THE IRISH COUNTY SURVEYORS
1834–1944

For Bernie

The Irish County Surveyors
1834–1944

A Biographical Dictionary

Brendan O Donoghue

[handwritten inscription]

FOUR COURTS PRESS

Typeset in 10.5 pt on 12.5 pt Ehrhardt by
Carrigboy Typesetting Services for
FOUR COURTS PRESS LTD
7 Malpas Street, Dublin 8, Ireland
e-mail: info@fourcourtspress.ie
and in North America for
FOUR COURTS PRESS
c/o ISBS, 920 NE 58th Avenue, Suite 300, Portland, OR 97213.

A catalogue record for this title is available
from the British Library.

ISBN 978–1–84682–063–2

SPECIAL ACKNOWLEDGMENT

This publication was grant-aided by the Heritage Council
under the 2007 Publications Grant Scheme.

AN
CHOMHAIRLE
OIDHREACHTA

THE
HERITAGE
COUNCIL

Printed in England
by MPG Books, Bodmin, Cornwall.

Contents

Illustrations

appear between pages 210 and 211.

CREDITS

Nos. 11 and 17, Waterford County Museum; Nos. 16, 19, 22 and 34, Dr R.C. Cox, Centre for Civil Engineering Heritage, Trinity College Dublin; Nos. 31–2, 62, 64–7 and 76–81, Institution of Engineers of Ireland; Nos. 33, 35, 59–61 and 63, author's collection; Nos. 40–2, Fr Browne collection, courtesy of Davison & Associates; Nos. 51 and 56, Irish Architectural Archive. Remaining items, including photographs from the Lawrence, Eblana, Valentine and Eason collections, are courtesy of National Library of Ireland; while specific dates are not available, the majority of these photographs were taken more than 100 years ago.

Abbreviations

BL	British Library
CC	*Cork Constitution*
CSORP	Chief Secretary's Office Registered Papers (NAI)
DB	*Dublin Builder*
DCE	Diploma in Civil Engineering (TCD)
HC	House of Commons
IB	*Irish Builder*
IBE	*Irish Builder and Engineer*
ICEI	Institution of Civil Engineers of Ireland
LCE	Licence in Civil Engineering (TCD)
NAI	National Archives of Ireland
NLI	National Library of Ireland
OPW	Office of Public Works
PRIAMCE	*Proceedings of the Incorporated Association of Municipal and County Engineers*
PRICE	*Proceedings of the Institution of Civil Engineers*
TICEI	*Transactions of the Institution of Civil Engineers of Ireland*
TNA	The National Archives, London (formerly Public Record Office)

Introduction and Acknowledgments

The first county surveyors took office in May 1834 after the failure of a previous attempt to place the work of the Irish grand juries under the supervision and direction of a corps of professional engineers. How were these surveyors selected for appointment? Who were they? How did they and their successors perform? What contribution did they make to the development of infrastructure throughout the country between 1834 and 1944? These are some of the questions to which answers may be found in this book.

The introduction of county surveyors to Irish local government was a major innovation, as dramatic a change in its day as the introduction of the city and county management system nearly one hundred years later. The appointments made in 1834 were, in fact, the first public appointments at local or central level in the United Kingdom which were required by law to be made on the basis of a competitive examination and an objective assessment of merit. Notwithstanding these novel features and the significant role the surveyors played in local development, the introduction and evolution of the county surveyor system and the careers of the men involved have received little attention in the history of public administration and of engineering in Ireland. This is not of course unique because, as Professor Joe Lee remarked in 1997, we have little systematic information on public servants, most of the major professions, academics or the many other categories of men and women who helped to build our state and our society.[1] In a similar vein, Diarmaid Ferriter has written that 'the Irish have shown an understandable penchant for celebrating and commemorating key historical revolutions, battles and the politics of high drama, but there have been too few attempts to chronicle the ordinary'.[2] In a sense, this book is just such a chronicle.

The fact that the establishment of the county surveyor system involved 'the earliest instance of a competitive examination for public office' would 'always be creditable to Ireland' in the opinion of the under-secretary, Sir Thomas Larcom.[3] The Board of Works also supported the introduction of the system, noting in their annual report for 1834 the beneficial effect likely to be produced on roads and bridges by the appointment of the surveyors; they commented also that while the surveyors' duties were 'very laborious and their remuneration small', the new men had 'very generally commenced their services with zeal and firmness, accompanied by a degree of skill in their business that will tend gradually to the improvement of the communications, and be productive

1 Professor Joe Lee, 'A brilliant start to filling the gaps in our biography', *Sunday Tribune*, 24 Aug. 1997.
2 Diarmaid Ferriter, *Cuimhnigh ar Luimneach, A History of Limerick County Council, 1898–1998* (Limerick, 1998).   3 NLI, Larcom papers, MS 7753.

of a considerable saving of expense to the counties'.[4] But the Select Committee on County Cess took a somewhat different view in 1836: they were 'deeply convinced that the county rate never can be as economically administered as by those who pay it' and they reported with regret that 'the magistrates and cess-payers do not give the same attention to the roads' as they did before the surveyors were appointed.[5]

As the calibre of surveyors improved, and as the system itself matured, it made a significant contribution to local development in many parts of Ireland although, as might be expected, performance was uneven, especially in the early years. In 1842, a royal commission recommended that the county surveyors should be replaced by district surveyors who would operate on the basis of the baronies and under the supervision of six circuit engineers employed by the Board of Works.[6] This recommendation was not accepted and, when the issue was reviewed again by a select committee in 1857, the outturn was more favourable: the committee reported that 'an able and efficient class of county officers has been formed' and that the system had been 'attended with great public advantage, both as regards the improvement of county roads and bridges and as regards the economizing of county funds'.[7] Thus, despite the initial misgivings of grand jurors about the value of appointing county surveyors and the doubtful competence of some of the early appointees, the system was to endure, with only minor changes, even after the grand juries themselves were replaced by the county councils in 1899.

Because of the absence of official records of the activities of the grand juries and of the first county councils it is difficult to attempt a complete evaluation of the county surveyor system or a detailed assessment of the contribution made by individual surveyors to the development of particular counties. What follows is based on a wide range of archival and printed sources but it cannot claim to be a definitive record; it is not all that I could desire, as Charles Gibson said of his history of Cork in 1861, 'but it is as good as I can make it'[8] and will, I hope, stimulate further research on the subject. Some of the biographical notes in Part III are relatively brief, either because the subject's career was a short one or because of the particular difficulty in turning up information on individuals who lived more than 150 years ago and rarely came to notice at national level. On the other hand, some of the notes have been expanded to include additional material on particular projects or controversies in which individual surveyors were involved but which may be of wider interest to students of the history of engineering and of local government in Ireland.

The research on which this book is based has been conducted intermittently over a period of more than twenty years and I am grateful to many people for their advice and assistance at various stages. Among these are the late Professor Sean de Courcy, Dr R.C. Cox, Dr Freddie O'Dwyer, Peter J. O'Keeffe, Dr Edward McParland, Professor Kevin B. Nowlan, Joss Lynam and Dr Paddy Caffrey. In the 1990s, a number of county managers, county engineers and county librarians were very helpful as was Mr W.J. McCoubrey, Chief Executive, Northern Ireland Roads Service. I am indebted also to a

4 *Third Annual Report from the Board of Public Works in Ireland*, HC 1835 xxxvi (76).   5 *Report from the Select Committee on County Cess (Ireland)*, HC 1836 xii (527).   6 *Report of the Commissioners appointed to Revise the several Laws under or by virtue of which Moneys are now raised by Grand Jury Presentment in Ireland*, HC 1842 xxiv (386).   7 *Report from the Select Committee on County and District Surveyors etc. (Ireland)* HC 1857 (Sess. 2) ix (270).   8 Charles B. Gibson, *The History of the County and City of Cork*, 2 vols (London, 1861), dedication.

large number of archivists and librarians: Julia Barrett, Architecture Library, University College, Dublin; John Callanan, Institution of Engineers of Ireland; Róisín Berry, Clare county archivist; Siobhan Fitzpatrick, Royal Irish Academy; Mary Higgins and Aisling Lockhart, Trinity College, Dublin; Mrs M.K. Murphy, former Archivist, Institution of Civil Engineers, London; Colum O'Riordan and Ann-Martha Rowan, Irish Architectural Archive; and members of the staff of The National Archives (formerly the Public Record Office), London. Special thanks are due to the staff of the National Archives of Ireland, especially Aideen Ireland, and to the ever-helpful staff of the National Library of Ireland, particularly Kevin Browne, Jim O'Shea, Tom Desmond, Brian McKenna, Sara Smyth, Aran O'Reilly, Thomas Fayne and Sandra McDermott, and to David Monahan for his photographic work.

Part I

Origins and Development of the County Surveyor System

Background

After the Act of Union (1800) Ireland was ruled by the government and parliament of the United Kingdom, based in London. One hundred members of the House of Commons at Westminster were returned by constituencies in Ireland, two for each of the thirty-two counties, two for each of the boroughs of Dublin and Belfast, and one for each of thirty-one other boroughs and for Dublin University. Four of the boroughs and the university gained a second seat in 1832. Twenty-eight Irish peers, elected for life from among all of the peers of Ireland, were entitled to sit in the House of Lords. A separate Irish executive continued to operate in Dublin headed by the lord lieutenant, the representative of the crown but appointed by the government of the day. While the situation varied with different personalities, the role of the lord lieutenant tended to be a ceremonial and representational one, rather than that of a head of government. The chief secretary, on the other hand, who was usually a member of the cabinet, was the effective political head of the Irish administration; he held the main parliamentary responsibility for Irish administration and legislation and had a myriad of other duties in Dublin and in London, involving constant travel between the two capitals. Day-to-day control of the executive was exercised by the under-secretary, the most senior civil servant in Ireland who, depending on his energy and ability, could play a crucial role in policy-making as well as in administrative affairs. Under this triumvirate, which was answerable to the cabinet in London and ultimately to parliament, the business of government in Ireland was carried on at national level by a semi-autonomous group of offices, boards and agencies, some of them reporting directly to the chief secretary's office in Dublin Castle, and some of them responsible to the Treasury or to other London-based departments of which they were part.

At sub-national level, the system of local government was not well developed. Municipal corporations, exercising relatively limited functions, existed in more than sixty boroughs until 1840 when their number was reduced to ten by reforming legislation. From 1828 onwards, separate boards of town commissioners were being established in a growing number of towns to provide the kind of services which growing urban populations demanded. Outside of these areas, the sole instruments of local government in the early decades of the century were the county grand juries, the forerunners of today's county councils.

THE GRAND JURIES

In addition to the judicial business for which they had originally been established, the grand juries had come to be responsible by the beginning of the nineteenth century for a wide range of fiscal (or civil) business in their counties. Under a series of acts passed during the previous two centuries, they had been assigned functions relating to the construction and maintenance of public roads and bridges; the provision and maintenance of courthouses, bridewells and gaols; the construction of minor marine works; and, in some counties, the maintenance of lunatic asylums, infirmaries and fever hospitals. Their annual expenditure on these functions and services amounted to about one million pounds in the early 1830s.[1]

The grand jury for each county was made up of twenty-three members of landed society, all selected for the purpose by the high sheriff at his discretion, subject only to having one member from each barony. It assembled twice each year at the county courthouse for the spring and summer assizes and transacted all of the fiscal business relating to the county in the few days before the assizes judges arrived to begin the criminal business. Until the 1820s, the procedure was that proposals for particular works were made on affidavit before a magistrate and submitted directly to the grand jury; if a presentment for the works was approved by vote of the jury – whose proceedings were not at this time open to the public – and fiated (in effect, confirmed) by the judge, the works could be carried out either by the person who obtained the presentment or his contractor. When the works were completed, another affidavit was sworn and submitted to the grand jury which then levied the amount of the presentment on the county cesspayers and recouped the person concerned.

The presentment system, as it was operated by the grand juries in the early decades of the nineteenth century, involved gross malpractice, inefficiency and waste. One of the leading critics of the system, speaking in parliament in May 1817, described it as 'twenty-three gentlemen, or it might be only twelve, in the midst of all the bustle and business of an assize, without any previous examination to guide them, taking upon themselves to levy money, to an indefinite amount, for various purposes, the items of which were almost innumerable; ... with a view of doing justice, they might just as well put the presentments and applications into a hat and draw out such number as they thought proper'.[2] In the same debate, another of the Irish members was equally bitter in his criticism of a system which, he claimed, had the people of Ireland groaning under the oppression caused by the abominable corruption practised under the grand jury laws.[3] Public money was frequently misapplied, according to one commentator,[4] much of it was either pocketed or wasted, according to another,[5] and the system as a whole had been 'productive of fraud, perjury and demoralization', in the words of a third.[6] Corruption and jobbery were widespread, with grand jurors themselves seeking presentments for

1 *Abstract of the accounts of Grand Jury Presentments in 1834*, HC 1835 xxxvii (220). 2 E.S. Cooper, MP for Sligo county, Parliamentary debates xxxvi, 2 May 1817, col. 115. 3 F.C. Ponsonby, MP for Kilkenny county, Parliamentary debates xxxvi, 2 May 1817, col. 115. 4 Thomas Newenham, *A View of the Natural, Political and Commercial Circumstances of Ireland* (1809). 5 Horatio Townsend, *A General and Statistical Survey of the County of Cork* (Cork, 1815). 6 Revd M.J. Keating of Limerick, *Suggestions for Revision of the Irish Grand Jury Law* (n.d.).

works of personal advantage and works which would benefit their friends and associates; as one writer put it, 'expensive bridges have been thrown over rivers, at places most unfrequented, merely to embellish the view from the windows of a neighbouring palace' and magnificent roads had been made at the public expense where there was no prospect of maintaining them.[7]

The granting of presentments to named individuals involved obvious favouritism and led to a situation in which, according to Alexander Nimmo,[8] road repairs were often carried out 'by a class of persons who make a trade of it, as a market for the labour of their poor peasantry; the latter were not, properly speaking, paid for what work they do, but have the amount allowed by their landlords as a set-off against the rent of their holdings'.[9] The grand jury itself was transient in nature, being dissolved at the end of each assizes, and it had no administrative or engineering organization for vetting the need for particular projects and the plans and cost estimates which were often inflated. Similarly, the arrangements for the supervision of work in progress and for the inspection of finished work were inadequate and unsatisfactory, with local landowners and others who might well have an interest in the works or the contracts acting as supervisors, overseers or conservators on behalf of the grand jury.[10] As a result, Daniel O'Connell could speak of families who had made fortunes and purchased estates out of their jobs in road-making 'though the roads they made were not now to be distinguished from the surrounding bog'.[11] The inefficiency of the system was compounded by the fact that appointments to offices under the control of the grand jury were made on the basis of a patronage system which was tightly controlled by local interests. There was serious dissatisfaction also with the system of local taxation through which the grand juries' operations were financed, as the county cess was levied in most counties on the basis of inadequate and out-of-date surveys and valuations;[12] besides, the cess was payable by the occupiers of land who had no say in the determination of the rate of tax or in the application of the proceeds. All of this led Gustave de Beaumont, the French writer who travelled in Ireland in the 1830s and who was scathing about government misman-agement, to remark that their membership of the grand juries provided the 'bad aristocracy' with a means of advancing their own affairs rather than the general interest, and allowed them to commit enormous abuses, gross frauds and monstrous excesses.[13]

7 W. Parker, *An Essay on the Employment which Bridges, Roads and other Public Works may afford the Labouring Classes, etc.* (Cork, 1819).　8 Alexander Nimmo (1783–1832), a Scottish engineer, came to work in Ireland in 1811 and was in charge of the publicly-funded programme of road, bridge and harbour construction in the western district of Ireland from 1822 until 1831.　9 Report of Alexander Nimmo, quoted *in Report of the Select Committee on the State of the Poor in Ireland (Summary Report)*, HC 1830 vii (667); evidence given by Nimmo to a House of Lords committee on the state of Ireland in 1824, quoted in *Report from the Select Committee on Grand Jury Presentments, Ireland*, HC 1826–1827 iii (555).　10 For a detailed commentary on grand jury law as it applied to road work, see William Greig, *Strictures on Road Police (sic)* (Dublin, 1818).　11 Parliamentary debates III xix, 11 July 1833, col. 565.　12 Report from the *Select Committee on County Cess (Ireland)*, HC 1836 xii (527).　13 Gustave de Beaumont, *Ireland: Social, Political and Religious*, edited and translated by W.C. Taylor (London, 1839); repr. with an introduction by Tom Garvin and Andreas Hess (Harvard, 2006), pp 134, 163–6.

FIRST EFFORTS TO APPOINT COUNTY SURVEYORS

With the large increase in local expenditure on mail-coach road construction and on public buildings in the first decades of the nineteenth century, much of it financed by loans and grants from the exchequer, government and parliament were forced to turn their attention to the need for greater efficiency, effectiveness and equity in the operations of the grand juries. William Vesey-Fitzgerald, MP for Ennis and Chancellor of the Irish Exchequer from 1812 until its abolition in 1816, took a particular interest in the subject but, while he noted 'the plunder of the public purse', he feared that 'a measure clashing with private interests' would encounter great opposition.[14] Nevertheless, he was in touch in 1814–15 with others interested in reform, including William Edgeworth, a civil engineer, Edward S. Cooper, MP for Sligo county, and Sir Henry Parnell, MP for Queen's County, who had a special interest in roads and other civil engineering matters and whose suggestions for legislation included a provision for the employment by each grand jury of a 'surveyor general of roads and other public works', with a deputy surveyor for each barony or half-barony.[15]

The first major effort to introduce new procedures began in 1816 when a select committee of the House of Commons, established in the previous year on foot of a motion moved by Cooper,[16] put forward proposals to reform the business of making and considering presentments for roads and other public works. It recommended that, as had been suggested by Parnell, 'there should be a surveyor of county works attached to each county and that no person should be appointed to that office who shall not have passed an examination before a board of civil engineers in Dublin, and have obtained a certificate of his qualification to discharge the duties of his office; the appointment of which board of civil engineers should be vested in the lord lieutenant'.[17] In a further report, the same committee proposed that there should be a new survey and valuation of all land 'under the direction of competent sworn surveyors appointed by the grand juries and valuators appointed by the parishes'.[18] In June 1816, after the select committee had reported, Cooper sought leave to bring in a bill to consolidate the entire body of grand jury law, with amendments to take account of the committee's recommendations, but he was urged by colleagues to proceed first with an amending bill which would deal with the most urgent issues.[19] Cooper, however, went ahead with his original scheme and, in May 1817, presented his draft of a consolidation bill to parliament.[20] This included provision for a system of circuit surveyors, who would be independent of the grand juries because, as Cooper put it, 'it was owing to the want of a responsible officer of this description that plans were bad, estimates inadequate, public works badly executed, and that in numberless instances private convenience was preferred to public advantage'.[21] Cooper

14 NLI, MS 7825, Letter Book of the Rt Hon. Wm Fitzgerald, Chancellor of the Irish Exchequer, containing copies of correspondence relating to his proposal to introduce legislative changes in the powers of Irish grand juries in relation to presentments and otherwise, 1814–15.   15 Henry Parnell (1776–1842) was one of the commissioners of the Holyhead road, MP for Queen's County from 1806 to 1832, author of *A Treatise on Roads* (London, 1833) and an honorary member of the Institution of Civil Engineers; for an obituary, see *PRICE* 2 (1842–3), p.11.   16 Parliamentary debates xxx, 7 Apr. 1815, col. 890.   17 *First Report from the Select Committee on Irish Grand Jury Presentments*, HC 1816 ix (374).   18 *Second Report from the Select Committee on Irish Grand Jury Presentments*, HC 1816 ix (435).   19 Parliamentary debates xxxiv, 6 June 1816, col. 1007.   20 Parliamentary debates xxxvi, 2 May 1817, cols 115–20.   21 Parliamentary debates xxxvi, 14 May 1817, cols 562–3.

was again urged by his supporters, including the chief secretary, Robert Peel, to withdraw the bill in favour of a shorter amending bill, and he agreed to do so.

The amending bill, presented by Vesey-Fitzgerald, was given a second reading in the commons on 13 June 1817.[22] It provided for the holding of special presentment sessions twice each year in the baronies throughout each county to consider proposals for public works before their submission to the grand jury. In addition, it proposed the appointment for each county (excluding Dublin city and county but including the other counties of cities and towns) of a county surveyor to report on the necessity for, and the expediency of, each proposal put forward at the sessions, and on the plans and cost estimates; the new surveyors were to be appointed by the lord lieutenant 'when and as soon as persons can be found properly qualified'.[23] In a radical departure from the patronage system, the bill went on to provide (as recommended by the 1816 committee) that no one was to be appointed to the new office until he had been examined by a board of at least three civil engineers set up by the lord lieutenant, and had obtained a certificate from the board of his fitness to hold the office.

The debates in parliament on the 1817 bill[24] exposed a wide divergence of view among the Irish members on the need for reform of the grand jury system. Vesey-Fitzgerald, in moving the motion for a second reading, argued that the bill was 'calculated to do more good to Ireland than any measure that had been adopted since the Union'. In his view, the provision for the appointment of county surveyors, and the other proposed reforms, would 'check the system of profligate jobbing and expense'. Sir Henry Parnell was concerned about the cost of the new appointments but felt that the surveyors would ensure that money levied on the cess-payers was well spent. But most of the other speakers were opposed to the appointment of surveyors. According to Sir George Hill, MP for Londonderry city, the bill was 'an abortion, or at best, a rickety offspring of the one abandoned'. It was neither wise nor just, he said, 'to revile and subvert a system which has worked so much benefit to Ireland'; the new system was 'calculated to deprive the gentlemen of the county of their proper influence and authority' and it would turn the grand jurors into 'insignificant cyphers'. Sir George went on to argue that to give the county surveyor 'paramount authority and control over every projected work, whether to mend or to make ... is degrading and intolerable' and far exceeded the proposition of the select committee which had reported in 1816. Finally, according to Sir George, 'it is notorious that a sufficient number, qualified as required, are not to be found' to fill the new positions. Sir Frederick Flood, MP for Wexford county, joined in the opposition because, he argued, the people of Ireland, who were in a state of starvation, ought not to be taxed with £15,000 to pay the considerable emoluments of the thirty-six officers who were to be appointed under the bill.[25] Another objection was raised by F.C. Ponsonby, the member for Kilkenny county, who was generally supportive of the bill but feared that the new appointments would give too much scope for political patronage. That same view

22 Parliamentary debates xxxvi, 13 June 1817, cols 959–60.   **23** An Act to provide for the more deliberate Investigation of Presentments to be made by Grand Juries for Roads and Public Works in Ireland, and for accounting for Money raised by such Presentments, 57 Geo III, c.107.   **24** Parliamentary debates xxxvi, 13 June 1817, cols 959–60; 30 June 1817, cols 1268–70; 10 July 1817, cols 1368–73.   **25** Thirty-six surveyors would have been needed if a separate appointment were made for each of the thirty-one counties (except Dublin) and the five counties of cities (again excluding Dublin), with the four counties of towns sharing the services of a surveyor with the adjoining county, as the act allowed.

was expressed in the House of Lords[26] by the earl of Donoughmore: 'the bill gave to the government a patronage of £15,900 a year and planted a portion of that patronage in every county'. Besides, the earl wondered, how could one man do the job in a large county like Tipperary? As against that, the earl of Limerick strongly supported the bill because it would 'abolish a system of consummate plunder which chiefly affected the lowest part of the people'.

In spite of the objections expressed by so many of the Irish members, the bill to provide for county surveyor appointments became law in July 1817 but this first effort to bring order and efficiency into the business of making presentments for roads and other public works was doomed to failure. The process began in December 1817 when applicants for the new positions were invited by public advertisement to present themselves for interview in Dublin before a panel of three distinguished practitioners in civil engineering and architecture: Thomas Telford, the Scottish engineer renowned for his extensive work on canals, roads, bridges and harbours, and who was to become first president of the Institution of Civil Engineers; Major Alexander Taylor, another Scotsman, who built the military road through the Wicklow mountains between 1800 and 1809 and had worked with his brother George and Andrew Skinner to produce the famous *Maps of the Roads of Ireland* in 1778; and Francis Johnston, architect to the Board of Works and designer of the GPO in Dublin and other important public buildings, churches and country houses. Johnston had been substituted at the last minute for Alexander Nimmo, who had accepted an invitation from the chief secretary, Robert Peel, to act as a member of the board.[27] It was Telford who suggested that Nimmo should be replaced 'without the least reflection upon his talents or integrity' because he was but a young engineer and was himself a candidate for a surveyorship; besides, in Telford's view, there should be one Irishman on the board and, as part of a surveyor's duty was to consist of attention to public buildings, it was desirable that one examiner should be an experienced architect.[28]

A list of those who presented themselves for interview does not seem to have survived among the chief secretary's office papers but, according to the chief secretary himself, 'no less than ninety-five persons had applied'.[29] Many of these, according to William Greig, who had worked as an engineer on the mail-coach roads for nine years, were men who, in the expectation of so many appointments taking place, had been induced 'to assume the title of engineer, who never dreamed of it before, as if a few months acquirements could fit them for such a situation'.[30] This may well have been the case because engineers, like surveyors and architects, had no uniform educational background or training at the time, and generally learned through a form of apprenticeship beginning at quite a young age.[31] Besides, there were no recognized tests of a man's entitlement to describe himself as a civil engineer, and no restriction on the use of the letters CE – in fact, as J.H. Andrews has written, 'there was nothing except conscience, or fear of ridicule, to prevent the initials being flaunted by anyone who took a fancy to them'.[32]

26 Parliamentary debates xxxvi, 10 July 1817, cols 1368–70.   27 BL, Peel Papers MS 40272 f 188, letter of 17 Dec. 1817.   28 BL, Peel Papers MS 40272 f 203, letter of 18 Dec. 1817.   29 Robert Peel, Parliamentary debates xxxviii, 6 May 1818, cols 539–40.   30 William Greig, *Strictures on Road Police* (*sic*) (Dublin, 1818), p. 231.   31 Study of engineering as an academic subject was not possible in England until 1838 and in Ireland until 1841 when TCD established an engineering diploma course.   32 J.H. Andrews, *Plantation Acres* (Belfast,

Regardless of their qualifications or lack of them, it seems that all of the candidates were called before the interview board which sat at the Paving House, 21 Mary Street, Dublin, from 10.00 a.m. to 4.00 p.m. and from 6.00 p.m. to 9.00 p.m. each day between 23 December 1817 and 19 January 1818. It 'spared no pains in fully and impartially investigating the qualifications of all those who presented themselves' but had to report that 'upon the whole of this painful and tedious investigation, finding so very few of the great number who presented themselves possessed of the necessary experience and practical knowledge in road-making and bridge-building, which form the chief portions of the duty of county surveyors, the board has not granted any certificates [of qualification for appointment]; and is of opinion that no advantageous proceedings can be taken under that clause of the act by which it is appointed'.[33] The whole episode left Telford with a poor impression of Dublin and of Ireland: he was later to write 'I ... spent six weeks in the metropolis of Pat's country; I confess it enables me to set a higher value on our own'.[34]

Faced with this embarrassing verdict on the calibre of the emerging civil engineering profession in Ireland, parliament was forced in February 1818 to suspend the 1817 act[35] and to replace it entirely by an act passed in the following June.[36] In presenting these measures, Robert Peel told the House of Commons that he found the result of the examinations 'most discouraging' and that, on mature consideration, he had been forced to change his opinion on the appointment of county surveyors – thirty-two fully competent men were necessary but while no less than ninety-five had applied with, on the face of it, 'the most respectable qualifications', their competence had not been borne out at the examinations.[37] Sir Henry Parnell regretted that only three of the candidates had been found competent[38] and wondered why sufficient qualified surveyors could not be recruited in England, given that reasonable salaries were on offer. Other speakers stressed the continuing need for reform, describing the existing grand jury system as an abominable system, a system of depredation and abuse, and a system productive of corruption, fraud and perjury. But Sir Frederick Flood stoutly defended the *status quo* and expressed his great pleasure that the 1817 Act would not come into operation because, as he put it, 'its tendency was to introduce into every county a set of men, total strangers to it, at salaries from £500 to £600 a year'; in his view, the grand juries were far better judges of what concerned the interests of their own counties than 'itinerant surveyors' could ever be and it was unjustifiable, absurd and impracticable to suggest that so many civil engineers were needed 'to guard, to overlook and to control the business of the grand juries'.[39]

1985), p. 253. **33** *Report of the Board of Civil Engineers which sat at No. 21 Mary Street, Dublin, from 23rd December 1817 to the 19th January, 1818*, HC 1818 xvi (2) **34** Quoted in L.T.C. Rolt, *Thomas Telford* (London, 1985), pp 144–5. **35** An Act to suspend, until the End of the present Session of Parliament, the Operation of an Act made in the last Session of Parliament, to provide for the more deliberate Investigation of Presentments to be made by Grand Juries for Roads and Public Works in Ireland, and for accounting for Money raised by such Presentments, 58 Geo III, c.2. **36** An Act to provide for the more deliberate Investigation of Presentments to be made by Grand Juries for Roads and Public Works in Ireland, and for accounting for Money raised by such Presentments, 58 Geo III, c.67. **37** Parliamentary debates xxxviii, 6 May 1818, cols 539–40. **38** Ibid. **39** Parliamentary debates xxxvii, 4 Feb. 1818, cols 145–6.

PRESSURE FOR REFORM IN THE 1820s

While forced to accept that the 1817 Act was now inoperable, parliament attempted, through the act of 1818, to preserve some of the expected benefits by requiring that all proposals made to the grand juries should in future be accompanied by maps, plans and estimates 'signed by some known surveyor, engineer or architect, or by some other competent person' and that these should be considered at special presentment sessions before coming up for decision by the grand jury at the assizes. It appears, however, that the operation of this revised system did not give general satisfaction and a parliamentary committee was established to review it. This committee reported in June 1819 that the principle of providing a strict and open investigation of applications for presentments 'by competent persons previous to the assizes ... is wise and judicious and calculated ... to provide for the better execution of public works'. They regretted however that 'the great benefits which were contemplated in the alteration of the grand jury laws have not been fully realised in the act of the last session' but attributed this failure not to any deficiency in engineering input, but rather to the limited number of persons who were qualified to take part in the presentment sessions.[40] Following this report, the 1818 act was repealed in July 1819, leaving it to the grand juries to carry on, more or less as they had done before. Presentment sessions (composed of the magistrates for each barony but with a somewhat less restrictive property qualification) were retained to screen applications for presentments, but there was no provision for a professional input to the making of these applications, or to their consideration at the presentment sessions, or when they came before the grand jury.[41] As one commentator described it: 'in avoiding the expense of civil engineers or inspectors, the public works ... are still left to local mismanagement and individual partiality'.[42]

Pressure for reform of the grand jury system continued in the 1820s. There were three reports in 1822 from a select committee suggesting changes in several aspects of the system but none of the recommendations dealt with the need for engineering input.[43] That issue was raised, however, in 1825 by a committee appointed by the Co. Cork grand jury which reported that the appointment of a 'properly qualified district engineer' for each of the two ridings into which the county had been divided in 1823 was essential if the execution of public works was to be improved.[44] That view was endorsed by a select committee in 1827 which noted that 'a considerable school of engineering' was said to be growing up in Ireland as a result of the various programmes of public works.[45] Again, in 1830, a select committee on the state of the poor in Ireland drew attention to the 'admitted necessity for a more effective reform than has been as yet applied to correct the acknowledged imperfections and abuses of the present system of grand jury presentments'; the need to put an end to 'the system of favouritism in the disposal of presentments' by

40 *Report from the Committee on Grand Jury Presentments in Ireland*, HC 1819 viii (387).   41 An Act to amend the Laws for making, repairing and improving the Roads and other Public Works in Ireland, by Grand Jury Presentments; and for a more efficient investigation of such Presentments and for securing a true, full and faithful Account of all Monies levied under the same, 59 Geo. III, c.84.   42 W. Parker, *An Essay on the Employment which Bridges, Roads and other Public Works may afford the Labouring Classes, etc.* (Cork, 1819). 43 *Reports from the Select Committee on Grand Jury Presentments, Ireland*, HC 1822 vii (353) (413) (451). 44 Reprinted in *Report from the Select Committee on Grand Jury Presentments, Ireland*, HC 1826–7 iii (555). 45 *Report from the Select Committee on Grand Jury Presentments, Ireland*, HC 1826–7 iii (555).

adopting an open system of contract; and 'the want of local engineers to direct and advise' on presentments.[46]

John Killaly (1766–1832), one of the most experienced canal and road engineers in Ireland, told the chief secretary, Lord Francis Leveson-Gower, in January 1830 that there was not, in his judgment, 'any one measure, as regards the interests of the country, more deserving of the attention of the legislature' than the reform of the grand jury system and, in particular, the elimination of 'the ills resulting from road jobbing'.[47] There were also some useful contributions to the debate in pamphlets published by public-spirited individuals and landowners. In 1831, for instance, Lord Carbery addressed a submission to the members of the imperial legislature in which he discussed the advantages of employing 'professed engineers', especially in the laying out of extensive lines of road or large bridges, and in furnishing estimates for substantial and difficult works'.[48] But he was not at all sure that engineers, whose education was expensive and whose time was valuable (a charge of up to three guineas a day was the norm) should be employed to act as undertakers or contractors for carrying out the ongoing work programmes of the grand juries. His preferred solution was the appointment by the government of a *county engineer* for each county (or more than one for larger counties) who would operate under the control of the grand jury and would have a 'general superintendence' of all roads and bridges; in addition, this corps of 'really active, efficient officers' would carry out periodic inspections, without notice, of all works being done by contractors under presentments, and certify the satisfactory completion of the work and its conformity with the presentment. A much more limited role for a new category of professional officers was suggested in a pamphlet published two years earlier by *A Country Gentleman* and which described in great detail how the work of making and repairing roads might be improved; the organizational arrangements would not be altered dramatically by this writer but 'properly qualified and instructed' men would be 'constantly going through the county' to 'point out to the overseers of roads the right system of making and repairing roads'.[49] A contrary view was that the taxation levied by grand juries was already excessive in the 1820s, chiefly because it had been extremely difficult to reduce the salaries bill for existing county officers; this writer's view was that roads should either be turnpikes, maintained from the proceeds of tolls, or roads maintained, as in England, by the parish which could do the work more cheaply and without being 'plundered by all it employs'.[50]

The failure of successive governments in the 1820s to respond to the many recommendations for the appointment of engineers at grand jury level can be attributed to the experience of 1817–18, the great difference of opinion that existed as to the remedy that should be proposed for the undoubted ills of the grand jury system, and reluctance to incur the wrath of the landed gentry by reducing their effective domination of the grand jury and its affairs. It is worth noting, however, that there had been a similar lack of action on calls by House of Commons committees in 1819, 1820 and 1821 (influenced to some extent by John Loudon McAdam) for the appointment in England

46 *Report of the Select Committee on the State of the Poor in Ireland (Summary Report)*, HC 1830 vii (667). 47 NAI, OP 972/62, letter of 16 Jan. 1830 from John Killaly.   48 Lord Carbery, *Observations on the Grand Jury System of Ireland with Suggestions for its Improvement* (London, 1831).   49 A Country Gentleman, *Hints on the System of Road Making* (Dublin, 1829).   50 John O' Driscoll, *Views on Ireland*, 2 vols (London, 1823).

of professional county surveyors to superintend and manage the large mileage of turnpike roads in each county.[51] The situation in Ireland changed with the formation by Earl Grey of a new cabinet in November 1830 and the appointment as chief secretary for Ireland of Edward Stanley whose policy objectives included administrative efficiency and economy and social aid of various kinds.[52] Stanley – the future 14th earl of Derby and three-times prime minister – was responsible during his three years in the Irish office for a considerable volume of reforming measures, including the establishment of a state-supported national education system in 1831; the establishment of the Board of Works in the same year to take over the functions of a number of exiting organizations and to administer a substantial new loan fund for public works; and measures to reform the tithe system. By 1833, he was ready to tackle the vexed question of grand jury reform believing, as he put it, that all classes were by then united as to the evils to which the system was subjecting the country. He came before parliament on 19 February with a statement of his proposals for the reform of the system, and gained general support for them.[53] The proposals included the revival of the county surveyor system, as provided for in the 1817 legislation, a step from which Stanley 'anticipated the most beneficial results'.

51 Sidney & Beatrice Webb, *English Local Government: Statutory Authorities for Special Purposes* (London, 1922).   52 Oliver MacDonagh, 'Politics, 1830–45' in W.E. Vaughan (ed.), *A New History of Ireland v: Ireland under the Union I, 1801–70* (Oxford, 1989).   53 Parliamentary debates III xv, 19 Feb. 1833, cols 955–63.

2

The First Appointments

THE STATUTORY FRAMEWORK

When the bill which was to become the Grand Jury (Ireland) Act 1833 was brought before parliament late in the 1833–4 session, most of the Irish members sought to have it postponed to allow more time for consideration of the new local representational arrangements it proposed.[1] Daniel O'Connell was among those who took this view, notwithstanding his dislike of the existing system under which 'favourites of the grand jury were fattened'.[2] But the government pushed the bill through against the wishes – and the votes – of many of the Irish MPs and the bill received the royal assent on 28 August 1833.[3]

The preamble to the new act[4] stated that the existing provisions for the consideration of presentments were not adequate 'to secure the needful investigation of the disbursement of the public monies and the due and economical performance of the works carried into execution' by the grand juries who had not 'sufficient time to deliberate upon and examine into the several presentments laid before them'. The act provided, therefore, that the presentment sessions in each barony would in future be composed of a number of the major cesspayers (not more than twelve and not less than five) in addition to the local magistrates, and that these sessions should receive and consider all applications for presentments for public works in the area, with county-at-large presentment sessions dealing only with the relatively small number of matters of concern to the county as a whole. All presentments would still have to be approved by the grand jury itself before any work could go ahead, but the jury was deprived of its earlier power to initiate projects and to alter presentments. In addition, the jury would have to sit in public when dealing with presentments, instead of retiring to the grand jury room from which the public had been excluded until then. To enable this somewhat more representative local government machinery to operate effectively, the act reintroduced provision for the appointment of a county surveyor for each county, except Dublin.[5] The appointment procedure was similar to that of the 1817 act: a panel of persons whose qualifications had been examined into and certified by a board of three civil or military engineers was to be set up, with appointments from the panel being made by the lord lieutenant.

---

1 Parliamentary debates III, xix 11 July 1833, col. 561.   2 Ibid., col. 564.   3 Edward Stanley resigned from the government in May 1833 and it fell to his successor as chief secretary, Edward Littleton, to manage the passage of the bill in the House of Commons.   4 An Act to amend the Laws relating to Grand Juries in Ireland, 3 & 4 Wm IV, c.78.   5 Provision for the appointment of district surveyors in Co. Dublin was made in 1844.

THE SELECTION PROCESS

The task of conducting the examination of candidates for appointment as county surveyor was assigned to the chairman of the Board Works, Colonel John Fox Burgoyne, his colleague, commissioner John Radcliffe, like Burgoyne a former army engineer, and the board's engineer and architect, Jacob Owen.[6] This arrangement subsequently attracted criticism in professional and other circles: according to a printed statement circulated in the mid-1850s,[7] one of the examiners – Owen – was 'merely an architect' and a significant number of the successful candidates were said (correctly) to have been assistants, apprentices, or private pupils of his in the Board of Works; besides, Radcliffe was said to be 'unknown to the public or to the profession as a civil or military engineer'.[8] Years later, Thomas Larcom, who was himself a commissioner of public works from 1846 to 1853 and under-secretary at Dublin Castle from 1853 to 1868, defended the arrangements made in 1834, arguing that because engineering talent in Ireland was not abundant 'it would not have been possible at the time to have found three other men combining the necessary qualifications who could have been called on to do the work unpaid' as the 1833 act provided.[9]

The full list of those who applied for county surveyor posts in 1833–34 cannot now be traced.[10] Burgoyne's recollection, more than twenty years later, was that about 50 candidates came forward for examination[11] although he had told the under-secretary in 1840 that there had been almost 100 candidates.[12] One of the candidates[13] thought that the number was higher still – 150 or upwards – and an even higher figure of 239 was quoted, but without identifying a source, in the *Tipperary Free Press*.[14] In any case, the process got underway in April 1834 when the chief secretary's office notified the candidates of the dates on which they should attend for interview, 'bringing specimens of your work and testimonials as to character and ability from professional and other gentlemen'.[15] Candidates were advised that they would have to undergo a very strict examination: they would have to show 'a correct knowledge and practical experience in land and road surveying and mapping; bridge-building, masonry and other works; the drawing up of specifications and estimates; as well as the details of the measurement and execution of the several descriptions of work connected with road engineering'. In addition, 'some knowledge of builders' business' was considered necessary.

Candidates duly attended at the offices of the Board of Works at the Custom House, Dublin, on 23 April 1834 and, according to Patrick Leahy, who was to be the first

---

6 *Returns of the Dates of Commissions issued by the Irish Government for the Examination of Candidates for the Office of County Surveyor in Ireland*, HC 1856 liii (335).   **7** NLI, Larcom papers, MS 7753, *Statement Respecting the Examinations for County Surveyorships Ireland*, undated; it was believed in official circles that the statement emanated from persons connected with the TCD engineering school.   **8** Radcliffe was, in fact, a peninsular veteran and military engineer who had been employed by the Directors General of Inland Navigation before his appointment as a commissioner of public works.   **9** NLI, Larcom papers, MS 7753, letter of 20 May 1857.   **10** The chief secretary's office file containing the list of applicants (NAI, CSORP 1833/5635) is not extant and while the CSORP index for 1833 contains entries for a large number of letters received that year from applicants for county surveyor posts, all of these letters are also included on the NAI list of non-extant items.   **11** Evidence given to the *Select Committee on County & District Surveyors etc. (Ireland)*, HC 1857 (Sess. 2) ix (270).   **12** NAI, OPW 1/1/4/19, Public Works Letters No. 19, p. 229, letter of 23 Jan. 1840.   **13** Evidence given by Henry Brett to the *Select Committee on County & District Surveyors etc. (Ireland)*, HC 1857 (Sess. 2) ix (270).   **14** *Tipperary Free Press*, 28 May 1834.   **15** NAI, CSO Unregistered

surveyor for the east riding of Co. Cork, underwent 'during a period of fourteen successive days, a most rigorous course of examination in the various departments of theoretical and practical science requisite to constitute a county engineer'.[16] Thomas Jackson Woodhouse, the first Antrim surveyor, provided a somewhat similar account of what transpired when he gave evidence to a House of Commons select committee in 1836[17] and several of the other surveyors corroborated this in 1857 when they appeared before another select committee:[18] Charles Lanyon, who was appointed to the Kildare post in 1834, confirmed that the examination went on every day for a fortnight; Henry Stokes, the first Kerry surveyor, asserted that the examination was 'an elaborate one'; and Henry Brett, who was appointed in December 1834 to King's County, recalled examinations which had gone on for some twelve days, with written questions and answers on education and previous experience, practical engineering matters, surveying, architecture, bridge-building, road maintenance and improvement, algebra, geometry, trigonometry, mechanics and hydraulics.

The examinations conducted in subsequent years were reported to be equally demanding. In his evidence to the 1857 select committee, Alexander Tate, a Dublin district surveyor, spoke of attending written examinations in the same room for ten days in 1847, from 10.00 a.m. to 4.30 p.m. each day; John Hill, appointed to Co. Clare in 1845, told of examinations spread over twenty-two days in 1841; and Horace Uniacke Townsend, a candidate examined for ten days in 1838 before his appointment to the north riding of Tipperary, confessed that he had 'become knocked up by the close confinement and the excitement – I had to stay away some time under a doctor's care, and go back again for further examination'. Burgoyne told the committee that the 1834 examinations 'were very elaborate and took a considerable time' and that the entire procedure was a written one. All of this led the 1857 committee to report that 'the examinations are stated to have been conducted with much care' – not a particularly decisive verdict on a system which was being strongly criticised at the time, and not a very convincing response to suggestions in some quarters that the business of the examinations had been conducted in a rather casual fashion.

THE FIRST SURVEYORS

Complete information on the outcome of the 1834 competition cannot now be obtained from primary sources as the original documents were either destroyed or mislaid due to the lax filing system in the chief secretary's office, forcing an embarrassed Larcom to report to parliament in July 1856 that 'there is no record of the list returned by the examiners'.[19] However, a memorandum which he prepared a year later stated that when the interview board had examined all-comers, they returned to the lord lieutenant, Marquis Wellesley, a list of those they had found qualified, 'in the order of their merit,

Papers 1834/720. 16 NAI, CSO Unregistered Papers 1834/184, letter of 10 May 1834. 17 *Select Committee on County Cess (Ireland)*, HC 1836 xii (527). 18 *Select Committee on County & District Surveyors etc. (Ireland)*, HC 1857 (Sess. 2) ix (270). 19 *Returns of the Dates of Commissions issued by the Irish Government for the Examination of Candidates for the Office of County Surveyor in Ireland, etc.*, HC 1856 liii (335); NLI, Larcom papers, MS 7753, letter of 20 May 1857.

from which the lord lieutenant was to select' but, he went on, 'it was not the practice of the lord lieutenant to select but to appoint the first on the list, and so on as vacancies occurred'.[20] There was, in any case, little time for deliberation on the results as the dates set for first series of special presentment sessions were rapidly approaching and the appointments were duly made with effect from 17 May 1834. There was some confusion initially about the position of the new surveyors in relation to the separate counties of cities and counties of towns (Belfast, Cork, Limerick, Londonderry, Waterford, Carrickfergus, Drogheda, Galway and Kilkenny) and some of the county surveyors and grand juries found it necessary to seek directions on the issue from Dublin Castle. The matter was soon sorted out and, generally with effect from 14 June 1834, the surveyors for the adjoining counties were appointed to be surveyors also for the counties of the cities and towns, with an entitlement to separate additional salaries.[21]

The appointments made in 1834 were the first public appointments in the United Kingdom which were required to be made on the basis of a competitive examination and an objective assessment of merit: the principle of competitive examinations was not adopted for civil service appointments until 1855 and was not extended to professional posts until the 1870s. Some questions about the appointments procedure appear, however, to have been raised at an early stage by Feargus O'Connor, the pro-repeal MP for Cork county from 1832 to 1835 (and later chartist leader), but when Burgoyne was asked to comment, he assured the chief secretary that he was 'very confident that no part of our conduct throughout that troublesome affair can be reasonably censured by any unprejudiced man'.[22] There was no vestige of patronage in the 1834 appointments, according to Larcom: rather than *selecting* from the pool of qualified candidates as the act would have allowed, the lord lieutenant offered appointment to the first person on the list and so on until all the posts were filled.[23] In the absence of any other system of ranking counties, this might be taken to mean that the person appointed to the Antrim post – Thomas Jackson Woodhouse – was first in order of merit, and so on down through the alphabetical list of counties.[24] But, as against this, Charles Lanyon who was appointed to Kildare (fourteenth on the alphabetical list) was believed to have achieved second place in the examination. Thus, the full order of merit can only be a matter of conjecture at this stage. In 1857, Burgoyne told a select committee that while he was chairman of the Board of Works (1831 to 1845), the lord lieutenant 'universally' acted on the order of merit and 'no influence was ever used on the part of the Irish government, nor any influence of any kind on the part of anyone else, to interfere with the examinations or the recommendations'.[25] However, the committee do not seem to have been convinced that there really was an established order of merit – their only conclusion was that the lists of candidates *are stated* 'at least since the examination of 1851, to have been returned to the lord lieutenant in the order of merit'.

20 NLI, Larcom papers, MS 7753, letter of 20 May 1857. 21 The counties of cities and counties of towns ceased to exist as separate entities under the Local Government (Ireland) Act 1898. 22 NAI, CSO Unregistered Papers, letter of 2 June 1834. 23 NLI, Larcom papers, MS 7753, letter of 20 May 1857. 24 A number of the provincial newspapers (for example, the *Leinster Express*, 24 May 1834) published lists of the newly appointed surveyors in alphabetical order by county, which suggests that the information was made available in that form by the chief secretary's office. 25 Evidence given to the *Select Committee on County & District Surveyors etc. (Ireland)*, HC 1857 (Sess. 2) ix (270).

As the 1833 Act did not apply to Dublin, thirty-two appointments were made in 1834, one for each of the two ridings in Co. Cork[26] and one for each of the other counties, excluding Dublin. The first surveyors (listed in the table below) were a disparate group. Some of them were relatively young men like Edmund Leahy who was appointed to Co. Cork (west riding) when he was only 20 years old, while others were much older and more experienced, including men like Leahy's 54-year-old father, Patrick, who had graduated to civil engineering when the business of land surveying went into decline on the commencement of the work of the Ordnance Survey in 1824. There were a number of Englishmen and some with railway engineering experience. A number of the new surveyors had been employed by the Board of Works on local projects in the years immediately preceding their appointment[27] and a significant number had been trained by Jacob Owen, the board's engineer and architect, and himself one of the examiners. Others had learned their engineering as part of the large team of assistant engineers, surveyors and valuers who worked with Alexander Nimmo until his death in January 1832. None of them, of course, had any formal engineering qualifications which did not become available at any of the universities in Britain or Ireland until the 1840s.

THE FIRST SURVEYORS

Antrim	Thomas Jackson Woodhouse	Londonderry	Stewart Gordon
Armagh	Henry L. Lindsay	Longford	James Bell (senior)
Carlow	Charles Gerard Forth	Louth	John Yeats
Cavan	Alexander James Armstrong	Mayo	Thomas Barclay
Clare	Richard Richards	Meath	Samuel Stephen Searancke
Cork (east riding)	Patrick Leahy	Monaghan	Edward V. Forrest
Cork (west riding)	Edmund Leahy	Queen's County	Alexander Harrison
Donegal	John Steedman	Roscommon	John Kelly
Down	John Fraser	Sligo	Francis Rawdon Hastings DuBourdieu
Fermanagh	Roderick Gray		
Galway	Henry Clements	Tipperary	Samuel Jones
Kerry	Henry Stokes	Tyrone	James Clarke
Kildare	Charles Lanyon	Waterford	William Johnston
Kilkenny	Samson Carter	Westmeath	Florence Mahony
King's County	Richard Boxall Grantham	Wexford	Barry Duncan Gibbons
Leitrim	Noblett Rogers St Leger	Wicklow	William Hampton
Limerick	John Walker		

When asked by a select committee in April 1835 if all of the first group of surveyors were sufficiently qualified, Burgoyne replied that 'we found many of them in some degree deficient in particular points'; he went on to say that the interview board had not rejected all who were inefficient but had 'recommended the best; we did not think all of

26 Co. Cork was divided into east and west ridings in 1823, but one grand jury continued to discharge all of the fiscal and criminal business of the county; Tipperary was not divided into ridings until 1838. 27 *A Return of the Names of Each Engineer Employed under the Commissioners of Public Works in Ireland ... in the years 1832, 1833 and 1834*, HC 1835 xxxvii (536).

them perfectly qualified to the extent we considered desirable'.[28] He had no doubt, however, that the examination process would have made those who had undergone it 'sensible' of their deficiencies and that they would endeavour to correct these. Asked if the establishment of a 'regular school for civil engineers, such as exists in Prussia and France' would provide a greater number of suitable appointees, Burgoyne hesitated, saying 'that would require much consideration'. However, William Bald, who had extensive experience of surveying and engineering in Ireland and on the continent, when asked by the same committee for his opinion on how the county surveyor system could be improved, had no doubt but that a school of roads and bridges should be established in Dublin, modelled on the École des Ponts et Chaussées in Paris, to impart both theoretical and practical knowledge to young men who wished to become assistant surveyors and to progress from there.[29] In what seems to have been a comment on Burgoyne's plan to establish the Civil Engineers Society of Ireland, Bald went on to say that while 'the formation of institutions and societies for learned purposes are very much to be admired and encouraged' these could not take the place of a school such as he had proposed. Sir Robert Kane, who had considerable influence on the development of science and education in Ireland, appears to have shared this view: in 1838 he had endeavoured in conjunction with Robert Mallet and others to organize from their own resources a training institution on continental lines but 'we were entirely in advance of our age; we met with no sufficient public support, and it could not be carried out'.[30]

TRANSITIONAL PROBLEMS

The new surveyors generally lost no time in taking up their positions as it was essential that they should attend the series of baronial presentment sessions which were to be held in preparation for the summer assizes. In making the appointments, it had been the policy that 'gentlemen ... should not be appointed to counties in which they have been accustomed to reside, or in which they have been chiefly employed'.[31] As a result, the new men found themselves working in unfamiliar territory as they grappled with the planning and supervision of a variety of schemes and projects. For some grand jurors, they personified the significant shift of power which the 1833 act had brought about at their expense, and in some cases they were looked on by grand juries 'not as their servants, but as spies appointed by hostile governments to look after them'.[32] Grand juries had been accustomed to running matters entirely their own way, and road contractors – who were often closely connected with the grand jurors and had been operating without effective supervision – had to adjust to a new situation in which all presentments were reported on by the surveyor and all completed works had to be

28 Evidence given on 8 Apr. 1835 to the *Select Committee on Advances made by the Commissioners of Public Works in Ireland*, HC 1835 xx (573); Burgoyne had expressed similar views in a letter to the chief secretary on 2 June 1834 (NAI, CSO Unregistered papers 1834/494). 29 Evidence given on 29 June 1835 to the *Select Committee on Advances made by the Commissioners of Public Works in Ireland*, HC 1835 xx (573). 30 Evidence of Sir Robert Kane, then president of Queen's College, Cork, in *Report from the Select Committee on Scientific Instruction; together with the Proceedings of the Committee, Minutes of Evidence and Appendix*, HC 1867–8 xv (432). 31 NAI, CSO Unregistered Papers 1834/720. 32 P.C. Cowan, *PRIAMCE* xxxii (1910–11), 25.

certified by him before payment could be made. In these circumstances, difficulties in making the transition to the new system could hardly have been unexpected. Relations between some of the surveyors and their grand juries were strained, sometimes to breaking point. Surveyors who attempted to rationalize and improve the road maintenance system provoked resentment among vested interests and their refusal of completion certificates to road contractors made enemies for them among interested parties. As late as 1893, R.B. Sanders, the King's County surveyor, was complaining that 'surveillance and check against jobbery makes the office of a county surveyor often very unpleasant ... when the county surveyor gives the honest and straightforward statement he should give ... he is rewarded by the strongest resentment from ... the very parties who have it in their power to most annoy and embarrass a county surveyor, both personally and in the discharge of his duties ... and very seldom will any effort to support him be made by those whose interests he has safeguarded, or whose money he has prevented from being spent for private purposes or squandered in useless and extravagant expenditures'.[33]

In addition to occasional disputes with the grand jury or with individual grand jurors about particular projects and on procedural matters, other factors contributed to the development of a less than harmonious relationship between the surveyor and the grand jury in some counties. The inability of the surveyors, due to the lack of any assistants or too few assistants, in the early years to supervise effectively the substantial road construction and maintenance programmes undertaken in those years led to regular complaints from some grand juries about the condition of the roads, and the extent of the private practice conducted by some surveyors was often a source of further friction. For their part, surveyors complained about the tendency of grand juries to substitute their own opinions for the professional judgment of their surveyor; as one of them put it, 'our present grand juries are now turned engineers and surveyors, by taking all the business out of the hands of the surveyor or surveyors, and pronouncing on the works themselves'.[34] A number of the early surveyors were dismissed by their grand juries and others who were in difficulties with their juries sought to be assigned to other counties where they could make a fresh start. In general, however, after the initial settling-in period, most of the surveyors seem to have developed a good working relationship with their grand juries, and with the authorities in Dublin Castle.

THE ROLE OF THE SURVEYOR

While it was in use until the 1940s (and in Northern Ireland until the 1970s), the title *county surveyor* was a misnomer from the very beginning and was regarded as an offensive one by many holders of the office.[35] In modern terminology, the office was that of county engineer and that title was being used by some of the surveyors themselves and even in some official quarters as early as 1835.[36] Each surveyor was legally responsible to his grand jury for a wide range of public works and services: his duties related predominantly to the

---

33 R.B. Sanders, 'Irish County Surveyorships and the Grand Jury System of Ireland', *PRIAMCE* xx (1893–4). 34 NLI, Smith O'Brien papers, MS 436, letter of 7 Apr. 1846 from Patrick Leahy.   35 J.H. Andrews, *Plantation Acres* (Belfast, 1985), p. 255.   36 See, for example, *Third Annual Report from the Board of Public Works in Ireland*, HC 1835 xxxv (76).

construction, maintenance and improvement of public roads and bridges but his responsibilities also extended to the construction and maintenance of various public buildings, including courthouses and gaols and, in some cases, district asylums. Given the size of many of the counties and the distances to be travelled, these duties were quite demanding. With the number of baronies ranging from twenty-three in Co. Cork to five in counties Leitrim and Monaghan, attendance at each of the baronial presentment sessions (often referred to as *the road sessions*) and the adjourned sessions, was itself a time-consuming business. The sessions generally took place between mid-January and mid-February, and again between late May and the end of June; and the adjourned sessions were spread over another month in each case.[37] In addition, the surveyors had to attend the spring and summer assizes – generally lasting for several days in March and July – to give professional advice and assistance in relation to the applications for presentments and to make applications for presentments themselves, where necessary. The grand jury had no direct labour organization but relied entirely on the contract system both for carrying out routine maintenance operations and for new works. This meant that the surveyor had to prepare specifications, plans and tender forms for all of the approved works before inviting competitive tenders and submitting them for approval to the adjourned presentment sessions or to the grand jury. If the works were to cost less than £50, a period of only thirty days was allowed for all of this activity to be completed; in other cases, the surveyor could wait until the next sessions or assizes before submitting the plans and specifications. Once tenders were approved, it was the surveyor's duty to supervise the performance of the works by the contractors, and to certify the payments due to them.

At each assizes, the surveyor was required to report to the grand jury on the progress of the county works, the performance of contractors, and the general condition of the roads, bridges and county buildings. In practice, he was also required to send regular reports and returns, each in a prescribed form, to the authorities in Dublin Castle who exercised fairly detailed surveillance over the affairs of the grand juries. The nature of the reporting requirements is seen in a letter of 9 May 1840 from the under-secretary's office to one of the surveyors demanding an explanation as to why a report was not written 'on half margin' as required by a circular from that office.[38] Bureaucratic requirements of this kind should not, however, be confused with central controls which were, in fact, very limited. A cursory review of the chief secretary's office registered papers for the years from 1837 onwards shows that a substantial number of memorials and letters of complaint about county surveyors were addressed to that office each year by individuals throughout the country and, while it was the practice of the office to call for a report from the surveyor in each case, the explanations offered seem to have been accepted without question in most cases. Although the lord lieutenant had power to dismiss a surveyor summarily, he and his officials were reluctant to intervene and generally left it to the grand juries to take action, where necessary. The official attitude is well illustrated by Burgoyne's response to a complaint made by Lord Clements in 1840

---

37 *Return of the Number of Days appointed by the Sheriff for Transacting the Fiscal Business in each County in Ireland*, HC 1844 xliii (130).   38 NAI, CSORP 1840/W/6004, letter of 9 May 1840 to Edmund Leahy and reply of 12 May.

about the performance of the Leitrim surveyor, T.D. Hall: in a letter to Thomas Drummond, the under-secretary, he advised that 'the government should not initiate an inquiry into assumed mismanagement on the part of a county surveyor on the representation of one, or even more individuals, however high in station and character they may be'; he went on to suggest that 'a fishing research into the general management of the county surveyor's business would hardly be fair' and that specific complaints should instead be brought to the attention of the grand jury.[39]

A form of supervision was indirectly exercised by the Board of Works arising from its functions in relation to the allocation of grants and loans to grand juries for major road and bridge projects and for the construction of courthouses and other public buildings. Where a project of this kind was submitted to the board, they arranged that the county surveyor's plans would be reviewed by one of their own engineers or, occasionally, by another county surveyor engaged on a fee basis for the purpose. Reviews of this kind often led to suggestions for modifications of plans which, on occasion, highlighted the inadequacies of some of the early appointees and were never welcomed by the surveyors. Burgoyne made no secret of his view that few of the early surveyors were not deficient in some respects and that a form of central supervision was essential: he told the under-secretary in 1840 that he 'was always of opinion that some general inspection and superintendence over the management of the surveyors in the execution of the county works would be very desirable, but the suggestion was overlooked either on account of expense or an objection to any such centralization' and they were all 'sent out uncontrolled in their proceedings'.[40] The Select Committee on Public Works expressed support in 1835 for Burgoyne's view that power to control and direct the county surveyors should be given to the Board of Works in order to ensure co-ordination and a measure of uniformity in standards[41] but no statutory provision for direct central supervision of the work of the grand juries or their surveyors was made until the enactment of the Local Government (Ireland) Act 1898.

The grand jury met only twice each year, had no legal existence between assizes, and was always under pressure to complete its administrative business in the few days before the judges arrived to deal with criminal matters.[42] Its secretary, although usually a very influential and well-connected official, had no supervisory role in relation to the other officers and there was no proper system of independent financial audit of the affairs of the grand juries until 1878. Thus, in practice, the county surveyors had a great deal of independence in going about their duties and the arrangements for providing and maintaining public works at local level depended very much for their effectiveness on the ability, integrity and energy of the individual surveyor who had personally to carry out all of the duties of the office in the first few years: there was no provision in the 1833 act for the appointment of assistants to the surveyor but only a provision for the employment of a clerk who was to be in regular attendance at the surveyor's office – generally located in the county courthouse.[43] However, having heard evidence that it was impracticable for

---

39 NAI, OPW 1/1/4/19, Public Works Letters No. 19, p. 229.   40 Ibid.   41 *First Report from the Select Committee on Public Works, Ireland*, HC 1835 xx (329).   42 The situation in Co. Dublin was different in that the law provided for the appointment of a finance committee of twelve members which had power to authorize payments between assizes.   43 Each surveyor was paid £50 a year to meet the cost of the office and the wages of his clerk: 3 & 4 Wm. IV, c.78, section 41.

a surveyor to carry out his duties effectively, even in the smallest county, without assistance, a select committee recommended in 1836 that the surveyors should be allowed to appoint assistants who would be paid out of public funds.[44] The government accepted this and provision for assistants, with a salary of up to £50 a year, was made by law in August 1836.[45] The number of assistants in any county was to be decided by the grand jury and the appointments themselves were to be made by the county surveyor from among persons certified to be fit and competent by the same board of engineers which examined candidates for county surveyorships. Appendix I deals with the issues arising in subsequent years in relation to the number of assistant surveyors, the appointment procedure and their qualifications and salaries.

EXTENSION TO DUBLIN

Bills were introduced at Westminster in 1838 and again in 1843 by the MPs for Co. Dublin providing for the extension of the county surveyor system to the county, but these did not reach the statute book. In August 1844, however, an act which passed without debate[46] authorised the lord lieutenant to direct the Dublin grand jury to divide their county into a number of road districts and to appoint a district surveyor for each, but there was to be no provision for assistant surveyors in Dublin until 1897. As in the case of the county surveyors, the new district surveyor appointments were to be made from a panel of candidates whose qualifications had been certified by a board of three civil or military engineers appointed by the lord lieutenant but the actual selection of those to be appointed was left to the grand jury. Initially, three road districts were established, and a board consisting of Colonel Harry Jones, chairman of the Board of Works, and two leading engineers, Sir John MacNeill and Thomas Rhodes, was established to examine the twenty-four candidates who came forward.[47] After scientific and practical tests conducted over three days, Alexander Tate was placed first in order of merit and was duly appointed early in 1845, together with Robert J. Hampton and Richard H. Frith.

SALARIES

While the 1817 Act had prescribed a salary of up to £600 for the surveyors of the larger counties, the 1833 Act fixed a standard maximum salary of only £300 a year, inclusive of all costs and travelling expenses, and without regard to the size of the county.[48] The small number of county surveyors who also served counties of cities and towns received

---

44 *Report from the Select Committee on Grand Jury Cess (Ireland)*, HC 1836 xii (527).   45 An Act to consolidate and amend the Laws relating to the Presentment of Public Money by Grand Juries in Ireland (Grand Jury (Ireland) Act 1836), 6 & 7 Wm IV, c.116, section 43; in Co. Dublin, however, provision for the appointment of assistant surveyors was not made until 1897.   46 An Act to Consolidate and Amend the Laws for the Regulation of Grand Jury Presentments in the County of Dublin, 7 & 8 Vict., c. 106.   47 Evidence of Alexander Tate in July 1857 to the *Select Committee on County & District Surveyors etc. (Ireland)*, HC 1857 (Sess. 2) ix (270).   48 3 & 4 Wm. IV, c.78, section 39; the Grand Jury (Ireland) Act, 1836, 6 & 7 Wm. IV, c.116, which consolidated grand jury law, restated the £300 salary limit in section 41.

additional salary payments, but these did not generally exceed £100 a year. Only in the case of Patrick Leahy, who was surveyor for Cork city as well as for the east riding of the county, was the maximum salary paid for the second post, bringing his total earnings up to £600 a year. The salaries fixed for the first three officeholders in Co. Dublin were only £130 a year, but, faced with evidence of the extent of the out-of-pocket expenses incurred by the surveyors in travelling, and an indication that they were considering defecting to the service of one of the railway companies, the grand jury soon obtained the consent of the lord lieutenant for higher salaries of £250 for two of the posts and £300 for the third and more demanding post in the southern district.[49]

The surveyors' salaries did not compare favourably with those fixed by law for other senior local officials whose duties were not so demanding; some of these enjoyed salaries of up to £500 a year, supplemented in some case by fees which they were entitled to charge for various services.[50] In earlier years, the holders of major engineering offices under other public authorities had also enjoyed much higher salaries than those of the surveyors. In the 1820s, for example, John Brownrigg and John Killaly, the engineers employed by the Directors General of Inland Navigation, earned salaries of £525 each.[51] The chairman of the Board of Works, the most senior engineering official in the country in the 1830s, earned £1,000 a year, the two other members of the board earned £600 each, and Jacob Owen, the board's engineer and architect, earned £800 a year at that stage.[52] Within a year of their appointment, surveyors were complaining that their salaries were far too low, bearing in mind the wide range of duties which they had to perform and the lack of any provision for the payment of travelling expenses.[53] The Select Committee on County Cess heard evidence in 1836 from several witnesses on this point, including Woodhouse, the Antrim surveyor, who stated that travelling expenses alone, including the keep of his two horses, cost him £200 a year or two-thirds of his salary.[54] But the committee was required primarily to concern itself with finding ways and means to reduce the county cess and so could not see its way to recommend any increase in salary. This led to the resignation later in 1836 of one of the surveyors, John Walker of Limerick,[55] and influenced the decision of Woodhouse to resume his earlier career in railway engineering.

The establishment in November 1840 of a royal commission to review grand jury law provided another opportunity for the surveyors to advance their case for higher salaries. Following a meeting in Dublin chaired by Charles Lanyon, a statement of their views was conveyed to the lord lieutenant by Henry Stokes, the Kerry surveyor, who declared that 'if any mischief has been done by bad surveyors, the cause of such being appointed is that better educated men could not be had for the salary of £300 per annum'.[56] The

49 *Abstract of presentments Dublin County Grand Jury, Easter 1846*; by 1857, two of the Dublin posts commanded a salary of £350, while the third man was paid £300. 50 Grand Jury (Ireland) Act, 1836, 6 & 7 Wm. IV, c. 116, section 110 and Schedule S. 51 *Civil Departments: Returns of the Names and Offices of all Persons whose Salary and Emoluments exceed £250 per annum*, HC 1830 xvii (480). 52 *Return ... specifying the Name, Salary and Situation of each Person receiving upwards of £800 per annum*, HC 1835 xxxvii (609). 53 See, for example, CSORP 1835/449, 562 and 1449, letters from Lindsay, St Leger and Walker. 54 Evidence of T.J. Woodhouse to the *Select Committee on County Cess (Ireland)*, HC 1836 xii (527). 55 NAI, CSORP 1835/1449; CSORP 1836/3245; *Report of the Commissioners appointed to Revise the Several Laws under or by virtue of which moneys are now raised by Grand Jury Presentment in Ireland*, Appendix B, HC 1842 xxiv (386). 56 Evidence of Henry Stokes in *Report of the Commissioners appointed to Revise the Several Laws under or by virtue of which*

absence of any provision for travelling and subsistence expenses was a particular grievance, given that a surveyor who regularly travelled around his county to inspect and supervise the various works was, in the final analysis, worse off than a colleague who rarely ventured outside his office in the county town. The surveyors must also have known that officials in other branches of the public service who received allowances for travelling and other expenses found that their incomes were considerably boosted by the generous rates at which these allowances were set. Anthony Trollope, for example, who served as clerk to one of the post office district surveyors in Ireland in the 1840s, received a salary of only £100 a year but found that his actual income was £400 after paying all of his expenses; his allowances of fifteen shillings a day for subsistence and six pence a mile for travelling expenses obviously included a large profit element.[57]

The report of the royal commission in 1842 had no effect on the salary level of the surveyors who, for another twenty years, maintained pressure on the government through letters, memorials and deputations for salary maxima which would vary with the size of the county, and for allowances to cover travelling and incidental expenses like postage and stationery.[58] In August 1857, a select committee concluded that the surveyors had 'a fair claim for some increased remuneration' and suggested that counties should be divided into three classes by reference to area, population, mileage of roads and other considerations, and that an appropriate salary should be fixed for each class.[59] In February 1858, the chief secretary, H.A. Herbert, presented a bill to give effect to this recommendation and to allow for the granting of pensions of up to two-thirds of salary to long-serving surveyors,[60] but he had been replaced by Lord Naas, on a change of government, before the bill could be enacted. There followed a three-year campaign by the surveyors to have the recommendation carried into law: deputations met Naas and his successor in London and Dublin and presented draft bills to them;[61] petitions were presented to parliament and to the lord lieutenant, some of them supported by substantial numbers of the Irish peers and MPs;[62] there were supportive leading articles in the press; and twenty-seven of the county grand juries were persuaded to vote in favour of allowing higher salaries. Eventually, in March 1861, another bill was brought forward by Edward Cardwell, who had replaced Lord Naas in June 1859, providing for increased salaries and for pensions for surveyors who were forced to retire on grounds of ill health, but this latter provision encountered serious opposition from MPs who argued that it would throw increased burdens on the county cess and lead to claims for similar treatment by other public officials. Finally, in July 1861, the bill, shorn of the superannuation provisions, was pushed through parliament by Sir Robert Peel, son of the former prime minister, who had just succeeded Cardwell as chief secretary.[63] It divided the counties into three classes for salary purposes, as recommended by the 1857 committee, and allowed for higher maximum salaries, at the discretion of the grand jury:

Moneys are now raised by Grand Jury Presentment in Ireland, HC 1842 xxiv (386).   57 Anthony Trollope, *An Autobiography*, 2 vols (London, 1883).   58 See, for example, NAI, Official Papers 1839/77/1, 1845/209 and 1848/17.   59 *Report from Select Committee on County & District Surveyors etc. (Ireland)*, HC 1857 (Sess. 2) ix (270).   60 HC 1857–58 ii (19).   61 NLI, Earl of Mayo papers, MS 11,035; Larcom papers, MS 7747.   62 See, for example, a memorial carrying the names of 'some of the principal members of the two branches of the legislature', published in the *Kilkenny Moderator*, 25 Jan. 1860.   63 An Act to enable Grand Juries in Ireland to increase the remuneration of County Surveyors and for other Purposes, 24 & 25 Vict., c.63.

Class 1–maximum salary £600
Cork (east riding), Cork (west riding), Antrim, Clare, Down, Kerry, Londonderry, Meath, Tipperary (south riding), Wexford.

Class 2–maximum salary £500
Armagh, Cavan, Donegal (northern division), Galway (western division), Galway (eastern division), Fermanagh, Kildare, Kilkenny, Queen's County, Roscommon, Sligo, Tipperary (north riding), Tyrone (northern division), Tyrone (southern division), Waterford, Wicklow.

Class 3–maximum salary £400
Carlow, Donegal (southern division), King's County, Leitrim, Limerick (eastern division), Limerick (western division), Longford, Louth, Mayo (northern division), Mayo (southern division), Monaghan, Westmeath.

Sir Thomas Larcom, as under-secretary, had been opposed to giving the grand juries any discretion in relation to the salary to be paid to an individual surveyor, as it could lead to 'jobbing presentments to please this or that grand juror' but he was over-ruled.[64] The proposed salary increases attracted criticism from ratepayers' representatives but the *Dublin Builder* hoped that this would be consigned to ignominious oblivion given 'the patent fact that the professional gentlemen holding the appointment ... are the worst-paid public officials, considering the onerous character of their duties and their position in society, of any in the Kingdom'.[65] There had also been criticism of the proposed increases in some newspapers which argued that holders of the office attracted lucrative private business, a criticism which led to the inclusion in the act of a provision allowing grand juries to prohibit private practice as a condition of any salary increase. Against this background, it cannot have come as a surprise to the surveyors to find that relatively few of them were automatically granted increased salaries. In 1863, only 17 of the 40 who were then serving had been granted the full increase; three had been awarded increases less than the maximum; and 20 others were still being paid at the rate set in 1834. On the other hand, none of the grand juries which had allowed salary increases opted to prohibit private practice.[66] Many surveyors had to work long and hard to convince their grand juries that they deserved to be paid the new maximum salaries or, indeed, to receive any salary increase at all; in 1877, the majority of them were still being paid at rates below the maxima provided for in 1861.[67]

64 NLI, Larcom papers, MS 7753, letter of 20 June 1857.   65 *DB*, 1 July 1861, p. 559.   66 *Return showing the Names of the County Surveyors in each County in Ireland, and their Assistants, with their respective Salaries, etc.*, HC 1863 l (277).   67 *Return of County Surveyors and their Assistants in each County in Ireland for the Year ending with the Summer Assizes 1877*, HC 1878 lxvi (218).

3

Changes in Personnel, 1834–69

The ranks of the original corps of thirty-two surveyors were reduced by illness and death within a few years of the first appointments. Of those who took up office in 1834, James Clarke of Tyrone was first to die (1838) and he was followed to the grave by William Johnston of Waterford (1840). Charles Forth died suddenly in Waterford in July 1845, only a year after he had moved there from Carlow, and William Hampton of Wicklow also died that year. Others who died in the following years were Alexander Harrison of Carlow (1851), Alexander Armstrong of Cavan (1852), Thomas Barclay of King's County (1855), Stewart Gordon of Londonderry (1860), and Samson Carter of Kilkenny (1861). Noblett R. St Leger, having moved from Leitrim to Sligo in 1836, continued in office there for thirty-six years until his death in 1872 when James Bell of Longford also died.

With dissatisfaction about their salary levels, the lack of provision for travelling and subsistence expenses, the absence of pension arrangements, and the limitations imposed by the grand juries on the number of assistants allowed in many counties, it is surprising that resignations by surveyors to pursue more lucrative professional business were not more common, especially in the years up to 1861 when salary increases were finally conceded. On the other hand, it must be remembered that a liberal regime applied in relation to private practice, that significant private sector engineering projects (apart from railway schemes) were few and far between in the second half of the nineteenth century, and that only a small number of the surveyors were skilled in architecture where greater opportunities for private practice arose. But there were some defections. John Walker of Limerick, who claimed that he had given up well-remunerated work at ruinous cost to himself to take on the post of county surveyor, resigned in November 1836 following the failure of the select committee which reported some months earlier to recommend an increase in salary; however, he was allowed to return to the service in the following year and worked in a number of counties until 1854. Thomas Jackson Woodhouse left his post in Antrim in 1836 to return to his railway engineering career. Barry Duncan Gibbons, the first Wexford surveyor, resumed his harbour engineering career in 1838, taking up a post as resident engineer at Kingstown and going on to become harbour engineer, and in 1857 principal engineer, in the Board of Works. Samuel Ussher Roberts, having served in Galway since 1855, followed Gibbons into the Board of Works in 1873, the only county surveyor to attain the rank of commissioner in that organization. The only transfer by a surveyor into the service of the Local Government Board arose in 1899 when P.C. Cowan moved from Co. Down to become the board's chief engineering inspector; before that, the board had little to do with road construction and improvement which were the areas in which the surveyors tended to specialise.

Some of the early surveyors resigned primarily to develop private practice of one kind or another. Edmund Leahy of Cork west riding resigned in January 1846, when he was refused leave of absence to pursue his parallel railway engineering interests. Robert Garrett of Donegal resigned in 1850 to pursue a career as consulting engineer, a move which was to cost him his life in 1857 at the siege of Cawnpore. Sir John Benson left his post in Cork east riding in the mid-1850s to concentrate on other engineering and architectural projects, mainly in the Cork area, and Charles Lanyon, the Antrim surveyor, did likewise in 1861 and, in addition, went on to become Mayor of Belfast in 1862 and MP for the city in 1866, for all of which he too earned a knighthood. Thomas Turner left his post in Cavan within a year of his appointment to return to his practice as an architect, although his partner, Richard Williamson of Londonderry, did not consider it necessary to resign his county surveyorship; Turner re-entered the ranks in 1883.

DISMISSALS

In contrast to those who virtually became institutions in their counties, other nineteenth-century county surveyors had relatively short terms of office because of the provision in the Grand Jury (Ireland) Act 1833 which allowed either the grand jury or the lord lieutenant to dismiss a surveyor.[1] The first to suffer this fate was Richard Richards whose removal from office was sought by the Clare magistrates only five months after his appointment; he elected to resign in November 1834 rather than face dismissal by the lord lieutenant. Others who suffered dismissal were Samuel Jones, Tipperary south riding (1852) and John Walker of Monaghan (1854). In some cases, incompetence (in the view of the grand jury) or a breakdown in relations between the surveyor and the jury was involved but there were cases too in which a grand jury was obliged to dismiss a man who had become totally incapable due to old age or ill-health, even though this left the unfortunate man with little to live on.[2]

In a number of cases where a surveyor resigned rather than suffer dismissal, or was actually dismissed, he was subsequently appointed by the lord lieutenant to a post in another county – an indication, perhaps, of the reluctance of the authorities in Dublin Castle to interfere with decisions made locally while upholding their own assessment of an individual's qualification to hold office. After Richards had resigned in Clare in November 1834, he re-entered the service in August 1840 when he was appointed as surveyor for Tyrone. Similarly, when Thomas Barclay of Mayo was facing dismissal in

---

1 Section 38, restated as section 40 of the Grand Jury (Ireland) Act, 1836. 2 Evidence of Henry Stokes, Kerry county surveyor, on 4 Aug. 1857 to the *Select Committee on County & District Surveyors etc. (Ireland)*, HC 1857 (Sess. 2) ix (270).

1836, he was transferred by the lord lieutenant to King's County. When W.H. Deane was dismissed by the grand jury of Tipperary south riding in July 1860 the chief secretary's office, having consulted Sir Richard Griffith at the Board of Works, decided that the reasons for his removal were not such as to debar him from further service and he was therefore appointed to Tyrone.[3] But the case of Patrick Leahy (Cork east riding) provides an illustration of the determination of the central authorities to allow the grand juries the freedom to dismiss their surveyors in circumstances where basic qualifications were not at issue. When Leahy was dismissed by his grand jury in March 1846 after his involvement in a questionable transaction relating to the repair of a quay came to light,[4] he attempted to avert dismissal by suggesting to Dublin Castle that he should be allowed to continue, at least for some months, so that he could finish the works in progress and settle the outstanding accounts. But the chief secretary's office was not prepared to entertain Leahy's pleas: he was advised that 'the law having invested the grand jury with the power of dismissal of any person holding the office of county surveyor, His Excellency the Lord Lieutenant can only proceed to appoint Mr Leahy's successor as soon as circumstances permit'.[5]

Questions about the insecurity of the surveyors' tenure of office continued to be raised as late as the 1880s. For instance, William Harte, one of the Donegal surveyors, told a royal commission in June 1887 that 'the notice to quit and capricious eviction from office has been used by those in power should the officer, in the honest discharge of his duty, render himself obnoxious to those into whose hands the law has delivered him, tied hand and foot'. Harte went on to quote from a resolution adopted at a well-attended meeting of county surveyors which spoke of their exposure to 'men of influence with whom they would frequently be in collision'; these were men, according to Harte, who could stop the salary of a surveyor 'and, when he is worn out in the public service, they can refuse or simply neglect to present for one shilling of retiring allowance for him'.[6] Notwithstanding criticism like this, the law remained unchanged until the end of the grand jury era and the position of a surveyor who lost favour with his grand jury continued to be a precarious one.

TRANSFERS

Extensive use was made between 1834 and 1869 of a provision in the Grand Jury Acts which allowed the lord lieutenant, from time to time, to transfer a surveyor from one county to another. In some cases, a surveyor was transferred to fill a vacancy caused by the death, resignation or dismissal of a colleague in another county. Occasionally, this involved only a single move, as in December 1845 when James Boyd of Clare was transferred to Wicklow to replace William Hampton who had died. In other cases, a chain of transfers was implemented: for example, when Charles Lanyon moved in October 1836 from Kildare to replace Thomas Jackson Woodhouse in Antrim, he was

3 NAI, CSORP 1868/8719.   4 *CC*, 14 and 17 Mar. 1846.   5 NAI, CSORP 1846 W 5222.   6 *Royal Commission on Irish Public Works, Appendix to the Second Report, Minutes, Evidence and Index* [C. 5264–1] HC 1888 xlviii.

succeeded in Kildare by John Yeats of Louth, who in turn was replaced by Francis Dubourdieu of Sligo, leaving that post to be filled by the transfer of Noblett R. St. Leger from Leitrim. It seems that these and most of the subsequent redeployments were arranged with the consent of the individual surveyors, or at their request; indeed, two of them told a select committee in 1868 that there had been 'only one single case of a removal ... which was done without the knowledge of the county surveyor'; this was probably a reference to the transfer in May 1867 of A.C. Adair from Clare to Tyrone (where a lower maximum salary rate applied) to make it possible for John Hill to return to that county.[7]

Most movements among the surveyors seem to have been initiated by the office-holders themselves who regularly applied for a transfer immediately a vacancy arose in another county, or set up direct exchanges with colleagues. There were numerous examples of this: John Walker of Carlow and Alexander Harrison of Monaghan exchanged places in August 1850; John Fraser of Down moved to Donegal in September 1852 replacing Henry Smyth who took over in Co. Down; and Frederick Deverell of Mayo and William A. Treacy, Cork east riding, took one another's places in April 1863. Personal reasons often led to such exchanges as in 1868, when Treacy, having moved to Mayo, found that he could not arrange proper tuition for his children there, and agreed to exchange places with W.H. Deane of Tyrone who had lost the confidence of his grand jury.[8] After 1861, some additional movements from one post to another followed the introduction of three levels of salary for counties in different classes, with a salary 50 per cent higher in Class 1 than in Class 3; by the 1870s, it was common practice when a better-paid post became vacant for up to five of the existing surveyors to apply for it.

In several cases surveyors returned after a period to a post they had held before. An extreme example was William Carroll who left Monaghan in September 1854 after only six months to take up duty in Waterford, but returned after another six months to Monaghan where he displaced Charles Tarrant who was transferred to the Waterford post. John Hill of Clare moved voluntarily in 1855, after ten years service, to King's County from which he returned to Clare in 1867 'at the request of local gentlemen'. In some cases, removal to another county was deemed necessary to protect the health or even the life of a surveyor. John H. Brett was transferred to Kildare in 1870 after six years service in the western division of Limerick because, having been fired on twice, he felt that his life was in danger there. James A. Dickinson had to leave Westmeath in 1869 after eight years service when he received letters threatening that he would be killed; he was under police protection for several months before his removal to the southern division of Tyrone. *The Times* reported in the same year that a third surveyor in a western county 'had to be sent North' for his own safety and commented that, in the case of all three, 'their only crime is that they will not give false certificates to road contractors for work which is not properly executed'.[9]

---

7 Evidence of Henry Brett and Alexander Tate in *Report from the Select Committee on the Grand Jury Presentments (Ireland)*, HC 1867–68 x (392). 8 NAI, CSORP 1868/8719. 9 *The Times*, 3 Nov. 1869.

RETIREMENTS

As was the case for most other public officials until late in the nineteenth century, no mandatory retirement age was prescribed for the county surveyors and some of them served to a great age before retiring. There was no provision for pensions in the early years and, despite regular representations and lobbying by surveyors themselves and recommendations of select committees of the House of Commons in 1857 and in 1868,[10] moves by the government to introduce even limited pension arrangements were resisted in the commons in 1861 and on a number of other occasions. Finally, in August 1875 the law was amended to allow a grand jury, with the consent of the lord lieutenant, to make pension payments to a retiring surveyor;[11] this was required, Lord O'Hagan told the House of Lords, 'by justice to certain old and deserving officers'.[12] Under the new act, a pension of up to forty-sixtieths of salary could be paid on retirement by reason of permanent infirmity of mind or body or where the surveyor was at least 60 years old and had at least ten years service *in the county*. In the case of a surveyor appointed after 1875, no pension was payable unless he had devoted his whole time to the post. There was no provision for aggregation of service in a number of counties for pension purposes, although the Antrim grand jury was allowed in November 1885 to add five years to the service of a man who had served there for twenty-five years; this, in effect, anticipated the 'added years' provision which became applicable to professional staff in the public service at a much later stage. When a private member's bill was introduced at Westminster in 1897 to allow for aggregation of service, with provision for pro-rata payments by the counties involved, it was rejected by the chief secretary on the basis that it 'violated all accepted notions of propriety and justice'.[13] However, provision to allow aggregation was made in the Local Government (Ireland) Act 1898 in the case of surveyors who were appointed after that act was passed.[14]

In the absence of provision for pension payments, a number of the grand juries allowed their surveyors to continue in office even where they had become seriously ill. In Galway, for example, Henry Clements 'was kept on for several years' before his eventual dismissal in 1858 because the grand jury 'acting with kindness and consideration' did not wish to 'throw a man upon the world'.[15] In the same way, the Donegal grand jury allowed John Steedman, their first surveyor, to continue in office until they were forced eventually in 1847 to dismiss him because of his inability to perform his duties; twenty years later, according to one of his former colleagues, the unfortunate Steedman was 'subsisting on the charitable contributions of a few surveyors and friends who endeavour to support him'.[16] In both cases, the grand juries had indicated that they would be perfectly willing to provide pensions if the law were amended to allow them to do so; according to under-secretary Larcom, both cases were brought to the special notice of the

10 *Report of the Select Committee on County & District Surveyors etc. (Ireland)*, HC 1857 (Sess. 2) ix (270); *Report from the Select Committee on the Grand Jury Presentments (Ireland)*, HC 1867–8 x (392).   11 An Act to enable Grand Juries in Ireland to grant Superannuation Allowances to County Surveyors in certain cases, 38 & 39 Vict., c.56.   12 Parliamentary debates III ccxxvi, 29 July 1875, col. 167.   13 County Surveyors (Ireland) Act Amendment Bill, 1897; Parliamentary debates IV xlvi, 12 Feb. 1897, col. 1897.   14 61 and 62 Vict., c.37, section 83 (13).   15 Evidence of Henry Brett, Wicklow county surveyor, in *Report from the Select Committee on Grand Jury Presentments (Ireland)*, HC 1867–8 x (392).   16 Ibid.

lord lieutenant on the basis that the surveyors had been 'discontinued from office solely in consequence of physical incapacity'[17] but the amendment of the law came too late to benefit the two men. Another of the original 1834 appointees was said in 1868 to have been 'entirely incapacitated for some years past';[18] this was probably Samuel Searancke of Meath who was forced by failing health to retire in July 1874, after forty years service, but with no pension. Within a year of the enactment of the 1875 act, Roderick Gray of Fermanagh, John Kelly of Roscommon, and Henry Stokes of Kerry, the only remaining 1834 appointees, retired on pension, each having served for more than forty years. Later examples of long-serving surveyors were Henry Brett of Wicklow who died in 1882 after forty-six years service; John Neville of Louth, who served for forty-six years before retiring, aged 73, in 1886; James Barry Farrell who retired in 1891 from his post in Wexford after a total of fifty-two years service; and John Hill of Clare who retired in March 1893 with over forty-seven years service.

NEW POSTS

The number of county surveyor posts increased gradually. In December 1838, Tipperary was divided into two ridings taking account, among other things, of the peripheral location of the county town of Clonmel and the extent of the county, and a second surveyor was subsequently appointed.[19] None of the other grand juries sought the division of their counties into ridings, as the law allowed, but in several cases arrangements were made to split a county into two or more divisions; this made it possible for the lord lieutenant to appoint a separate surveyor for each division, relying on the power in the Grand Jury (Ireland) Act 1836 to appoint one or more surveyors for any county. Such arrangements were introduced in response to frequent complaints by the surveyors of some of the larger counties about the burden of work they had to carry. Counties affected were:

- *Limerick*, divided into eastern and western divisions in 1837.
- *Galway*, divided into eastern and western divisions in 1838.
- *Tyrone*, divided into eastern and western divisions in 1847 and, in the 1860s, into northern and southern divisions.
- *Donegal*, divided into northern and southern divisions in 1851.
- *Mayo*, divided into northern and southern divisions in 1856, after a period during which there had been a second surveyor for the county as a whole.
- *Down*, divided into northern and southern divisions in 1862.
- *Cork east riding*, divided into northern and southern divisions in 1862.

With the seven new posts resulting from the establishment of county divisions and the creation of two ridings in Tipperary, the total number of surveyor posts had increased from the original 32 to 40 by 1862. Appointments to new posts and to fill vacancies which

17 NLI, Larcom papers, MS 7747. 18 Evidence of Henry Brett, Wicklow county surveyor, in *Report from the Select Committee on Grand Jury Presentments (Ireland)*, HC 1867–8 x (392). 19 Donal A. Murphy, *The Two Tipperarys* (Nenagh, 1994).

arose because of deaths, resignations and removals from office, meant that the panel of qualified candidates which had been established in 1834 was exhausted after a few years. In all, 38 appointments were made from that panel, 32 in May 1834 and 6 others in the following three years.

THE 1838–41 EXAMINATIONS

In April 1838, after the original examiners, Burgoyne, Radcliffe and Owen had been commissioned to hold a new series of examinations, a circular was issued to about thirty men who had expressed interest in county surveyor positions notifying them that the examinations would begin on 23 April and directing that they should bring testimonials with them and specimens of their drawings and other work. They were also advised that the examinations would be 'in the practice and principles of surveying, road-making, bridge-building and ordinary civil engineering work; these with a knowledge of the most advantageous manner of employing workmen, and of the nature and modes of applying different materials are the essential articles of information required; but it would be advantageous if the party also possessed some knowledge of building in general and architecture, likewise of drawing and mathematics'.[20] Having examined thirteen candidates, the examiners presented the names of two qualifiers in mid-May to fill vacancies in Galway and Wexford.[21] In June, six of the other candidates were advised that they were found not to be competent and a number of others were told that they had been placed on a list of second-class candidates who might qualify after some further experience.[22]

The examinations then appear to have taken on a rolling character. In December 1838, three of the second-class candidates were invited to attend for interview on the basis that the board 'not having been satisfied with the answering at the last examinations of the majority of the candidates … have thought it advisable to examine new candidates but they are willing, if you choose to attend, to put some further questions to you with a view to ascertain whether in the interim you have acquired more knowledge of the different branches of engineering, but especially in that of bridge-building'.[23] Having conducted examinations from 10 to 20 December of these candidates and sixteen others, the board reported on 26 March 1839 that three of them could be appointed to fill any vacancies that might arise but only on the basis that they were 'if not so perfectly qualified as we should desire, are as capable as are likely to be obtained or expected for the employment'.[24] Three others were placed in a second class comprising 'those whose attainments would induce us not to recommend them for the present but who may make themselves in a moderate space of time equal to the appointment'; two of the men involved (John Neville and John Benson) were later to rank among the most able and effective county surveyors of the nineteenth century. The remaining thirteen 'whose acquirements we consider very inferior' were declared not to be qualified. This less than satisfactory outcome prompted

20 NAI, OPW 1/1/4/15, Public Works Letters No. 15, p. 3.   21 NAI, CSORP 1838/1033, letter of 18 May 1838 from the Board of Works.   22 NAI, OPW 1/1/4/15, Public Works Letters No. 15, p. 178.   23 NAI, OPW 1/1/4/16, Public Works Letters No. 16, p. 338.   24 NAI, OPW 1/1/4/17, Public Works Letters No. 17, p. 239.

Burgoyne to tell the under-secretary in January 1840 that 'the office, by degrees, is getting into hands unequal to a proper performance of the duties', adding that since the first examination 'a very inferior class have presented themselves from which the board of examiners have selected the best, but I have never deemed them to possess the qualifications that they considered desirable'.[25]

When the first three candidates had received appointments, those in the second class were invited to attend again for examination in August 1840 and of these John Neville, being found to be considerably improved, was recommended for appointment; it appears that Benson did not take up the invitation to be re-examined at that stage.[26] Fresh applications were received later that year but examinations were not held for some months. In August 1841, when a vacancy arose in Queen's County, the Board of Works told the lord lieutenant that they would shortly send him the results but, in the meantime, they felt able 'to report decidedly that Mr Henry Owen stands first on the list' and recommended him for appointment; Owen was a son of one of the examiners and worked with him at the Board of Works.[27] A list of a further eight qualified candidates was produced in December 1841;[28] by March 1846, four of these had received appointments and the others were not available for one reason or another. The lord lieutenant therefore directed that new county surveyor examinations should take place but, because of the additional work already arising in the organization of famine relief works, the Board of Works advised that they could not arrange a general examination 'in the present state of the office business' but would send for 'the gentlemen they think likely to be suitable and examine them'.[29] Having examined John Benson (who had been examined previously in 1838), the board advised that they were now satisfied as to his competence[30] and a few days later they reported that they had also re-examined William Augustus Treacy and recommended him for appointment.[31] This brought the number of appointments arising from the 1838–41 examinations to thirteen.

THE 1847 EXAMINATIONS

Due to pressure of work arising from the famine, a fresh group of up to forty candidates was not summoned for examination until June 1847. In September, the examiners – now comprising Colonel Harry Jones (who had replaced Burgoyne as chairman of the Board of Works), together with Radcliffe and Owen – produced an initial list of four candidates who appeared to be 'the best qualified of all that attended the examination'; they advised that these should be sufficient to fill the places that were vacant and 'when the Board shall have time to examine some others of those who attended, they will forward a complete list stating the qualifications of all the candidates'.[32] The organization of these 1847 examinations seems to have been an administrative disaster. Due to what was described as 'an error merely clerical', the wrong name was given for one of the first four qualifiers,

25 NAI, OPW 1/1/4/19, Public Works Letters No. 19, p. 229. **26** NAI, OPW 1/1/4/21, Public Works Letters No. 21, pp 7, 41, 104. **27** NAI, OPW 1/1/4/23, Public Works Letters No. 23, p. 331. **28** *Returns of the Dates of Commissions issued by the Irish Government for the Examination of Candidates for the Office of County Surveyor in Ireland, etc.*, HC 1856 liii (335). **29** NAI, OPW 1/1/4/33, Public Works Letters No. 33, p. 338. **30** Ibid., pp 336, 352. **31** Ibid., p. 394. **32** NAI, 2D/57/33, OPW Letter Book, letter of 9 Sept. 1847.

forcing the lord lieutenant reluctantly to cancel an appointment he had already notified to the individual concerned and the relevant grand jury.[33] On top of this, the examiners discovered that many of the candidates had not been given an opportunity of answering 'several of the questions which were of the greatest importance' and were forced to recall these for a further examination at the end of October.[34] To make matters worse, the recall notice was misdirected in at least one case and yet another examination had to be arranged in early November for the candidate concerned. The examiners moved quickly after that and produced their final list, which included five additional qualifiers, on 30 November. All nine of those who qualified were appointed in the following four years.

THE 1851 EXAMINATIONS

Radcliffe and Owen were again members of a board set up in April 1851 to conduct examinations, with Richard Griffith (who had become chairman of the Board of Works in 1850) as the third member. On this occasion, there were said to have been as many as seventy-four candidates, even though no publicity was given to the competition. A list of sixteen qualifiers was sent to the lord lieutenant in December 1851 after examinations which went on for ten days[35] and seven appointments were made in the following five years.

It seems to have been generally accepted, following the 1838 and subsequent examinations, that offers of appointment were made strictly in accordance with the order of merit decided by the examiners. According to Larcom only one exception had arisen – when the Cavan grand jury pressed in 1853 to retain as county surveyor a man who had been acting satisfactorily as deputy for more than a year but whose name was not next on the list of qualified candidates.[36] The general belief seems also to have been, as *The Engineer* reported in 1869, that 'once a person did not take advantage of his turn he lost all subsequent claim to any vacancy that might take place: when his turn came, to use a common phrase, he must either take it or leave it'.[37] The record shows, however, that some qualifiers were allowed to decline offers of appointment in particular counties while continuing to have their names retained on the list; the outstanding example of such a case – but not the only one – was that of Samuel Ussher Roberts, who qualified in 1841 but did not take up an appointment until the mid-1850s.

THE 1856–57 CONTROVERSY

Jacob Owen held office as principal engineer and architect in the Board of Works from 1832 to 1856 when he was succeeded by his son, James Higgins Owen, and he had been a member of every county surveyor examination board during that period. Soon after his

33 Ibid., letter of 20 Oct. 1847.   34 Ibid., letter of 22 Oct. 1847.   35 Evidence of Revd J.A. Galbraith and J.L. Worrall to *the Select Committee on County & District Surveyors etc. (Ireland)*, HC 1857 (Sess. 2) ix (270); *Returns of the Dates of Commissions issued by the Irish Government for the Examination of Candidates for the Office of County Surveyor in Ireland, etc.*, HC 1856 liii (335).   36 NLI, Larcom papers, MS 7753; Alexander Tate, in his evidence to the 1868 Select Committee, suggested that there had been one other case, but could not recall the details.   37 *The Engineer*, xxviii, 29 Oct. 1869, 291.

retirement, a major controversy broke out about the manner in which the examinations had been conducted over the years and there were specific allegations about Owen's influence on the results. He had, in fact, succeeded not only in establishing an architectural dynasty for his family and protégées in the Board of Works but had also exercised considerable influence in favour of those of his family and of his assistants who competed for county surveyor posts. Among these were his son-in-law, Charles Lanyon, appointed in 1834 and transferred to the more prestigious Antrim post in 1836; and his son, Henry, who was recommended for appointment in Queen's County in August 1841, on the basis that he 'stands first on the list' even though the full list was not completed until the following December.[38] Some years later, when Henry Owen and another surveyor were applicants for a transfer to Waterford, he was again favoured by the board although on this occasion they 'felt some delicacy in making any precise recommendation'.[39] Pupils or assistants of Owen who gained appointments included Richard Lanauze (1840), Richard Williamson (1847) and Thomas Turner (1852). In all, it was claimed that at least thirteen of the candidates who qualified at the examinations of 1838, 1847 and 1851 were pupils of Owen[40] but this seems not to have attracted public attention before 1856. Even then, Burgoyne asserted that while he and Radcliffe, his fellow examiner, had been 'perfectly satisfied as to the honour of Mr Owen as an examiner', they had taken the precaution, 'in order to prevent any appearance of partiality', of having the papers of candidates who were associated with Owen examined by two of the Board of Works' engineers and by Thomas Rhodes, all of whom 'reported most favourably upon them'. Many years later, however, Robert Cochrane, a successor of Owens's and who had himself worked at one time as an assistant county surveyor, noted that there was 'a run on certain offices in which young men had opportunities to train for county surveyorships and for which a fee of £300 for each pupil was usually charged'; the posts he had in mind must surely have been those in Owen's office.[41]

The immediate cause of the 1856 controversy was an examination of candidates for a vacant district surveyor position in Co. Dublin in which Jacob Owen had participated in March 1856 shortly before his retirement. His co-examiners were Colonel Sir John Graham McKerlie (who had become a commissioner of public works in July 1855) and Radcliffe who had been a commissioner since 1831. No public notice was given of this examination which qualified seven candidates, one of whom (R.A. Gray, a drainage engineer with the Board of Works) was duly appointed. When the original appointments as Dublin district surveyors were being made in 1844, a suggestion that the positions should be offered to those who had qualified at the last county surveyor examinations was rejected[42] because the view was taken that the relevant legislation required candidates for Dublin posts to be assessed separately. The view that there were quite separate appointments systems had been maintained in 1845 when Robert Hampton, a Dublin district surveyor, applied to replace his father, the Wicklow county surveyor, who had died suddenly; on that occasion, the Board of Works advised that as Robert had not been

38 NAI, OPW 1/1/4/23, Public Works Letters No. 23, p. 331.   39 NAI, OPW 1/1/4/32, Public Works Letters No. 32, p. 237.   40 NLI, Larcom papers, MS 7753, Statement Respecting the Examinations for County Surveyorships Ireland, undated.   41 Robert Cochrane, 'Remarks on some Examinations, Competitive and Qualifying, as Avenues of Employment for Engineers', *TICEI* xxxi (1906), p. 7.   42 NAI, OPW 1/1/4/31, Public Works Letters No. 31, p. 204.

examined for a county surveyor post, and notwithstanding the circumstances of the case, the lord lieutenant should 'not depart from the principle which has been laid down and which has been universally followed'.[43] Notwithstanding these earlier rulings, the examiners at the Dublin competition in 1856 took the view that the examination was 'sufficiently rigid to fit those that passed it for the office of county surveyor of counties generally' and their names were therefore added to the existing panel of qualifiers for those posts.[44] When one of them was nominated for a provincial surveyorship, an anonymous printed statement was circulated alleging irregularities and deficiencies in the examinations and procedures followed since 1834 and highlighting the success-rate of Owen's pupils; it was believed within the Board of Works that the authors were associated with the school of engineering in Trinity College none of whose graduates had succeeded at that stage in qualifying for a surveyorship.[45] Radcliffe subsequently noted that a clever college graduate had been passed over, although well qualified, and Larcom conceded that with hindsight Owen ought to have been prohibited from taking pupils on the grounds that 'it was not possible but that the same turn of mind, thought and habit should prevail in the questions put by the same man as teacher and examiner'. He felt also that professional examiners ought to have been appointed rather than 'a practical man like Owen'. Radcliffe hoped to retrieve the situation by demonstrating clearly that the questions and answers at the Dublin examination were such as to qualify a successful candidate for any surveying post, but when he looked for the papers, he was furious to discover that they had been burned or destroyed during the clearing out of Owen's office in the Custom House following his retirement.[46]

The controversy led to questions and debates in the House of Commons. In July 1857, Larcom was obliged to make a return giving details of all the competitions held since 1834 (in so far as these were available), details of the appointments made on foot of each competition and a list of the qualified candidates who had yet to receive an appointment.[47] It is clear from the surviving records of the chief secretary's office and of the Board of Works that there were numerous errors in the return – for example, names and dates were incorrectly stated in several cases, and other names were omitted entirely. Remarkably, however, this seems to have been the first occasion on which official information of any kind on the outcome of the various examinations was made available publicly or even made available to candidates, successful or otherwise.[48] In light of the return, the commons decided in July 1857[49] to order the establishment of a select committee to 'inquire into the duties, functions and mode of remuneration of county and district surveyors in Ireland; and also as to the best mode of examination … with a view to establish a system of competition, and secure to the public the services of the best qualified candidates'. In making a case for the inquiry, G.A. Hamilton, MP for Dublin

43 NAI, OPW 1/1/4/33, Public Works letters, No. 33, p. 32.   44 NLI, Larcom papers, MS 7753, OPW letter of 5 June 1856.   45 NLI, Larcom papers, MS 7753, Statement Respecting the Examinations for County Surveyorships Ireland, undated.   46 NLI, Larcom papers, MS 7753, letter of 3 July 1857 from OPW. 47 *Returns of the Dates of Commissions issued by the Irish Government for the Examination of Candidates for the Office of County Surveyor in Ireland, etc.*, HC 1856 liii (335).   48 A policy of non-disclosure seems to have originated in 1834: when Henry Brett inquired from the chief secretary's office in May of that year, a few weeks after the initial appointments had been made, if his name was on the list of other qualified candidates, he failed to get any information even though his name was on the list and he obtained an appointment six months later. 49 Parliamentary debates III cxlvi, 14 July 1857, col. 1501.

University, conceded that 'efficient and useful men' had been appointed but argued that the examination system should be transformed into a more competitive and transparent one so as to open it up to 'an educated class above the lowest class' and to stimulate 'an advantageous and a useful system of education' – a reference hardly unrelated to the interests of the engineering school of Trinity College Dublin which was turning out growing numbers of graduates in engineering in the 1850s. Sir Denham Norreys, MP for Mallow and a member of the Cork grand jury, alleged that the examination system was disgraceful without spelling out his reasons, but John Vance, MP for Dublin city, was not so circumspect: there was great dissatisfaction, he said, because 'the commissioners, it was stated, had in many instances, appointed their own private friends'. Vance went on to suggest that there should be due notice of examinations, that the names of candidates, successful or not, should be published, and that 'no examiners ought to be permitted to examine their own pupils'. In face of complaints like these, the chief secretary remarked that 'it would be a desirable thing that the parties who were examined should not be examined by those with whom they had been private pupils'.

The select committee under Hamilton's chairmanship completed its task in record time and reported in August 1857.[50] They heard oral evidence from a number of surveyors who told of their experience of the examination system, gave details of their onerous duties, and argued their case for substantial salary increases. Revd Joseph A. Galbraith, registrar of the school of engineering at Trinity College Dublin, complained that no advance publicity was given to the date, time and place of any of the examinations conducted since 1834, a claim which was supported by the evidence of several witnesses; it emerged that the Board of Works simply kept a list of those who applied and notified these when examinations were due to be held. There were complaints from candidates at the various examinations about the lack of information on the results. Alexander Tate, a Dublin district surveyor since 1845, had assumed that his name was on the list of qualifiers at the 1847 examinations but was shocked to find it omitted from the returns presented to parliament in 1856; Larcom subsequently assured him that he had indeed qualified in 1847, and that his name had been omitted in error. J.L. Worrall, a candidate in 1847 and 1851, complained that he had not been able to get any information about the 1847 results and had not learned until 1854 that his name was included on the 1851 list.

Although it had been asserted in the evidence given to the select committee that the lists of qualified candidates had been returned to the lord lieutenant in order of merit in the 1834–51 period, the committee did not seem to have been convinced of this and they strongly criticised the fact that the lists, and the system generally, were 'wanting in the publicity, both before and after the examinations, which is desirable in order to constitute a proper competitive system'. They found fault also with the constitution of the board of examiners which, in their view, should be changed occasionally and should comprise a civil engineer, a county surveyor, and a gentleman of high scientific attainments. A bill to give effect to the recommendations of the committee was drafted quickly in consultation with some of the surveyors[51] but, partly because of changes in personnel at the Irish office, legislation to establish a new selection and appointment system was not enacted

---

50 *Report from Select Committee on County & District Surveyors etc. (Ireland)*, HC 1857 (Sess. 2) ix (270).
51 NLI, Mayo papers, MS 11,035.

until 1862. Larcom had advised that 'to do away with the existing list would give great dissatisfaction … but the question is whether the object is not worth braving that'.[52] In the event, the act expressly provided that persons who had been certified following the controversial 1856 examination for Dublin surveyorships, and those who had earlier been certified under the law relating to county surveyorships generally, would continue to be eligible for appointments in Dublin or elsewhere. As a result, the influence of Jacob Owen and of the Board of Works on appointments was to continue until 1869 when these lists were exhausted.

According to the return made to parliament in 1856, nine candidates certified in December 1851 had not yet received an appointment but all of those who were still available had received appointments by 1862. In addition, Samuel Ussher Roberts, who had qualified in 1841 and had declined several offers of appointment in the following years, was appointed to west Galway in 1858, having already acted as surveyor for the county of the city of Galway since 1855. In an effort to give retrospective legitimacy to the outcome of the Dublin examinations, McKerlie, Radcliffe and Owen had been appointed as examiners in March 1857 and instructed to review 'the applications of those who had been lately examined for Dublin surveyorships'. The outcome later that month was a report certifying that all seven of these were fit for the office of county surveyor generally[53] and all of them were appointed by 1867, including the 'clever and well-qualified graduate', Bernard B. Murray, whose failure to gain a position after the 1856 examinations had fuelled the controversy. These appointments brought to thirteen the number of appointments made between 1856 and the commencement of the new appointments system in 1869 and created a grand total of eighty county surveyor appointments under the original and, by then, discredited selection procedures.[54]

52 NLI, Mayo papers, MS 11,035, letter to Lord Naas, 17 June 1858.   53 NAI, CSORP 1857/1909.   54 The total of eighty excludes four appointments made under the separate Co. Dublin system between 1844 and 1856.

4

A New Appointments System, 1869–99

THE AGE OF EXAMINATIONS

Following the Northcote-Trevelyan report of 1854 on the organization of the permanent civil service, an autonomous new organization – the Civil Service Commissioners – was established in London in 1855 to begin the process of substituting recruitment by open competitive examination for the former patronage system; fifteen years later, when a statutory order made competitive examinations mandatory for all civil service posts with only limited exceptions, *The Times* commented that 'the principle of unlimited competition has triumphed'. In this period, described by Gladstone as 'the age of examinations', the County Surveyors (Ireland) Act 1862[1] was enacted, despite what Sir Robert Peel, the chief secretary, described as the 'determination of some Irish members to offer every opposition to the measure'.[2] The Civil Service Commissioners were given the responsibility of certifying the qualifications of applicants for appointment as county surveyor or as Dublin district surveyor; they were required, whenever a vacancy occurred, to hold an examination in Dublin after due public notice, and the lord lieutenant was required to appoint 'one of the persons who shall be certified'. The approval of the grand jury was required before an appointment became effective – but in practice this procedure seems to have been a mere formality; there is no record of the rejection by a grand jury of the name submitted to them.

Even before the act came into operation, the proposed arrangements had been criticised by some of the existing surveyors and by others. There were fears that men who had considerable experience of grand jury work would be reluctant to submit themselves to an examination of the kind that the Civil Service Commissioners were likely to arrange. An experienced surveyor, in evidence to a select committee in 1868, argued that the post required 'not mere book knowledge, or the knowledge which is to be obtained of theoretical matters, but practical knowledge and experience in dealing with men, and the management of works ... and you do not want to get for county surveyors persons fresh from Trinity College, or any other college'.[3] Another surveyor told the same committee that he could not see how the commissioners 'are capable of judging the fitness of a candidate for the office of county surveyor' and that a 'practical man', such as an engineer, should therefore be associated with the selection process.[4] It was argued also that there was real advantage in the traditional system under which a panel of

1 An Act to amend the Law relating to the Appointment of County Surveyors in Ireland, 25 & 26 Vict., c. 106. 2 Parliamentary debates III, clxvii, 7 July 1862, cols 1565–6.   3 *Report from the Select Committee on the Grand Jury Presentments (Ireland)*, HC 1867–8 x (392), evidence of Alexander Tate.   4 Ibid., evidence of Henry Brett.

qualified candidates was maintained, allowing the authorities to fill vacancies as soon as they arose. This last point led the Select Committee on Grand Jury Presentments in its July 1868 report to point out that the new system would 'necessarily create considerable delay in filling up the appointment' and it recommended that there should always be a panel of two qualified candidates available for filling vacancies immediately they arose.[5]

ARRANGING THE EXAMINATIONS

Because of the provision which allowed for the appointment of persons whose names were already on the list of qualified candidates, vacancies which required the 1862 Act to be activated did not occur until 1869. On 16 September of that year, the under-secretary at Dublin Castle told the Civil Service Commissioners that the list was then exhausted and he requested them to set about making arrangements for the first examinations as there were already two vacancies for which speedy provision was needed.[6] The letter went on to say that while the act allowed any of the persons certified by the commissioners to be selected, the lord lieutenant 'would wish to exercise this choice by appointing on each occasion those persons whom the commissioners shall report to him to be the best qualified'.

The format of the first – and subsequent – competitions was heavily influenced by suggestions made by the under-secretary who pointed out that:

> The post of county surveyor differs from most others for which the selection is made by competitive examination in as much as a county surveyor does not enter on his duties in a subordinate capacity, but has from the first the whole responsibility of the works of his county. On this account, His Excellency [the lord lieutenant] thinks it undesirable that candidates should be deemed eligible who are not at least twenty-six years of age and I am accordingly to suggest that twenty-six and forty be the limits of age. It is also essential that the commissioners should be satisfied as to the character of candidates and their health. Each candidate should be required to show that he has been engaged in the practice of his profession on adequate works for a sufficient time; or has had in some other way sufficient opportunities of becoming acquainted with the practice of his profession in some one of its branches listed below ...

A two-part programme of examinations was agreed between the Dublin Castle authorities and the commissioners. Part I was to comprise seven separate three-hour papers on mainly theoretical and academic subjects carrying a total of 340 marks while Part II, comprising ten three-hour papers on more practical matters, was to carry 660 marks; there was to be an interval of a fortnight between the two sets of examinations 'so that candidates may not be worn out by the lengths of the examination in the last days'.

5 *Report from the Select Committee on the Grand Jury Presentments (Ireland)*, HC 1867–8 x (392).   6 Letter of 16 Sept. 1869, published in the *Fifteenth Annual Report of Her Majesty's Civil Service Commissioners*, [c. 197] HC 1870 xix (1).

No minimum pass mark was prescribed in any subject so as not to exclude a deserving candidate who might 'break down in some out of so many subjects' and, so that lack of scholastic attainments would not be unduly penalised, candidates would be deemed eligible to proceed to Part II if they showed *some* proficiency under *at least one* of the Part I subjects. According to William R. Le Fanu, an experienced railway and consulting engineer who had become a commissioner at the Board of Works in 1863, the civil service commissioners had first proposed an examination which was so scientific, and put so little emphasis on practical matters, that only young men of no experience, but with a university education or its equivalent, would get through.[7] Le Fanu disclosed in 1877 that, on behalf of the lord lieutenant, he had suggested some changes of emphasis – and these, presumably formed the basis of the proposals in the under-secretary's letter referred to above. However, even with these amendments, the net effect of the new system was to create a group of surveyors very different in terms of education and experience to those who had attained the rank in the previous thirty-five years.

THE FIRST EXAMINATIONS

Public notice of the dates of the first examinations and particulars of the examination subjects was duly given, specifying that the first part was to take four days, commencing on 2 November, and the second part was to begin on 16 November and last for five days. There were to be seventeen papers in all, as follows, with three hours allowed for each of them:[8]

Part I
1. Geometry, Algebra, Trigonometry
2. Differential and Integral Calculus and Geometrical Optics
3. Statics and Dynamics
4. Hydrostatics, Hydraulics, Pneumatics and Heat
5. Electricity and Magnetism
6. Chemistry
7. Geology and Minerology

Part II
8. Knowledge of Materials
9. Theory of Construction
10. Roads, Drainage and River Works
11. Architecture
12. Sewage and Irrigation
13. Water Supply
14. Sea and Reclamation Works
15. Harbour and Dock Works
16. Canal Engineering
17. Railway Engineering

The announcement of the examinations brought a mixed reaction. A leading article in *The Engineer* welcomed the fact that the appointments were being thrown open and that so severe a criterion of professional ability was being demanded, and hoped that before long similar arrangements would apply to situations in England.[9] At the same time, the

7 *Report of the Committee appointed by the Treasury to Inquire into the Constitution and Duties of the Board of Works in Ireland*, [c. 2060] HC 1878 xxiii (1), evidence of W.R. Le Fanu.   8 The papers were subsequently printed in full in the *Fifteenth Annual Report of Her Majesty's Civil Service Commissioners*, [c. 197] HC 1870 xix (1); examination papers for the competitions held in 1870 and 1871 were published in the subsequent reports. 9 *The Engineer* xxviii, 29 Oct. 1869, 291.

journal was critical of several features of proposed arrangements: the fact that nine days were to be 'appropriated to the ordeal' of the examinations as against three for the majority of university examinations; the cost and difficulty for anyone from England having to travel backwards and forwards twice or stay in Dublin for three weeks; the remarkable similarity between the examination subjects and the curriculum prescribed for the Dublin University school of engineering; and the attempt to test the practical ability and professional experience of candidates by written papers rather than by reference to their past employment and employers' references. Another view of the examination programme, which might well have represented that of many of those involved in Irish engineering, was expressed by 'A Retired C.E.' in a letter to the *Irish Builder*.[10] The writer had two sons, one of whom was 26 years old, had been educated at Queen's College, Cork, and had just finished an apprenticeship with a county surveyor, while the other, aged 39, had eleven years experience of engineering, including a period as an assistant county surveyor. The younger son was quite capable of passing the examination but felt he would need ten years experience in the profession before he could competently act as a county surveyor; the other son was unwilling to enter the list with 'schoolboys fresh from their studies' against whom his superior knowledge and practical experience of county works would gain him insufficient marks. In the writer's opinion, the whole examination ought to relate mainly to a knowledge of county works and the valuation of land and, to demonstrate what could happen when theoretical science took precedence over practical considerations, he instanced the case of a bridge designed by a county surveyor who, at one time had been a mathematical tutor (and hence enjoyed the soubriquet *Old Square Root*) and which had to be replaced at considerable cost by a new bridge of metal girders.

Notwithstanding whatever misgivings there may have been about the nature of the examination, and the relatively short notice, there were 35 applicants. Of these, 9 withdrew or were deemed ineligible, leaving 26 to contend with the formidable series of written papers. In all, 24 of the candidates were declared to be unsuccessful and the remaining 2 were duly certified just before Christmas 1869. First in order of merit, with 515 marks out a possible 1,000, was Leveson Francis Vernon-Harcourt, the 30-year-old scion of an English family which was prominent in the church and in politics, and was himself a graduate of Balliol College, Oxford; Vernon-Harcourt's term of office as county surveyor in Westmeath was to last for less than a year – he returned to private practice, going on to become a consulting engineer and, in 1882, professor of civil engineering at University College, London. The second-placed candidate was another 30-year-old Londoner, Henry Temple Humphreys, who had considerable practical experience of engineering in England and in India; he gained 475 marks out of 1,000 and was appointed to serve in the western division of Co. Limerick. In reporting the results, *The Engineer*, which had commented so favourably on the new appointments system only a few months before, hoped that both gentlemen had insured their lives, congratulated them on 'no small display of courage' and suggested that 'a surveyorship in Sierra Leone would be the safer of the two posts'.[11] A letter to the editor of the same journal some weeks later from 'an unsuccessful candidate' was very critical of both the selection

10 *IB*, 1 Nov. 1869, 252. 11 *The Engineer* xxix, 7 Jan. 1870, 11.

process itself and the outcome; it suggested that the system was not the open and competitive one which candidates had been led to expect and that the practical questions were so simple that they could have been answered by someone just out of engineering school, thus failing to compensate the experienced practical engineer for having to contend with calculus, chemistry etc. The writer went on to suggest that the appointment of a candidate who was known to be a near-relative of the chief secretary, Chichester Fortesque,[12] had led most of the candidates to suspect that it was merely a waste of time and money preparing for the examination unless one had a friend in court.[13]

IMPACT OF THE NEW SYSTEM

Of the eighty county surveyors who had been appointed before 1869, the vast majority were Irishmen who had no formal third-level engineering qualifications but had learned their engineering on the job. In 1854, thirteen years after the school of engineering was established at Dublin University, not one of the seventy-eight students who had been awarded the school's diploma had obtained a county surveyor post, although one had served for a time as an assistant county surveyor in Clare;[14] the same was true in 1857 by which time 101 young men had passed through the school.[15] Between then and 1867 when the last appointments were made under the old system, only three graduates of the school joined the ranks of the surveyors; the first of these, in 1857, was J.B. Pratt, who had graduated in 1844, and he was followed in 1861 by J.A. Dickinson, and in 1862 by Bernard B. Murray. The Queen's Colleges at Belfast, Cork and Galway had also begun to award diplomas and degrees in engineering in the mid-1850s but none of their graduates obtained a county surveyor post under the original appointments system. Against this background, the results of the 1869 examinations must have come as a considerable shock to Irish engineers generally, and particularly to those who had entered the profession by the more traditional routes rather than by university education and had built up solid practical experience of public works. A London-born graduate of Dublin University (J.H. Moore) took first place at the 1870 examinations and there was a shift in favour of Irishmen at examinations held from 1873 onwards. But the dominance of the young university graduate who could score heavily in the theoretical subjects was to continue until 1899. This was in line with the more general experience of the new selection systems for public appointments which, according to G.M. Trevelyan, 'showed a belief in higher education which was something new in England'.[16]

The reports of the Civil Service Commissioners for the years up to 1883 gave full details of each county surveyor competition, setting out the names of the successful candidates, the marks gained by each man in each subject, and often the examination

---

12 Chichester Fortescue, chief secretary from 1868 to 1871, could hardly be described as 'a near relative' of the first-placed candidate; he became the fourth husband of Countess Frances Waldegrave in 1863, after the death in 1861 of her third husband, George Granville Vernon-Harcourt who was the candidate's uncle; the candidate's cousin, Sir William Vernon-Harcourt, was elected MP for Oxford in 1868.   13 *The Engineer* xxix, 4 Feb. 1870, 67.   14 Robert V. Dixon, *A Farewell Address to the Students of the School of Engineering*, Appendix B (Dublin, 1854).   15 Evidence of Revd J.A. Galbraith on 28 July 1857 to the *Select Committee on County and District Surveyors etc. (Ireland)*, HC 1857 (Sess. 2) ix (270).   16 G.M. Trevelyan, *British History in the Nineteenth Century* (London, 1922).

papers themselves. In subsequent years, the published reports included only the essential facts – the number of competitions held, the number of applicants, and the number certified to be qualified.[17] In the early years, the practice after each examination was to certify only as many candidates as were needed to fill the vacancies that existed, even when other candidates had obtained relatively high marks. A suggestion made by a select committee in 1868,[18] and supported by the authorities in Dublin, that two 'supernumeraries' should be named to fill the next two vacancies – and thus avoid delays – was rejected by the commissioners on the grounds that the act appeared to require that a fresh examination of candidates should be held *after* a vacancy had occurred. It was agreed, however, in 1874 that all candidates who reached the prescribed standard would in future be certified but only for the purpose of providing a panel of qualified candidates who could act temporarily as deputy county surveyors in the absence through illness of an incumbent, or pending the filling of a vacancy.[19] Those on the panel had no claim to a permanent post and had to sit the examination again if they wished to gain one. The Civil Service Commissioners suggested repeatedly in the period 1879–91 that the law should be changed to obviate the need for a new examination each time a vacancy occurred[20] but no action was taken. As a result, applicants who had been awarded marks well above the qualifying standard at previous examinations were obliged to sit the full examination again if they wished to be considered when a subsequent vacancy arose; to do otherwise would be a step backwards, according to one of the existing surveyors, as 'lower grade candidates' could then wait around 'hanging on to any and every sort of occupation, or perhaps remain idle, until a vacancy occurs'.[21]

The appointments system established in 1869 continued with relatively few changes until 1899. In all, 23 competitions – 2 in one year on a few occasions – were held during that period for one, 2 or occasionally 3 or 4 posts. The number of candidates was generally in the range 15 to 20 and the men who gained the highest marks were invariably offered the appointments available. Only those who were adjudged to have the necessary practical experience were allowed to sit the competitive examinations, the basic two-part structure of which remained unchanged.[22] The number of separate papers was, however, reduced in the 1870s (from 7 to 4 in Part I and from 10 to 5 in Part II), but without affecting the balance of the marking scheme or the range of theoretical and practical subjects covered.[23] Experience led the Civil Service Commissioners to suggest in 1870 that 'with the view of saving both time and expense, it would be desirable to eliminate by preliminary tests all candidates who were unable to write English legibly, correctly and grammatically', but the authorities in Dublin rejected this because travel to Dublin for separate examinations would entail great expense for some candidates.[24] The commissioners

17 Detailed results are available on Civil Service Commission files for some years, for example TNA, CSC 10/30, 10/870, 10/1002, 10/1117, 10/1196.   18 *Report from the Select Committee on the Grand Jury Presentments (Ireland)*, HC 1867–8 x (392).   19 *Nineteenth Annual Report of Her Majesty's Civil Service Commissioners*, [*c*.1330] HC 1875 xxii (1).   20 See, for example, *Thirty-Sixth Annual Report of Her Majesty's Civil Service Commissioners*, [c. 6760] HC 1892 xxvii (1).   21 R.B. Sanders, 'Irish County Surveyorships and the Grand Jury System of Ireland', *PRIAMCE* xx (1893–4), 251.   22 Parliamentary debates IV xiv, col. 1826, records a question on 18 July 1893 by the MP for Dublin University who was told that a graduate of the engineering school had not been allowed to sit the examination because he lacked the necessary practical experience.   23 The papers for the September 1898 examination can be found in *TICEI* xxxi (1906), 124–43. 24 *Seventeenth Annual Report of Her Majesty's Civil Service Commissioners*, [c. 672] HC 1872 xix (1).

also suggested that no one should be admitted to Part II of the examination unless he had shown 'something more than superficial knowledge' (instead of just 'some proficiency') under two at least of the more theoretical papers which comprised Part I; this idea found more favour with the Dublin authorities but, so as not to put too much weight on the theoretical part, they suggested that candidates should be required to qualify in only one of the subjects involved. In the event, neither of these ideas was implemented.

The fact that the two successful candidates at the first examinations in 1869 were Englishmen prompted the writer of a letter to one of the newspapers in 1870 to suggest that 'surely there still are some Irishmen sufficiently proficient in their profession to be able to take charge of a county, as in the present list of surveyors all, with two or three exceptions, are natives'.[25] In the event, only 9 of the 42 appointments between 1869 and 1899 went to Englishmen and one to a Scotsman, the last of them in 1891. More than half of those who became county surveyors during that period were 30 years old or less when appointed, and only 5 of them were more than 35 years old. At least 36 of the appointees had third-level engineering qualifications, most of them from institutions in Ireland. The school of engineering at Dublin University, the longest-established in the country, accounted for 16, or almost half of the successful Irish graduates. Of the remainder, 10 were graduates of Queen's College, Belfast, 4 had graduated at Queen's College, Galway, and one at the Cork college; and there were 2 associates of the Royal College of Science for Ireland where a three-year course in engineering had been established in 1867.[26] Of the new surveyors who appear not to have had third-level qualifications, 4 had gained considerable engineering experience in England or further afield on both public and private sector projects, and another was an Irish engineer with at least thirty years experience, much of it with the Board of Works.

25 *Express*, 8 Oct. 1870.   26 *Royal College of Science for Ireland, Directory*, 1897–8; *Royal University of Ireland*, Calendars, 1883–1909; TCD School of Engineering, *A Record of Past and Present Students* (Dublin, 1909).

5

Working with the County Councils, 1899–1922

Under the Local Government (Ireland) Act which became law in August 1898, the administrative and fiscal business of the grand juries was transferred to county and rural district councils which were to be elected on a much wider franchise than that which applied at parliamentary elections and which included women for the first time. When the votes were counted after the first elections in April 1899 it was clear that a dramatic shift in the location of political power had taken place in the counties, with the unionists swept from office in three of the four provinces. The political position of the large Irish landowners was undermined and nationalist councillors, many of them 'farmers, shopkeepers and publicans' as F.S.L. Lyons put it,[1] became the dominant force in local affairs in most parts of the country.

Reform of local government at county level had been on the political agenda intermittently for many years before 1898. More than sixty years earlier, William Smith O'Brien was calling for the establishment of a three-tier system of representative local government in rural areas, based on parish councils, baronial district boards and county fiscal boards which would ultimately be responsible for the management of all county affairs.[2] Bills designed to replace the grand juries by popularly elected representative boards were introduced by Irish members in the House of Commons at intervals from the 1850s onwards but were not enacted and, for a variety of reasons, successive governments failed to tackle the question themselves. Instead, as the century advanced, government and parliament looked to the poor law guardians rather than the grand juries to assume additional responsibilities; new functions relating to sanitary and environmental matters, public health, welfare, nuisances, burial grounds and labourers' housing were assigned to the guardians which had a greater measure of popular acceptance and democratic legitimacy than the grand juries, had better-developed administrative organizations, and operated under an elaborate system of central direction and control. In 1898, the total expenditure of the guardians well exceeded that of the grand juries which continued to be responsible for roads and bridges, courthouses and minor additional matters.[3]

Henry V. White, Queen's County surveyor, in a paper presented to the Institution of Civil Engineers of Ireland in 1885, spoke of 'the certainty of ... the establishment of

1 F.S.L. Lyons, *Ireland since the Famine* (London, 1971), p. 212.   2 *Limerick Star & Evening Post*, 23 Feb. 1837.
3 *Returns of Local Taxation in Ireland for the Year 1898*, HC 1899 lxxxiii (pt. 1) (360).

county boards to replace grand juries' and suggested that his colleagues should forward practical recommendations to the government in relation to the legislation which was expected to emerge.[4] Ten years later, it was clear that the days of the grand jury system were numbered and there was ample time for the county surveyors and other grand jury officers to assess the implications for them of a radically different system. As permanent and pensionable officers, the surveyors were to be transferred by the 1898 Act to the new county councils but with no special provision for early retirement as was made for county secretaries.[5] Many of them were naturally worried about the impact of the new legislation and several publicly expressed their views on the subject at a meeting in Cork in August 1898. John Horan of Limerick spoke of the apprehension of his colleagues about the proposed new councils but went on to say that 'we as loyal men will do our best to work under them and we can only hope that the new bodies will rise to a sense of the increased responsibilities placed upon them'. And R.W. Longfield, then serving in Donegal but shortly to move to Co. Cork, looked forward not only with a considerable amount of apprehension but also with 'curiosity as to what these new bodies will do'; he anticipated a considerable amount of friction between the district and county councils, a considerable amount of trouble, and a considerable amount of odium thrown on the county surveyors.[6] He was not far wrong.

Before the coming into operation of the act in April 1899, four of the longer-serving surveyors decided to retire – two of the three Cork surveyors, Nat Jackson and A. Oliver Lyons; Frederick R.T. Willson of Fermanagh; and A.C. Adair of Derry. Jackson and Adair each had forty-four years service. Adair based himself on the Book of Exodus when he announced his retirement to his grand jury at their last meeting in March 1899, stating that he would prefer to have them fix the rate of his pension, because in a short time there might be a new king in power who knew not Joseph! The majority of the surveyors remained at their posts, though not without some misgivings as to what the new system would bring. The transition cannot have been an easy one for men who, in many cases, would probably have been seen by their new political masters as an integral part of a debased *ancien régime*. Apart from having to work with new and generally inexperienced local representatives, there were other significant implications for the surveyors: their duties became more demanding with the establishment of the rural district councils which alone had power to put forward proposals for roads expenditure; their position in relation to salary levels was considerably worsened; the way was left open for the introduction of direct labour work on the roads with an obvious impact on the surveyors' workloads; and the right to private practice was increasingly challenged and was prohibited in the case of a growing number of new appointees.

4 Henry V. White, 'County Work in Ireland considered in relation to Grand Jury Laws', *TICEI* xvii, (1887), p. 17.   5 For details of the surveyors and assistant surveyors serving in 1898, see *Return of Officers in the Service of the Grand Jury in each County in Ireland*, HC 1898 lxxiv (237); for details of those transferred to the county councils, see *Return showing the Names … of all Officials transferred by virtue of the provisions of the Local Government (Ireland) Act, 1898 to County Councils, District Councils or Boards of Guardians etc.*, HC 1901 lxiv (331).   6 *PRIAMCE* xxv (1898–9), 35–6.

APPOINTMENTS

Those who aspired to a county surveyor post after 1899 had to face an appointments system which was radically different, with canvassing and local patronage becoming essential for success and a knowledge of Irish being demanded in a number of counties. Under the 1898 act, each county council (rather than the lord lieutenant) was given the duty of appointing a county surveyor but the approval of the Local Government Board was required for each appointment and for the dismissal of a surveyor. The provisions of the 1862 act under which the qualifications of candidates were to be certified by the Civil Service Commissioners continued to apply as did the requirement that a fresh examination of candidates was to be held whenever a vacancy arose; the law was changed, however, in 1900 to allow a person to be appointed to a vacant post if he had been certified at any previous examination.[7] The net effect of these amendments was that the open competitive system which had operated since 1869 no longer applied; instead, whenever a vacancy arose, applications were invited by the county council by public advertisement, candidates whose qualifications had not already been certified were required to sit an examination arranged by the Civil Service Commissioners and the councillors then proceeded by majority vote to elect a surveyor from among those who qualified at this or any previous examination. In effect, therefore, the examination became a qualifying one only, leaving councillors free to elect any of the qualified applicants regardless of the marks attained. This was described as 'an evil and retrograde step' by the Kildare surveyor, Edward Glover, in his presidential address to the Institution of Civil Engineers of Ireland in December 1900[8] and fears were expressed that the candidates who had 'most friends at court' and could 'most effectively pull the wires' would gain future appointments;[9] experience of the system in the subsequent years was quickly to prove that these fears were well-founded.

The first new appointment made under the 1898 act was that of Charles Boddie who was elected by Derry county council in March 1900 although he had been placed fourth at the qualifying examinations held at the end of 1899.[10] W.F. Barry – the only applicant – was appointed to the Monaghan post in June 1900.[11] Three years later in Wicklow, Stephen G. Gallagher, who had obtained lower marks than any of the other three qualified candidates, was appointed.[12] Results like these prompted the Local Government Board's chief engineering inspector, P.C. Cowan, himself a former county surveyor, to state publicly that it was regrettable that the candidate who obtained first place in the examinations would not necessarily have any better chance of appointment than any of the other candidates who reached the qualifying mark, and he noted wryly that he – a Scotsman – would probably never have had his pleasant experience as a county surveyor in counties Mayo and Down if the new arrangements had applied in previous decades.[13] In 1905, Cowan returned to the theme in an address to engineering students at Trinity College, remarking that 'now, the candidate who has the strongest local influence is

---

7 The County Surveyors (Ireland) Act, 1900, 63 & 64 Vict., c. 18.   8 Edward Glover, *Presidential Address*, *TICEI* xxviii (1902), 114.   9 *IB*, 1 May 1900, 353, letter from AMICE to the editor.   10 TNA, CSC 10/1759.   11 *Northern Standard*, 16 June 1900.   12 TNA, CSC 10/2219.   13 Remarks at a meeting of ICEI, 4 Nov. 1903, *TICEI* xxxi (1906), 32.

usually appointed, although he may be last in the list of men who qualified at the examination'; this was a very serious and retrograde step, in his view, as it meant that county councils could no longer hope to get 'the services of the picked men amongst young engineers' as they had done in former years.[14] By 1907, the *Irish Builder and Engineer* was reporting agitation to restore the former competitive system 'under which such a splendid service was constituted', noting that the new appointments system had lowered the dignity of the office and the status of the profession; it was 'wholly based on canvassing' according to the journal, and secured 'not the best engineers, but those who can boast of the best wire-pullers amongst their supporters'.[15] 'At present', the *Builder* added, 'the fact of being able to speak Irish, to be a native of the district or related to inhabitants thereof, or to be a supporter of the dominant party on the council, are all mentioned as high qualifications for the post of county surveyor'. At the same time, the editorial conceded that open competition might not be the best way of securing the most practical and capable men, as against those who are 'cleverest at passing examinations', and this view was endorsed by the *Roscommon Herald* some years later: appointing the man who gained the highest examination marks could have unfortunate results as 'the men crammed for examinations very often turned out to be useless for real practical work'.[16] A somewhat similar view was held by J.H. Moore, Meath county surveyor, who had been appointed initially in 1870 on foot of gaining first place in the examinations that year; Moore 'always had his doubts as to the propriety of these competitive examinations, and whether they got the best men', but he still thought that the old system was better than the new one under which 'it was practically known for certain who would be appointed' as soon as a vacancy occurred.[17]

In their operations generally, the new county councils failed to bring an end to the small-scale corruption which had characterised the old regime.[18] On the contrary, many of the nationalists who dominated these councils acquired a reputation for opportunism and graft and quickly developed a reputation for jobbery and petty corruption which was not entirely dispelled by comparisons with the old regime;[19] in the colourful words of Professor Joe Lee, they 'dutifully attended to the three Fs of popular politics – friends, family and favours'[20] and, while there were honourable exceptions like Cork county council, a patronage system of appointments, often involving an over-riding preference for local candidates, was just one demonstration of this. As early as 1905, Arthur Griffith was denouncing the fact that efficient administration and moral standards were being impaired by this system and suggested that patronage should be exercised in the interests of the nation rather than the individual.[21] Sinn Féin itself went on to adopt policies to rid local government of corruption in its employment practices and otherwise but, on the ground, canvassing for county surveyor posts – and for positions generally – continued to be the order of the day in most counties in the first two decades of the century. Appointments generally went to local men, graduates of the Queen's Colleges or of the

14 P.C. Cowan, 'Presidential Address to the Engineering Students Association TCD', *IBE* 30 Dec. 1905. 15 *IBE*, 9 Mar. 1907, 174.   16 *Roscommon Herald*, 25 Oct. 1919.   17 Remarks at an ICEI meeting on 4 Nov. 1903, *TICEI* xxxi (1906), 36.   18 Eunan O' Halpin, *The Decline of the Union* (Dublin, 1987), p. 16.   19 Patrick Maume, *The Long Gestation: Irish Nationalist Life 1891–1918* (Dublin, 1999), p.30.   20 J.J. Lee, 'Centralisation and community', in *Ireland: Towards a Sense of Place* (Cork, 1985), p. 84.   21 Arthur Griffith, *The Resurrection of Hungary* (Dublin, 1905), p. 155.

Royal College of Science in Dublin, rather than Trinity College graduates who had gained many of the posts on offer between 1869 and 1898. Opposition to the appointments system came from many quarters. In 1919, when a vacancy arose in Roscommon, Fr Michael O'Flanagan – by then a vice-president of Sinn Féin – preached a special sermon attacking not just the taking of money bribes for votes but, as he put it, the subtle insidious form of bribery which was commonly described by the expression 'scratch me and I'll scratch you'.[22] Anyone who voted for base personal motives should be removed from office at the first opportunity, according to Fr O'Flanagan – retribution, incidentally, of a temporal nature being preferred to the invocation of hellfire or other interdiction by the church. And just to show that canvassing could sometimes lead to surprising results, one might note that the local man who was eventually appointed to fill the vacancy in Roscommon had just returned from service in France as a lieutenant in the Royal Engineers.

Notwithstanding continuing complaints about their severity and their lack of relevance to the practical every-day work of a county surveyor, the structure and marking scheme of the Civil Service Commissioners' examinations, as established in the 1870s, were maintained for ten years after 1899. In May 1900, the *Irish Builder* strongly endorsed the view of a correspondent who argued that the examination was 'adapted for a high-flown Master of Arts than for a practical every-day engineer, whose chief duty is to attend to roads and courthouses'.[23] In response, perhaps, to criticisms of this kind, the questions set at the examinations held in 1903 over nine consecutive days were said to have been 'of a more advanced practical character than on former occasions'[24] and in 1911 the structure of the examinations was altered significantly: a candidate was then required to pass in only one of the four theoretical subjects (mathematics, mechanical philosophy, experimental sciences, and geology and mineralogy) while still having to pass in county works, railway and canal engineering, marine engineering and hydraulic engineering.[25] In 1914, the Local Government Board framed a new set of regulations which appeared to shift the balance again in favour of the more theoretical subjects;[26] candidates now had to pass in two of the three Part I subjects (mathematics, mechanics and geology) and in three of the five more practical Part II subjects but, surprisingly, a pass in the paper on county works was still not essential.[27] These changes prompted a potential candidate to complain in the columns of the *Irish Builder and Engineer* that the net effect was to make the examinations even more extensive than they had been; he went on to argue that 'if a young engineer be engaged in charge of important works, he has enough to do from early morning to late into the evening … it is very difficult to see how such a person is to read through and make up the extraordinarily extensive amount of mathematical, scientific and engineering matter which the programme demands'.[28] Not unreasonably, the writer suggested that if there were to be separate examinations of candidates, the majority of whom were by then university graduates in engineering, the emphasis should be shifted from 'the morass of theoretical engineering' to roadmaking which constituted about 75 per cent of the expenditure for which the surveyors were

22 *Roscommon Herald*, 1 Nov. 1919.   23 *IB* xlii, 1 May 1900, 346.   24 *IBE* xliv, 13 Aug. 1903, 1907.   25 TNA CSC 10/3398, table showing results of examinations held on 11 July 1911 and following days.   26 *Annual Report of the Local Government Board for Ireland, 1913–14*; County Surveyors Qualifications Order Ireland, 1914.   27 TNA, CSC 10/4045.   28 *IBE*, 18 July 1914, 438.

responsible but which had been the subject of only one question out of 128 in the most recent examinations. It seems, however, that all suggestions for further changes in the examination system were ignored by the Local Government Board and the system continued as it was until the board went out of existence in 1922.

In the first decade after their establishment, some county councils decided that a knowledge of Irish should be required of all candidates for local appointments. When Cork county council notified the Local Government Board of this policy in 1908, the board replied that Irish could not be included as a qualifying subject at the county surveyor examinations but told the council that they were free to give preference to candidates who had both passed the qualifying examination and had demonstrated that they had a knowledge of Irish. The council agree to go ahead on this basis but insisted that in no case would they appoint a person who was ignorant of the language, no matter how high his other qualifications might be. In the event, the two candidates with the highest marks were appointed; both of them had satisfied the three priests who were appointed by the council to act as examiners that they had a competent knowledge of Irish.[29]

Qualifying examinations were held in 1899, 1900, 1903 and in each year from 1906 to 1911. Examinations were held twice in 1913 and again in 1914, 1915, 1917 and 1920. The final examination held under British administration began on 18 February 1921 and, while 4 of the 5 candidates qualified, the 3 vacancies which existed were filled by men who had qualified at earlier examinations.[30] A pseudonymous article in the *Irish Builder and Engineer* in July 1921 noted that the list of candidates at the last examinations 'was surprisingly small, considering that two of the three vacant positions [Down and Armagh] were always held to be plums of the profession'.[31] This was advanced as evidence that 'something was radically wrong' with the system but, in reality, it seems more likely that the small candidature was caused by the disturbed conditions of the time and, in particular, the fact that most local authorities had withdrawn recognition from the Local Government Board. The writer, nonetheless, went on to claim that the examination, as a test of qualification, or even as a means of discovering the best man, was quite useless and that only 'those a few years out of college and who are in minor positions' had any chance of competing successfully; there was no oral examination or interview which could help in assessing a candidate's general deportment and ability to deal with employees and councillors, and academic and technical subjects were allotted too high a proportion of the total marks with, for example, only 100 marks for eleven answers to questions on county works, as against 200 for eight questions on mechanics.

In all, 38 county surveyor appointments were made between 1899 and 1921, five of them involving the transfer of a surveyor from one county to another. All but a few of the remaining appointments went to men who held third-level engineering qualifications, but Trinity College graduates, who had taken nearly 50 per cent of the positions filled in the previous twenty years, now took only 3. Galway graduates took 12 positions, Cork graduates took 5, and Belfast graduates took 3, with holders of the ARCSI taking another 5. A total of 8 positions went to men who were less than 30 years of age, 15 appointees were between 30

29 *IBE*, 31 Oct. 1908, 679; 14 Nov. 1908, 712; 12 Dec. 1908, 765.   **30** TNA, CSC 10/4646.   **31** *IBE*, 30 July 1921, 509; the third vacancy was in Kerry, southern division.

and 39, and the remaining 8 were aged 40 or more, the oldest being 54 at the date of his appointment.

SALARIES

The act of 1861 which specified maximum salaries of £600 for county surveyors in the larger counties was repealed in 1898 and it was left to the new county councils to determine the salaries to be paid to new appointees. Several councils awarded substantial increases to their surveyors, others declined to grant any additional remuneration notwithstanding the very real increase in workloads, while others still felt that existing salaries were too high and should be reduced when next a vacancy arose. As a protection for existing staff, the 1898 act had provided that, if under the new system the duties of an individual surveyor were increased, the Local Government Board could fix a new rate of salary for the post where agreement was not reached at local level. Within a few months of the commencement of the act, the board received applications from a number of surveyors seeking increased salaries and the general body of surveyors submitted a detailed memorandum setting out the additional duties devolving on them and the increased travelling expenses they were incurring in attending meetings and in making quarterly inspections of works; they also argued that their income from private practice must necessarily fall off because of their more demanding public duties. Rather than deal with individual cases as they arose, the board adopted a range of salary scales for the surveyors, differentiating among counties primarily on the basis of road mileage, and notified these to the county councils in February 1900.[32] Some councils accepted the new scales but they were considered excessive by many others who, with the support of the General Council of County Councils, conducted a bitter campaign against their implementation, arguing that the figures had been determined in an arbitrary manner and without giving councils an adequate opportunity of presenting their side of the case. The dispute was brought to the courts by a number of councils but judgment was given against them in the queen's bench division in December 1900.[33] However, lord chief baron Palles ruled in the court of appeal in February 1901 that the board had indeed erred in the procedure they adopted and they were then obliged to reconsider each case on its merits, sometimes holding local inquiries to hear both sides.[34] The board, however, defended itself vigorously against the assertion that they were 'raising salaries broadcast and without any justification'; they pointed out that the increases they had allowed related to the transferred officers only, that they had a duty to protect these officers, that the increases would add to rate poundages by no more than 'an imperceptible fraction of one penny in the pound', and that, leaving aside the special cases of the surveyors, the effect of their actions was 'in the nature of a drag, or veto, as regards increases of salaries proposed by local bodies'.[35]

The opposition of county councillors to high salary levels was even more strongly expressed when vacancies came to be filled in county surveyor posts in the early years of

32 *Annual Report of the Local Government Board for Ireland, 1899–1900.*   33 *The Times*, 3 Dec. 1900.
34 *R (Wexford County Council) v. The Local Government Board for Ireland (Webster's Case)*, [1902] Irish Reports 2, p. 348.   35 *Annual Report of the Local Government Board for Ireland, 1900–01.*

the century. In many cases, the salary decided on by particular councils before advertising a vacancy was much less than had been paid to the outgoing officeholder – and was sometimes even below the maximum salary levels fixed by law in 1861 – the view being taken in some cases that the burden on the ratepayers arising from pension payments to a retiring surveyor would have to be offset, at least in part, by the payment of a lesser salary to his successor. When the first vacancy which came to be filled under the new arrangements arose in Derry in 1899, the council proposed a salary of only £250 a year, instead of the £600 paid to the retiring officer, but they accepted with some reluctance the view of the Local Government Board that this would be insufficient to attract experienced men to take on the more demanding duties and extra travel that would be involved in servicing the new county and district councils; as a compromise, a salary of £350 a year was fixed.[36] Later that year, the salary for a new appointee in Co. Down was fixed at £800 but this was exceptional. The county councils of Monaghan, King's County and Wicklow appointed new surveyors in the following few years, all of them at salaries considerably less than their predecessors had been receiving. Similarly, Sligo county council decided in 1906 to offer a salary of only £400 to the successor of a man who had been paid £500 a year even after a survey conducted on their behalf had shown that first-class counties paid their surveyors about £700 a year on average, while second-class counties (of which Sligo was one) averaged about £600, with only a small number of counties paying less than £500 a year to existing office-holders.[37]

As further posts came to be filled in the years after 1906, the level of salaries continued to fall. The Local Government Board regularly protested about this, arguing that the salary levels were inadequate to attract candidates of good calibre, and their chief engineering adviser spoke publicly in 1911 about the fact that salaries had fallen below a desirable level even though the duties of the office had been increased.[38] But while the board withheld approval in some cases until better conditions of appointment were offered, their arguments had little effect in most cases. When they told Kerry county council in 1913 that a salary of £300 a year – the level set for the first surveyors eighty years earlier – was not enough, one councillor was moved to wonder 'how could a man spend that much at all'? Many of his colleagues around the country were obviously of the same view because relatively few of the posts advertised in the years up to 1915 carried salaries of more than £300 and only a very small number – generally in Ulster – offered salaries that approached or exceeded £500. This led the council of the Institution of Civil Engineers of Ireland to resolve in March 1914 that in no case should the initial salary for a county surveyor post be less than £400 a year, with provision for reasonable increments and separate travelling and subsistence allowances[39] but this intervention was no more effective than that of the Local Government Board had been.

TRANSITION YEARS

At the county council elections in June 1920 Sinn Féin candidates won majorities on 29 of the 33 councils. In the months that followed, many councils began to transfer their

36 *IB*, 1 July 1899, 73.   37 Report of meeting of Sligo County Council, *Sligo Champion*, 2 June 1906.   38 P.C. Cowan, address at Congress of Royal Sanitary Institute, Belfast, *IBE*, 19 Aug. 1911, 565.   39 *TICEI* xl (1914), 110.

allegiance from the Local Government Board to the underground Dáil Department of Local Government which called on them to take stern action against any of their officials who refused to carry out instructions or who continued to work with the board.[40] The authorities retaliated by cutting off grants for roads and other purposes and this, coupled with the virtual collapse of rate and rent collections in many areas, and the higher wage rates which applied after the world war, forced councils to adopt drastic cost-cutting measures, including the laying-off of workers in many areas. As the war of independence intensified, roads were trenched or obstructed, and bridges damaged, to frustrate the movement of crown forces.

For many county surveyors and other local officials, especially those who had been appointed in grand jury days, these developments must have created serious difficulties and a real crisis of conscience, and some of them, in turn, must have been regarded with suspicion by the newly elected Sinn Féin councillors. While a number of local officials, including a few county secretaries, resigned or were suspended or dismissed rather than carry out instructions emanating from the Dáil Department,[41] none of the county surveyors suffered this fate although the retirements of E.A. Hackett of Tipperary south riding and Singleton Goodwin of Kerry, both in May 1920, and R.H. Dorman of Armagh in March 1921, were probably influenced by the disturbed conditions of the time. Most of the county surveyors, however, including fourteen whose appointments preceded 1898, appear to have managed to maintain reasonable working relationships with their councils during the difficult 1920 to 1922 period and were still in office when new governments came into office in both parts of Ireland.

40 Arthur Mitchell, *Revolutionary Government in Ireland* (Dublin, 1995), p. 160.   41 Mary E. Daly, *The Buffer State: the Historical Roots of the Department of the Environment* (Dublin, 1997), pp 56–8.

6

New Structures, 1922–44

INTRODUCTION

When the county surveyors and several hundred county councillors and other delegates assembled in April 1910 in the theatre of the Royal Dublin Society in Leinster House, Dublin, to hear the lord lieutenant, the Earl of Aberdeen, open the first Irish Roads Congress[1] they, like Irishmen of all political opinions, would probably have been amazed if they could have foreseen the partition of Ireland and the establishment of two separate administrations twelve years later. By March 1912, however, the surveyors felt it necessary to send a deputation to the Local Government Board to discuss how their interests might be protected in the event of home rule legislation[2] and the idea of partition was being discussed in the House of Commons a few months later. New structures were eventually decided on and given statutory effect by the Government of Ireland Act in December 1920 but the act came into force only in the six northern counties where powers were progressively transferred to the new parliament and government during the second half of 1921. The northern surveyors then found themselves and their councils responsible to a new Ministry of Home Affairs while the others continued to work under the general supervision of the Local Government Board until April 1922 when the formal transfer of power to the government of the Irish Free State took place.

County surveyor vacancies existed in a number of the southern counties in 1922 and, under an order made later that year by the Minister for Local Government and Public Health, new qualifications were prescribed for the office: the requirement that a candidate must have been examined and certified by the Civil Service Commissioners in accordance with the 1862 act was maintained but there was still no requirement to hold a degree in engineering; instead the candidate had to show that he had 'been regularly trained as a civil engineer', was engaged in the practice of his profession and, for a period of not less than four years, had charge of important works.[3] Examinations were eventually held in Dublin over a period of seven days beginning on 20 September 1923. Candidates who held a degree in civil engineering were exempt from Part I which comprised papers in mathematics, mechanics, and geology/mineralogy. In Part II, there were papers on structures, materials, railway and canal engineering and marine engineering together with papers on hydraulic and sanitary engineering and three separate papers

---

1 *Record of Proceedings of the Irish Roads Congress* (Dublin, 1910).   2 *Annual Report of the Local Government Board for Ireland,* 1911–12.   3 County Surveyors Qualifications Order, *Iris Oifigiúil,* Nov. 1922.

on county works. In addition, there was a paper on 'Gadhilg' (*sic*) in which a qualifying mark had to be attained before a certificate could be issued. Seven candidates qualified at these examinations and three others, who had previously passed in the professional/ technical subjects, successfully took the examination in Irish in order to maintain their eligibility for appointment. Examinations were again held in April 1924 with a similar range of subjects but the requirement to obtain a qualifying mark in Irish was dropped. On that occasion, 10 candidates qualified, bringing the panel of qualified candidates up to 20; of these, 7 were appointed in 1923–4, filling vacancies in counties Clare, Kilkenny, Laois, Longford, Mayo, Meath and Roscommon. Examinations were held over five days in October 1927, structured as in 1923 and 1924, but with Irish restored as a qualifying subject, and examinations in 1930, 1932 and 1935 led to the qualification of 23 additional candidates. The need for further examinations was eliminated in 1938–9 when it was decided that competitions for county surveyor posts would no longer be confined to those who had qualified at these examinations.[4] However, a certificate issued on foot of any of the 1923–35 examinations continued to be accepted in lieu of a university degree in engineering until the late 1940s.

THE LOCAL APPOINTMENTS COMMISSION

In spite of the declared policy of Sinn Féin, canvassing for county surveyor posts – as for posts at all levels in local government – continued after 1922. In 1926, however, the Local Appointments Commission was established to select and recommend persons to be appointed by local councils to professional, technical and higher administrative posts, including county surveyor positions.[5] The commissioners approached their task on the basis that the best-qualified candidate should always be appointed and so they adopted the practice of recommending one person only for each vacancy, thus depriving county councillors of the power which they had since 1899 to select the person to be appointed. This generated resentment and controversy, especially in the early years. When Cornelius Murphy, a chief assistant county surveyor in Cork since 1925, was recommended to Westmeath county council for appointment to the vacant post in that county in March 1928 following the first competition held by the commission for a surveyorship, some councillors argued that the recommendation should be treated with contempt, and referred back, while others criticised the sending of 'planters into Westmeath over the heads of natives';[6] the council eventually voted to make the appointment but only under protest.

Fianna Fáil, in opposition, aligned themselves with those who argued that local knowledge and the wishes of local councillors should be taken into account in making appointments[7] and a modification of the system was inevitable when the party took office in February 1932. Éamon de Valera, as President of the Executive Council, wishing 'to give effect to the fullest practicable extent to the principle of decentralization by devolution of responsibility to local authorities', requested that the commission should

4 *Annual Report of the Department of Local Government and Public Health*, 1938–9.   5 Local Authorities (Officers and Employees) Act, 1926 (No. 6).   6 *Midland Reporter*, 15 Mar. 1928.   7 Mary E. Daly, 'Local Appointments' in Mary E. Daly (ed.), *Country and Town: One Hundred Years of Local Government in Ireland* (Dublin, 2001), pp 50–1; Mary E. Daly, *The Buffer State: the Historical Roots of the Department of the Environment* (Dublin, 1997), pp 132–3; 155–7.

in future 'submit a panel of three names to the local authority concerned, except in cases where that authority expressly indicates a desire that only one name shall be submitted'. However, the 'three names' procedure which was introduced in June 1932 was not a success, with some local authorities refusing to make appointments where the name of a particular candidate was not among those submitted to them. It was claimed also that bribery and canvassing continued to prevail in some parts of the country[8] and the system naturally favoured the local candidate with the most influence. For these reasons, Mr de Valera, told the commissioners in July 1933 that they should revert to the practice of recommending only one person for each vacant position and that procedure was followed consistently thereafter. No county surveyor was appointed as a result of the 'three names' procedure.

A UNIFIED ENGINEERING SERVICE

When the local government system in the Free State was remodelled in 1925 rural district councils were abolished and full responsibility for roads (except those in boroughs and urban districts) was vested in the county councils whose sanitary, housing and health functions were assigned to separate Boards of Health and Public Assistance. These boards, as well as the urban councils, generally engaged consulting engineers, or employed their own engineers and architects, when building and engineering projects were to be carried out. In the 1920s, therefore, county surveyors' duties continued to be primarily related to roads and bridges in rural areas so much so that at national level they were treated as part of the roads establishment but the Department of Local Government and Public Health was not happy to allow this situation to continue: in its report for 1928–9, it pointed out that it had been felt that 'in confining the activities of the county survey staff to road work only, full advantage was not being taken of their professional quali- fications and training'.[9] One year later, the department was drawing attention to the need for better co-ordination of engineering services at local level, pointing out that the 27 county councils employed about 30 county surveyors, 128 whole-time assistant surveyors and 10 part-time assistant surveyors, while the boards of health, also operating on a county basis, employed up to 100 engineers and clerks of works, several of whom were also assistant county surveyors.[10] All of this, according to the minister, General Mulcahy, resulted in a very considerable amount of overlapping and a very considerable want of co- ordination.[11] He went on to say that 'we are getting at present county surveyors ... [who are] first-class all-round engineers ... it would be very wrong that men with the possibilities of these men should get into the groove of simply being roads men and would be denied ... the possibility of developing generally on the engineering side ... however, the present position is that the county surveyor is simply responsible for his roads'.

To remedy this situation, the approach initially decided on by the department was that the county surveyor, in additional to his traditional roads functions, should be regarded as the chief engineering officer of the county and should have general

8 Nicholas Mansergh, *The Irish Free State: Its Government and Politics* (London, 1934), p.233.   9 *Annual Report of the Department of Local Government and Public Health*, 1928–9.   10 *Annual Report of the Department of Local Government and Public Health*, 1929–30.   11 Dáil Debates 34, 30 Apr. 1930, cols 1205–6.

responsibility for water and sewerage schemes, housing and the other services administered through the boards of health. The surveyor would not necessarily be expected to prepare engineering plans for these services himself but would act as adviser to the board and, with the assistant surveyors, undertake the technical supervision of schemes. Stephen G. Gallagher – himself a county surveyor – in his presidential address to the Institution of Civil Engineers of Ireland in November 1931 spoke of the anxiety which was initially caused by these proposals but it seems that his real concern was that engineers in private practice would suffer under the new arrangements. Thus, Gallagher was happy to assure the institution that while the county surveyor 'would be placed in the position of county engineer', with functions of an advisory and supervisory nature, he would have the right to advise that an architect or consulting engineer should be engaged except where schemes were of a minor character. It was obvious, Gallagher went on, that such a change would not in any way interfere with the practice of members specializing in those public works usually carried out by county councils and boards of health.[12]

Steps towards the development of unified engineering services on these lines were initiated in the 1930s as vacancies arose.[13] For example, when a position in Tipperary north riding came to be filled in 1935, the duties prescribed for the new appointee included not just the traditional duties of the county surveyor but also a requirement 'to advise the Tipperary north riding Board of Health and Public Assistance on engineering matters and to exercise supervision'. It was made clear also that 'the drainage and all other engineering work of the county' would be under the supervision of the new appointee and would 'form part of his ordinary duties'. The last competition for a county surveyor post of this kind was held in 1942 but the vacancy in Co. Sligo was filled by the appointment of P.J. Haugh, who had been Leitrim county surveyor since 1936. The last person to attain the rank of county surveyor in the twenty-six counties was, therefore, J.T. O'Byrne, who had been appointed to the Wicklow post in July 1940 and who served until March 1964. The era of the county surveyor did not end, however, until the retirement in 1974 of J.G. Coffey, who had been Kilkenny county surveyor since March 1940.

In Northern Ireland, county surveyors continued to be appointed until the 1970s. However, following the report of a review body in June 1970, legislation was enacted to replace the county councils by a new system of district councils and to transfer roads and other major functions to the Ministry of Development. The county surveyors and other road staff became civil servants in October 1973 when the reorganization scheme came into effect; the office of county surveyor was then abolished, serving surveyors became divisional road managers in six road divisions and one – T.A.N. Prescott – became first director of a new centralised roads service.

THE FIRST COUNTY ENGINEERS

After the abolition of the boards of health in 1942 and the transfer of their functions to the county councils, reform of the engineering services was taken a stage further. Under

12 *TICEI* lviii (1931–2), 5–6.   13 *Annual Report of the Department of Local Government and Public Health,* 1930–1.

a series of ministerial orders made from 1943 onwards, individual county councils were required to appoint a 'chief engineering officer' to give them expert engineering advice and assistance in the performance of their functions. However, as the specific statutory duty to appoint a county surveyor had not then been repealed,[14] each order provided that the new office and that of county surveyor 'shall be amalgamated to form for all purposes one office, to be called the office of '———— County Engineer'. Qualifications were declared by the minister for each new office on a case by case basis as vacancies arose;[15] instead of the earlier requirement that a candidate should have been 'regularly trained as a civil engineer', candidates were now required to hold a university degree in engineering or possess equivalent professional qualifications, but those who held certificates of qualification as a result of examinations held by the Civil Service Commissioners continued to be eligible. Initially, in 1943–4, the age limits (except in the case of existing permanent local officers) were 26 to 45 years, as had been the case for county surveyors in the 1930s but the lower limit was raised to 30 in 1944, at which stage six, rather than four years experience 'in charge of important civil engineering works' was required. The duties of each new county engineer post were stated to be 'to give to the county council such engineering services of an advisory, supervisory or executive nature as are required by the county council in the exercise and performance of any of its powers and duties and include the duties of county surveyor'. The latter was necessary because, until the enactment of the Local Government Act, 1955, there were still a number of statutory provisions under which specific functions were assigned to county surveyors, rather than to their county councils.[16] The first competitions for posts of county engineer were held by the Local Appointments Commission in 1943 and resulted in three appointments – Claud Warner to Leitrim, Michael G. Ahern to Longford, and Thomas Kelly to Wexford; Warner took up office at the end of May 1944, a few days before the others, and was therefore the first person to serve in the new grade.

14 The provision was repealed by the Local Government (Repeal of Enactments) Act 1950 (No. 26).   15 The first national qualifications for county engineer posts were declared on 21 Apr. 1948.   16 Local Government Act, 1955 (No. 9), section 22.

Part II

The Work of the County Surveyors

7

Road Construction and Maintenance

Before the county surveyors came into office in 1834, considerable progress had been made by the grand juries in building new roads and improving the condition of the existing ones, while turnpike trusts had provided a network of trunk roads linking the main population centres.[1] With greater appreciation of the economic importance of better means of communication and with the application of more scientific principles to the business of road-making, major improvements and extensions of the main road network were made under the Mail Coach Road Act 1805;[2] competent surveyors were appointed to plan new routes and improvements of existing ones and over £450,000 in loans and grants was provided to assist the grand juries in implementing these plans.[3] By 1816, over 2,000 miles of roads had been surveyed in this way[4] and a considerable additional mileage was developed in the following ten years. In the 1820s, another important contribution to road-building, especially in the more remote areas, was made by a small team of engineers, Alexander Nimmo, Richard Griffith and John Killaly, who had been appointed in 1822 by the lord lieutenant to carry out programmes of public works in distressed areas with funds provided by the government. Over 1,000 miles of roads were constructed under these arrangements, mainly in counties along the western seaboard from Mayo to Cork and in Tipperary, Roscommon and Leitrim,[5] and were maintained by the Board of Works until 1854.

Against this background, and because of their determination in the 1830s and 1840s to minimize increases in the county cess, most grand juries were unlikely to be easily convinced by their new professional advisers that there was need to continue, or to expand, the costly roads programmes of earlier years. Nevertheless, a total of about 1,500 miles of new roads had been built under the direction of the surveyors by 1841.[6] Expenditure on new road and bridge construction had increased to about £150,000 a year in the early 1840s but the average cost per mile of new road was only £264.[7] It is clear from this and other evidence that many of the new roads were minor roads, often in

---

1 J.H. Andrews, 'Road Planning in Ireland before the Railway Age', *Irish Geography*, 5:1 (1964), 17–41; Peter O' Keeffe, *Ireland's Principal Roads, AD 1608–1898* (Dublin, 2003); David Broderick, *The First Toll Roads* (Cork, 2002). 2 An Act to amend the Laws for improving and keeping in Repair the Post Roads of Ireland, and for rendering the Conveyance of Letters by his Majesty's Post Office more secure and expeditious, 45 Geo. III c. 43. 3 *Return Showing all Sums of Money ... voted or applied ... in aid of Public Works in Ireland since the Union etc.*, HC 1839 xliv (540). 4 J.H. Andrews, 'Road Planning in Ireland before the Railway Age', *Irish Geography*, 5:1 (1964), 26. 5 *Twentieth Annual Report of the Commissioners of Public Works in Ireland*. 6 *Report of the Commissioners appointed to revise the Several Laws under or by virtue of which Monies are now raised by Grand Jury Presentment in Ireland*, Appendix B, HC 1842 xxiv (386); total adjusted to include mileage in Co. Tyrone. 7 P.C. Cowan, chairman's address in *The First Irish Roads Congress: Record of Proceedings* (Dublin, 1910).

remote rural areas and coastal districts, and that they can have consisted of little more than a cleared track, with limited drainage and a few inches of gravel or broken stone laid directly on the subsoil.

The most significant development in the years after 1834 was a dramatic increase in the mileage of roads under maintenance by the grand juries. According to a report presented to parliament in 1842, the total mileage was then 38,400,[8] nearly three times the total of 13,191 miles under maintenance in 1834,[9] and returns from county surveyors in 1868 indicate that the mileage was then about 50,000, almost four times the 1834 total.[10] The increased mileage was due mainly to the effective 'taking in charge' by the grand juries of existing roads, rather than the construction of new ones, but the extent to which the surveyors may have been responsible for the increased mileage by making applications themselves for presentments for road maintenance is not clear. Under grand jury law, the cost of maintenance of any road could be presented for and, if a question arose as to whether it was a 'public road', the matter could be argued before the judge at the assizes and decided by him. If a road was repaired at *any time* by a grand jury, that was accepted as *prima facie* evidence that it was a public road, and might still be so regarded even if, for one reason or another, the road had dropped out of the maintenance system for a period.[11] With legal provisions like these, the dramatic increase in the mileage of roads taken in charge – in effect, public roads – is easy to understand.

Like all other grand jury activity, road construction and maintenance was required by law after 1834 to be carried out on a contract basis. The road contractors, often local farmers, were generally appointed for three-year periods after a competitive tendering process but these contractors were altogether different from other contractors, according to one of the surveyors: 'one begs payment as a favour; the other, having performed his work, requires it as a right. The latter (the ordinary contractor) is almost always a capitalist; the former (the road contractor) is mostly a pauper, or a small farmer'.[12] Contractors regularly submitted very low bids in an effort to gain access to what was one of the few sources of off-farm income available at the time; moreover, sections of road for which tenders were sought were often chosen to suit the convenience and capacity of contractors who provided their own horses and carts, sometimes employed their own relatives, and did little more than spread prescribed quantities of hand-broken stone or gravel on road surfaces, leaving the material to be compacted and ground in by the passing traffic. In this situation, the annual cost of maintenance of the contract roads was very low – seldom more than £10 per mile in 1834 – but the standard of the work was uneven and not always acceptable. Besides, ensuring satisfactory completion of the work was a perennial problem for the surveyors, especially where contracts were awarded on the basis of unrealistic prices.

Whether the standard of maintenance on county roads improved in the early years due to the efforts of the surveyors is a moot point. Some of the grand juries did not

8 *Report of the Commissioners appointed to revise Several Laws under or by virtue of which Monies are now raised by Grand Jury Presentment in Ireland*, Appendix B, HC 1842 xxiv (386); total adjusted to include mileage in Co. Tyrone.   9 *Report from Select Committee on County and District Surveyors etc. (Ireland)*, HC 1857 (Sess. 2) ix (270)   10 *Returns from County Surveyors in Ireland ... of the Number of Miles of Roads under charge etc.*, HC 1867–68 lv (446).   11 *Annual Report of the Local Government Board for Ireland*, 1903–4.   12 John Neville, 'Grand Jury Laws and County Public Works, Ireland', *Dublin University Magazine*, xxvii: clix (1846).

believe that it had. The Armagh jury, for example, declared in 1837 that the roads were then in a worse condition than in 1834 and, in Co. Down, the surveyor was admonished because the Board of Works had drawn attention to the state of some of the post roads in the county.[13] For their part, surveyors drew attention to a number of factors to explain why, as one of them put it in 1846, the system of road maintenance introduced in 1834 had been a failure to some extent: 'the whole powers of the surveyor at present are passive – on a neglected work, to refuse a certificate. He is, therefore, in a position between Scylla and Charybdis. The public and the contractor are sometimes both dissatisfied; the former that it is aggrieved, the latter that he is not paid; and without a knowledge of the real cause, the fault is laid on the surveyor'.[14] Some surveyors campaigned unsuccessfully in the 1840s for modifications of grand jury law to make the system a more efficient one; it was suggested, for example, that the surveyor himself should be allowed to apply for presentments for all maintenance work which he considered necessary and that tenders should then be invited and accepted only on the basis of a predetermined scheme of road districts and sections, instead of leaving it to the presentment sessions and the contractors to decide how the work and the contracts should be organized.

As part of their road construction and improvement activity, the early surveyors built a considerable number of bridges but the absence of grand jury records and the fact that relatively few bridges carry plaques indicating the date of their construction, or the names of their designers or builders, makes it difficult in many cases to establish with certainty which bridges may be attributed to a particular surveyor. Excluding Northern Ireland, it has been estimated that there are some 16,000 masonry arch bridges in Ireland, many of which were erected in the late eighteenth century or in the first half of the nineteenth century.[15] The available records suggest that several thousands of these bridges, including many which are still in use, were either designed by the county surveyors between 1834 and 1844 or rebuilt under their supervision in those years. The early surveyors also built bridges in wood where resources for more substantial masonry structures were not available, but wooden bridges rarely survived for long and most had been replaced by the end of the nineteenth century.

The work of the county surveyors and the mileage of roads for which they were subsequently to have responsibility were directly affected by the Famine. Relief measures introduced in March 1846 following the failure of the previous year's potato crop in many districts involved the holding of special presentment sessions at which projects costing almost £250,000 were sanctioned.[16] The surveyors were required to examine and report on these projects – most of which involved road construction and improvement – and in many cases to organize and supervise the works. The numbers who applied for employment were very large and proper supervision became almost impossible in most areas. Under schemes administered directly by the Board of Works, employment reached

13 C.E.B. Brett, *Court Houses and Market Houses of Ulster* (Belfast, 1973).   14 John Neville, 'Grand Jury Laws and County Public Works, Ireland', *Dublin University Magazine*, xxvii: clix (1846).   15 *Reports on inspection, assessment and rehabilitation of masonry arch bridges and of concrete bridges* (Department of the Environment, Dublin, 1988 and 1990); on stone bridges generally, see Peter O' Keeffe & Tom Simmington, *Irish Stone Bridges: History and Heritage* (Dublin, 1991).   16 A.R.G. Griffiths, 'The Irish Board of Works in the Famine Years', *Historical Journal*, xiii (1970), 638.

a peak of nearly 100,000 before the works were closed down in August 1846 by direction of the Treasury, with perhaps another 30,000 employed on schemes for which the county surveyors were directly responsible.[17] With the realization that the failure of the 1846 potato crop would be even more extensive than in the previous year, public works schemes were again authorized in August 1846.[18] From early September, special presentment sessions to prepare lists of possible projects were held in public; these were variously described as tumultuous, turbulent and even riotous, with huge numbers of hungry labourers and others demanding that even the most outlandish and badly prepared schemes should be approved. Notwithstanding the primacy of the Board of Works and the large team of inspectors and supervisory staff it built up, many of the county surveyors also had to play a central – and not very popular – role in the administration of the 1846–7 relief works programme, reporting at the presentment sessions on the costs and the benefits of individual schemes and on those that might be approved, and organizing and supervising works scattered throughout their counties. After presentment sessions in Limerick, one of the surveyors was 'hunted like a mad dog by the whole country population'[19] while violence was common also on the relief works themselves, with supervisors often in considerable danger and liable to suffer even assassination attempts. In Clare, one official reported that strong nerves were needed and John Hill, the county surveyor, agreed, telling the chief secretary's office that 'these people are dangerous'.[20] Task-work was unpopular and efforts to initiate it often led to dangerous situations for the county surveyors who, in some cases, appear to have decided unofficially to apply a less rigorous regime than the central authorities had laid down so as to allow workers to earn more realistic amounts. The majority of the relief works undertaken in 1846–7 again consisted of the construction and improvement of roads even though most areas already had more than adequate networks. An Irish MP told parliament in January 1847 that the new roads were 'not alone useless, but in most cases, positively mischievous' and ran 'from nowhere to nowhere'.[21] Another spoke of 'works worse than idleness … public follies … works which will answer to no other purpose than obstructing the public conveyances'.[22] The cost of the work generally exceeded contract rates and many projects were left unfinished, causing frequent accidents and serving as 'a constant unsightly monument of a disastrous period'.[23] In some cases, it was alleged that even major roads on which relief schemes had been undertaken were left in a worse state than they had been before.

In the aftermath of the famine, the finances of the grand juries were under pressure due to difficulties in collecting county cess from individuals who were in financial difficulties, the demand for repayment of monies advanced by the government for relief works, and the need to complete the more important road schemes which were still in progress when the relief works were terminated. In parallel with these developments, long-distance passenger and freight traffic, including the mails, was beginning to transfer

17 Mary E. Daly, *The Famine in Ireland* (Dundalk, 1986).   18 An Act to Facilitate the Employment of the Labouring Poor, for a limited period, in the Distressed Districts in Ireland, 9 & 10 Vict., c. 107. 19 *Correspondence relating to measures for relief of distress in Ireland (Board of Works Series), July 1846 to January 1847*, HC 1847 1 (764)   20 NAI, CSORP 1846/1390, report of 6 Oct. 1846.   21 Col. F. French, Parliamentary debates III lxxxix, 25 Jan. 1847, col. 456.   22 Lord George Bentinck, Parliamentary debates III lxxxix, 4 Feb. 1847, col. 775.   23 *Final Report of the Board of Public Works, Ireland, relating to measures adopted*

to the railways from the main roads on which, notwithstanding improvements in the roads themselves and in vehicle design, loads of more than 1½ tons were rarely carried. These factors led to a decline in interest in road construction and improvement and impelled the juries to attempt to stabilize roads expenditure as a whole. Although the financial situation improved in 1853 when the government decided to eliminate completely the repayments due under the Labour Rate Act and to begin the transfer of functions from the grand juries to the poor law guardians or to central level, investment in improving the road system was no longer seen to be as important for economic and social development as it once had been and even expenditure on basic road maintenance continued to be severely restricted: average expenditure per mile in 1854 was much less than it had been twenty years earlier although the cost of labour had nearly doubled since then.[24] The attitude of the juries was fully consistent with the policy of the authorities in London from where Sir Charles Trevelyan, the powerful permanent head of the treasury, wrote to the Board of Works in March 1850 suggesting the promotion of a bill to amend grand jury law to take account of the fact that extensive expenditure on roads would be unnecessary in the future because of the development of the railway network.[25] Further evidence of the declining government interest in road development was the enactment of legislation to allow the Board of Works to divest itself entirely of the roads obligations it had assumed over the previous thirty years; in consequence, full responsibility for more than 1,000 miles of roads built by the board and its predecessors – and which they had been maintaining by direct labour – was transferred to the grand juries with effect from July 1854.[26] The repeal of the last of the Turnpike Acts in 1857[27] and the transfer of responsibility for the remaining 325 miles of turnpike roads to the grand juries in April 1858 marked the final phase of this process and confirmed that the impetus had gone out of road-making.

As individual surveyors began to adopt a more strict approach to the issue of the completion certificates without which road contractors could not receive payment, the contractors became more cautious, the number of tenders received fell substantially in some areas, and there were many cases where no tenders were submitted for particular works. For example, John Hill, King's County surveyor, was forced to tell his grand jury in July 1856 that no tenders had been submitted for 184 miles of roads, including 63 miles of the most important roads in the county, and that the neglect of these roads in the coming winter would be 'a most serious public evil'.[28] Situations like this led to the enactment in 1857 of legislation[29] designed 'to remedy the great loss and inconvenience arising from the want of proper contractors'; under this act, direct responsibility for maintaining considerable lengths of roads in some counties was taken over by the surveyors in the following years where no tenders, or unsuitable tenders, were received for necessary works. Hill, who had pressed strongly for this change in the law, was among the first to develop a well-organised direct labour system to cope with what came to be

for the relief of Distress, July and August 1847, HC 1849 xxxiii (1047). **24** *Report of the Commissioner appointed to Inquire into the Turnpike Trusts, Ireland, HC 1856 xix (2110)*. **25** NLI, Larcom papers, MS 7746. **26** Grand Jury (Ireland) Act, 1853, 16 & 17 Vict., c. 136. **27** Turnpikes (Abolition) Ireland Act, 1857, 20 & 21 Vict., c. 16. **28** Quoted in Justitia, *Observations on the General Grand Jury Act* (Dublin, 1857), p. 12. **29** Grand Jury (Ireland) Act, 1857, 20 & 21 Vict., c. 15.

known as 'works in the charge of the county surveyor'; in all, over 400 miles of roads in his county were said to have been almost impassable before this.

Where, as in most counties, contract work was still predominant, difficulties continued to be experienced until the end of the century with small contractors – some of them 'broken down farmers', as one surveyor described them – who still took jobs at recklessly low prices, left larger jobs unfinished, or neglected their responsibilities in the later years of a contract. Surveyors themselves were not generally allowed to prosecute defaulting contractors in the courts, and many of them complained about the delays caused by the need to get approval from the grand jury at each stage of a prosecution, and about the refusal of some juries to support action of this kind. Consolidation of small contracts to reduce the number of separate contractors involved and to provide a minimum road length of a few miles in the hands of each contractor was still an issue, with only limited success being achieved by those surveyors who attempted to enforce this. Difficulties continued to be experienced in ensuring adequate supervision of a multiplicity of contractors, especially in those counties where the surveyor was allowed to employ only a small number of assistants. The general practice was to require contractors to begin operations by placing specified quantities of materials on the side of the road for measurement by the surveyor or his assistants because, as one surveyor explained, 'there is no other mode by which it can be ascertained whether the contractors put on the specified quantity of material they are bound to use to repair the roads'.[30] But even when this was done, problems continued to arise, with heaps of materials sometimes being moved from one road section to another to be measured again. With all of its faults, however, the contract system of road maintenance was to survive in some counties into the middle of the twentieth century.

In the absence of any official relief agency, the county surveyors in a number of the western counties were called on at intervals from the 1870s onwards to take responsibility for programmes of relief works financed by the government. When distress arose in parts of the west in 1879–80, funds of £271,000 were allocated for road works but according to one senior official[31] much of the money was wasted and failed lamentably to achieve the objective. This experience brought about a hardening of official attitudes[32] and so, when relief works were decided on in subsequent years, different arrangements were put in place. In 1886, a separate body of Piers and Roads Commissioners was established with a fund of £20,000 to be spent on the provision of work for unskilled labourers in coastal areas in Galway and Mayo but it was still found to be necessary to call on the county surveyors to work with the commissioners in selecting and designing suitable schemes, and to act as their agents at construction stage. By August 1887, over forty piers and landing places, with new access roads, bridges and causeways had been constructed or improved mainly in west Galway and Mayo, and the care and devotion of the surveyors concerned to the work deserved 'most ample acknowledgment' according to a report from Colonel Fraser RE to parliament in 1889.[33] Notwithstanding this endorsement of the value

---

30 James Bell, letter to the *Longford Journal*, 18 Jan. 1839.   31 Sir Henry Robinson, *Memories: Wise and Otherwise* (London, 1923), p. 23.   32 Mary E. Daly, *The Buffer State: The Historical Roots of the Department of the Environment* (Dublin, 1997), pp 22–3.   33 *Piers and Roads: Counties of Galway and Mayo, Report by Colonel Fraser RE* [c.5729] HC 1889 lx.

of county surveyor involvement, when further relief funds were allocated in 1890 following a potato crop failure, Arthur Balfour, the chief secretary, insisted that the works should be tightly controlled by the central authorities in order to avoid what he described as the 'gross abuses' and reckless administration' of earlier years; Colonel Peacocke of the Royal Engineers was therefore brought in to advise the county surveyors – whom Balfour considered to be 'unreliable' – in the selection of works'. These ad-hoc short-term relief measures were followed in 1891 by the establishment of the Congested Districts Board with wide powers to promote economic and social development in districts along the western seaboard. The efforts of the board to improve communications and to develop the fishing industry regularly involved the surveyors, particularly in the construction of roads and bridges and the improvement and extension of fishing piers. By 1919, over £418,000 had been allocated by the board for roads and bridges and more than £108,000 for piers and other marine works.[34]

In the thirty years before 1885, there was very little change in the total mileage of roads for which the county surveyors were responsible. In that year, basic maintenance of the network of some 50,500 miles absorbed the not inconsiderable sum of about £630,000 but expenditure per mile was very low, in some counties even below the 1834 level.[35] Few new roads were constructed but some major bridges were built to replace earlier wooden structures; in the 1860s, for example, the bridges at Wexford and New Ross were replaced by iron bridges designed by the Wexford surveyor, J.B. Farrell; Samuel A. Kirkby was responsible in the early 1880s for the construction of long bridges on cast-iron piles to replace bridges at Youghal and at Kinsale, Co. Cork; and an iron swing-bridge was completed in 1887 to link Achill island to the mainland. After its introduction to Ireland about 1860, concrete had slowly come into use for bridge work, initially in mass concrete structures such as those constructed by the west Cork surveyor, Nat Jackson, from the late 1870s onwards. By the early years of the twentieth century, as specialist companies and consulting engineers emerged, reinforced concrete structures, using wrought-iron or steel for added strength, were gradually introduced.

While the lack of interest on the part of the public authorities in Ireland as well as in Britain in road development continued into the 1880s – and seems to have been matched at engineering level[36] – a serious debate about the manner in which road maintenance work could best be carried out by the grand juries began in 1885. The county surveyors were still very much divided at that stage on the central issue of whether the contract system or a direct labour system was best.[37] The majority were still opposed to any fundamental change in a system which, in their view, had served the country well and provided a network of roads which was tolerably well maintained. Direct labour would be financially unacceptable and operationally impracticable, according to many surveyors: there would be substantial extra costs arising from the purchase of plant and equipment

34 W.L. Micks, *History of the Congested Districts* Board (Dublin, 1925), pp 85–6.   35 J.O. Moynan, 'On the Maintenance and Repairs of Country Roads in Ireland', *TICEI* xvii (1887), 180–204.   36 In the index to sixty octavo volumes of the *Proceedings of the Institution of Civil Engineers* for the sessions 1879–80 to 1893–4, thirty-two pages are taken up by items of railway interest, with less than two covering papers on road subjects; the index to the *Transactions of the Institution of Civil Engineers of Ireland* lists only three papers on roads matters between 1845 and 1885.   37 See papers by H.V. White, F. Willson and J.O. Moynan, and subsequent discussion, in *TICEI* xvii (1887) 19–81; 144–204.

and in respect of overseers, time clerks, and pay clerks, particularly as the main object of the hired labourer as against the small contractor, was 'to put in his day', as one surveyor argued.[38] Besides, a direct labour system would interfere with the engineering duties of the county surveyor and reduce his status to that of a 'check clerk'. However, some of the surveyors in the 1880s took a different view, including several of those who, because of the lack of contractors in their areas, already had responsibility for the maintenance by direct labour of significant road mileages. There were some who called for the complete replacement of what one described as the 'crude, antiquated system' introduced in 1834 and the introduction of legislation under which responsibility for the carrying out of road works would be assumed directly by the county surveyors. Others called for a change of this kind but only in relation to the main roads and those other roads where, because of heavy traffic, poor subsoil conditions, or local scarcity of materials, the contract system was not capable of delivering acceptable standards of maintenance. Another suggestion was the introduction of a 'half-contract' system under which materials would be supplied under contract, in accordance with prescribed specifications, with the surveyors carrying direct responsibility for the recruitment and organization of labour for the conduct of the road work. While the policy debate continued, some surveyors took the law into their own hands by introducing new methods and higher standards of maintenance even though the statutory basis for their actions was uncertain. In a number of counties, *de facto* systems of direct labour were introduced. In 1890, Peter Cowan, the surveyor for Co. Down, secured the support of his grand jury for the introduction of steamrolling, and hired – and later purchased – an Aveling & Porter roller for service on the roads nearest the city of Belfast. He was soon followed by John H. Brett, the Antrim surveyor. Some years later, William Collen of Dublin, having initially failed to convince his grand jury of its merits, acquired a steamroller at his own expense to demonstrate its effectiveness; by 1900, he had a force of eleven rollers at work throughout the county.

The shortcomings of the traditional system of road construction and maintenance became even more apparent from the 1890s onwards with the use on the roads of new forms of recreational and work-related transport. While horse-drawn vehicles still predominated, heavy steam-propelled traction engines towing a number of wagons were being used intensively for freight traffic in some areas[39] with devastating effects on road surfaces, bridges and culverts. By 1896, the cycling boom was at its peak, and the rapidly growing cycling fraternity, many of whom were drawn from the middle and upper classes, were complaining loudly about the condition of road surfaces which caused frequent accidents as well as punctures in their relatively expensive tyres;[40] the grand juries were regularly berated by cycling magazines for 'the disgraceful state of the Irish roads' and county surveyors and their assistants were criticised for 'neglect of the ordinary duties for which they are paid'.[41] The pressure for better roads intensified after the first petrol-engined motorcar made its appearance in Ireland in 1896. Motoring grew

---

38 P.J. Lynam, *TICEI* xvii (1887), 68.   39 In 1896, the Locomotives on Highways Act raised the speed limit for these vehicles from 2 mph to 14 mph and put an end to the red flag requirement which had applied since 1865; the speed limit was raised again to 20 mph in 1903.   40 Brian Griffin, *Cycling in Victorian Ireland* (Dublin, 2006). 41 The *Irish Cyclist*, founded in 1885 by Richard and Alex Mecredy, played a prominent part in this campaign, as did the weekly, the *Irish Wheelman*; see, for example, *Irish Wheelman*, vii:8 (22 May 1894).

rapidly in popularity and 3,668 cars and motorcycles had been registered by September 1905.[42] By then, the Irish Roads Improvement Association had developed into a powerful lobby group representing the influential motoring community whose vehicles raised great clouds of dust on poor road surfaces in dry weather and were prone to skidding when surfaces were wet and muddy.

Although these developments led to intense discussion of the organizational and practical arrangements for road maintenance and pavement improvement, there was little recognition of the need for change in the Local Government (Ireland) Act 1898. The power of initiative in relation to annual road works programmes was given to the rural district councils, rather than the county councils, and the total expenditure that could be incurred on roads in any county was capped at 125 per cent of the average spending of the grand jury in the previous three years. The new county councils were authorised to purchase or lease quarries, and to purchase or hire steamrollers, scarifiers and stone crushers, but only for the purpose of making quarried materials and machinery available to road contractors.[43] Thus, the traditional legislative bias in favour of contracting was maintained and was confirmed by an order of the Local Government Board in January 1899[44] which provided that all public works undertaken by the county councils should normally be executed by contract except where it was not possible to obtain suitable tenders or contractors.

In the year ended 31 March 1901, 54,063 miles of public roads were maintained by the county councils at a cost of £721,923. In the following ten years, expenditure on road improvement and maintenance increased substantially and the first steps were taken towards revolutionary change in the manner in which road work was carried out – a change which was to have a dramatic effect on the role and responsibilities of the county surveyors. Following an amendment of the law in 1900, the Local Government Board made an order which allowed a county council to introduce direct labour operations in any district where the council, after considering a report from the county surveyor, thought fit.[45] A number of councils had pressed strongly for this amendment, influenced more by the need to provide employment for rural labourers than by considerations of efficiency and effectiveness in road-making or by pressure from the roads lobby, but councils generally were slow to introduce new arrangements. Limerick was first to change over fully to direct labour: under pressure from the labourers and their supporters, a direct labour scheme covering all 1,830 miles of road in the county was adopted in October 1902 and approved by the Local Government Board in March 1903.[46] In Tipperary north riding, where there had also been pressure from labour interests and difficulties arising from lack of contractors, a scheme covering 394 miles of main roads was approved at the same time. In the following three years, schemes covering some of the roads in ten other counties came into effect. In all of these cases, the councils raised short-term loans to buy steamrollers, scarifiers and other road-making plant and the county surveyors directly employed teams of gangers, labourers and pay-clerks to implement the new system.[47]

42 *Annual Report of the Local Government Board for Ireland*, 1905–6. 43 61 & 62 Vict., c.37, section 12.
44 The Local Government (Procedure of Councils) Order, 1899, articles 22 and 24. 45 Local Government
(Ireland) (No 2) Act, 1900, 62 & 63 Vict. c. 41; Local Government (Procedure of Councils) Order, 1901.
46 *Annual Report of the Local Government Board for Ireland*, 1902–3. 47 *Annual Reports of the Local*

In 1906, however, there was still a strong divergence of views among county surveyors on the direct labour issue. Of the eight who replied to an enquiry from the Irish Roads Improvement Association that year, three still supported the contract system: P.J. Lynam of Louth believed that road labourers would not work well without a level of supervision which could not be afforded; Henry Webster of Wexford claimed that labourers were scarce and in great demand in his county; and James Heron of Down believed that reorganizing and perfecting the contract system was the quickest and best way to improve the roads of his county. Two others, J.W. Leebody of Tyrone and S.A. Kirkby of Cork, believed that direct labour would give better results only on the more important roads.[48] William Collen of Dublin was the strongest supporter of direct labour operations, describing the contract system as a farce whose object appeared to be to subsidise the small farmers of the county. R.H. Dorman of Armagh was another strong advocate of the new system as was E.A. Hackett of Tipperary south riding.

By 1908, with direct labour schemes in operation in thirteen counties, the Local Government Board was able to report that the new method 'continues to find favour' because of the improvement in the standard of road maintenance which resulted; it noted also that in Clare 'the decrease in pauperism' in some areas was being attributed to the employment provided on the roads.[49] By 1910–11, seventeen counties were operating direct labour schemes covering 8,792 miles of road out of the total of 54,086, but only Limerick and Tipperary north riding relied entirely on direct labour at that stage.[50] As the move towards direct labour accelerated in line with the growth in the volume of mechanically propelled vehicles and the increased use of motor lorries for freight movement, the surveyors became responsible in most cases for large direct-labour organizations. By August 1927, a small army of more than 20,000 road workers had come into existence[51] and employment levels rarely dipped much below that level until the 1960s. When percentage grants (financed out of the proceeds of new duties on motor spirit and on motor car licences) were introduced in 1910 for steamrolling and other improvements of main roads, many councils were reluctant to take them up, preferring to continue to allocate the available local resources to labour-intensive work on purely local roads. High levels of spending on these roads and an emphasis on the creation and maintenance of employment continued to be dominant features of road policy and financing for many years, causing surveyors regularly to complain that job creation came first and real transportation needs a poor second.

The limited road improvement grants which were available from 1910 onwards enabled some of the surveyors to carry out substantial improvement projects on main roads and roads which had come to be heavily used by motorised tourist traffic. During the world war, however, when grants were severely curtailed and borrowing was restricted, road improvement operations were again confined to work which was urgently necessary.[52] Ordinary maintenance activity was also reduced because of increases in the rates of road workers' wages and in materials costs, and because of the release of large

Government Board for Ireland, 1903–4, 1904–5 and 1905–6.   **48** *Important Articles on Road-Making and Maintenance, published for the Irish Roads Improvement Association* (Dublin, 1906).   **49** *Annual Report of the Local Government Board for Ireland*, 1907–8.   **50** *Annual Report of the Local Government Board for Ireland*, 1910–11.   **51** *Annual Report of the Department of Local Government and Public Health*, 1925–7.   **52** *Annual Report of the Local Government Board for Ireland*, 1916–17.

numbers of road labourers, and even some road machinery, for tillage and harvesting operations to meet the wartime food production crisis. By March 1921, the IRA campaign in various areas led to the trenching of roads and the destruction of bridges in order to hinder the movement of crown forces. Moreover, with continuing pressure on local finances, only 400 out of about 8,500 road workers were still employed, road maintenance work had come to a virtual standstill, and the condition of the roads was rapidly deteriorating.[53] Trade unions criticised the fact that labourers were dismissed while salaried officials like the county surveyors generally remained on full pay[54] but it appears that only one of them, E.K. Dixon of Mayo, in whose county all road work had been suspended, was given notice of dismissal after he had rejected a proposal that he should accept only 50 per cent of his normal rate of pay. Even in this case, however, the council found it expedient to withdraw the notice and to ask Dixon to continue in office.

The huge challenge facing the government of the Irish Free State, the local authorities and the county surveyors in providing an adequate road network was clearly set out in a report by the Commission on Reconstruction and Development which was published in May 1923.[55] The commission was concerned primarily with the problem of unemployment which was being exacerbated by demobilization from the army but, noting that road work had generally been adopted in the past as the most suitable in an emergency to absorb surplus labour, it recommended that a long-term road improvement programme should be adopted immediately. Such a programme was justified by the fact that less than half the network of 8,000 miles of trunk and main roads was fit to carry motor traffic; many of these roads had never been steamrolled and the network as a whole had suffered in the previous four years due to lack of maintenance, the abnormal military traffic, the transfer of traffic from the railways due to the dislocation of that network and the destruction of many roads and bridges during the war of independence and the civil war. To strengthen, recondition and improve the trunk and main roads was estimated to cost, on average, £2,000 a mile, and a first five-year programme to deal with 3,600 miles was recommended. This, it was estimated, would provide employment for 10,500 men working at quarrying, stone-breaking and spreading material on the roads.

In 1922, practically all roads were of the unbound type, with weak foundations and poor alignment. Ten years later, following the allocation of more than £5 million in state grants to the county councils, substantial lengths of what were then being described as 'the National Roads' had been reconstructed in asphalt, concrete, tar or bitumen macadam and other dustless and impervious surfaces. According to the Department of Local Government and Public Health, 'all the main traffic lines and the important streets of every town have been brought to a standard which compares favourably with the roads and streets of any other country' and this fact stood out as 'a testimony to the energy, enthusiasm and efficiency of the survey and administrative staff of the local authorities'.[56] Investment in the improvement of the main roads continued throughout the 1930s but complaints were frequently made that value was not being obtained for the expenditure

53 Michael Laffan, *The Resurrection of Ireland: The Sinn Féin Party, 1916–1923* (Cambridge, 1999), pp 330–1.
54 Mary E. Daly, *The Buffer State: The Historical Roots of the Department of the Environment* (Dublin, 1997), p. 81.   55 Commission on Reconstruction and Development: Interim Report on Reconditioning and Improvement of Roads (Stationery Office, 1923).   56 *Annual Report of the Department of Local Government and Public Health*, 1931–2.

incurred on direct labour maintenance work and there was some support for the view that, for small roads, with light traffic, the contract system was 'adequate to meet the necessities of the case'. There were fluctuations in the numbers employed on roads in the late 1920s and 1930s but the average was generally between 16,000 and 20,000 in those years, not including almost equal numbers engaged on employment schemes; these schemes, for which the surveyors were also responsible, generally involved additional road and footpath works.[57] In 1939–40, as the era of the county surveyor was coming to an end, and as county surveyors were gradually assuming new responsibilities for housing and water and sewerage projects in addition to their roads functions, their combined teams of road workers had increased to almost 40,000. However, some of the surveyors were still grappling with the problems which their nineteenth-century predecessors faced arising from the employment of small contractors on road works: over 10,000 miles of roads were still being maintained by contractors at that stage at an average cost of only £10 per mile[58] – little more than the figure which applied 100 years earlier!

57 *Annual Reports of the Department of Local Government and Public Health*, 1927–8 to 1939–40.   58 *Annual Report of the Department of Local Government and Public Health*, 1939–40.

8

Public Buildings and Other Public Works

The Antrim surveyor, Thomas Jackson Woodhouse, told a parliamentary committee in 1836 that 'extensive architectural works seldom occur or come under the charge of the surveyor' and when the need arose, as in the case of hospitals, gaols and courthouses, 'other parties would be called in' because the county surveyor could not be expected to undertake such work.[1] Woodhouse agreed that a man could be a very good engineer and a very bad architect and he offered the view that engineers 'know but little of architecture'. While a small number of the early surveyors distinguished themselves as architects – generally on foot of work which they undertook for private clients – developments in relation to the building of courthouses, gaols, lunatic asylums and the Queen's Colleges can all be taken as evidence that the grand juries, as well as the authorities in Dublin, had little confidence in the design skills of the average county surveyor.

County courthouses
In the decades immediately before the appointment of the surveyors, the grand jury system had made 'its last and, perhaps, its finest contribution in the sphere of public buildings: the majestic series of courthouses which adorn the provincial towns'.[2] County courthouses had been built or rebuilt in most of the county towns and maintenance work, rather than new construction, was all that was then required in most areas. Where, exceptionally, a new county courthouse was required after 1834, or major alterations were needed, only a few of the surveyors were called on to provide designs. In Belfast, Charles Lanyon designed the building at Crumlin Road, and at Coleraine a new building was provided by the Londonderry surveyor, Stewart Gordon. But a lack of confidence in the architectural capacity of their county surveyor was shown by the Down grand jury in 1838 when a new building was required at Newry; in this case, Thomas Duff was brought in to design the building, much to the annoyance of the surveyor, John Fraser. Again, when it was decided in the 1840s to replace Gandon's building at Waterford, the assignment went to John B. Keane, rather than the local surveyor. It was Keane who also won the commissions to provide new buildings at Nenagh (1840) and at Limerick (1850), while a new building at Ennis, also completed in 1850, seems to have drawn heavily on plans presented by him ten years earlier. Where alterations or extensions to earlier

1 Evidence given on 17 May 1836, in *Report from the Select Committee on County Cess (Ireland)*, HC 1836 xii (527). 2 Maurice Craig, *The Architecture of Ireland from the earliest times to 1880* (London, 1982); see also Niall McCullough, 'Courthouses – The Mirror of Society', in Mildred Dunne and Brian Phillips (eds), *The Courthouses of Ireland* (Kilkenny, 1999).

buildings were decided on, the county surveyor was more likely to be called on to provide plans. Thus, John Neville extended the Dundalk building in 1846 and J.B. Farrell carried out the last of a long series of alterations to the Wexford building in 1862. Substantial reconstruction work was undertaken at Downpatrick (1855) and Newry (1862) by the Down surveyor, Henry Smyth, and Arthur C. Adair was responsible for additions and modifications to the courthouse in Londonderry in the late 1890s. But there were many cases in which modifications or improvements were put in the hands of outside architects; for instance, William Barre was commissioned to extend the building at Omagh in 1853 and in Sligo, J. Rawson Carroll was engaged for the rebuilding of the courthouse in 1876.

Sessions houses

In addition to the large and often monumental county courthouses, changes in the operation of the legal system also necessitated the building of new sessions houses, or local courthouses, in numerous towns around the country in the decades before the county surveyors came into office. However, there was still a need for additional accommodation and the Board of Works was in contact with surveyors about plans for no less than fifteen new buildings in a number of counties in 1837.[3] In these cases, the board provided detailed specifications setting out the accommodation required and while they recognised that design work of this kind may not have previously been attempted by many of the surveyors, they made it clear that they would be expected to undertake it in the future.[4] In some cases, surveyors appear to have been confident enough to go ahead and design the new buildings themselves, using patterns which were then familiar. In other cases, when a surveyor declined to undertake the work himself, or when his own plans were rejected, an outside architect with relevant experience was brought in. In Kilkenny, for example, W.D. Butler, one of the most prominent architects of the period, provided plans for new courthouses at Callan and Urlingford and shared the 5 per cent fee for superintendence of the construction work with the surveyor, Samson Carter. Other examples of sessions houses completed by surveyors in the early years are Arklow and Tinahely, Co. Wicklow (William Hampton); Balbriggan and Swords, Co. Dublin (Alexander Tate); and the building at Ballymena by Charles Lanyon. Henry Smyth, the Down surveyor, provided a new building at Banbridge in 1872, the Louth surveyor, Patrick Lynam, completed a new courthouse at Drogheda in 1889, and as late as 1916 the Westmeath surveyor, A.E. Joyce, completely rebuilt the courthouse at Athlone. But architects in private practice also won a substantial number of commissions for the design of relatively modest courthouse buildings after 1834. W.D. Butler of Dublin and his pupil, William Caldbeck, were pre-eminent in this respect while George Richard Pain was engaged to design a substantial number of courthouses in Munster.

County gaols

Most of the county towns had been provided by the grand juries with substantial new county gaols or extended gaols before the appointment of the county surveyors. In all, there were forty-one of these buildings in 1822,[5] many of them designed by the same

---

3 NAI, OPW 1/1/4/12, Public Works Letters No. 12, pp 252–4.   4 Ibid.   5 *First Report of the Inspectors General on the General State of the Prisons of Ireland*, HC 1823 x (342).

group of architects who built the county courthouses, including Richard Morrison, John Hargrave, and the Pain brothers. Between 1822 and 1832, ten new gaols were provided and substantial additions and improvements had been made at sixteen of the existing facilities, leaving only two of them in need of improvement or replacement in the view of the authorities.[6] With the availability of so many relatively new buildings and the decline in the prison population generally, only a few of the grand juries had to undertake the construction of new gaols in their areas after 1834 and only two of them seem to have called on their surveyors to provide designs. In Belfast, Charles Lanyon designed the gaol at Crumlin Road (1843–5) to replace a building at Carrickfergus on which the government inspectors found it 'painful' to have to report again in 1832,[7] and John Neville completed a new county gaol at Dundalk in 1854. In other cases, outside architects were called in, for instance at Nenagh, where a new building designed by J.B. Keane was completed in 1842. Some additions and modifications to earlier gaol buildings (such as those carried out at Londonderry by Stewart Gordon in the 1850s) were made by various surveyors but, in the majority of cases, it seems unlikely that the surveyors made any significant contribution to prison architecture before they and the grand juries were relieved entirely of responsibility for the prisons in 1877.

Bridewells

In addition to the county gaols, many of the smaller towns had bridewells or local gaols to which petty offenders, debtors and prisoners in transit were committed, and these were also the responsibility of the grand juries and their surveyors. In all, 112 bridewells were in use in the first decades of the nineteenth century, many of them damp, unsanitary and insecure. By 1832, however, most of the bridewells in use were newly built, according to Inspectors General of Prisons; no less than eight of them were opened in Kerry alone in 1831.[8] Architects in private practice continued to be engaged in the majority of cases after 1834 when the need for a new bridewell arose (sometimes in conjunction with a sessions house) with W.D. Butler and William Caldbeck appearing to have a near monopoly of the business in the 1840–60 period. Occasionally, individual surveyors were very critical of being bypassed and disclaimed all subsequent responsibility for a building designed by an outsider. Bridewells designed solely by county surveyors were very much the exception and most surveyors would have carried out no more than minimal maintenance work at the bridewells before these buildings, like the county gaols, were transferred to the new General Prisons Board in 1877 – and, in nearly 50 per cent of cases, promptly closed down.[9]

Lunatic asylums

In the 1820s and early 1830s, under an extensive building programme undertaken by the Commissioners for the Erection of Lunatic Asylums, nine district asylums were provided throughout the country, from designs by Francis Johnston and William Johnston Murray.[10] When a second series of seven new asylums was commissioned in the mid-

6 *Tenth Report of the Inspectors General on the General State of the Prisons of Ireland*, HC 1831–32 xxiii (152). 7 Ibid. 8 Ibid. 9 R.B. McDowell, *The Irish Administration, 1801–1914* (London, 1964). 10 *Report from the Select Committee of the House of Lords appointed to consider the State of the Lunatic Poor in Ireland*, HL 1843 (193).

1840s, none of the surveyors was engaged to provide designs for the buildings. However, at Ballinasloe, where only an extension of the existing asylum was required, the commission went to the east Galway surveyor, James F. Kempster, who was responsible also for some further additions to the complex in the 1880s. County surveyors were more prominent in implementing a third programme of asylum building which began in the 1860s and which involved sixteen new buildings. Henry Smyth, one of the Down surveyors, was the architect of the asylum at Downpatrick (1865–9); Arthur C. Adair, then the Clare surveyor, was partly responsible, with William Fogerty, a Limerick architect, for the asylum at Ennis (1870); and J.B. Farrell and James Bell, both county surveyors, collaborated in the design of the large new building which was completed at Enniscorthy in 1868.

The Queen's Colleges
Legislation enacted in 1845 provided for the establishment of Queen's Colleges at Belfast, Cork and Galway. Although the Board of Works was authorised to employ the 'county surveyor, or other competent surveyor or architect', to design the necessary buildings, Charles Lanyon was the only surveyor to be awarded a commission; his buildings for Queen's University Belfast, opened in 1849, are still in use.

Workhouses
Some of the early surveyors were prepared to operate as building and civil engineering contractors in the 1830s and even in the 1840s, when the Royal Institute of the Architects of Ireland was seeking to prohibit its members from undertaking building contracts. Patrick and Edmund Leahy, the Cork surveyors, won two of the workhouse construction contracts allocated by the Poor Law Commissioners in 1839–40[11] and seem to have completed the buildings at Lismore and Mallow satisfactorily. At Listowel, where the Leahys had contracted to provide another workhouse at a cost of £5,980,[12] they failed to complete the contract, forcing the commissioners to take over the works and complete the building themselves by direct labour.[13] The Leahys were the only county surveyors who were awarded workhouse contracts, although Henry Stokes, their Kerry colleague, submitted an unsuccessful tender for the buildings at Tralee.

Piers and harbours
A large number of piers, quays, small harbours, landing places, and associated coast roads, were constructed at various locations on the coast in the 1820s with the aid of grants from the Commissioners of Irish Fisheries and later from the Board of Works, usually at a cost of £500 to £800.[14] Under a new programme of marine works initiated in 1846 as part of the response to the famine,[15] forty-five projects were completed between then and 1852; the individual schemes, which generally cost between £500 and £3,000, were designed and supervised in most cases by Board of Works engineers and

11 *Sixth Annual Report of the Poor Law Commissioners*, 1840; *Seventh Annual Report*, 1841.   12 *Seventh Annual Report of the Poor Law Commissioners*, 1841; *Eighth Annual Report*, 1842.   13 *Ninth Annual Report of the Poor Law Commissioners*, 1843.   14 *First and Second Reports of the Commissioners of Inquiry into the state of the Irish Fisheries*, HC 1837 xxii (77) (82).   15 Fisheries (Ireland) Act, 1846, 9 & 10 Vict., c. 3; Piers and Harbours (Ireland) Act, 1847, 10 & 11 Vict., c.75.

carried out by direct labour.[16] Grand juries were authorized by law in 1836 to construct piers and quays, or to repair or enlarge any existing pier or quay, but they appear to have made little use of these powers. In the 1850s, however, most of the facilities for which the Board of Works had been responsible were vested in the grand juries[17] leaving the county surveyors with responsibility for the maintenance of more than 200 piers, harbours and landing places.[18] Apart from basic maintenance and occasional storm repairs, few of the surveyors undertook significant works on any of these piers except where development was initiated and financed by the Congested Districts Board in the first decades of the twentieth century.[19]

16 *Return of All Piers and Harbours Built under the Board of Works in Ireland since the passing of the Act 9 Vict.,* *c.3*, HC 1884–85 lxx (167).   17 Grand Jury (Ireland) Act, 1853, 16 & 17 Vict., c. 136, second schedule. 18 *Return of all Piers and Harbours built under the Board of Works in Ireland since the passing of the Act 9 Vict.,* *c. 3*, HC 1884–85 lxx (167).   19 Reports from county surveyors on the condition of these piers were submitted to parliament in 1885 (*Statements by County Surveyors on the Condition of Piers and Harbours etc.*, HC 1884–85 lxx (266)); more detailed reports were prepared in 1886 (*Royal Commission on Irish Public Works, Appendix to the Second Report, Minutes, Evidence and Index* [C.5264–1], HC 1888 xlviii).

9

Private Practice in Architecture and Engineering

County surveyors were precluded by law from receiving fees or payments from contractors to the grand jury but there was nothing to prevent them from taking on other appointments, from engaging in private practice, or from accepting fees for assignments from other grand juries or public bodies. The majority of the early surveyors appear to have taken full advantage of this liberal regime and a number of them became better known for their private work than for their public duties. A small minority of the surveyors were more than competent architects, with private practices which extended to a wide variety of projects. On the engineering side, John Fraser, surveyor for Co. Down, freely admitted in 1841, that he earned as much from his private practice as he was paid for his official duties;[1] and in Cork, Patrick and Edmund Leahy, together with Patrick's two other engineer sons, developed a major civil engineering practice, the income from which must have far exceeded their official salaries.[2] Charles Lanyon, after he took up duty in Antrim in 1836, went on to become one of the most important and prolific architects in Ireland[3] and was responsible for many of Belfast's prominent buildings while John Benson, who was appointed to Cork east riding in 1846, also had an extensive private architectural practice.

The extent to which the early surveyors engaged in private practice often gave rise to criticism by their grand juries. Henry Lindsay, in Armagh, was the subject of complaints in 1837 when it was alleged that his private business was leading to neglect of his public duty, but he insisted on his right to continue his private practice.[4] John Fraser, in Co. Down, also found himself in difficulties with his grand jury on this score, as did the Leahys in Cork where a magistrate, in sworn evidence to a royal commission in 1844, claimed that 'the taxation of the county will always be increasing, so long as the engineer of the county has the privilege or the power of a private office'.[5] The fact that so many of the surveyors engaged in private practice in the early years is not surprising, when account is taken of their dissatisfaction with their rate of pay and with the lack of any provision for payment of travelling and subsistence expenses to them or for pensions. In addition, allowance must be made for the fact that there was a scarcity of private engineering and architectural practices in the Ireland of the 1830s and 1840s, especially

1 *Report of the Commissioners appointed to revise the Several Laws under or by virtue of which Moneys are now raised by Grand Jury Presentment in Ireland*, Appendix B, HC 1842 xxiv (386).   2 Brendan O Donoghue, *In Search of Fame and Fortune: The Leahy Family of Engineers, 1780–1888* (Dublin, 2006), pp 107–9.   3 Frederick O'Dwyer, 'The Architecture of the Board of Public Works 1831–1923', in *Public Works: The Architecture of the Office of Public Works 1831–1987* (Dublin, 1987).   4 C.E.B. Brett, *Court Houses and Market Houses of Ulster* (Belfast, 1973).   5 Evidence of Robert A. Rogers, in *Evidence taken before the Commissioners appointed to Inquire into the Occupation of Land in Ireland*, HC 1845 xxi (657).

outside the capital, and that Dublin-based architects and engineers were unlikely to find it profitable to undertake the design and supervision of relatively small isolated projects around the country. As late as 1857 it was suggested by one of the surveyors that as one could not expect to have an engineer or architect of eminence in private practice in country districts, 'it would be a very great convenience and benefit to the country at large to be able on the spot to obtain professional services of such value' as could be provided by a county surveyor.[6]

By the 1860s, the fact that the official salary of several of the county surveyors accounted for only a small part of their total incomes was attracting critical media and other comment; indeed, a *Dublin Evening Post* editorial went so far as to suggest that the reason the office was anxiously sought after was 'because it is generally the means of obtaining a very large amount of private practice' resulting in a situation in which, in several of the counties, 'salary is only a small part of the income'.[7] However, the surveyors themselves strongly defended the *status quo* in their evidence to a select committee which gave 'anxious attention' to the issue in 1857 and which concluded that, on the whole, the grand jury should have discretion to permit or to prohibit private practice.[8] Charles Lanyon told the committee that he saw 'many advantages connected with a private practice' and argued that it helped 'to get a better class of men to undertake the duties of county surveyor'; Henry Brett, then serving in Wicklow, claimed that 'an efficient officer is decidedly improved by allowing him to have private practice, provided it is not to a very great extent'; and Alexander Tate, a Dublin district surveyor, maintained that 'where private practice is enjoyed to a large extent, you will find concurrently with it county works efficiently and satisfactorily managed'. When increased salaries for county surveyors were authorized in 1861, the law was amended to allow a grand jury to direct that its surveyor, as a condition of receiving an increased salary, 'shall not engage in any private professional practice'[9] and when pensions for the surveyors were authorised in 1875, those appointed after that were required to have given their whole time to their posts if they were to qualify. But many of the grand juries were slow to prohibit private practice, and many of the surveyors seem to have found it more advantageous to continue with private practice than to serve in a pensionable office.

In the early years of the twentieth century the boards of guardians and the new rural district councils brought about a significant expansion of activity in the provision of water and sewerage schemes and in the construction of labourers' cottages – areas in which the newly elected councillors were obviously more willing to respond to the needs and demands of their constituents than their predecessors had been. Although extensive private practice by county surveyors in other fields was no longer the norm at that stage, these expanded programmes provided fresh opportunities for a significant number of the surveyors – and some of their assistants – to supplement their salaries by seeking to be engaged to design small water and sewerage schemes in rural areas and to design isolated labourers' cottages or groups of cottages. A few of the surveyors also continued to attract

6 Evidence of Alexander Tate to the *Select Committee on County & District Surveyors etc. (Ireland)*, HC 1857 (Sess. 2) ix (270). 7 *Dublin Evening Post*, editorial, 22 June 1861. 8 *Select Committee on County & District Surveyors etc. (Ireland)*, HC 1857 (Sess. 2) ix (270). 9 An Act to enable Grand Juries in Ireland to increase the Remuneration of County Surveyors and for other Purposes, 24 & 25 Vict., c.63, section 6.

commissions for new building and other work at the lunatic asylums and from school authorities, and to undertake architectural work for private clients.

As the first decade of the twentieth century was a difficult period for the building industry and the allied professions, with work for architects in particularly short supply, the Royal Institute of the Architects of Ireland began to campaign publicly in those years against the surveyors who continued to engage in private architectural practice. The council of the institute in June 1901 called on county councils to make arrangements which would enable the surveyors to devote all of their time to their official duties.[10] Some months later, a general meeting of the Architectural Association of Ireland resolved that county surveyors' private practice constituted 'a serious grievance and injury to the members of the allied professions of architecture and engineering'; members of the association felt that the surveyors were in an unfairly favoured position to secure private work and that this state of affairs was an injustice to private practitioners and to the ratepayers; the *Irish Builder* actively supported the architects' case, criticized the institute for its 'years of inaction' on the subject and listed ten surveyors who had been actively engaged in architectural work.[11] In the years that followed, the architects' organizations regularly protested when engagements went to persons who were not among their members, declaring this to be 'a serious menace, not alone to the interests of architects in private practice' but also 'to the development of the art of architecture itself'.[12] In 1909, the institute stepped up the campaign, declaring that it was contrary to the public interest for county surveyors and their assistants to engage in private practice in competition with other architects and engineers and calling on all county councils to require their officers to devote all of their time to their official positions.[13] The county councils gradually responded by making it a condition of appointment of new surveyors that they should devote the whole of their time to their official duties but nothing could be done in the case of existing office-holders, some of whom continued to have private practices into the late 1920s.[14]

The more important architectural and engineering projects carried out by individual surveyors in the course of their private practice are noted in the biographies in Part III. What follows is intended to give an indication of the nature and extent of the work involved:

Country houses and villas
Numerous country houses as well as urban residential development projects were designed by Charles Lanyon. Thomas Turner was also a prolific designer of country houses in Ulster, and John Neville and Samuel Ussher Roberts were prominent in similar work. A number of the surveyors continued to engage in house extension or improvement work at least until the end of the nineteenth century but there are only a few examples of new houses designed by surveyors after the 1860s.

---

10 *IB*, 20 June 1901, 771.   11 *IB*, 18 July 1901, 803; 1 Aug. 1901, 818.   12 *IBE*, 1 Dec. 1906, 964.   13 *IBE*, 9 Jan. 1909, 26.   14 A survey conducted in 1918 established that 99 of the 185 assistant surveyors then serving had private practice rights (*IBE*, 2 Nov. 1918, p. 466).

Churches, convents and schools

Although Catholic church building gained momentum in the period immediately after emancipation in 1829 and was to become the dominant architectural activity in Ireland until well into the twentieth century,[15] little of this work was carried out by county surveyors. Charles Lanyon, however, was a prolific church architect, designing buildings for Church of Ireland and Presbyterian communities, mainly in Ulster. Others who designed churches were Stewart Gordon of Londonderry, John Benson of Cork and Samuel Ussher Roberts of Galway. The fact that two of John Neville's daughters became nuns has been linked to 'his prodigious output in the field of convent design, which spread far beyond the boundaries of Co. Louth to his native Limerick and to Roscommon and Tyrone'.[16] Thomas Turner designed schools in the northern counties as did Lanyon and Richard Williamson of Derry.

Banks, commercial buildings and offices

Competition among the emerging banking companies in the nineteenth century led them to engage some of the major figures in Irish architecture to design impressive buildings for their headquarters and for branches in the larger towns, resulting in buildings which are among the most important survivors from Victorian and Edwardian Ireland.[17] County surveyors obtained commissions from some of the banks and contributed notable buildings such as those by Charles Lanyon and Thomas Turner in Belfast, Armagh and Derry. James Bell (junior) designed the Ulster Bank in Longford in 1860 and Henry Brett of Wicklow and his sons undertook much of the National Bank's building work in the 1870s. John Neville is thought to have designed many of the shops/stores in Dundalk's main streets while John Benson provided the Butter Market and the Princes Street (English) Market in Cork. In Belfast, Lanyon's Custom House, completed in 1857, is noteworthy while Thomas Turner contributed council offices in Derry and the Town Hall in Coleraine.

Railway stations

Like the banks, many of the early railway companies set out to provide imposing buildings of some architectural merit at their stations and termini, thus generating some substantial architectural commissions. The surveyor who profited most from this was Charles Lanyon who has been described as the 'prince of railway architects'.[18] Lanyon's stations included buildings in Belfast, Coleraine, Cookstown, Larne and Bangor. Thomas Turner was responsible for an imposing Italianate terminus at Foyle Road, Derry (now demolished), while John Benson designed the station at Penrose Quay, Cork, for the Great Southern & Western Railway, the Albert Quay head office building for the Cork & Bandon Railway and the nearby Albert Street terminus of the Cork & Passage Railway.

Miscellaneous projects

Surveyors attracted a variety of unusual design commissions. John Benson provided the buildings for the Cork Exhibition of 1852 and earned a knighthood for his design of

---

15 Sean Rothery, *A Field Guide to the Buildings of Ireland* (Dublin, 1997).   16 Christine Casey, 'John Neville: Louth County Surveyor', *County Louth Archaeological and Historical Journal*, xxi:1 (1985), 23–30.   17 Emmet O Brien, 'The Architecture of Bank Buildings in Ireland, 1726–1910' (PhD, UCD, 1991).   18 Alistair Rowan, *The Buildings of Ireland–North West Ulster* (London, 1979).

Dublin's Great Exhibition buildings in 1853; his Berwick Fountain still adorns the Grand Parade in Cork city but his Turkish Baths in Bray have long since been demolished. Charles Lanyon contributed the Palm House at the Botanic Gardens in Belfast, the Campanile at Trinity College, Dublin, and Scrabo Tower, in Co. Down, a memorial to the Marquis of Londonderry. Other surveyors who designed public monuments included Charles Tarrant who designed the clock-tower on the quay in Waterford, and Carter Draper who designed the memorial in Wicklow town to Captain Halpin, one-time master of the *Great Eastern*.

Civil engineering

Apart from railway engineering and minor land reclamation and other projects for local landowners, the most significant engineering projects associated with the private practices of the surveyors involved harbour improvement work, such as that carried out in Wexford by J.B. Farrell and at Dundalk by John Neville. In Clare, John Hill was responsible for major land drainage and reclamation work and in Cork, Edmund Leahy provided outline plans for a ship canal to allow steamships drawing fourteen feet of water to berth near the centre of the town of Bandon – a project which failed to reach construction stage.[19] In the later decades of the nineteenth century and in the early years of the twentieth century, the boards of guardians, and subsequently the rural district councils, undertook a growing number of water supply and sewerage schemes and labourers' cottage schemes and in many cases employed county surveyors to design these.

Railway engineering

Several of the first group of surveyors had experience of railway engineering before taking up their appointments: Patrick Leahy (Cork east riding) had worked under Nimmo on surveys for the company which was authorised in 1826 to build a line from Limerick to Waterford and on some early English railway surveys in the early 1830s,[20] while Charles Forth and Thomas Jackson Woodhouse had worked with Charles Vignoles on the Dublin & Kingstown Railway in the early 1830s. When the climate for railway investment improved in 1835–6, generating what came to be known as the little railway mania, it was not surprising that some of these men became heavily involved in a number of the twenty-six separate railway projects which were put forward in Ireland during those years. In the north, Woodhouse, in partnership with William Bald, surveyed the route for a Belfast–Ballymena railway in 1835 and prepared plans for the Ulster Railway Company's line from Belfast to Portadown and Armagh, while at the other end of the country, the Leahys prepared plans for a Cork–Passage railway in 1836.[21] When the cyclical trend in railway investment reached a new peak in 1844–5, prospectuses were published for up to 150 new railway schemes in Ireland, many of them involving county surveyors.[22] In the north, Charles Lanyon was engineer to a new company which revived the scheme for a Belfast to Ballymena railway and was engaged also by the Londonderry

19 *Bandon Navigation: Report of the Proceedings of the Committee appointed on 17 December 1841*, Cork Archives Institute, U 140/D. 20 7 Geo. IV, c. xxxix; see Brendan O Donoghue, *In Search of Fame and Fortune*, pp 25–6, 37–8. 21 *CC*, 7 May 1836. 22 *Irish Railway Gazette* i:5 (2 Dec. 1844); i:10 (6 Jan. 1845); ii:4 (22 Nov. 1845); ii:6 (8 Dec. 1845).

& Coleraine Railway. The 1825–6 scheme for a Limerick–Waterford Railway was revived, with Charles Forth heading the team which was given responsibility for completing the route surveys and preparing the detailed plans. Henry Brett, then the Mayo surveyor, worked under Sir John MacNeill on surveys for railways in Connacht, including an abortive Irish West Coast Railway scheme, and Barry Duncan Gibbons of Wexford was another who became involved. By far the most spectacular and speculative venture into railway engineering, however, was made by the Cork surveyors, Edmund and Patrick Leahy, who promoted the Cork & Bandon Railway in 1844, followed up in 1845 with proposals for extensions and branches to serve Bantry, Kinsale, Macroom, Killarney and Tralee, and developed a number of other substantial proposals which brought their planned railway mileage to more than 500 by the end of 1845.[23]

The involvement of county surveyors in the promotion, design and construction of railways created a direct conflict of interests since it was their public duty to ensure that the safety and carrying capacity of the roads were not prejudiced by railway construction works. In Co. Louth John Neville opted not to engage in railway work himself and had no hesitation in the 1840s in taking strong action against the Dublin & Belfast Junction Railway and its formidable engineer, Sir John MacNeill, when he found that some proposed bridges would cause inconvenience and endanger public safety.[24] But who was to protect the public interest where the surveyor was himself the railway company's engineer? The authorities in Dublin seem to have allowed the situation to drift in 1844–5 when many of the surveyors were heavily involved in railway engineering even in their own counties; they may have been influenced by the relative scarcity of engineering talent in many parts of the country, the huge workloads already being carried by the small number of consulting engineers available for railway work, and the general desire at the time to press on rapidly with railway development. However, when Edmund Leahy sought in early 1846 to appoint a deputy to carry out his county surveyor duties while he himself worked to consolidate his various railway interests, the Board of Works took the view that matters had gone too far and advised that 'the interests of the counties will not be properly guarded, having reference to the injury that their public works may receive from the construction of railway works passing over or near them, if the county surveyors are not in their own counties unconnected with the management of the latter;[25] Leahy's request to appoint a deputy was therefore rejected and he was told that the lord lieutenant could not sanction his involvement in any way with railways passing through Co. Cork while he continued to act as county surveyor there.[26] Other surveyors, however, managed to maintain their involvement in railway engineering, even in their own counties, into the 1850s and 1860s. Charles Lanyon, for example, was engineer to several lines constructed in the north during those years; John Bower, who took up duty as surveyor in Carlow in 1861, was engineer to railway companies in Donegal; and Henry Brett, while serving as surveyor in Wicklow, worked with John Hill, then the King's County surveyor, on plans for the Tullamore–Athlone line which opened in 1859. Brett

23 See Brendan O Donoghue, *In Search of Fame and Fortune* (Dublin, 2006), pp 125–140. 24 Christine Casey, 'John Neville: Louth County Surveyor', *County Louth Archaeological and Historical Journal*, xxi:1 (1985), pp 23–30; see also Canice O'Mahony, 'Iron Rails and Harbour Walls: James Barton of Farndreg', *County Louth Archaeological and Historical Journal*, xxii:2 (1990), 134. 25 NAI, CSORP 1846 W 676. 26 Ibid.; *Irish Railway Gazette* ii:13 (26 Jan. 1846); CC, 27 Jan. 1846.

was also employed on the engineering of the Athenry–Tuam line in 1860 and a few years later, he became heavily involved in planning and promoting a Dublin–Baltinglass railway which failed with heavy loss to himself.

Light railways and tramways

When a new phase of light railway and tramway development, mainly in the south and west of Ireland, began under a series of acts passed between 1860 and 1896 county surveyors were again involved. Initially, there was no provision for state or grand jury financial support for new undertakings but from 1880 onwards loans could be made on the security of a baronial guarantee and, after 1883, grand juries had power to guarantee the payment of dividends of up to 5 per cent on the capital where the net operating receipts were insufficient. These developments opened up new opportunities for county surveyors, some of whom engaged actively in the design of particular schemes. For the majority, however, light railway development involved significant additional duties: a surveyor was required to provide detailed reports on all aspects of a proposed scheme as a basis for a decision by the grand jury on whether to support the scheme, and at construction stage the surveyor was obliged to exercise general supervision over the works which often involved sections of lines running along the margins of public roads. When new railways and tramways became operational, the surveyor had to audit the gross and net receipts to establish whether a baronial contribution towards the dividend was justified and, in extreme cases, when hopelessly uneconomic tramways or railways had to be taken over entirely by a committee of management appointed by the grand jury, the county surveyor became, in effect, general manager of the line; the Schull and Skibbereen and West Clare lines were among the better-known examples of such lines.

Part III

Biographical Notes on County Surveyors

ADAIR, Arthur Charles (*c*.1824–1903), trained in architecture and engineering as a pupil in the Belfast office of **Charles Lanyon**. He was subsequently employed by the Board of Works on the supervision of various building projects and spent some years in Canada where he worked on railway construction. He returned to Ireland in the early 1850s and became Clare county surveyor on 27 August 1855, taking up residence at Bindon Street, Ennis.

Adair carried out some improvement works at the county courthouse in Ennis which had been completed to a design by J.B. Keane in 1850. His major building project, undertaken in conjunction with the architect, William Fogerty, was the Clare lunatic asylum at Ennis (later Our Lady's Psychiatric Hospital) which was constructed in an Italianate style between 1866 and 1870 to accommodate 260 persons. Fogerty was primarily responsible for the architectural aspects, with Adair taking responsibility for the water and sanitary arrangements, heating, lighting and services generally; he also superintended the construction works with Fogerty's nephew, Joseph. The buildings cost about £35,000 and were so large by the standards of the day that a commentator in the *Irish Builder* wrote, soon after their completion, that 'one would imagine this establishment could accommodate all the indoor and outdoor idiots, madmen, women, and children in the whole kingdom of Ireland'.

In Ennis, Mill Road Bridge, a single large arch with massive rusticated voussoirs and heavy piers, was completed in 1857 probably to a design by Adair, and he invited tenders in 1866 for the erection of a new iron superstructure on the bridge at Bindon Street, since reconstructed in reinforced concrete. Like many other surveyors of the period, he was a severe critic of the contract system of road maintenance as it operated under the law then in force. His refusal to grant certificates to many contractors where there was 'gross violation' of their contracts drew criticism from the lord chief baron who declared at the spring assizes in 1863 that if a grand jury found a county surveyor acting harshly towards the contractors, they ought to dismiss him. Adair had acted with a little too much stringency, according to the judge, and should in future concern himself only with the question of substantial compliance: 'persons must look rationally into the affairs of men' he said 'and it is not necessary that a road be in a beautiful line to entitle the contractor to a certificate'. Adair, however, continued to press for stronger action against contractors, complaining at the spring assizes in 1867 that 'a system has arisen of persons taking contracts at a low price without any intention of completing them; the consequence is that many of the important roads in the county are in a bad state of repair, and likely to remain so if steps be not taken to compel such contractors to carry out their work'. He assured the grand jury that, if their solicitor were directed to seek court decrees for damages against the defaulting contractors and their sureties, he would give him all possible assistance in bringing the proceedings to a successful conclusion. Adair, however, was transferred to the western division of Co. Tyrone in May 1867 before anything could be done to put this policy into effect; the transfer was believed to be the only one which had been made since 1834 without the consent of the surveyor concerned and it meant that the maximum salary available to Adair was reduced from £600 to £500 a year.

Adair lived at Lisnamallard House, near Omagh, during his term of office in Tyrone and had his offices in the town. At the spring assizes in 1874, the grand jury agreed to increase his salary but this was blocked on a legal technicality by some dissenting cesspayers. Adair then sought a transfer to a better-paid post and this was arranged in the following May when he was appointed to replace **Richard Williamson**, the Derry surveyor, who had died; he was allowed the maximum salary of £600 despite suggestions that some overburdened cesspayers stood aghast at such reckless expenditure. One of Adair's nine assistants in Tyrone, W.J. Robinson, also moved to Derry in 1874, beginning what was to be a thirty-five-year term of office as city surveyor under the borough corporation; Adair, however, filled the separate post of surveyor to the city grand jury in addition to his county duties.

Adair was responsible for a water supply scheme at Limavady (1877); additions and improvements at the lunatic asylum (1888–90); extensive additions in 1896–8 costing some £5,500 at the Greek Revival courthouse in Bishop Street which was designed by John Bowden and completed in 1817; and new county offices which were completed in 1899. He carried on a private practice as an architect in Derry from 1875 onwards – a period of great prosperity and rapid development in the city – and engaged in other activities which attracted attention and criticism in parliament. On 9 July 1880, J.G. Biggar, the Parnellite MP for Cavan, asked the chief secretary for Ireland if it was consistent with his official duties for Adair to act as a land agent in Tipperary but the chief secretary chose not to become involved: he replied that he did not know the facts of the case and did not think he was duty bound to make enquiries, as Adair had been appointed twenty-five years earlier at a time when private practice could not be forbidden. Biggar was not satisfied with this reply and, in asking a further parliamentary question on 29 July, succeeded in putting it on the record that Adair had been refused salary increases in his two previous counties; that the public works in Derry were in such a state as to cause serious public agitation; and that Adair, as well as acting as a land agent and as secretary to a club, was conducting a private practice as architect, engineer and land surveyor. Again, however, the chief secretary responded that it was not for him to intervene and Adair was allowed to continue in office until the end of the grand jury era.

At the assizes in March 1899, a month before the new county council was to take office, Adair tendered his resignation because, as he put it, he was getting old – he was then 75 years of age – and would not be able for the extra work that would arise in serving the new county council and rural district councils; besides, quoting loosely from the Book of Exodus, he thought it appropriate that his pension should be determined before the county council came into office as 'there might then be a new king in power who knew not Joseph'. Because of a dispute with the Local Government Board about the level of salary for the office, a successor was not appointed until March 1900 and Adair continued on an acting basis until then, drawing both his salary of £600 and his pension of about £350 to the great annoyance of some members of the county council.

Adair married Sarah Magee and had at least one son, Arthur John, born in Ennis in March 1867. In Derry, he lived at Crawford Street, a street of large houses built for the well-to-do by a speculative developer between the late 1850s and the early 1870s. After a long illness, he died there on 15 December 1903, aged 79, and was buried in the city cemetery. An obituary in the *Derry Standard* described him as 'an upright and unswerving public official of high ability, jealous of the dignity of his office, and determined never through fear, favour or affection to be swerved from the strict path of duty … His work will live in many an important public building and engineering work in the county'.

AHERN, Michael Gerard (1907–92), was born on 5 April 1907 in Cork, the son of an assistant county surveyor. He was awarded the BE degree at University College, Cork, in June 1927 and qualified at the last examinations for county surveyorships held in 1935. His first appointment in 1927–8 was as assistant county surveyor in Cork where he was engaged mainly on road construction. Like so many of his contemporaries, he then worked for a period on the Shannon Scheme and was subsequently employed by the Valuation Office and the Turf Development Board. He was employed as contractor's engineer by the South of Ireland Asphalt Company between 1937 and 1941, working mainly on water supply schemes. He joined the Department of Local Government and Public Health in April 1941 as a temporary engineering inspector and was given responsibility under **T.C. Courtney** for supervising the scheme of large-scale turf production by local authorities which was in operation during the emergency years.

Following the first county engineer competition held by the Local Appointments Commissioners in December 1943, Ahern was recommended for appointment to the Longford post. He took up duty there on 1 June 1944 but transferred to Co. Limerick in July 1948. He held strong views on a range of local government issues and was never slow to express them. He believed that not all county managers were 'suitable persons', that there was a danger that they would become 'yes men' who would sacrifice local autonomy in favour of supporting the government line, and that their appointment had downgraded the position of the county engineers. He was an outspoken critic of road policy in the 1950s and early 1960s, when a high proportion of the total resources available were devoted to the black-topping of minor roads rather than the improvement of the main arteries. In addition, he was a staunch defender of the local authority direct labour system, arguing that it enabled services to be provided efficiently and at less cost than would arise if the work were to be done by those he dubbed 'privateers'. Ahern was in office in Limerick until November 1974 when he retired. He died on 15 January 1992.

ARMSTRONG, Alexander James (fl. 1829–52), probably a native of Donegal, became the first Cavan county surveyor on 17 May 1834. He had been employed as an assistant draftsman at the Boundary Survey Office in Dublin in 1829–30 at a salary one guinea a week and was subsequently employed on a number of Board of Works projects.

In his evidence in 1836 to the commissioners who were inquiring into the condition of the poorer classes, he said that while the turnpike and other principal roads in his county were in good condition the bye-roads were in very bad repair 'in consequence of a system of jobbing by contract'; this was having an adverse effect on the development of tillage and mountain farming because, with improved roads, one horse could carry nearly double the quantity of produce to the market. Armstrong set about redressing this situation and built long stretches of new roads in the Killeshandra and Bailieboro areas and in other parts of the county. He was responsible for the construction in 1836–7 of the three-arch masonry bridge over the river Erne at Belturbet, for which the Board of Works had advanced a loan of £1,700 in the previous year; the bridge still carries the N3 national primary road running northwards from Cavan to the border with Northern Ireland at Aghalane Bridge.

In 1841 Armstrong told an official inquiry that he had been obliged to give up his limited private practice as it interfered with the performance of his roads duties. By then, some 3,000 miles of roads were in his charge and he had three assistants. His workload increased still further during the famine period, when he was obliged, like the other surveyors, to assist the Board of Works in the selection, planning and supervision of relief works on which 17,500 persons were employed throughout the county in December 1846; at one stage, early in 1847, he received a letter through the post advising him that his life would be in danger if he proceeded with a certain line of new road.

Armstrong served in Cavan until his death at Bath on 1 February 1852. His son, Edward James, educated by a Mr Moore in Cavan, entered Trinity College, Dublin, on 2 November 1855, aged 15, but did not graduate.

BARCLAY, Thomas (fl. 1834–55), was the first county surveyor for Mayo, appointed on 17 May 1834. At the spring assizes in 1836 a major dispute arose between him and some members of the grand jury when he insisted that a particular project had not received the necessary prior approval at the presentment sessions. Notwithstanding an apology by the surveyor, the grand jury resolved that his conduct was 'marked by the most persevering disrespect' and that he should be dismissed immediately. However, the foreman of the jury later told the chief secretary's office that, because Barclay had a wife and large family, they would not object to his appointment to another county. In addition, the Board of Works advised that as they had 'frequent testimony of Barclay's good conduct and qualifications, he should not suffer an extent of stigma that he does not deserve'. It was arranged, therefore, that Barclay would be transferred to King's County on 11 April 1836, in an exchange with **Henry Brett**.

Barclay went to live at Gayfield, Tullamore, and seems to have had an uneventful career in his new county where, in addition to his official duties, he conducted a limited private practice in surveying. He was in his usual good health when he set out on his biannual circuit of the county in May 1855 to attend the various presentment sessions in preparation for the summer assizes. He collapsed in the street in Portarlington on 24 May on his way to the courthouse, and died almost immediately, leaving a large family to mourn him. According to the *King's County Chronicle* he was highly esteemed in the county and considered to be a man of considerable professional attainment.

BARRY, William Fitzgerald (1863–1951), was born in Dublin in 1863. Having earned a diploma in civil engineering at the Royal College of Science for Ireland in 1887, he trained in mechanical engineering at the Great Southern & Western Railway's works at Inchicore. He was engaged on the construction of the West Clare Railway in 1887 and later worked as an assistant in the county surveyor's office in Portarlington under **Henry V. White**. He was resident engineer on water supply schemes at Enniscorthy and Ferns before moving in July 1895 to Kilkenny as assistant county surveyor under **A.M. Burden**. He transferred to the staff of the new county council in 1899 at a salary of £130 a year.

Barry was a candidate at county surveyor examinations held from 1893 onwards and, although his marks exceeded the qualifying standard on each occasion, he had to wait until 1900 to obtain an appointment. He became acting county surveyor in Monaghan in 1899, after **James Heron** had moved to Co. Down and when the substantive post came to be filled by the county council in June 1900, after the usual advertisements, Barry was the only applicant and was duly elected to the post which carried a salary of £400 a year. It seems, however, that he cannot have been very happy in Monaghan, as he was a regular applicant for county surveyor posts in other counties in the early 1900s. In December 1906, he was elected by the casting vote of the county council chairman to the Sligo surveyorship, and he resigned his post in Monaghan in the following January in prospect of taking up his new duties. However, it emerged soon afterwards that one of the councillors who had voted for him was an undischarged bankrupt – and therefore ineligible for membership of the council – leaving the way open for Barry's opponent in the final council vote, **R.J. Kirwan**, to claim the position. Barry had to carry on in Monaghan for five years more before securing another appointment. He was an unsuccessful candidate for the Wexford post in June 1909 when **Stafford Gaffney** was appointed but when the latter gave notice in November 1911 of his intention to resign, the post was advertised again. Barry applied once more and after a relatively short campaign, was elected unanimously from five applicants at a council meeting on 11 January 1912. The post carried a salary of £500 a year.

Barry continued the substantial programme of bridge works begun by his predecessor. Good progress on the construction of a new ferro-concrete bridge at Ferrycarrig had been made by the contractors, Messrs Colhoun of Derry, before he took up duty and the bridge opened for traffic in November 1912. It was 350 feet long, 25 feet wide, with 11 spans of approximately 30 feet, and a cast-iron opening span; it was constructed on the Hennebique principle, and was only the third of its type in Ireland. The bridge was not, however, a great success: it had to be closed for major repairs as early as 1932 when it was 'in an unsound condition' according to Barry, and investigations by consultants in the early 1960s disclosed severe spalling of the concrete, with bracing detached from the piles, and severe vibrations. Weight restrictions followed, with the limit reduced to three tons by 1972. A new bridge consisting of eight spans of precast prestressed concrete beams on tubular steel piles, designed by J.B. Barry & Partners, was completed by Ascon Ltd in 1980 beside the 1912 structure at a cost of over £800,000.

By the end of 1912, Wexford county council had approved plans for another new ferro-concrete bridge, with an opening span, to replace an 1844 wooden bridge over the river Slaney at the Deeps, near Killurin; the plans in this case were prepared by Messrs Delap and Waller of Dublin, and the cost was not to exceed £5,000. The bridge was quite similar to that at Ferrycarrig: 360 feet long in all but only 18 feet wide, and with spans of about 30 feet. The British Reinforced Concrete Engineering Company was awarded the contract to build the bridge at a price of £3,713, with the bascule opening section being provided by the Cleveland Engineering Company at a price of £1,250. The bridge was completed in 1915 and is still in use, although in need of some rehabilitation work.

Yet another new bridge, this time an inter-county bridge over the river Barrow, was authorized by Wexford county council in August 1915 when they allocated £2,425 to meet their share of the cost of replacing the wooden bridge which had been constructed at Mountgarret, just above New Ross, by the American engineer, Lemuel Cox, in 1794–6. By April of the following year, Barry and his Kilkenny colleague, **A.M. Burden**, had received preliminary plans from Delap and Waller for a reinforced concrete bridge, with a Scherzer-type opening span of 40 feet and 10 other spans of about 30 feet, and Kilkenny county council had agreed to support the scheme. By the time tenders were invited in June 1916, Burden had taken over primary responsibility for the project but, for cost reasons, the scheme did not go ahead at that stage. In 1920, Barry was warning his council that the old bridge was 'absolutely dangerous' but the two councils could still not raise the cost of a new one, then estimated to be up to £18,000. After the almost complete destruction of the wooden bridge by fire in 1921, the new reinforced concrete bridge, with some design alterations, was eventually completed and opened to traffic in November 1929 under the general supervision of **R.F. Bowen**, who had become Kilkenny county surveyor in 1923.

The iron bridge at Wexford (completed by one of his predecessors, **J.B. Farrell**, in 1866 to replace the original wooden bridge of 1795) continued to be a cause of concern for much of Barry's time in Wexford. As early as 1914, he advised the county council that there was need for a new bridge which he estimated would cost over £37,000. In 1918, the admiralty offered to pay £5,000 towards reconstruction of the bridge if the county council would put up the same amount, but the offer was withdrawn when it became apparent that £10,000 would not be nearly enough to put the bridge into a proper state of repair. In July 1922, in the early stages of the civil war, some of the bridge spans were blown up and contractors had to be engaged urgently to carry out essential repairs; the contractor's men were fired on more than once and, in 1923, the army had to complete the restoration work. The bridge was to survive for another thirty-five years, with speed limits and weight restrictions applying from 1927 onwards, and the placement of an array of tar barrels to force the single-line traffic to move from one side of the bridge to the other as it crossed the waterway. In 1938, Messrs Delap and Waller expressed the opinion that the bridge was not strong enough to carry its own weight but there were to be many more years of debate before construction of a new bridge was authorized by the Minister for Local Government in 1952, ten years after Barry's retirement. The new seven-span structure, 1,250 feet long and with clearance of up to 20 feet above high water level, was the product of the first major design and construct contract in Ireland; it was completed in 1959 by Ascon Ltd at an overall cost of some £300,000 on the site of the original wooden bridge. At the official opening in September of that year the Minister for Local Government, Neil T. Blaney, noted that the prestressed concrete superstructure was the first example of the use of this new technique in a major Irish bridge and went on, according to his official script, to express confidence that the bridge would long endure. But this third Wexford Bridge was fated to have a much shorter life than either of its predecessors, inadequate though they may have been. Expensive repairs became necessary in 1978 and structural defects which could lead to the ultimate collapse of the bridge became evident in 1994. The entire superstructure was replaced in 1997 by a composite steel-girder bridge, with precast concrete deck slabs, at a cost of some IR £6 million.

The only substantial building project for which Barry was responsible seems to have been the reconstruction in 1930 of the old county gaol at Wexford to form a new courthouse and county offices. However, he made a significant contribution to the modernization of the Wexford road network and of road construction systems in the 1920s. By 1924, he had six portable Acme stone-breakers in operation around the county as well as other forms of mechanization. After what were described as heroic efforts to improve the road from Wexford to Enniscorthy by traditional means in 1924–5, he arranged the resurfacing in concrete of over eleven miles of the road in 1927–8 by the Pioneer Road Construction Company of Dublin. This was one of the earliest concrete road projects carried out in the country and the new section of road was believed to be the longest continuous stretch of concrete road in Great Britain or Ireland at the time. It consisted of a 16-foot-wide carriageway, formed of 6 inches of unreinforced concrete, with a 2-foot-wide section of macadam on each side, and it was completed at a cost of 7s. 6d. per square yard. Four concreting units were employed, using 17,000 cubic yards of stone drawn from quarries near Wexford and at Brownsbarn, near Enniscorthy, as well as 8,500 cubic yards of sand dredged from the Slaney by local cotmen.

According to his county manager, Barry had served the county with outstanding courtesy and efficiency when he retired on 1 December 1942 at the age of 78. He died in December 1951.

BEATTY, Joseph Nicholson (1868–1939), was born on 5 December 1868 at Donacloney, Co. Down. He was educated at Lurgan college and graduated from Queen's College, Belfast, with a BE degree in 1893. He was employed as clerk of works and then as assistant county surveyor under **P.C. Cowan** in Co. Down before going to work in England in 1896 with Naylor Brothers, engineers, on the Ashbourne & Buxton Railway. When **David Megaw**, the special assistant to the Antrim county surveyor, was obliged to take extended sick leave in 1903, Beatty was appointed from twenty-four applicants to take his place temporarily as supervisor of the county's steamrolling programme which was being carried out on a direct labour basis using five steamrollers purchased from borrowed funds. After Megaw's return to duty, Beatty continued in the service of the county council and worked under Megaw when the latter became county surveyor in 1914.

Following Megaw's sudden death in August 1927, the vacancy was advertised in February 1928 at a salary of £800 a year, with £150 for travelling expenses. Beatty was unanimously elected to the post in the following month, having been acting as county surveyor in the interim period. He was prominently associated with the work of laying out long lengths of concrete roads in the Antrim, Ballymena and other areas of the county; a total of thirty-three miles of these roads was laid between 1925 and 1931. Beatty served for seven years as county surveyor and retired in 1934. He subsequently devoted a good deal of time to farming at his native Donacloney although he continued to live in Belfast. He died on 11 April 1939, aged 70.

BELL, James (*c*.1794–1872), practised as an architect and builder in Trim where, in 1817, two years after the battle of Waterloo, he designed and constructed the town's Wellington Monument – a massive structure comprising a Corinthian column on a square base and surmounted by a statue of the Iron Duke whose family home at Dangan Castle was not far from the town. The statue itself was the work of Thomas Kirk (1781–1845) whose statue of Nelson stood on the pillar in Dublin's O'Connell Street until 1966.

Bell was responsible for the preparation of land surveys and maps in Co. Armagh in the 1820s. In May 1828 when he came to Edgeworthstown House, Co. Longford, to advise on repairs to a chimney flue and to the library, he was described by Maria Edgeworth as 'a plain, practical, chubby, innocent-looking, ruddy rough-faced youngish man, with hair roughed up anyhow'. He subsequently advised on a new house being built at Black Castle, Navan, for friends of the Edgeworth family and in April 1829, Maria wrote that 'Mr Bell, the most obliging of builders and architects' was buying timber and slates in Dublin for two houses which she proposed to build in the village of Edgeworthstown and for which Bell provided the plans. In the same year Bell designed the market house which still stands at the end of the village main street – a two-storey five-bay building, with a limestone pedimented centrepiece bearing the Edgeworth arms. Together with James Pettigrew, he was awarded a £780 contract by the grand jury of Meath in 1828–9 'to build the Meath side of Clonard Bridge on the Dublin to Mullingar Road', now known as Leinster Bridge.

When he was appointed to be the first county surveyor for Co. Longford on 17 May 1834, Bell seems to have hoped to hold the post while continuing to live in Trim. However, as Maria Edgeworth suggested in a letter of 10 June 1834 'the grand jury of the county of Longford will insist upon Mr Bell's residence in this county if he accepts the superintendence of our roads, and his scheme of a clerk and an office in Edgeworthstown and visits from time to time as the case might be will never do'. Bell did indeed take up residence in Co. Longford soon after his appointment, and lived at Newtown Forbes during most of his period as county surveyor. Although he complained of the inadequacy of his salary out of which he 'had to support two horses and a servant to attend them', he seems to have carried on very little private business, earning by his own account only £100 between 1834 and 1841 from small commissions undertaken in Co. Westmeath. He does not seem to have carried out any architectural work for the grand jury, although it is possible that some works at the county courthouse in Longford in 1859–60 may be attributable to him, perhaps in conjunction with his son, **James Bell (junior)**.

Bell had served as county surveyor for thirty-eight years when he died at his residence in Newtown Forbes at the age of 78 on 10 May 1872. An obituary appended to a grand jury report recorded that he had discharged his duties with much satisfaction to the several grand juries and all others concerned up to the time of his death. For some strange reason, however, the local coroner decided to hold an inquest into Bell's death, even though he had been in conversation with the surveyor's medical attendant on the evening of his death and could easily have satisfied himself that death was due to natural causes. This earned the coroner a reprimand from a committee of the grand jury which also disallowed his fee for the inquest.

BELL, James (junior) (1829–83), was born at Newtown Forbes, Co. Longford, the eldest son of **James Bell**, Longford county surveyor from 1834 to 1872. Although still only 18 years of age, he was sufficiently competent in engineering to hold an appointment as assistant engineer under the Board of Works on the famine relief works in Longford in 1847–8. He then became an articled pupil in the office of **Charles Lanyon**, the Antrim county surveyor, who had an extensive private practice as an engineer and architect in the north of Ireland at the time. The young Bell was thus able to gain a wide variety of experience in engineering and architecture and was said to have displayed unusual ability, especially as a draftsman, in advance of the recognized standards of the day. In addition, he was said to have demonstrated 'an original and somewhat daring genius for architectural design, which if pursued and chastened by more experience, might have resulted in conspicuous success in the profession'. Unfortunately, Bell's work at this period, like that of so many other able young engineers and architects who worked in Lanyon's large office, is difficult to differentiate from that of the master himself. It is recorded, however, that Bell worked as resident engineer and clerk of works under Lanyon on the reconstruction of the Belfast lunatic asylum in 1850–4 and was involved with him in 1856–7 in the preparation of plans for the National Gallery in Dublin; these plans were subsequently set aside in favour of a scheme completed in 1864 to a design by Captain Francis Fowke, an Irish engineer who had worked on the Victoria and Albert Museum at South Kensington, London.

In the late 1850s, Bell moved to Dublin where he took up a post as chief draftsman with the Board of Works. Here again, although he was involved in the design of many important public projects, his name was not directly associated with any of them. He left the board around 1860 to work in Longford with his father and, having qualified at county surveyor examinations held in 1857, sought unsuccessfully to be appointed to vacancies which arose in the following years. Eventually, in 1862, after **Alexander Tate** had left his Dublin district surveyor post to replace Lanyon in Antrim, Bell competed successfully for the vacancy against seventy candidates. He was to serve as district surveyor for the northern division of Co. Dublin for over twenty years.

Described by the *Dublin Builder* in February 1863 as 'a young and rising member of the profession' of architecture, Bell had delivered an obituary to the Royal Institute of British Architects following the death of Benjamin Woodward on 24 June 1861. He was one of about thirty architects who attended a meeting in Dublin in January 1863 to protest against the conduct of a competition for the design of the Dublin Exhibition Palace and Winter Garden, which was to be built on the site at Earlsfort Terrace which is now occupied by the National Concert Hall and was formerly occupied by University College, Dublin. Although the protest failed to prevent the award of the commission to Alfred G. Jones who completed the building in May 1865, it led to the revival of the moribund Royal Institute of the Architects of Ireland in which Bell was active in subsequent years, being elected a fellow in May 1863 and auditor in 1869. Although he resigned his membership at one stage, he went on to become a member of the council of the institute in 1877. He became a member of the Institution of Civil Engineers of Ireland in 1864, a member of its council in 1867, and a member of the

Institution of Civil Engineers (London) in 1879. Elected a member of the Royal Society of Antiquaries of Ireland in 1868, he conducted a survey for the society of the round towers and crosses at Monasterboice and presented drawings, plans and photographs of the monuments to the society at Kilkenny in April 1871; he also submitted a report on the condition of the monuments and proposals for their conservation, following which a subscription list was opened to finance the works.

As Dublin district surveyor, Bell seems to have discharged his routine roads duties in unexceptional fashion and he carried out some works on sea walls and harbours in the county. In 1877, he found himself having to cope with a new problem – the breaking up of the roads in the Drumcondra area by the Dublin Tramways Company's contractor who left them in a state dangerous to the public; Bell prosecuted the company which was fined £20 and the contractor undertook to have the track-laying completed forthwith. In 1881, he was called on to provide an independent report on charges of defective construction of the new Dublin city sewers. His finding that many of the complaints were justified led to a sworn inquiry conducted by the Local Government Board's chief engineering inspector, **Charles P. Cotton**, whose report of June 1881 basically supported Bell's conclusion that insufficient engineering staff and clerks of works had been engaged for the effective supervision of the project, leading to problems in some areas with levels and manholes. Cotton, however, absolutely rejected allegations that there had been some kind of conspiracy to allow faulty work to be certified and found that many complaints had been put forward as part of an organized attack on the contractor.

Bell continued to practice occasionally as an architect exhibiting what an obituary described as 'the powers and vigour of design which his professional contemporaries and intimate friends credited him with'. His works included an unsuccessful competition entry for the Unitarian Church, St Stephen's Green, Dublin (1861); additions and alterations at Lissbrack House, Co. Longford, also in 1861; and Church Street Orange Hall, Longford (1861–2). His Ulster Bank building at Longford (1863) was described at the time as an unbroken and recessed Italian facade in limestone, with a rusticated and pseudo–arcaded ground floor and segmental pedimented windows on the first floor. With **J.B. Farrell**, the Wexford county surveyor, he was joint architect of the Wexford lunatic asylum at Enniscorthy, completed in 1868; it is possible that he had carried out preliminary work on this project during his service with the Board of Works ten years earlier. In Dublin, he was responsible for a clerk's residence for the Balrothery union (1873); substantial alterations, improvements and additions in 1870 at St Columba's Church of Ireland, Swords, completed in 1818 to a design by Francis Johnston; a new sexton's lodge and attractive gates and pillars at the entrance to the Swords churchyard; and a new school at Malahide, completed in 1879. Plans and superintendence were given voluntarily by Bell for several of these latter projects.

Bell was said to be a man of 'taking appearance and genial manner' who inspired the personal regard of many friends who were 'attracted by a character singularly affectionate and humorous' and he was 'much respected by those who had business relations with him'. During his service in Dublin, he lived initially at Balbriggan and later at Windsor Terrace, Malahide. He resigned from his position as surveyor in early 1883 due to failing health and died at the age of 54 on 24 October of that year. He was buried at Mount Jerome cemetery with two of his children who had died in infancy; it seems likely that he himself had designed the attractive carved portland headstone and limestone grave-cover with its decorative inscription.

BENSON, Sir John (1812–74), was the first county surveyor to receive a knighthood. He was born at Collooney, Co. Sligo, the son of John Benson, in what was described as 'a comfortable one-storey thatched house'. He taught himself architecture and engineering with the support of the local landowner and MP, Edward Joshua Cooper (1798–1862), who sent him to the Dublin Society's Drawing School where he earned a premium in 1832. Benson worked initially in Sligo and elsewhere in the west of Ireland and spent some time in the service of the Board of Works. His early architectural projects included a large two-storey Italianate house at Lisheen, Ballysodare (1842), dismantled 100 years later; the small Church of Ireland church at Knocknarea (1843), and the Dominican church and convent in Sligo (1845) which was destroyed by fire around 1870. Benson also designed the Catholic Church of the Assumption at Collooney, an English Gothic composition described in the 1850s as 'a sermon in stone, uplifting us to Heaven'; the church was substantially completed in 1843, but the tower was not raised to its full height of 150 feet until 1878. He was also involved in the 1850s in the redesign and enlargement of Markree Castle, the family home of his patron, which had been remodelled into a Gothic castle by Francis Johnston fifty years earlier. He collaborated with **Noblett St Leger**, the Sligo county surveyor, in the design of Victoria Bridge (now Douglas Hyde Bridge) over the river Garavogue in Sligo town; the five-arch bridge, with nicely detailed masonry, opened in 1852.

Benson sat the county surveyor examinations held in December 1838 but failed to qualify although the examiners reported that he was one of a number of candidates 'who may make themselves in a moderate space of time equal to the appointment'. When he was examined again in March 1846 the examiners were 'satisfied as to his competence'; he was immediately appointed as surveyor for the west riding of Cork and transferred at the end of the month to the east riding, allowing **W.A. Treacy** to be appointed to the west riding. The two new surveyors immediately had to face a particularly difficult situation: the failure of the 1845 potato crop had led to serious distress in many parts of the county, legislation had been enacted in early March 1846 to allow programmes of relief works to be approved at special baronial presentment sessions, and the Cork grand jury had decided to hold these sessions for all twenty-three baronies between 13 April and 13 May. Although a high proportion of the applications coming before the sessions were rejected, there were still very large numbers of projects to be examined and reported on by the new surveyors. Both men had to tell the final county sessions on 13 May that they had been unable to give sufficient attention to the individual projects and this, coupled with the reluctance of the grand jury to spend substantial additional monies on works which would have to be funded, in part, from the county cess, led to the rejection of more than half of the schemes which had been put forward. In spite of the difficulties, relief works began under the supervision of Benson and Treacy soon afterwards, and considerable numbers were employed before the authorities directed that all works should be closed down in August 1846.

The two surveyors, with augmented engineering organizations, and under the direction of the Board of Works, were again heavily involved in relief works later in 1846 and in 1847, and from time to time came under criticism from various quarters. There were urgent demands for the speedy initiation of relief schemes in the autumn of 1846, with large crowds attempting to influence the decisions of both the magistrates and the surveyors when they came together at the presentment sessions. On one occasion in October, it was reported that Benson and other officials had to seek refuge in the local RIC barracks from a mob which had pursued them from the sessions. In the following month, Benson told the Board of Works that the funds which had been allocated for the relief works were inadequate as the number of destitute had been found to be much greater than could have been foreseen. A total of 5,715 men were already employed on the works in his area, all of them on task work which he had introduced with great difficulty; as a result, the public were getting 'near double the value' for the same outlay, and the men were able to earn up to 1*s*. 2*d*. each day. As the numbers employed built up, Benson and Treacy and their assistants were heavily censured – probably unfairly – on account of the design, execution and cost of various schemes. But it seems clear that they did their best to cope with unprecedented difficulties and that there were conflicting pressures on them: on the one hand, demands for the employment of the maximum possible number of starving and penniless workers, often on worthless and unproductive schemes, and on the other hand, the demand for reasonable efficiency, value for money, and expeditious completion of schemes. Overall, Benson's labours in 1846–7 were said to have been 'trying and arduous' and, when the emergency was over, it was generally agreed that he carried out his assignment on the relief works 'most satisfactorily'.

The finances of the Co. Cork grand jury were in poor shape as a result of the famine and they adopted a parsimonious approach towards all proposals for new expenditure in the immediately succeeding years. Nevertheless, Benson's achievements as county surveyor in the field of road and bridge building were very considerable. He built over 130 miles of new roads in the east riding and took over responsibility, on behalf of the grand jury, for nearly 150 miles of turnpike roads and some 85 miles of roads constructed and formerly maintained by government engineers. By his own account, he erected over thirty new bridges at a cost of about £13,000, some of them replacing bridges (including those at Carrigadrohid and Macroom) damaged or swept away in the serious floods of 1853, and he supervised bridge repair works costing more than £2,000. His lasting works in the county include three large bridges over the river Blackwater (including Rathcoole Bridge and a replacement bridge at Mallow) and a skew bridge over the Lee at Carrigrohane, built in the late 1840s. At the end of 1854, Benson gave notice of his intention to resign from his post as county surveyor and did so on 24 March 1855, after the spring assizes. Fulsome tributes were paid to him at the various presentment sessions held during January and at the assizes, and his departure was much regretted by the magistrates and grand jurors. The 'upright and efficient' manner in which he had performed his onerous and important duties was recorded in numerous resolutions and his talents and achievements were the subject of generous comment.

While still serving as county surveyor, Benson was commissioned to provide the buildings for the Cork Exhibition of June to September 1852 which was inspired by the Great Exhibition at the Crystal Palace in London the previous year. Sir Thomas Deane, the most prominent and senior Cork architect of the day, was heavily involved in the planning and organization of the exhibition and it is possible that he may have had some input also to the design process. To cater for the many exhibits and the large crowds which were expected, Benson successfully adapted the great hall of the Corn Exchange to form a large barrel-vaulted hall, 225 feet long, 52 feet wide and 45 feet high, with extensive use of glass, iron columns and timber frames. In addition, he constructed a new Fine Arts Hall at the rear, 152 feet long by 52 feet wide. The reconstructed Corn Exchange building became the nucleus of Cork City Hall in the 1890s and survived until its destruction by the Black and Tans in December 1920. The new fine arts building was dismantled after the exhibition and re-erected in 1854, under Benson's supervision, as a centre for the arts in Cork on the site now occupied by the Cork Opera House. The Athenaeum, as it was first known, could seat some 2,000 people in its main hall which measured 110 feet by 52 feet, exclusive of the galleries. It was sold under the Incumbered Estates Acts in 1863, subsequently reconstructed, and continued in service as the Cork Opera House until December 1955 when it was destroyed by fire. The last traces of Benson's building were removed some years later when the present Opera House, designed by Michael Scott, was constructed on the site.

In August 1852 Benson was declared the winner (ahead of Deane & Woodward and Richard Turner, among others) of the competition for the design of temporary buildings to be erected on the lawns around Leinster House, Dublin, to accommodate the Industrial Exhibition which the Royal Dublin Society, with the financial support of William Dargan, had decided to organize in the following year. Due to the limited time available, it was arranged that the work would be carried out by direct labour and Benson undertook to supervise it himself. The demand for space grew to such an extent that the floor area finally provided was almost double the 140,000 square feet originally intended; the great central hall alone was 425 feet long, 100 feet wide and 108 feet high at the centre, with a roof carried on semi-circular laminated timber ribs, each weighing 9 tons and supported on cast-iron columns. In all, some 2,300 tons of wood, 500 tons of cast-iron and 70,000 square feet of fluted glass were used in the work. Despite a very severe winter, the buildings were completed in time to allow the exhibition to be opened with elaborate ceremony on 12 May 1853; Benson was knighted by the lord lieutenant, the Earl of St Germains, on the occasion in recognition of his services. The exhibition was visited by Queen Victoria, her consort and their two sons in August 1853 and a total of over one million people had passed through the halls before the exhibition closed at the end of October. The buildings, which had cost a total of nearly £60,000, were taken down in 1854.

Following his resignation from his county surveyor post, Benson concentrated on his private architectural practice, the post of city surveyor to which he had been appointed in January 1851, and his positions as architect to Cork corporation and engineer to the Cork harbour commissioners to which he had been appointed at the end of 1854. Under his direction, the river navigation was greatly improved, a new pier was built at Monkstown, and a new deep-water Victoria Quay was built in the city. He was responsible for the construction of two of the city's main bridges. In 1861, he completed St Patrick's Bridge, with three masonry arches, to replace an earlier bridge which had been swept away in the floods of 1853; this was among the widest bridges in Ireland or in England at the time. His single-span cast and wrought-iron North Gate Bridge, opened in 1864 and demolished in the late 1950s, was one of the longest bridges of its kind constructed in Ireland. His most important engineering work in the city was the waterworks, with its imposing brick and limestone buildings, and steam-powered pumping engines and water-powered turbines which drove water to concrete reservoirs on a hillside high above the level of the city; the works were in partial operation by 1858 and completed in 1861. He was involved also in the 1860s in the preparation of a preliminary report on the main drainage of the city.

Although he was unsuccessful in an 1851 competition for the design of a city hall, Benson must have presented a serious challenge by then to the position of Sir Thomas Deane as the leading architect in Cork. He is, in fact, best remembered for his contribution to the architecture of the city in the 1850s and 1860s. Many of his buildings were constructed in red sandstone with contrasting dressings in local limestone. They include the Butter Market, Shandon (1852) with massive triumphal archway and columns; Princes Street English Market (1862); and the Berwick Fountain on the Grand Parade (1860). Although he was for a time worshipful master of the local masonic lodge, his ecclesiastical commissions crossed the religious divide in Cork. He added a tower and west front to St Mary and St Anne's (North) Cathedral in the 1860s but plans which he prepared in 1868 at the request of Bishop Delaney for a total reconstruction of the cathedral were never carried out. He designed St Vincent's Church, Sunday's Well, constructed in 1856 in a Gothic style (but completed ten years later by the English architect, George Goldie) and two small churches in St Patrick's cemetery. His last works were the enlargement of St Patrick's Catholic Church (1872–3), and the design of St Luke's Church of Ireland, a commission gained after a limited competition in 1873 and completed after his death by W.H. Hill. He was believed to have given professional advice and assistance gratuitously for a number of these ecclesiastical and charitable projects.

Outside of Cork and his native Sligo, Benson's architectural projects were quite few; for William Dargan, he designed the Turkish Baths in Bray, now demolished, and in Ballina, he was responsible for the addition of a spire to the Catholic cathedral in 1858. His expertise in engineering and architectural matters was, however, widely recognized and he regularly acted as expert witness and as arbitrator in major land acquisition and contract cases around the country. He was one of the main witnesses before the select committee of the House of Commons dealing with the Dublin Waterworks Bill in 1861.

Benson also engaged in railway engineering. Having been actively involved since the early 1850s in the controversies surrounding the selection of a route for a Cork–Macroom railway, he served as engineer to the Cork & Macroom Direct Railway Company which constructed its line between 1863 and 1866. He was a director of the Cork & Kinsale Junction Railway which opened in 1863 and was engineer for the Cork & Limerick Direct Railway (from Charleville to Limerick), completed in 1862, and the Cork Blackrock & Passage Railway which he converted to narrow-gauge operation in 1872. He designed the Penrose Quay terminus in Cork in 1858 for the Great Southern & Western Railway, with mass-concrete foundations and a colonnade of twenty Doric columns, and was involved in the engineering of the long tunnel which brought the company's Dublin line from a temporary station near Blackpool into the new terminus nearer the city centre. The attractive limestone head-office building for the Cork & Bandon Railway at Albert Quay (1852) is also attributed to Benson while the nearby Albert Street station, completed for the Cork, Blackrock & Passage Railway Company in 1873, was his last contribution to the railway architecture of Cork.

Benson married Mary Clementine Pyne in 1849 but had no children. He was in poor health from 1872 onwards and was obliged to retire from his various public positions in the following year. He died at the age of 62 on 17 October 1874 in London, where he had been receiving medical attention, and was buried at Brompton cemetery. His death was 'accelerated by overwork' according to one account, and brought on by 'excessive attention to his professional duties' according to another.

BODDIE, Charles Littleboy (1861–1924), was born at Straffan, Co. Kildare, on 8 October 1861, the son of George Boddie of Aberdeen. He was educated in Belfast and studied arts and engineering at university level. He trained as a civil engineer with the Belfast & Northern Counties Railway Company and became that company's district engineer at Coleraine in 1889. He was the recipient of presentations in April 1900 when he left the service of the company to take up the position of Derry county surveyor – he received a gold watch and chain, an illuminated address and a purse of sovereigns, and his wife was given a solid silver tea and coffee service and an oaken tray.

Boddie's appointment by the new county council on 16 March 1900 was the first new appointment to be made under the provisions of the Local Government (Ireland) Act 1898 which left it to each county council to determine the salary to be paid to their county surveyor and allowed them to appoint any person who had passed the qualifying examination, instead of being required to appoint the man who was placed first; Boddie was placed fourth of the five candidates who sat the examination in November 1899. While his predecessor had enjoyed a salary of £600, the new surveyor's starting salary was only £350; the county council had originally intended that the salary should be less again – £250 a year – but they accepted the view of the Local Government Board that this would be insufficient to attract experienced men to take on the more demanding duties and extra travel that would be involved in servicing the new county and district councils.

Modifications and re-roofing at the county courthouse in Derry were planned and carried out by Boddie in 1902–3 at a cost of some £3,500 although he complained of being overburdened with the 'ordinary work'. When difficulties arose with the contractors, it was suggested in a letter from an anonymous architect to the *Irish Builder* that matters would have been different if an experienced architect had been engaged for the project. Nevertheless, Boddie went on to enlarge and renovate Coleraine courthouse in 1908 and was responsible in 1915 for a new courthouse at Limavady, described by C.E.B. Brett as 'oddly toy-like and Germanic'.

Boddie seems to have been more successful in performing his roads duties and he undertook a steamrolling programme at an early stage. As the programme developed, he prevailed on his council to prohibit the assistant surveyors from engaging in outside occupations or in private practice. There was particular difficulty with one assistant in the Coleraine area who was employed also as an income tax collector – work which, it was argued, was certain to make both the assistant and the county surveyor unpopular with the 'respectable ratepayers' in the district. Boddie also found himself in difficulty with contractors from time to time. On one occasion in 1916, he was sued at Derry assizes for libel by three brothers who had been carrying on a road contract held by their late father; Boddie had reported to the council that materials had been stolen from the roadside after measurement by one of his assistants and alleged that representatives of the contractors were privy to this. His report was held to be a privileged one, submitted without malice, and the action against him failed.

A cloudburst on 24 July 1916 swept away many bridges in Co. Derry and caused serious damage to roads. Boddie's response won him great praise: a loan of £10,000 was raised by the county council to cover the cost of making good the damage and eleven bridges were replaced in as many months at a cost of £700 to £800 each. In the early 1920s, he accompanied county and district councillors on an extended study tour of England and Scotland to gain first-hand acquaintance with the most modern methods of road construction. A scheme which he subsequently drew up for the development of the road network in Derry was approved by the council and while implementation had to be postponed for financial reasons, it remained, according to the *Irish Builder and Engineer*, 'as a documentary memorial to the professional skill of Mr Boddie and his ability to realize the present needs and foresee the future requirements of road transit in the North of Ireland'.

After an illness lasting over six months during which he had been given leave of absence from his duties, Boddie died at his residence in Coleraine on 6 June 1924. His professional ability, his devotion to the duties of his office, and his sterling qualities as a man were lauded in obituaries. With his wife, the former Elizabeth Anne Edwards, he had one daughter and five sons; one of his sons, Donald George was a group captain in the RAF medical service and another, Ronald Charles, served in the Royal Navy and ended his career as a rear admiral.

BOWEN, John Kingston (1888–1944), was born on 18 April 1888, the second son of a large farmer, Patrick Bowen of Ballyorban House, near Monkstown, Co. Cork. He was educated at

Christian Brothers' College, Cork, and graduated from Queen's College, Cork, with a BE degree in 1908. A younger brother, **Richard F. Bowen**, who was to follow the same academic route ten years later, became Kilkenny county surveyor in 1923.

Bowen's first professional appointment (1908–11) was as engineer to the Timoleague & Courtmacsherry Extension Light Railway Company. He later became an inspector with the Congested Districts Board, based at Claremorris, and worked on a variety of development projects in Co. Mayo. When he was appointed Waterford county surveyor on 1 April 1915, having taken top place at a qualifying examination held in the previous January, Bowen was not yet 27 years of age. Initially, the position was not a well-paid one: the county council had fixed the salary at £350 a year, with £100 towards the purchase of a car, even though the Local Government Board took the view that this level of salary was inadequate to attract well-qualified applicants and that an allowance of £100 would not enable a surveyor to provide and maintain a good car. With the rapid increase in living costs brought about by the war, Bowen was soon able to persuade his council to allow him £250 a year for travelling expenses and his salary level was also improved in the following years.

Like his contemporaries, Bowen had to cope with a substantial shortage of funds and serious damage to roads and bridges in the early 1920s; by 1922, only 70 men were employed on roads in the county as against 600 a year earlier, and 32 important bridges had been destroyed. His most notable achievement at this time was what he described as the salvage of the main road bridge at Ballyvoyle, near Dungarvan. This large masonry-arch structure, completed in 1862, was badly damaged by an explosion during the civil war on 4 August 1922 but Bowen had it open again to traffic one year later on completion of some rather unusual operations to prevent the collapse of further arches (as had happened in the case of the adjoining railway bridge) and to rebuild the collapsed central arch.

Bowen's success as a county surveyor was said to be due in no small measure to his skill as an administrator and his capacity for organization. He was an early advocate of the use of machinery on a large scale, both for quarry work and road construction, and he succeeded in bringing the Waterford roads up to a high standard in the 1920s. A major report which he presented to his council in 1926 contained some observations on current and likely future developments in relation to road maintenance which continued to be relevant for decades afterwards. The report noted that while road expenditure came to something less than one quarter of the council's expenditure, it attracted the great bulk of the criticism of ratepayers, notwithstanding the benefits to the public, and the large amount of employment it provided. He complained that there was a tendency on the part of the council, however, to provide first for other services and to achieve economies by giving whatever was left to roads, even though wages had doubled since pre-war days and materials of all kinds cost more; in this situation, he asserted, it was only by improved methods, plant and management that the cost of road-making had been contained. With motor traffic increasing rapidly and increased speeds adding to the damage being done to road surfaces, Bowen demanded strict enforcement of the speed limits and weight limits on lorries. He was ahead of his time in calling for a system of vehicle taxation based on lorry weights and the relative damage done by them to roads. He strongly advised that the use of heavy lorries on roads which were not capable of carrying them should be prohibited, at least until funds were made available to reconstruct the roads. Finally, he pleaded for a better understanding of the need for more expenditure on roads: 'we must settle down to hard facts, and clear our minds of cobwebs, obsessions, prejudices and nonsense, and calmly recognize that when we purchased, and insist on operating, thousands of motor lorries, we also purchased and became liable for a road expenditure of millions of pounds, just as much a part of the cost of these vehicles as are themselves, their tyres, engines and supplies. They go together, the vehicle and the highway on which it must run, and all costs must be paid'.

Bowen was said to have an imposing presence and a fine speaking voice and always took an outstanding part at conferences, on deputations and at meetings of professional societies. Having filled his official position with distinction for nearly thirty years, he died at his residence, Ballinaparka, Cappoquin, on 30 November 1944, aged 56. His wife, Nancy, whom he had married in 1920, was a noted horsewoman; she died in 1955 and is buried with him in St Mary's churchyard in Dungarvan. His younger daughter, Muriel (1926–2000), was a renowned journalist and author, working mainly in London, while his son, Vincent, and grandson, John (both civil engineering graduates of University College, Cork) maintained the family connection with the development of Co. Waterford's infrastructure through their involvement with Bowen Construction and its antecedents.

BOWEN, Richard Francis (1895–1938), a younger brother of **J.K. Bowen**, Waterford county surveyor from 1915 to 1944, was born in Cork on 29 September 1895. He was educated at Christian Brothers' College, Cork, and at University College, Cork, from which he graduated with a BE degree in 1918. He held various temporary posts between 1918 and 1920 when he joined the British Reinforced Concrete Engineering Company by which he was employed for three years. When the position of Kilkenny county surveyor was advertised in March 1923 at a starting salary of £600, Bowen was one of seventeen applicants. He was elected to the office in October of that year and took on the additional post of borough surveyor in July 1930. He proved to be an able successor to the long-serving **A.M. Burden** whose daughter he was to marry.

Bowen was regarded in Kilkenny as a man of outstanding qualities and professional ability. He was respected and admired by staff and councillors alike, and there was general agreement that he made a substantial contribution to the development of the county. A report which he presented to the county council in the autumn of 1933 neatly summarized the work that had been done in his first ten years. Following the destruction caused during the war of independence and the civil war, 105 bridges had been repaired or replaced. In addition, the major bridge at Mountgarrett had been replaced by a new concrete structure, built between 1927 and 1929 at a cost of £24,000. Tarred road surfaces had been introduced in the county in 1924 and the mileage of such roads had since grown to 242. The steamrolling of another 424 miles of roads had been completed, a mile and a half of concrete roads had been built, and the streets of Kilkenny and Thomastown had been concreted. Overall, more than 1¼ million tons of stone and other materials had been applied to the 1,611 miles of roads in the county in the ten-year period, and an enormous amount of work had been done in the building and reconstruction of gullets and walls, widening and strengthening roads, easing bends, and signposting. By 1933, 1,223 miles of roads were being maintained by direct labour – about 76 per cent of the total – and the council's workforce on roads had increased to 1,500 from the 1923 figure of about 300. Finally, Bowen pointed out that when direct labour was introduced in the county in 1903, the county surveyor had five assistants and a clerk to assist him – and the only addition to the staff in the intervening thirty years was a typist!

In his role as Kilkenny borough surveyor, Bowen was associated with a substantial programme of road works, public housing and other development. In light of the reputation gained by the city in more recent years for its approach to heritage and conservation matters, it is worth noting what the *Kilkenny People* wrote about Bowen's approach over seventy years ago. Remarking that the city had been transformed since he took office, the paper went on:

> He recognised, as fully as any man could, that Kilkenny with its historical background, reaching back many centuries, was in very truth 'no mean city', and although progressive and forward-looking in his views, his steady purpose was that modern improvements should not be out of harmony with its ancient traditions. So far as his opportunities allowed him, he planned wisely and well.

But there was obviously criticism from some quarters of Bowen's activities for the paper noted that, while not afraid of legitimate criticism, he never pandered to 'the noisy clamour of pseudo protectors of public interests or the raucous vapourings of baseless and malicious innuendo'.

Bowen was conscientious and scrupulous in carrying out his professional duties and 'service' was said to be his dominant characteristic. He played a part in many community enterprises and was one of those who were instrumental in establishing in Kilkenny the Moccasin boot and shoe company whose modern factory he designed. From the mid-1930s, however, he was in indifferent health and was obliged to take regular breaks at continental health resorts. On each occasion he returned to his duties with new vigour and a determination to resume the rather hopeless struggle against what was to be a fatal illness. He died at his home at Sheestown, Co. Kilkenny, on 15 May 1938 at the early age of 42. His funeral procession to St Kieran's cemetery assumed vast proportions, according to the report in the *Kilkenny People*, reflecting his splendid record of loyal and devoted service to the public interest.

BOWER, John C. (1818–86), was born at Downpatrick, Co. Down, on 8 December 1818, the son of William Bower, a Scottish engineer who had settled there and had married Sarah Graham, daughter of an eminent mathematician. John Bower was the third child of the marriage, a remarkably quick, active and clever boy who was educated by his grandfather and served a pupilage under his father. After some unsuccessful mining speculations which depleted his finances, William Bower died in 1839, leaving his 20-year-old son to take on full responsibility for the support of his mother and her large young family. According to an obituary, this responsibility was nobly and willingly undertaken and carried out by Bower, who 'devoted himself wholly to it, never marrying so that he might the more efficiently supply his father's place, his devoted affection to his widowed mother and sisters being one of the strongest traits in his fine character'.

Bower worked initially as assistant to **John Fraser**, the Down county surveyor, but soon joined the staff of Sir John MacNeill, the leading Irish railway engineer, as the youngest of his assistants. One of his major assignments, begun in 1841, involved a detailed survey, valuation and rearrangement of the earl of Leitrim's estates comprising some 1,100 acres in Co. Donegal, and the carrying out of improvement works funded by a loan of £21,000. Bower was engaged in a variety of other work for public bodies and private clients in Donegal and neighbouring counties throughout the 1840s and was employed by the Board of Works in connection with the relief works carried out in the county in 1846–7. He was living and working in the Glasgow area in 1849 but was back in Ireland by 1851 when he entered into partnership in Dublin with Charles Sweeney. His reputation as a consulting engineer grew rapidly, as did the number and variety of his engagements, including surveys and valuations under the Incumbered Estates Acts. It was said of him, however, that he cared far more for the scientific completeness of his work than for his own pecuniary advantage so that, even with abounding work, he made little profit. Either as promoter or opponent he was eagerly sought after by various clients for advice, for evidence and for arbitration, and he became involved also in some unusual projects. One of these, undertaken with W.G. Plunkett of Boyle, involved experiments in the manufacture of paper, textiles and fabrics from plants such as coltsfoot, mangolds, turnips and beet; a patent was taken out on this process in 1856.

As a relief from the mental tension and anxiety of his extensive private practice (or so it was said), Bower sought an appointment in the quieter pastures of local government in the 1860s, having passed the county surveyor examinations in December 1851. When **W.H. Deane** was dismissed from his post as county surveyor in the south riding of Tipperary in July 1860, the *Dublin Builder* reported that Bower who was 'favourably known to the profession and to the public' was to replace him, but Bower's appointment as county surveyor did not occur until the following December when he was assigned to Carlow. He seems to have carried on the routine duties of his new post with reasonable success but was not involved in any significant new projects except some security and other works costing about £600 which he undertook at Carlow courthouse by direct labour in 1867. He maintained an extensive private practice during his time in Carlow, operating initially from offices at South Frederick Street, Dublin, later from a base at Sackville Street, and later still from an office at St Stephen's Green North.

Bower was involved in railway engineering from the 1840s onwards. In November 1846 he presented a paper to the Institution of Civil Engineers of Ireland describing a new method of setting out railway curves and, in later years, he was not afraid to clash with some of the major figures of the railway world and to challenge the accepted wisdom. In May 1855, in another paper presented to the institution, he argued trenchantly that considerable extra costs had been incurred in constructing double lines of railway where this was not justified by safety or traffic considerations. In March 1857, in a third paper, he denounced what he described as 'The Monster Contract System' and severely criticized the powerful 'monster' railway contractor (clearly intended as a reference to William Dargan) and his contract prices compared to those prevailing on county road and other works. Bower pleaded for the letting of railway construction works in small contracts, and against the taking of shares by large contractors – 'the corruption of railway contracts' – and the dominance of the contractor over the engineer. Three years later, he was again engaged in public controversy and in what the *Dublin Builder* described as 'a smart correspondence' with Sir John MacNeill, his former master, about the curves and other design features of their respective lines.

During his term of office in Carlow, Bower continued his involvement in railway engineering, mainly in the north of Ireland, and engaged in parliamentary work and other private practice. From 1860 to 1863 he was engineer to the Finn Valley Railway Company whose fourteen-mile line from Strabane to Stranorlar was constructed between 1861 and 1863, and he continued as engineer to the company until 1869. He was engineer also to the Londonderry & Lough Swilly Railway in the early 1860s and engineer to the Dublin & Antrim and the Letterkenny Railway Companies from 1864 to 1875. He was involved in the mid-1860s with a proposal to link Rosslare harbour with Waterford, a link which was not finally put into place until 1906. An action brought by Bower in the courts in 1867 for the recovery of £510 in fees which he claimed to be due to him illustrates the nature and scale of his operations at that time. He had been engaged in December 1866 by a group in Sligo to oppose two bills which were being promoted in parliament to authorize improvement works by the Sligo town commissioners and the harbour commissioners. Bower's evidence in court told of his visits to Sligo for consultations with his clients, visits to Westminster over a period of several months to attend meetings of the committee dealing with the bills, and consultations with his senior assistant, then based in Derry, and his several juniors, based at his Dublin offices. All of this activity was charged out to the client at rates of five guineas a day in Bower's case, three guineas in the case of his chief assistant, and one guinea each for the juniors.

Bower was engaged in 1869 by the Newry town commissioners to make proposals, in conjunction with **Bernard B. Murray**, surveyor for the southern division of Co. Down, for a new water supply scheme to serve the town. He put forward a scheme, estimated to cost £12,250, based on a reservoir at Glen Head within a mile of the town centre and which he claimed would provide eighty days supply. He was given to understand that these proposals had been accepted but learned sometime later that an alternative scheme, devised by the English consulting engineer, Thomas Hawksley, was favoured. Bower felt obliged to defend his proposals publicly and did so in a 112-page pamphlet published in Dublin in October 1870.

The drainage of the river Barrow had been engaging the attention of landowners and engineers for many years before Bower began his association with Carlow town, parts of which were regularly inundated by the river, as is still the case. Sir John MacNeill had prepared plans for the drainage of the upper Barrow in the 1830s and Board of Works engineers returned to the matter in the 1840s. In 1868, a committee representing a large number of the affected landowners engaged Bower to prepare fresh plans and costings. His scheme involved the formation of a drainage district of about 50,000 acres and the carrying out of works costing almost £200,000, but many of the landowners shrank from committing themselves to so large and expensive a project and suggested instead the carrying out of less expensive works to relieve flooding in the upper reaches of the catchment. Bower argued strongly in 1870 against this approach, claiming that these works would simply exacerbate the problems in other areas and lead to ruinous inundations downstream. In 1871 his proposals were still being considered by local landowners and were the subject of deputations to the authorities in Dublin in support of claims for financial assistance from public funds. By 1877, works on the first and second divisions of the original project were in progress, under a different engineer, James Dillon, but despite the attention of two royal commissions in the 1880s, no further action was taken, mainly because the estimated cost of the works involved had by then risen to more than £500,000. Bower had been gone from Carlow for fifty years before work on the Barrow drainage scheme was begun by the Board of Works under special legislation enacted in 1927; the project was completed in 1934.

In 1877 Bower was heavily involved in the planning of the Belfast, Hollywood & Bangor Extension Railway, necessitating his presence in London as a witness before a House of Commons committee which was considering the bill. The grand jury's tolerance of his extern activities seems to have run out at this stage and, following a unanimous vote of censure at the summer assizes, he resigned from his county surveyor post in July 1877. Another of his railway engagements – the Letterkenny Railway – was responsible for bringing an end to his career as a consulting engineer. This assignment was the cause of much trouble and expense to Bower and, having obtained new acts of parliament and helped to re-establish the company after a collapse, Bower found himself in conflict with the directors. His health suffered and in 1884 he retired from all work and went to Scotland, hoping the rest might in time restore him. However, his health continued to deteriorate and he died in Edinburgh on 31 December 1886 at the age of 68. An obituary concluded:

Few could have recognized in the gaunt old man – old before his time – who daily was to be seen reading in the Philosophical Institution, the strong, ardent, energetic and successful engineer of some years back. The testimony of one who knew him best was that he was widely acquainted with current literature; a man of great poetical gifts, having left several unpublished works; wonderfully beloved by his friends and relatives; of a most generous and noble nature, a sanguine and enthusiastic temperament, and a helper – far beyond prudence sometimes – to any brother in need.

Bower was a member of the Kilkenny and South East of Ireland Archaeological Society from 1862 onwards and was active in the Institution of Civil Engineers of Ireland of which he was vice-president in 1859.

BOYD, James (*c*.1781–1853), a 55-year-old civil engineer based in Belfast, was recommended for appointment as Mayo county surveyor in March 1836 when it appeared that the incumbent was about to be dismissed by the grand jury. That vacancy did not arise but Boyd became Clare county surveyor on 15 September 1836 and went to live in Ennis, the county town, where he set up his office. He was appointed also as surveyor for the county of the city of Limerick for which he received an additional salary of £60 a year, and was engaged by the Board of Works, on a fee basis, in surveying and superintending the construction of government roads in adjoining counties. He was willing also to do work for private individuals but reported in 1841 that 'this remote district does not offer any'. In November 1836, soon after the law had been amended to allow for the appointment of assistant surveyors, Boyd sent his son, Campbell, to be examined for the position. The Board of Works decided to issue a certificate of qualification in respect of young Boyd, but only on condition that his father would submit evidence that he had been gaining 'proper knowledge of the business'.

Victoria Bridge, Ennis, a large shallow limestone arch over the river Fergus, completed in 1840, is probably attributable to Boyd. Three years earlier, he prepared plans for small courthouses at Killaloe and Tulla. In 1840, he presented the grand jury with plans for a new county courthouse in Ennis which had been prepared by J.B. Keane, architect of several other county courthouses. Boyd told the jury that the design would combine 'solidity with convenience and utile beauty', but the project lapsed for lack of resources and was not taken up again until 1846.

Boyd was transferred at his own request to the Wicklow county surveyor post on 13 December 1845 and soon afterwards was heavily involved in dealing with famine conditions in the county. While there had been some limited problems arising from the effects of the potato blight on the 1845 crop, the complete destruction of the following year's crop brought severe famine to Wicklow in the winter of 1846–7, especially in the more remote mountainous areas. Reports from Boyd to the relief commissioners brought graphic accounts of the deepening crisis and of the distress of the urban and rural populations. As county surveyor, he had to take charge of a programme of relief works on which 1,225 people were employed in mid-December 1846. One month later, the numbers had risen to more than 5,000, but with the continuing scarcity and high prices of food, and a heavy snowfall in February, Boyd's reports continued to tell of death and suffering on a large scale. The numbers employed on the relief works in Wicklow reached a peak of 6,678 in early March 1847 when a decision was made to close down the works and to replace them by outdoor relief and food supply measures, leaving Boyd free to return to the more normal duties of a county surveyor.

Boyd served in Wicklow until his death at the age of 72 on 9 July 1853 at East Hall, Delgany, where he had lived for some years. Campbell Boyd, who had become an assistant county surveyor in Clare in 1836, transferred with his father from Clare to Wicklow in 1846 and served as assistant county surveyor there until his retirement in 1904, shortly before his death; in all, he had the remarkable total of sixty-eight years service in the grade.

BRAZILL, Thomas (fl. 1849–82), was listed as an associate of the Institution of Civil Engineers of Ireland in 1849, with an address at 15 Wentworth Place, Dublin, and his name first appears in Dublin directories in 1852 when he was in practice as a civil engineer with offices at Holles Street. By 1855, he had become auditor of the institution and was a member of its council from 1863 onwards. He seems to have practised during those years as a consulting engineer.

Brazill was certified to be fit to fill a county surveyor position in March 1857. He was subsequently told that he was placed fifth on the list of qualifiers but did not receive an appointment until October 1865 when he was assigned to Kildare. At his first assizes in March 1866 he advised the grand jury that he had found the 1,013 miles of roads in the county to be in generally poor condition and that more active supervision was needed to ensure continuous attention by the contractors to repair work. Although reluctant to incur additional expenditure, the grand jury therefore allowed him to recruit a third assistant at a salary of £60 a year, but his own salary was fixed at £350, well below the maximum of £500 allowed by law.

In 1867, Brazill was successful in the court of queen's bench in an important action which he brought against the Great Southern & Western Railway Company and which resulted in the imposition on them of an obligation to maintain bridges over the railway and the approach roads to these bridges. Apart from this, Brazill's service in Kildare seems to have been relatively uneventful until 1869. On the proposal of the Marquis of Kildare, he was reprimanded by vote of the grand jury at that year's spring assizes for having certified a payment for a road that had not been maintained in proper condition and his approach generally to defaulting road contractors was also heavily criticized by some of the other jurors. As the assizes were drawing to a close, Brazill announced that he would resign his post at the next assizes. But when the grand jury came together again in July, he told them that he had arranged to exchange places with either **J.H. Brett**, one of the Limerick surveyors, or **J.A. Dickinson** of Westmeath. The jury strongly objected to the fact that these arrangements had been made without reference to them and they insisted that Brazill should immediately resign, as he had committed himself to doing. Brazill complied with their wishes on 28 July and Brett was appointed in December 1869 to replace him. Brazill himself was never afterwards appointed to another county surveyor post notwithstanding a number of applications to the chief secretary's office in the 1870s.

While his term of less than four years in Kildare was not marked by any significant engineering achievement, Brazill deserves to be remembered for his contribution in the 1850s to the debate on the problem of providing a proper water supply for Dublin. The city and its suburbs had a population of some 350,000 at that stage, but its water supply – taken largely from the canal basins within the city – was seriously inadequate and believed to be the cause of widespread disease, especially among the less well-off. A succession of prominent Irish and English engineers put forward proposals for new water supply schemes involving a variety of sources, including the canals, the river Vartry, Lough Dan and the upper reaches of the river Liffey at Ballysmuttan, and Parke Neville, the city engineer reported in 1854 on proposals to abstract water from reservoirs to be formed on the river Dodder. Brazill, however, considered that the Dodder was not the solution – he dismissed it as a mere mountain stream – and he alone suggested that a reservoir formed from the Liffey just below the waterfall at Poulaphouca would be the most suitable means of meeting what he described as 'the pressing necessity of procuring a copious supply of pure water' for the city. In his *Report to the Corporation of Dublin on the Proposed Supply of the City and Suburbs with Pure Water at High Pressure* (1854), he suggested that the river could readily be impounded at that point, flooding the valley to a depth of 50 feet and creating a reservoir of up to 250 million gallons at an elevation of 430 feet above sea level; from there, water would be conveyed by gravity in an open masonry conduit via Ballymore Eustace, Kill, Rathcoole and Saggart to holding tanks at Crumlin, providing a supply of as much as 300 gallons per head per day, or nearly eight times the expected immediate requirement. All of this could be accomplished, according to Brazill's estimates, for an outlay of some £105,000.

The question of how best to improve Dublin's water supply was ultimately referred to a royal commissioner, Sir John Hawkshaw, one of the most prominent English civil engineering consultants of the day, who held a four-day inquiry in the city in August 1860 to hear the arguments advanced by proponents of the different schemes. The corporation's consulting engineer, Thomas Hawksley of London, felt that the two canals would provide a sufficient supply 'of unobjectionable waters for domestic use' within the cost limits the corporation had in mind, and he criticized Brazill's scheme on the grounds that it would involve a great deal of unnecessary additional expense. In the event, the commissioner recommended a scheme put forward originally in 1854 by a Dublin-based engineer, Richard Hassard (1820–1913), which involved impounding the Vartry near Roundwood, some 700 feet above sea level. This was adopted unanimously by the corporation in October 1860, mainly through the influence of Sir John Gray, and was approved by parliament in July 1861, thus resolving for the time being what the *Irish Builder* described as the 'vexata questio ... one of the most memorable and protracted that ever occupied the consideration of a municipal body'. The Vartry scheme was constructed between 1862 and 1868 at a cost of about £650,000, providing a reservoir of 2,500 million gallons. Time, however, was to prove that Brazill's approach was the best long-term solution even though his proposals had been rejected in 1860 by Sir John Hawkshaw who felt that 'if on no other ground, yet on the score of colour' it would not be advisable to adopt Poulaphouca as a source of supply. To cater for the water needs of the expanding city and to provide a source of hydroelectric power, construction of a dam on the Liffey at Poulaphouca began in 1938 and, after delays caused by the shortage of materials during the Second World War, work was competed in 1945, creating a 5,000-acre lake at Blessington, larger even than the reservoir Brazill had contemplated. Water drawn off from the new lake now provides the bulk of the supply to Dublin city and county, while the Vartry, even with a second 1,250 million gallon reservoir, constructed between 1908 and 1923, contributes less than 20%.

Brazil's career after his resignation in Kildare in 1869 is not documented. *Thom's Directory* listed him as a civil engineer and architect with an address at Charlemont Mall, Dublin, in 1881 and 1882 but he also had a Manchester address in those years. There are no entries for him in the directory in the subsequent years.

BRETT, Henry (1805–82), was born at Carrowreagh, near Tobercurry, Co. Sligo, son of John Brett (1765–1844). His younger brother, John (1807–71), continued to live in the area and was engaged in farming and as a road contractor; he later became a large landowner and chairman of the local board of guardians. In the 1820s, Henry was employed by the tithe

commissioners in valuing the parish of Achonry, Co. Sligo. He worked also for the Waste
Land Reclamation Company on land improvement projects along the Mayo–Sligo border and
was for a time a civil assistant with the Ordnance Survey and an overseer of roads and other
grand jury works. He was living at Tobercurry when he was appointed county surveyor for
King's County on 9 December 1834 and was required to set up his office at Tullamore which
had replaced Philipstown (now Daingean) as the county town under legislation enacted the
previous year.

Brett transferred to Mayo on 11 April 1836 and performed his duties there so successfully
that he was accorded regular votes of thanks by successive grand juries. He claimed that
'nothing could be worse than the state of the roads in Mayo' when he took up duty there but
that, after six or seven years, he had brought them to a good state. Because of the size of the
county, he was forced to travel up to 6,000 miles each year on official business, even though
he had four assistants by 1841 to assist him in supervising work on the 1,400 miles of roads in
his charge. Travel expenses alone cost him about £150 a year, and he was forced to spend up
to eighty nights each year away from his base in Castlebar, staying at inns and hotels. To
compensate for this, and to take account of his small net salary, the grand jury lobbied
successfully to have Brett appointed by the Board of Works as superintendent of the
government roads and bridges in north-west Mayo, a post which earned him an additional £1
10s. each week. In that capacity, he continued the road and bridge construction programme
in the Ballina and Belmullet areas which was begun by Alexander Nimmo in the 1820s; a
number of the plans and drawings which he produced in those years have survived in the
collection of OPW Architectural and Engineering Drawings held at the National Archives.
The Board of Works were obviously impressed by Brett's efforts in their service and, in their
annual report for 1837, quoted extensively from a detailed report of his on the success of the
work programmes in the half-barony of Erris and on the need for further road investment
there. Brett noted that much had been done in the 1820s to improve the district by opening
government roads, allowing wheeled vehicles to penetrate to Belmullet for the first time; as a
result, the town was 'a neat and bustling place of business ... with all the appendages of a
rising and prosperous seaport' and a population of about 900. But there were still districts
'exceeding ten miles square in which there are no roads that can be travelled by a wheeled
carriage'; these areas were 'excluded from intercourse by land with the grain market and the
inhabitants are obliged to have recourse to illicit distillation to meet the demands of their
landlords for they have no other means of converting their grain into money, or rendering it
portable for transport over the mountains'.

During the famine, Brett was responsible under the general direction of the Board of
Works for supervision of the relief works in Mayo, an undertaking which was said to have
'taxed to the utmost all his ability, energy and zeal to prevent their becoming useless and
wasteful'. An obituary in the *Irish Builder* recorded that 'with a staff of 200 engineers and
clerks, and administering the employment and payment of 80,000 labourers, he toiled
indefatigably at the task, while he strove hard to induce the Government to adopt a more
economic and wholesome method of meeting the difficulty of the moment. To the system of
relief works, he entertained the strongest objections, both for their extravagant and wasteful
expenditure and for the demoralizing consequences for those relieved'. While the official
record shows that the numbers employed on the works in Mayo peaked at about 62,000 in
March 1847 (rather than 80,000) they were certainly large enough to justify Brett's concern,
expressed in a letter to the under-secretary in December 1846, about the neglect of farm
operations and the fearful consequences that would follow the failure to provide for the next

year's crop. In the autumn of 1847, when his fears were borne out, he reported that in some areas deaths from starvation were almost as frequent as they had been in the previous year and the poor houses were full; moreover, there was every reason to expect that Mayo could become a desolate county unless some action was taken by providing funds for 'employment of a useful and productive character' such as the enclosure and reclamation of waste lands, drainage and the construction of lime-kilns to fertilize the land.

Brett prepared plans for courthouses at Belmullet and Swinford which were constructed around 1838, and may also have been responsible for the design of the courthouses built at Balla and Ballina in the early 1840s. In his private capacity, he carried out some survey and valuation work and was involved in railway work from the 1840s onwards. In 1843–4, under Sir John MacNeill, he carried out surveys for lines in Connacht and, in the late 1850s, he worked with G.W. Hemans, the former chief engineer of the Midland Great Western Railway, on the planning of the Athenry & Tuam line which was completed in 1860. He was engineer to the Dublin, Rathmines, Rathgar & Rathfarnham Railway in the 1860s, a project which never took off, and some years later he worked with **John Hill**, Clare county surveyor, on the planning of the Midland Counties & Shannon Junction line, which was to link Clara and Banagher but which was not constructed until 1884. In 1863–4, he was engineer for a proposed railway from Sallins, on the Dublin–Cork line, to Baltinglass; this scheme had to be abandoned at the time, with heavy financial loss to Brett, but it was eventually carried through in 1885.

After the famine years the Mayo grand jury terminated the appointments of Brett's assistant surveyors in the interests of economy and, according to himself, the post then 'became so laborious and onerous' that he was glad to leave the county. He was transferred to Waterford on 8 October 1849 and moved again on 15 July 1853, this time to Wicklow, where he was responsible for some significant projects. A replacement bridge over the river Dargle at Bray, completed in 1856, was among the first of these and the foundation stone of his three-arch Seskin Bridge over the river Slaney, near the Glen of Imaal, was laid in the same year. The old seven-arch masonry bridge at Wicklow was widened on the upstream side in 1862–3 after more expensive schemes to replace it by a new stone bridge or an iron bridge were abandoned. A new single-span bridge over the river Dargle at Enniskerry followed in 1865 and yet another costing £3,000 was completed at Avoca in 1869. The Avoca Bridge, described at the time by the *Irish Builder* as a handsome structure, carried a roadway 24 feet wide, it had 3 arches, each with a span of 45 feet, and was built of granite taken from quarries at Aughrim. The bridge replaced an older one which was carried away in the floods that followed the thaw in early 1867 and its foundations were said to have been carried down eight feet below the bed of the river. Brett's bridge was itself seriously damaged in the floods that accompanied 'hurricane charlie' in August 1986 but was subsequently reconstructed.

Following the failure of the Wicklow harbour commissioners to raise the necessary funds to carry out ambitious plans for the development of their harbour which had been prepared by **Barry Duncan Gibbons** in 1852, Brett supervised the implementation of a more modest scheme in 1855–6; this involved the construction of a 460-foot-long, 45-foot-wide pier at the end of the Murrough, some dredging work, and new quay walls, and was estimated to cost £8,500. After James Barton had replaced Brett as engineer in June 1857, further payments to the contractor were withheld and the contract was terminated because Barton alleged that the masonry and materials were defective and that the pier was 50 feet short. The contractor sued the harbour commissioners for damages and, after a nineteen-day hearing, the jury held in his favour in April 1858 and awarded costs against the commissioners; as no less that seven engineers, including Brett and **John Neville**, had appeared for the contractor, and six, including Gibbons and **Charles Lanyon** for the commissioners, the overall cost of the proceedings must have been very considerable indeed.

Brett went on to carry out further harbour works costing some £2,000 at Wicklow in 1869 and also carried out works at Arklow harbour which, in the view of Charles Vignoles FRS, president of the Institution of Civil Engineers of Ireland in 1864, amounted to 'a successful contest … with the sands which threatened to close up the harbour'. In Bray, where he was later to become surveyor to the township, Brett was responsible for works at the esplanade in 1861, a new water supply scheme and a new sewerage scheme which cost £10,000 in 1869. He was one of a number of engineers who submitted plans in 1867 for the construction of a harbour at Bray but these were not accepted and a harbour was not built there until the mid-1890s. He led the opposition on behalf of the town commissioners in 1875 when the railway company proposed to construct a tunnel to bring Quinsboro Road under their line to the esplanade in Bray. At Wicklow, he carried out alterations and improvements in 1866 at the county courthouse – said to be in a disgraceful condition a few years earlier – and he may also have been responsible for the provision of a new courtroom in 1876, most likely in association with his son, **John Henry Brett**.

As county surveyor in Wicklow, Brett had a continuing involvement with the development between 1862 and 1868 of Dublin Corporation's waterworks near Roundwood. He was one of those who petitioned unsuccessfully against the waterworks bill when it came before parliament in 1861 and, when construction began, he kept in close touch with all stages of the work. With two English engineers he was engaged by Daniel Tighe of Rosanna, a local landowner, in 1866 to assess the risk that the large earthen embankment might fail, causing a catastrophe such as had occurred in March 1864 at Sheffield where large loss of life had resulted from the collapse of a much smaller dam. Brett and his colleagues reported that a failure of the Roundwood embankment 'would sweep away the village of Ashford' and when a partial failure occurred in January 1867 there was considerable alarm in the area. Tighe brought proceedings against the corporation some months later, claiming damages of £25,000 arising from devaluation of his property by reason of the danger to which it was exposed and, when two arbitrators (one of whom was **Sir John Benson**) failed to resolve the matter, the case went for trial before a special jury at Wicklow courthouse in October 1867. Giving evidence for the plaintiff at the arbitration and in court, Brett was extremely critical of the methods used in constructing the foundations of the embankment and of the lack of proper provision for storm overflows; he concluded that 'the embankment as it at present stands is an object of permanent apprehension'. He was forced to admit in court, however, that he had no experience himself of reservoir construction and, in opposition to his view, a number of other engineers gave expert evidence on the safety of the works. Nevertheless, the jury awarded compensation of £12,051 to the claimant. The *Irish Builder* felt that the corporation had been harshly dealt with and commented that the case was one of considerable importance from a legal and an engineering point of view because damages had never previously been awarded in anticipation of an event which might never come to pass.

Another court case in March 1868 gave Brett an opportunity of renewing his criticism of aspects of the Vartry scheme. A bakery owner in Newtownmountkennedy claimed damages arising from the destruction of his premises in October 1867 when a large water main burst in the street outside. Brett gave evidence that these pipes had been thrown out of wagons onto the road and that he had complained at the time about this and about the obstruction which had been caused. The judge warned the jury (without wishing to disparage Brett's statements) about the expert evidence of engineers generally: 'though he entertained great respect for those gentlemen, he found that when a matter is favourable to the party for whom they were examined, they looked into the point minutely while giving other matters, perhaps equally important to the case, only a cursory examination'. The case was settled in favour of the plaintiff who was awarded damages of £300.

During his years of service in Wicklow, Brett worked also on the valuation of land which was to be disposed of under the Incumbered Estates Acts. He was engaged as architect and surveyor to the National Building and Land Investment Company of Ireland which was set up in 1865 to purchase large estates for subdivision and to advance loans for farm development and for urban working class housing. He maintained the interest in the reclamation of waste land – bogland, slob land, mountain land and foreshore – which he had developed in Mayo in the 1830s, and engaged in controversy with Mayo landowners on the issue in 1875; in a letter published in *The Times* on 16 February, he maintained that, in the previous 100 years, vast areas had been transformed by the 'labour and energy of the peasantry' rather than by capitalists, and insisted that 'under careful arrangement and judicious management reclamation by the people is not only practicable, but is the only safe and economic mode by which it can be made remunerative and a source of wealth to the community at large. Reclamation by the state, by companies, or by proprietors on a large scale, I assert to be inapplicable, but reclamations by the occupiers, under proper control and with reasonable assistance can be made a source of undoubted profit to the proprietors and occupiers of the soil'. He set out his arguments on the subject more extensively in a pamphlet *The Reclamation of the Waste Lands of Ireland* which was published in 1881.

Said to be 'an incessant and laborious worker', Brett carried on an extensive private civil engineering and architectural practice under the title Henry Brett & Sons in the 1870s and early 1880s, with offices at St Stephen's Green and later at Dame Street, Dublin. The sons involved were John Henry Brett (a county surveyor from 1863 to 1914) and Henry Charles Brett, who had been awarded the LCE at Trinity College, Dublin, in 1870 and was to become chief inspector of the land loans service of the Board of Works in 1884. After the death in 1872 of William Caldbeck who had been architect to the National Bank for nearly twenty years, the Bretts gained much of the bank's business in the following years. They provided new or reconstructed bank premises at Castlebar, Claremorris, Carrick-on-Suir, Waterford and Millstreet, the latter two illustrated in the *Irish Builder* in September 1877 and April 1878, respectively. Their other architectural projects included St Joseph's convent chapel, Kilkenny (illustrated in the *Irish Builder* in January 1881), while their engineering projects included sewerage works at Ballaghadereen and Naas and a water supply scheme at Greystones.

Brett had a serious accident in November 1871 in the course of his duties: he was standing on an outside car, preparing his wraps for a lengthened drive over the roads when the horse, which had just been taken out fresh from the Woodenbridge hotel stables, started suddenly and caused him to be flung violently to the ground over a pile of stones. However, he recovered well from his injuries and was able to resume his activities. After forty-six years service as a county surveyor, he died at his home, Rosemount, Booterstown, Co. Dublin, on 13 May 1882 at the age of 77 and is buried in Deansgrange cemetery in a large family grave. According to an obituary 'he retained to the last all the energy of his earlier years' but his death certificate gave 'exhaustion' as the cause of death. The obituary went on to say that Brett was 'well informed upon nearly every subject and possessed a remarkable memory for incidents which, added to his generous and kindly nature, made his society most attractive and agreeable, and gained him a host of friends'. He was survived by his wife, Mary, who died in 1894, and by three daughters and three sons, John Henry and Henry Charles, referred to above, and Joseph P. Brett BL, who was the author of *Grand Jury Law* (Dublin, 1894) and *The Law and Procedure relating to Compensation for Criminal Injuries* (Dublin, 1899).

BRETT, John Henry (1835–1919), was born in Tobercurry, Co. Sligo, the eldest son of **Henry Brett**, who had become King's County surveyor in December 1834. He was educated at Waterford academy and by private tuition but, while an obituary in the *Irish Builder and Engineer* stated that he was a graduate of Trinity College, Dublin, there is no record of him in the registers of students and graduates of the university. After training by his father in engineering and surveying, he was employed on railway work by William Dargan and subsequently by a number of railway companies. He is recorded in 1857 as one of those who tendered unsuccessfully for the construction of a bridge over the Grand Canal to bring the Harcourt Street railway line to its terminus in Dublin and for the terminus itself. In 1855 he had been an applicant for a position under the East India Company and in 1858 he applied unsuccessfully for a post as road engineer in Jamaica.

Brett successfully sat the examinations for Dublin district surveyor posts which were held in March 1856. He was advised a year later that, on foot of these examinations, he was also deemed to be qualified for an appointment as county surveyor but he had to wait another six years before his name reached the top of the list of qualifiers. He was assigned to the western division of Limerick on 15 July 1863 and went to live at Rathkeale. Shots were fired at him in November 1868 and, following a dispute with a road contractor in October 1869, pellets fired through his parlour window lodged in his side. Although Brett was not seriously wounded, he believed that his life was in real danger following what he described as these 'murderous attacks' and he was given a police guard. According to *The Times*, the only crime committed by Brett and two other colleagues who had also been attacked was that 'they will not give false certificates to road contractors for work which is not properly executed' and Sir Thomas Larcom, the retired under-secretary, commented that 'no man can do his duty honestly without danger of the assassin's bullet'.

When a county surveyor vacancy arose in Kildare in July 1865, Brett applied unsuccessfully for a transfer to that county. He applied again when the post fell vacant in July 1869 on the resignation of **Thomas Brazill**, and confidently expected to be transferred on that occasion for his own safety. However, there was a long delay by the authorities in Dublin in making the Kildare appointment, leading the *Leinster Express* to declare in October that the roads had been allowed to deteriorate so much that, if something was not done soon, a balloon would be the only satisfactory means of locomotion in the county. Brett's eventual appointment in December was welcomed by the *Express* – 'a better selection we believe could not have been made' – but the journal was concerned that the ratepayers would be mulcted to make good the damage that had been done to the roads in the interregnum.

Initially, Brett's remuneration in Kildare was set at a gross total of £550 a year, out of which he had to pay the salaries of his three assistant surveyors, his clerk, and his travelling expenses. At the summer assizes in 1871, when he complained that his net salary was only about £180 a year after all of these expenses had been paid, the grand jury reluctantly agreed to fix his salary at £400 a year and to meet the salaries of £70 for each of the three assistants from county funds. The remuneration issue came up again in 1875 when the question of carrying out improvements at the county infirmary arose. Brett took the view that the preparation of plans for these works did not come within his sphere of responsibility as county surveyor and claimed to be entitled to extra remuneration if he were to undertake the task. This view was subsequently confirmed by counsel's opinion obtained by the grand jury, forcing the governors of the infirmary to retain their own architect for the purpose. The grand jury, however, had their revenge in the following year when Brett sought to have his salary increased to £500, the maximum set by law in 1861 for the county; a motion to increase the salary was declared lost, after one influential member had pointed out that the increase granted in 1871 was on the understanding that the surveyor's duties would extend to all public buildings in the county.

Brett was one of a number of engineers who prepared plans in 1879 for a new sewerage scheme at Naas but the board of guardians considered his plans too costly although the plans they favoured – prepared by an English engineer – were rejected by the Local Government Board. In the early 1870s, he was responsible for alterations to the courthouse at Maynooth and the county courthouse in Naas which had been constructed before 1805 to a design by Richard Morrison. At the spring assizes in 1871, the judges had complained strongly about the wretched accommodation at Naas and the need for painting and general repairs, and while they recognized that it would not be possible to reconstruct the building to conform with the internal arrangements at Carlow courthouse – another of Morrison's buildings completed in 1834 – they advised Brett to inspect that building with a view to adopting some of its features at Naas. While there were some who felt that no amount of expenditure could convert the building into a commodious and convenient courthouse, Brett and his grand jury went ahead with alterations in 1871–2. With this experience, Brett seems to have considered himself to be something of an expert on the subject and contributed a paper to the Architectural Association of Ireland setting out in some detail the considerations to be taken into account in designing county courthouses and gaols; the paper was published in the *Irish Builder* in 1875.

Together with his father and his brother Henry Charles, Brett carried on a substantial private practice in engineering and architecture in the 1870s and early 1880s initially under the name of Henry Brett & Sons. After the death of his father, the style John H. & H.C. Brett, civil engineers and architects, was adopted and the firm operated from offices at 49 Dame Street, Dublin, and later at 12 South Frederick Street. The brothers worked together on the planning and construction of new lunatic wards at Naas workhouse in 1881. John Henry worked with W.M. Mitchell on the design and construction of the new Ledwich wing at Mercer's Hospital, Dublin, in the early 1880s, the plans having been chosen from schemes submitted by four of Dublin's most prominent firms of architects in November 1879.

Formal notice of Brett's transfer to Antrim, to succeed **Alexander Tate**, was read to the county's grand jury at the spring assizes in March 1886 but he had actually taken up duty in the county in the previous November. He acknowledged the advice Tate had given him on local circumstances and on road management systems 'which were not familiar to me from previous experience'. Five years later he introduced steamrolling, beginning with roads on the outskirts of Belfast – the second of the county surveyors to adopt this new technique. Brett served in Antrim until his retirement at the beginning of 1914 when he was approaching 80 years of age. He acted as a justice of the peace for the county until his death at his home at Alexandra Gardens, Belfast, on 26 December 1919, aged 84. He is buried at Deansgrange cemetery in Dublin in a large family grave with his father and mother, his wife Mary Josephine (daughter of James Brady MD of Dublin), his only daughter Maud, his son Hubert, his three sisters and two brothers, Joseph who died in 1910 and Henry Charles who died in 1926. Two of his five sons, Henry James and John Aloysius, both graduates of Trinity

College, Dublin, served in the British foreign and colonial service, the former in China and the latter in Pakistan.

BREWSTER, Henry (1820–83), possibly a native of Carlow, was born on 6 January 1820. He was employed by the Board of Works on the supervision of relief works in Co. Limerick in 1846–7 and was forced to suspend the works in the Kilfinane area in March 1847, throwing a large number of men out of work, when dissatisfaction with the task-work system led to the appearance on the roads of a man armed with pistols who threatened the pay-clerks and overseers. On another scheme, on the road between Ballylanders and Mitchelstown, he reported that a party of five armed men had forced the overseers to kneel on the road and had discharged pistols over their heads, while the workers were 'laughing and chuckling the whole time'.

Brewster was still based in Limerick when he qualified to become a county surveyor at the examinations held in November 1847 and he was appointed as second surveyor for Co. Mayo on 23 August 1848. When **Henry Brett**, the other Mayo surveyor, transferred to Waterford in October 1849, Brewster took over responsibility for the whole county but because of the heavy workload two divisions were again established on 31 March 1856. From then onwards, Brewster served in the northern division. He was an active road-builder, having contributed a total of 250 miles to the county's stock by 1868, and always had a good relationship with the grand jury, a situation which may have been helped by the fact that he was said to be connected by marriage with 'one of the oldest and most respected families in Mayo'.

In his last report to the grand jury in March 1883, Brewster told them that, after thirty-four years service, he thought it high time to retire on pension, but he seems to have continued in office for some time after that to allow for a successor to be appointed. He died of a heart attack at his home in Church Street, Castlebar on 28 June 1883 and was buried in the cemetery attached to the old Church of Ireland church in the town. At the summer assizes a few weeks later, the grand jury paid tribute to him for his long years of efficient and diligent service to the people of Mayo and the *Connaught Telegraph*, in recording his death, spoke of his 'highly honourable and distinguished career'.

Brewster married Eliza O'Malley (1825–88) and one of their daughters, Sarah Mary, married his south Mayo colleague, **Edward Glover**. His only son, Edward (b. 1860) was awarded the BAI degree at Trinity College, Dublin, in 1883 and worked with Glover in south Mayo in 1884–5 and subsequently with Glover's successor, **P.C. Cowan**. In 1886–7, he acted as resident engineer under Cowan on the construction of the Achill viaduct which he had helped to design while working with Glover, and he subsequently assisted **A.E. Joyce**, the Westmeath surveyor, in a number of surveys for new railways which were proposed in the west of Ireland under the Tramways and Light Railways Acts. Subsequently, the younger Brewster was in private practice in Castlebar until his death in 1897.

BURDEN, Alexander Mitchell (1864–1923), was born on 12 October 1864 at 8 Alfred Street, Belfast, where his parents lived. He was named after his great-grandfather, Alexander Mitchell (1780–1868), the Dublin-born blind engineer who was the inventor and patentee of the screw-pile which became widely used in the foundations of marine structures in the second half of the nineteenth century. Mitchell's daughter, Margaret, married William Burden, first professor of midwifery at Queen's College, Belfast. Their son, Henry Burden MD (1835–93), who worked as a pathologist in Belfast and was president of the Ulster Medical Society in 1888–9, married Anna McCormack and was the father of the future county surveyor.

Burden was educated at the Royal Belfast Academical Institution from 1875 to 1882 and at Queen's College, Belfast, where he was a gold medallist and was awarded a BE degree with first-class honours in 1885 as well as a pass in the BA examination in biological science. He worked in 1885–6 as assistant to the Antrim county surveyor, **Alexander Tate**, and his successor, **John Henry Brett**, and was subsequently an assistant in the Belfast offices of Lanyon and Lynn and of Messrs Young & Mackenzie, architects. From 1888 to 1892, he worked in Argentina, first on survey work for the Buenos Aires Great Southern Railway and later as resident engineer in full charge of the construction of some fifty miles of that line. For the following two years, he was an assistant surveyor on the civilian staff of the royal engineers, a much sought-after appointment at the time and one which was gained only after 'severe examinations'.

In November 1894 Burden was appointed county surveyor for Co. Kilkenny where he was to serve for over twenty-eight years. Although he was a Unitarian in his younger days and later a member of the Church of Ireland, he married the daughter of a prominent Catholic Kilkenny brewing family, the Sullivans of Lacken Hall, who was a great-granddaughter of Daniel O'Connell. Burden came to be highly regarded in Kilkenny for his ability, his helpfulness, his quiet charm, and a host of other characteristics which were said to mark him out as 'a very perfect gentleman'.

Burden transferred to the service of the new county council in 1899 but was not very happy when they fixed his salary at £500. Nevertheless, he went on to develop an excellent working relationship with the council to which he was to give well over twenty years loyal service; he was the best officer that ever served under the council, according to one of its members. He advised them at an early stage of the need to spend more money on strengthening the formation of the county's roads, instead of wasting money on superficial maintenance – an argument that was to continue to be made by roads engineers for most of the following century. Direct labour was introduced in the county in 1903 and steamrollers and other machinery was brought into use.

In 1907 Burden together with a barrister and an architect, was appointed to constitute a bridge commission to review the case for rebuilding or replacing St John's Bridge in Kilkenny city. A new bridge was proposed in the following year but the plans ran into some difficulty with the Local Government Board and the Board of Works where 'important technical questions' arose about the appropriate borrowing term for works in ferro-concrete of which there was then little experience. Once these issues were resolved, a fifty-year loan term was agreed, and construction work went ahead in 1910 on a bridge designed by Burden in association with L.G. Mouchel & Partners; this was a ferro-concrete arch of 140 feet clear span, then the longest span of its kind in Britain or Ireland. Testing of the bridge in November 1910 seems to have been a major event in Kilkenny, with a large number of Irish and English engineers and a large number of the local population turning out to witness the process. Three rows of sandbags were piled high along the thirty-four-foot-wide bridge, and traction engines and loaded wagons were driven across repeatedly between them but the maximum recorded deflection of the arch – which was thirty inches thick at the centre – was found to be only one-fifth of an inch. The bridge is still in service although remedial and strengthening works were carried out in the 1970s when a new concrete slab was laid over the original decking.

Burden was heavily involved also in the Waterford bridge controversy which raged from 1907 to 1910. When eventually a ferro-concrete bridge on the Hennebique system was decided on, he was appointed to act as engineer-in-chief for the joint committee of the local authorities which had overall responsibility for the project. Tenders were invited in July 1910 for the bridge which was to have 13 spans, each 40 feet wide, with an 80-foot-wide steel opening span, all supported on cylindrical steel piers constructed around clusters of concrete piles up to 65 feet long and filled with concrete. The main contract was awarded to Messrs Kinnear, Moodie & Co. of Glasgow in September at a price of £64,311 and the opening span was provided by the Cleveland Bridge and Engineering Co. of Darlington. The bridge was opened to traffic by John Redmond MP in February 1913; it survived for over seventy years until replaced by the present bridge, constructed between 1982 and 1989.

In June 1920, Burden advised his council that Mountgarrett Bridge over the river Barrow, on the boundary between Wexford and Kilkenny – a wooden bridge built in 1794–6 by Lemuel Cox – was so dangerously defective that it would be a mistake to spend money on its repair. (His colleague, **W.F. Barry**, was advising his council in Wexford at the same time that the bridge was 'absolutely dangerous'). According to Burden's estimates, a new bridge would cost up to £18,000 which would have to be shared by the two county councils as the Ministry of Transport had said that they had no funds from which a grant could be made. Neither of the councils could raise their share of the cost in 1920 due to their depleted finances, and Burden did not live to see his plans for a replacement bridge come to fruition; the wooden bridge was almost totally destroyed by fire in June 1921 and a new bridge was finally built between 1927 and 1929.

Burden was joint engineer for a major water supply scheme for Kilkenny city, completed in 1905 at a cost of about £23,000, and he also carried on a limited private practice in the area. In 1906, he prepared plans for the renewal of the derelict canal from Kilkenny to the tidal waters of the river Nore at Inistioge at an estimated cost of £78,400; the canal was to be 35 feet wide at the surface, up to 6 feet deep and with 18 locks, and was to follow closely the line of the river for nearly 20 miles. Burden's plans and estimates were reviewed by the Royal Commission on Canals and Waterways which found no fault with the scheme but reported in 1911 that, because the greater part of the benefit would be of a local character 'the work should not be undertaken unless the county and city of Kilkenny are prepared to guarantee the interest on two-thirds of the required capital, the balance being provided by the state as a free grant'. In the event, the canal project did not proceed.

In early 1922, Burden had to absent himself from his duties due to ill-health and he was forced to tender his resignation in February 1923. He died at his home, Bellevue, Kilkenny, on 9 March 1923, aged 59, and was buried in the grounds of St Canice's cathedral. In a generous tribute, the mayor of Kilkenny said that Burden was a man in whom the whole city and county took a just pride, because not only was he a courteous gentleman, but also a very distinguished engineer; they prided themselves on having such a man in their midst, and there was universal regret at having lost such a valuable asset to the city and county.

BURKITT, James Parsons (1870–1959), was born at Killybegs, Co. Donegal, on 20 August 1870, son of Thomas Henry Burkitt, a Presbyterian minister, and his wife, the former Emma Eliza Parsons. He was a graduate of Queen's College, Galway, where he was awarded a BA degree in 1891 and a BE in 1892; he won awards in each year of his three-year engineering course and achieved first-class honours in both degree examinations. He served as an assistant county surveyor in west Galway under **James Perry** in the early 1890s, supervising the construction of a pier and a swing-bridge over an estuary of the sea. In 1893 he was assistant engineer on the Westport–Mulrany extension of the Midland Great Western Railway and in the following year worked on the construction of the Collooney and Claremorris line. He was subsequently engaged on the construction of the Belfast waterworks and, under **P.C. Cowan**, on the Downpatrick waterworks.

Burkitt's appointment as Fermanagh county surveyor was announced in December 1898. In the early 1900s he began building the first section of a hilly zigzag road officially known as Marlbank but which was nicknamed 'Burkitt's Folly'; today, this forms an important part of a scenic route heavily used by tourists visiting the Marble Arch caves. He introduced tarred roads to Fermanagh in the 1920s and was one of the pioneers in the use of concrete in bridge works. In 1909 he was responsible for the construction for the Lisnaskea rural district council of one of the earliest ferro-concrete bridges in Ireland – a single span of forty-two feet over the river Tempo at Drumlone, designed on the Hennebique system by Mouchel & Partners and still in service.

There had been pressure for a bridge or ferry to link Boa Island in Lough Erne with the mainland from early in the twentieth century but there was no agreement among the 137 families on the island as to where the link should be made: some wanted a link with Pettigo, others wanted a connection towards Kesh, and a few wished to have a connection in the direction of Beleek. Nothing was done until the 1920s when, after partition had come into effect and the nationalist-controlled Fermanagh county council was dissolved, Northern Ireland's first prime minister, Sir James Craig, later Lord Craigavon, promised the protestant people of Beleek and the western part of Fermanagh that he would have them properly linked by road with the rest of Northern Ireland without having to pass through the Free State. To achieve this objective, Burkitt was authorized to construct three large reinforced concrete bridges in lower Lough Erne, using the patented Kohn system which was promoted by the Trussed Concrete Steel Company of London. Captain Westropp George of Dublin was appointed resident engineer in January 1924, tenders were invited in the following September and the contractors began work in January 1925. In just over two years, they had completed the Rosscor viaduct, three miles from Beleek (two spans of 31 feet, and seven of 35 feet), and two somewhat similar bridges at each end of Boa Island, one with three spans and the other with five spans of around 30 feet each. The bridges and ancillary roads through the island cost a total of about £90,000 and were opened by Lord Craigavon in February 1927. In 1935, and with the same general objective as the Boa Island bridges, Lady Brooke and Lady Craigavon bridges were completed in upper Lough Erne, to link Lisnaskea via Trasna Island to the south of the county; these are fine examples of early reinforced concrete beam and slab bridges designed on the Hennebique system, with fourteen spans and ten spans, respectively.

Like his colleagues, Burkitt found himself engaged in a constant struggle to secure the proper performance by road contractors of their duties in relation to what Winston Churchill described as 'the muddy by-ways of Fermanagh'. He seems to have been one of the few surveyors who were actually assaulted in the course of this duty. In May 1915, on a tour of inspection in his horse and trap, a road contractor attacked him during the lunch-hour, called him a rotter and struck him with a stick. Although Burkitt said at the subsequent court hearing that he was not hurt and did not want to press the case, the contractor was fined ten shillings and ordered to pay the prosecution's costs.

Burkitt has been described as the most original of all the ornithologists of the first half of the twentieth century. Using a small meat safe, propped up with a stick which was removed by means of a piece of string, he trapped robins in his garden and in the surrounding countryside and marked their legs with aluminium rings, using different patterns rather than colours because he was colour-blind himself. This enabled him to work out territory sizes for individual birds, to study their distribution and behavioural patterns and to estimate their ages. He was the first to prove that hen robins sing, unlike most female birds, and claimed that each bird greeted him with a song as he made his rounds of their territories. Burkitt wrote that he was 'thankful to have had this little piece of research into a corner of God's garden' but he confessed to a colleague that he sometimes worried that he might be more interested in the creation than the Creator. His work on robins and other birds was published in *British Birds* in the 1920s and in the *Irish Naturalist* between the 1920s and the early 1950s.

Burkitt married Gwendoline Hill, daughter of the distinguished Irish architect, William Henry Hill, and the couple lived at Lanakilla, near Enniskillen. Small in stature, a man of integrity and few words, and devoutly religious, he served Fermanagh until his retirement in April 1940. He died on 30 March 1959, aged 89, having spent his last years reading his bible and working in his garden. He was buried in Trory churchyard on the shores of Lough Erne. One of his sons, Denis Parsons Burkitt (1911–93), having begun the study of engineering at Trinity College, Dublin, transferred to medicine and became a well-known surgeon, researcher, and epidemiologist, working initially in Africa and later in London in the field of cancer and its treatment. Burkitt's Lymphoma, a lethal form of predominantly childhood cancer of the lymphatic system, is named after him and he played a substantial part in identifying the causes of the disease and suitable treatments. In later years, his research suggested that various 'diseases of civilization' were linked to a lack of fibre in western diet and his book *Don't Forget Fibre in Your Diet* (1979) rapidly sold over 200,000 copies in many languages.

BURTCHAELL, Peter (1820–94), was born on 31 October 1820 at Brandondale House on the banks of the river Barrow near Graiguenamanagh, Co. Kilkenny. He was the second son of David Burtchaell (agent for the earl of Clifden and JP for counties Carlow and Kilkenny) and his wife, the former Jane Dames of Dublin. Peter was educated in Dublin at the Feinaiglian institution and, as a 17-year-old young man, was apprenticed to the Waterford county surveyor, **William Johnston**. He was sent forward in July 1837 to be examined by the Board of Works with a view to appointment as one of Johnston's assistant surveyors and was allowed take up the position. However, a special report by Jacob Owen, the board's chief engineer and architect, noted that Burtchaell had 'little experience, and that chiefly in office duties; he is, however, an intelligent and promising young man, but his permanent appointment to the situation of assistant to the county surveyor should perhaps be made contingent on his improvement in a practical acquaintance with the duties he will be required to perform'. By 1846–7, Burtchaell had moved on to become engineer in charge of the Board of Works drainage and land improvement schemes in Co. Carlow. He gained experience also of railway surveys and worked as contractor on roads and bridges in counties Waterford, Kilkenny and Carlow. He was appointed Carlow county surveyor on 29 March 1851 and was transferred to his native Kilkenny at his own request in December 1860, following the death of **Samson Carter**.

Burtchaell worked with **J.B. Farrell**, the Wexford surveyor, on the design of New Ross Bridge, constructed in 1868–9 at a cost of just over £40,000. One of his principal concerns during his years in Kilkenny, where he also held the separate post of surveyor to the county of the town, was St John's Bridge over the river Nore – an attractive three-arched masonry structure which had been completed in 1767 to replace a late medieval bridge which was carried away in a great flood in 1763. The steep gradient of the approaches to the bridge were considered to be difficult and dangerous for horse-drawn traffic and the large piers were thought to be interfering with the flow of water in the river to an extent which increased the risk of flooding in the city. It was reported in 1861 that a new stone bridge was to be built according to Burtchaell's plans and, in 1866, tenders were invited to take down the old bridge and to replace it by what was described as a new 'flat' bridge at a cost not to exceed £3,000. Nothing came of this. In 1871, Burtchaell recommended the construction of a bridge with three cast-iron arches on stone piers and abutments, having abandoned for technical reasons the idea of a stone bridge; the latter would have cost less and would give more local employment but the overriding consideration in Burtchaell's view was to maximize the waterway. When it emerged at the spring assizes in 1872 that no suitable tender had been received for the cast-iron bridge, the grand jury appointed a committee to review the question of whether a traditional stone structure would suffice. In the event, the 1767 bridge survived until 1910 when it was replaced by the single-span concrete structure which is still in use.

Burtchaell was always very critical of the legal provisions under which contracts for road construction and maintenance works had to be awarded to the lowest bidder. In his first report to the Carlow grand jury in 1851, when he found the condition of the county roads to be indifferent, he complained that contractors seemed to believe that it was sufficient 'to sprinkle a few loose broken stones over the road surface' a few days before the presentment sessions or assizes at which they expected to receive payment. More than twenty years later, he was pointing out to the Kilkenny jury that he was being forced to allow 'almost ignorant men' to take large contracts for low prices, thus 'driving away respectable contractors of experience'. Plans and specifications were seldom looked at or understood by contractors, he alleged, until after they had commenced work, and delay and great public inconvenience was being caused because of the inexperience and incompetence of those who submitted the lowest tenders.

By April 1894 Burtchaell had been in office as county surveyor for forty-three years. He had served for over thirty-three years in Kilkenny and had seen generations of grand jurors come and go. But while he appeared to be in robust good health, he was in reality suffering from an acute form of heart disease and this, coupled with some serious disagreements with the grand jury, induced him to give notice of his intention to retire from his post at the end of the year. Before he could do so, he was taken suddenly ill on 21 June 1894 and died later that day at his home at Larchfield, near Kilkenny. The *Kilkenny Moderator and Leinster Advertiser* wrote that he was a clever engineer 'who was held in the highest esteem by all classes for the ability and rectitude with which he performed his official duties'. However, 'he was not versed in the ways of diplomacy; he never stooped to flatter or to court the favour of any man; on the contrary, he was outspoken, impulsive occasionally, quick to resent but honest and upright as the day'. In his private life, he was said to have all the qualities of a Christian gentleman, and was a loyal member of the masonic order.

Burtchaell was elected in 1870 to be a trustee of the Royal Historical and Archaeological Association of Ireland – the successor of the Kilkenny Archaeological Society, founded in 1849, and now the Royal Society of Antiquaries of Ireland. He married Maria Isabella, elder daughter of Lundy Edward Foot, a Dublin barrister, in 1852 and the couple had three sons and two daughters. The eldest son, George Dames Burtchaell MA KC LLB MRIA (1853–1921), who died as a result of a motor accident on a Dublin street, was Deputy Ulster King of Arms, a barrister, an antiquarian and an expert genealogist. The second son, David Edward (1859–1910), graduated in engineering from Trinity College, Dublin, in 1881, served his time to his father and assisted him for some years in the 1880s; he worked also as resident engineer on the Tralee & Dingle, Farranfore to Valencia, and Waterford to New Ross railways and on a variety of civil engineering projects around the country. The third son, General Sir Charles Burtchaell, had a distinguished career in the Royal Army Medical Corps, including service as principal medical officer in France and in India.

Peter Burtchaell used Alexander Nimmo's theodolite throughout his engineering career – 'a small but wonderfully handy and accurate little instrument' according to a report in the *Irish Builder and Engineer* in 1906. The instrument was then the property of David Burtchaell and was being used on work connected with the Greystones main drainage scheme. The subsequent fate of the instrument with which Nimmo carried out so much valuable work on roads, piers and harbours all around the Irish coast is unknown.

CAFFREY, John (1873–1947), was born on 16 October 1873, son of John Caffrey and Sarah McKee, and was educated in Belfast at Hardinge Street school and the Municipal Technical Institute. He worked initially as a teacher and attended various summer courses at the Royal College of Science in Dublin and in London. He began his engineering training in Belfast in 1898 under James Munce, assistant city surveyor, and worked with him in 1902–3 on surveys of Belfast river courses, and under H.A. Cutler, Belfast city engineer, on street surveys for tramway electrification.

Caffrey was resident engineer on the construction of inverted siphons on the Armagh–Keady railway in 1907 before taking up an appointment in Armagh as principal of the municipal technical school. In 1911, he was responsible for the design and construction of extensions (including the addition of a third storey) to the 1815 market house in Armagh to accommodate his school, making extensive use of reinforced concrete in floors and beams; the whole reconstruction was planned so sensitively that the new work was said to be practically indistinguishable from the old. Some years earlier, he had reconstructed the Tontine rooms at Armagh to create a new city hall, now demolished, and he designed a number of other buildings in and around Armagh, including a substantial bakery premises.

When a vacancy in one of the two county surveyor posts in Co. Donegal arose in July 1912, one of the assistant county surveyors immediately asked the council to postpone the filling of the post for a year, by which time he hoped to be qualified for the position. In the following November, the council proposed to the Local Government Board that the man in question, P.J. Kelly, should be appointed county surveyor for the southern division of the county but the board refused to sanction this because Kelly did not meet the prescribed qualifications relating to training and experience. Members of the council reacted angrily, some proposing that their man should be given further time to qualify and others suggesting that a committee of the council itself should consider the board's response and review Kelly's qualifications. In the following January and before the report of the committee was completed, Kelly withdrew from the contest and the council took steps to fill the post in the normal way. Two months later, in March 1913, John Caffrey was appointed at a salary of £400 a year, £100 less than had been paid to his predecessor. He had qualified at the civil service commissioners examinations in 1907 and was an unsuccessful applicant for a county surveyor post in Cork in 1909.

Maintaining the large network of roads in Donegal, even with the limited grants which became available from the Road Board after 1910 for the main roads, was always a difficult and expensive operation for the county council and its surveyors. The increasing use of cars and heavy lorries from 1914 onwards, and the destruction of bridges and cratering of roads during the civil war exacerbated the problems and led the council to seek to introduce relatively low weight restrictions and to prohibit the heavier vehicles from using all but a few hundred miles of roads in the county. The Department, however, took the view that such provisions would virtually amount to a repeal of the existing law relating to mechanically propelled vehicles entering the county, and would preclude the allocation of Road Fund grants for its roads. By

1937, Caffrey was still complaining about the damage caused by heavy lorries to the network of over 3,000 miles of county roads, much of which was still being maintained under the contract system: the railways, he said, had diverted much of their traffic to these roads, lorries were using even the weak county roads and the bog roads, and the one-ton lorry had been replaced by vehicles carrying three tons or more. 'If a limit is not put on the weight of vehicles using some county roads', he reported, 'the amount required to keep them in order will be considerably increased in the future'. But no such restrictions were imposed and it was left to Caffrey's successors to grapple with the problem of bringing the roads of Donegal up to modern standards.

Caffrey had to undertake a considerable amount of bridge work during his years in Donegal. In 1915, he told the county council that the old wooden bridge over the river Eske at Donegal was dilapidated and dangerous and he obtained approval to build a new ferro-concrete structure there at a maximum cost of £200. He was responsible also for the construction of a girder bridge over the river Drowes at Bundrowes on the Donegal-Leitrim border, and ferro-concrete bridges at Stragally and Breenagh. Another ferro-concrete bridge on the Hennebique system was built at Cruit Island in 1917, replacing a timber structure.

Following the retirement of **J.H. Steadman**, the northern division surveyor in 1933, the Department of Local Government and Public Health pressed the county council to appoint a single surveyor for the county, as most other counties had already done, and to introduce a unified engineering service in the interests of greater efficiency. Caffrey agreed to take over the whole county but only if he were allowed two senior assistants and a salary of £900, plus an allowance for board of health work, but the council were not convinced that this would be an efficient or effective arrangement or suitable for a county of the size and topography of Donegal. Eventually, in October 1934, with threats from the Minister that grant allocations for unemployment relief works and for roads would be withheld or cancelled if the matter was not resolved, the council gave in and accepted his proposal; Caffrey then received the salary of £1,000 he had sought and was allowed two senior assistants, one of whom was his son, John J. Caffrey BE BSc (1901–76), who was to continue in office until March 1966.

After a long illness, during which his son acted as county surveyor, Caffrey was obliged to retire in March 1938 and died in 1947. Following his retirement, the council attempted to restore the former arrangement by appointing his son and the other senior assistant as county surveyors but the Minister and the Department maintained their earlier position. The stand-off continued until the appointment of a new county surveyor in October 1939.

CARROLL, William (*c*.1823–97), was the son of William Carroll, a merchant, of 11 Eccles Street, Dublin. He served a pupilage to Jacob Owen in the Board of Works and subsequently worked under Isambard Kingdom Brunel as resident engineer on various railway schemes, including the Oxford, Worcester & Wolverhampton Railway. He returned to live at the family home in Eccles Street in the 1850s and became Monaghan county surveyor on 11 March 1854 after the dismissal of **John Walker**.

In his first report to the grand jury in July Carroll wrote that 'it is not in my power to congratulate you upon the condition of the county roads, but I trust in being able to do so on a future occasion'. He drew attention to the fact that contracts were being taken at prices which were wholly insufficient to complete the works by persons who 'trusted to chance for any payment they may receive, knowing that they can lose nothing owing to the laxity of the authorities in prosecuting them as defaulters'. However, Carroll's plans to improve the situation were interrupted when, for reasons which are not easily explained, he was transferred to Waterford on 12 September 1854 to replace **Alexander Schaw** who had died. He was reappointed to the Monaghan post on 30 March 1855 and by the following July was able to tell the grand jury that he had already taken steps to see to it that contracts were let at prices which were fair and remunerative and subject to more strict supervision; he admitted that costs would increase in the short-term, but argued that the new arrangements would make more sense in the longer-term. He was concerned, however, about the large mileage of byroads for which the grand jury had accepted responsibility but his suggestion that these should no longer be maintained from public funds fell on deaf ears. More than thirty years later, in March 1886, he was still criticizing the continuing increase in the county's public road mileage, pointing out that the figure had increased by about 100 since 1870; the additional costs arising on foot of the extension of the contract repair system to these minor roads had caused the total roads expenditure to increase to some £14,600 each half-year.

Carroll served in Monaghan without incident for thirty-six years. His resignation on grounds of ill-health was accepted by the grand jury at the summer assizes in July 1891 and he was awarded a pension of £267 a year. The jury expressed their regret at the loss of an official who had faithfully and conscientiously performed his duties over a long period and, in a tribute to him, the *Northern Standard* noted that he had been a master of his work and a courteous and obliging official while maintaining an independent attitude. Carroll died on 27 July 1897, aged 74, and was buried at Deansgrange cemetery in Dublin. His wife, Lucy Agnes, died on 17 April 1917. His brother, Frederick Arnold Carroll, who was also an engineer, was active from the 1850s to the 1880s on railway and other work.

CARTER, Samson (*c.*1804–60), died in unusual and tragic circumstances. Between one and two o'clock on the afternoon of Saturday, 1 December 1860, a country gentleman, believed to be of high position, committed suicide in Dublin by placing his silver double-barrelled pistol in his mouth and blowing his brains out. The incident took place in the upstairs front bedroom of a house of ill-repute at 68 Montgomery Street – the street which gave its name to *Monto*, Dublin's notorious red-light district. The police were quickly on the scene and took little time to establish from papers in the pockets of the dead man's clothes that he was Samson Carter, the Kilkenny county surveyor. The *Freeman's Journal* reported 'the fearful and tragic occurrence' in great detail on the following Monday, with eye-witness accounts of the condition of the bedroom and of the body, and full reports of Carter's three-day stay in the house with an English woman who gave her name as Ellen Seymour, alias Montezuma. The Tuesday edition carried verbatim accounts of the proceedings at the inquest at which a verdict of suicide 'while labouring under temporary insanity' was returned. On Wednesday 5 December, both the *Kilkenny Moderator* and the *Kilkenny Journal* reprinted the story, the former protesting that it would most gladly have spared the afflicted family the deep pain which publication of the details would cause them if it were not for the fact that the Dublin papers had already made these public!

Carter was a son of Major Samson Carter (1777–1854), a distinguished soldier and one-time resident magistrate at Nenagh, who was awarded the freedom of the city of Dublin in 1828 for the 'effective and gentlemanlike manner' in which he had preserved the peace in Co. Waterford while serving there as a magistrate. He became county surveyor in Kilkenny in May 1834 at the age of 30, and was surveyor also for the separate county of the town of Kilkenny for which he was paid an additional salary of £50. At the time of his death, he was a 56-year-old widower with six children, his wife, Mary Pasley, daughter of a Co. Galway clergyman, having died in 1850. Although a sensation of the deepest horror was said to have swept through Kilkenny when the news of his death reached the city, his erratic behaviour during the previous year had led many who knew him to believe that Carter was indeed bordering on insanity. An extravagant lifestyle had led to financial difficulties and a degree of intemperance which, according to one of the local papers, had 'completely prostrated his intellect'. He had gone to Dublin to negotiate a solution to his financial problems, staying initially at the Prince of Wales Hotel and moving to Montgomery Street in a depressed state of mind when a settlement failed to emerge. But he was never to return to Kilkenny, even in death: he was buried with his father at Dublin's Mount Jerome cemetery.

From 1828 to 1830 Carter was employed by Alexander Nimmo as surveyor and superintendent on the central Erris road. In 1832, he was engaged by the Board of Works to report on a Wicklow copper mine for which a loan was being sought and about the same time reported to the board on proposals for a new Waterford–Tramore road. With **Noblett St Leger**, another of those who worked with Nimmo and who became Leitrim county surveyor in 1834, he was engaged on road surveys in Co. Monaghan and together they were engaged in 1834–5 to prepare a survey of Waterford harbour for the Commissioners for Improving the Port – a partnership which Carter marked by naming his second son Noblett St Leger Carter.

Carter was seen to be a man of considerable ability whose talents were fully appreciated in his public and private capacities and who seems to have had little difficulty in carrying out his basic functions in relation to road maintenance in Kilkenny. Within a year of his appointment in 1834, he had prepared proposals for substantial stretches of new roads between Inistioge and Waterford and from New Ross to Waterford. By 1841, he had been allowed five assistants to assist in supervising work on the 800 miles of roads which were then in charge; only three of his colleagues, many of whom had much larger mileages to deal with, had greater numbers of assistants. This made it possible for him to take on whatever private business he could get. In addition, he was able to act as surveyor to the trustees of the thirty-two miles of turnpike road from Carlow to Kilkenny and onwards to the Tipperary boundary, and the thirty-four-mile-long Waterford to Shankill turnpike, running northwards through Co. Kilkenny; he earned an additional £70 a year from these posts but resigned from them in 1852 when the decline in income from tolls left no funds available either for maintenance work or to meet salary payments.

In 1841 Carter became embroiled in controversy about the involvement of county surveyors in architectural work, particularly the design of courthouses and other county buildings. As he explained it himself to the commissioners who had been set up in 1840 to review the grand jury laws, he had an arrangement with W.D. Butler, one of the most prominent architects of the period and at the time 'a personal intimate', as to 'all the architectural business which I might obtain for him'. Thus, when his own plans for new courthouses at Callan and Urlingford were rejected by the Board of Works in 1836, he naturally turned to Butler. The latter's plans were approved by the authorities in Dublin and by the grand jury, and the two identical courthouses were completed in 1840 with Carter, apparently, sharing with Butler the 5 per cent fee for superintendence of the work. The arrangement between the two men broke down, however, later in the year when plans were required for improvements and alterations at Kilkenny gaol. Having obtained the approval of his grand jury to his outline proposals in the summer of 1840, Carter asked Butler to prepare detailed plans for the £10,000 scheme on the basis that if these were put forward 'without allowing the intervention of any other competitor', Carter would be given one-third of the fee. This seems to have been too much for Butler, who reported the matter to the Royal Institute of the Architects of Ireland which in turn lodged complaints with the commissioners who were reviewing grand jury laws. But Carter was quite unapologetic about his conduct, claiming that the affair had been brought forward 'from a persevering, personal malignity, than with any view to the furtherance of the public service'. He wrote to Butler asking 'Do you imagine I will, or would, tamely submit to Keane, Darley, Morrison, or even yourself stepping in between me and the body to whom I am legally appointed the professional officer and adviser, and permit them to usurp my prerogative? No! Sooner than do so, I would expend one-half of my salary in obtaining plans from an architect of eminence, if even from the other side'.

On taking up duty in Kilkenny, Carter lived initially at Castleview, where his office was located, and later at St John's Quay in the city. He was elected to membership of the Royal Irish Academy in 1837 and was also a member of the Kilkenny and South-East of Ireland Archaeological Society.

CHADWICK, William J. (1895–1951), was born on 9 May 1895 at Borrisoleigh, Co. Tipperary, where his father was clerk of the petty sessions, and was educated at Rockwell College and at University College, Dublin, from which he graduated with a BE degree in 1916. He was an assistant county surveyor in Kildare from 1917 to 1919 and worked in England as a contractor's engineer and as resident engineer, mainly on water and sewerage schemes, until 1928.

On his return to Ireland in 1929, Chadwick worked briefly on the Shannon Scheme. Later that year he became Nenagh town surveyor and resident engineer on a water supply extension scheme for the town. He became an assistant county surveyor with Tipperary north riding county council in May 1931 and acting county surveyor in October 1934 when **T.C. Courtney** left the post to return to the Department of Local Government and Public Health as chief engineering adviser. The county council then resolved to fill the post themselves by direct election from among candidates who were serving officers of the council or of other local councils, but they were prevailed upon to face the reality that the appointment could only be made on the recommendation of the Local Appointments Commissioners. In the event, it was Chadwick, the local candidate, who was duly recommended for appointment to the permanent office in June 1935.

Chadwick died in office on 2 September 1951. Unusually, there is a memorial to him in the form of a statue of Our Lady of Fatima in the grounds of the Catholic church in Nenagh – erected by the road-workers to the memory of one 'whom they will always remember as a respected official and kind friend'.

CHARLTON, James Wallace MC OBE (fl. 1914–56), served with distinction with the rank of captain in the First World War, gaining the Military Cross and the Croix de Guerre. He graduated in engineering in 1922 and became an assistant county surveyor in Co. Fermanagh in December 1925 under **James P. Burkitt**. After a period as deputy county surveyor he was appointed county surveyor in 1940. In the late 1940s and 1950s, he worked closely with the central authorities in the preparation and execution of major works on the more important roads in the county and undertook some experimental work on the removal of peat by explosives on the Enniskillen–Brookeborough road in the 1949–52 period. Charlton retired in 1956.

CLARKE, Eugene O'Neill (1867–1943), was born on 8 August 1867, son of Eugene Clarke, proprietor of a general drapery business at West Street in Drogheda, and his second wife, Frances M.J. O'Neill. Having gained a BE degree from the Royal University of Ireland, as well as the Royal College of Science diploma in 1887, he worked until 1893 as an assistant engineer with Robert Worthington, a major railway contractor, on the construction of railways in the south and west of Ireland under the Tramways and Light Railways Acts. He was living at Adelaide Road, Dublin, immediately before his appointment as Leitrim county surveyor in 1897.

Hartley Bridge, which carries a county road over the river Shannon a few miles north of Carrick-on-Shannon, was constructed under O'Neill Clarke's supervision in 1912–13. The bridge and approach roads were built by direct labour at a total cost of about £1,200 – some £500 was contributed locally and a grant of £500 was made by the Congested Districts Board. It has a rather strange profile and is not very pleasing aesthetically due in part, perhaps, to modifications made at the insistence of the Board of Works to increase the waterway, but it deserves recognition as one of the earliest surviving reinforced concrete structures of its kind in Ireland. There are six main spans, the longest just over forty feet, with two additional shorter spans over the flood plain at the western end. The deck slab is carried on beams spanning between two longitudinal parapet beams which are supported on piers made up of two columns with horizontal tie beams and a bracing strut. A five-tonne weight restriction was imposed on the bridge in 1984 when it became apparent that the structure had deteriorated seriously, with extensive spalling of the concrete and widespread and deep corrosion of the reinforcing moss-bars which had minimal concrete cover. Examination of the structure at that stage also suggested that the bridge was grossly under-reinforced and otherwise seriously deficient by modern standards of design and construction. Nevertheless it still stands, ninety-five years after its completion, as a tribute to the enterprise and initiative of a county surveyor who, when others were still debating the risks involved in using reinforced concrete for bridges of this kind, decided to go ahead on his own with a novel form of construction.

Clarke introduced steamrolling in the towns of Manorhamilton and Dromahaire in 1905 and some years later in the Carrick-on-Shannon area, but found that the rural district councils in Leitrim, as in other counties in the west, still clung to the old labour-intensive system of maintaining roads and were critical of all attempts to introduce change. Writing in the *Irish Builder and Engineer* in April 1910, he explained that there was still fear of the cost of steamrolling, even when it was spread over a term of years, but the most important factor was the influence of the road contractors on the members of the district councils. Particular roads had remained in the hands of individual contractors for many years, with contracts being renewed every few years, and often handed on to sons of the earlier contractors. So much was the contract looked on as a family asset, that it sometimes formed part of a marriage settlement when a son was being settled in life. Unless, therefore, special efforts were made to influence the district councils, Clarke argued that the roads of Leitrim would continue to be inadequate for the requirements of the day. Matters did not improve in the subsequent years and, with the withdrawal by the county council of recognition of the Local Government Board, the cancellation of state grants, and anti-rates agitation, local services generally were in a sorry state in Leitrim by the autumn of 1921; the roads had come to resemble quagmires and were often impassable, and there were attempts by the local Sinn Féin/IRA organization to take control of the organization of the works. There was little the county surveyor could do until more settled conditions returned in 1922–3.

The stone clock-tower in the Market Square, Carrick-on Shannon, erected in 1905 to commemorate Owen McCann, the first chairman of Leitrim county council who died in 1901, was designed by Clarke. He retired in February 1936 after thirty-nine years service, with glowing tributes from the council to his work and worth. He was one of the last of the county surveyors who had served under the grand jury system as well as under the county councils. He continued to live in Carrick-on-Shannon for some years after his retirement and carried on a limited private practice.

CLARKE, James (*c*.1774–1838), described himself in 1818 as a land surveyor who had 'served a regular apprenticeship to an eminent civil engineer' – he was referring to the post office surveyor, William Larkin – and had afterwards been employed 'in making county surveys, and in superintending extensive lines of post roads, for the conveyance of His Majesty's mails'. He went on to explain that he had been 'constantly engaged in the different departments of his profession' and could refer to gentlemen of respectability for his conduct and qualifications.

Living in Carrick-on-Shannon, Clarke was one of the assistant engineers who worked with Alexander Nimmo on road planning and construction in counties Roscommon, Leitrim and Sligo in the late 1820s. He was paid one guinea a day and an additional 2*s*. 9*d*. a day for the forage of his horse. Subsequently, he advertised his services as a land surveyor in the *Roscommon and Leitrim Gazette* and practised as a land surveyor and engineer at Belturbet. In 1832, when he prepared plans for the Roscommon grand jury to submit to the Board of Works for a seven-mile stretch of road from Tarmonbarry on the Shannon to Lung bridge on the Roscommon–Mayo boundary, he was said by the board to have had 'much practical experience' in such matters; the road in question was originally laid out by Richard Edgeworth and Alexander Nimmo, and was considered by the board to be of great importance because it would open up to the province of Connacht the transport facilities of the Royal Canal. Clarke's plans were approved and a loan of £7,111 was allocated for the project; contractors were at work on the road in 1833 under Clarke's supervision and the project was completed in 1836.

In his *Remarks on the Maintenance of Macadamised Roads*, published in 1843, Sir John Fox Burgoyne, then chairman of the Board of Works, remarked that 'the importance of rolling roads, either newly constructed, or when subjected to extensive repairs, seems never to have been duly appreciated' and noted that the practice of rolling had been rare in Ireland. On foot of this publication, Burgoyne has often been credited with being the first and most strenuous advocate in Ireland or England of horse road-rolling as a measure of economy and efficiency in road-making. However, James Clarke's *Practical Directions for Laying Out and Making Roads*, published in 1818 but not, perhaps, circulated very widely after its publication, gives him a better claim to this distinction for it strongly recommended that 'before any new road is opened … the contractor should be bound to roll it frequently with a cast-iron roller; this will press down any small stones which may rise above the surface, close the tracks of wheels as soon as formed, and keep the whole road smooth and hard'. Clarke's book, one of the first to discuss the subject of road-building in the Irish context, is notable in other respects, including the extent to which his directions are quantitative, something which was rare in the literature of the time on road-making. It included a section on stone bridges, recommending the use of a single low arch as against the several small, high, semicircular arches which formed most country bridges at the time.

In 1833, with Thomas, his brother, who was also living in Belturbet at the time, Clarke prepared surveys and maps of a number of estates, including the Dopping estate of 1,860 Irish acres which straddled the Dublin-Kinnegad road and the Royal Canal where they pass through Co. Meath. Their attractive maps of the estate were on a scale of twenty-four perches to one inch and the survey data included details of each tenancy and information on the quality of the lands involved.

James Clarke must have been up to 60 years of age when he became the first county surveyor for Tyrone on 17 May 1834. He went to live in Omagh but, with his wife, Frances (Fanny) Williams, continued to have extensive property interests in and around Belturbet which had been the subject of a complex settlement in 1819 before the marriage took place. Clarke died in office on 4 September 1838 as a result of a fall from his gig, leaving five young children. His widow, in her will made in March 1856, divided the family property between her three sons (one of them, James, apparently a Church of Ireland clergyman) and her brother-in-law, Thomas Clarke, who received the interest on £600, her two farms and her household furniture and chattels.

Following the death of his brother, Thomas Clarke acted as county surveyor until the following spring assizes when **James B. Farrell** was appointed to the post. Clarke then sought another county surveyor position 'so that he would be able to render assistance to the widow and children of his late brother'. The Tyrone grand jury were satisfied that he had given diligent and efficient service, but the Board of Works advised that they had sanctioned his acting appointment only because of 'the peculiar circumstances of the case'. They believed that Clarke, on a strict examination, would not be found to be duly qualified for a permanent appointment and his application for a position in another county was therefore turned down.

CLEMENTS, Henry (fl. 1790–1864), was a son of Hill Clements, a land surveyor and civil engineer, who was in partnership for many years with David Aher (1780–1842), a leading surveyor, cartographer, geologist and engineer, who worked for the Grand Canal Company and for various mining companies on the development and operation of collieries near Castlecomer where he lived. Both Aher and Clements worked for the Bogs Commissioners between 1809 and 1813 and were engaged until 1824 on the survey and mapping of Co. Kilkenny on foot of a contract from the grand jury. They also undertook road design, construction, survey and mapping projects in Kilkenny and Carlow and further afield for the grand juries and the postmaster-general. Hill Clements worked as an assistant to Richard Griffith while the latter was engaged on a major public works programme in Cork and Kerry from 1822 onwards, and published *A Plan for Improving the Port and Quays of Limerick* in 1836.

Hill Clements married Mary Aher, his partner's sister, and two of their sons, Henry and Hill junior, followed their father into engineering careers. They were both engaged as surveyors by the Bogs Commissioners and subsequently worked with their father on the Kilkenny grand jury survey. Henry's initial engagement by the commissioners in 1810 had been on the basis of half a guinea a day, as against a standard fee of one guinea for most other surveyors, which suggests that he must have been either very young or inexperienced, or both, at that stage. He was probably in his forties when he became the first county surveyor for Galway on 17 May 1834. He acted also as surveyor to the county of the town of Galway which had its own grand jury, and received an additional salary of £100 for this. Clements lived in Galway city and apart from his public duties did 'some few small things', as he put it, for private clients, mainly plans and maps for persons seeking presentments for new works. He had some involvement with improvement works at Galway docks in the late 1830s and while he provided plans for a new courthouse in Ballinasloe in 1838, the building which was completed there a few years later appears to follow a standard design attributed to William Caldbeck. The Board of Works was in touch with him in 1837 about the need for further new courthouse buildings at Clifden and Oughterard but the identical courthouses provided in the two towns in 1843 also appear to follow the standard Caldbeck plan.

Clements was allowed two assistants in 1837 and two more some months later, but disagreements with the grand jury as to the division of responsibility between him and the assistants led the jury to seek to have the county divided and to have a second surveyor appointed. When this was conceded by the authorities in the spring of 1838, Clements was assigned to the new western division, where he had to carry responsibility, initially without the services of any assistants, for some 900 miles of roads. He complained that his area was so extensive that he had to keep extra horses and that his travelling expenses cost him up to fifteen shillings a day out of his own pocket. In July 1839, perhaps to compensate for the absence of paid assistants, he advertised for a pupil who, for a moderate fee, would have an opportunity of acquiring a knowledge of the various branches of engineering and surveying as well as 'the practical part of building and architecture'.

In 1846–7 Clements was heavily involved in the organization of famine relief works and, like several of the other surveyors, found the assignment to be a somewhat hazardous one. He was assaulted on the works in 1846 and Richard Griffith, then deputy chairman of the Board of Works, thought it necessary to ask the police to send two men to guard his house. Later that year, his office in Galway was forcibly entered by men demanding employment on the works.

Clements was in difficulty with the Galway town grand jury at the assizes in March 1855 when they complained that he had awarded a contract to a person who had been found on a previous occasion to be guilty of 'underhand arrangements'. He escaped with a formal caution on that occasion but was heavily criticized again at the summer assizes, this time for having certified payments for work which had not been properly carried out. Soon afterwards, in August 1855, he was replaced by **Samuel Ussher Roberts** as town surveyor. He continued in office as county surveyor throughout 1856, carrying on his normal duties in relation to the roads of west Galway and endeavouring, as he told the jury in the summer, to 'promote the extirpation of noxious weeds on the sides of the roads' in response to complaints which has been expressed by the assizes judge. He was allowed by the county grand jury in March 1857 to appoint Roberts as deputy county surveyor 'in consequence of his own increasing infirmities'. This arrangement continued until July 1858 when Clements was removed from office and replaced by Roberts; according to one of his colleagues, he had been kept on by the grand jury, acting out of kindness and consideration and 'not wishing to throw a man upon the world' until it became impossible to continue. He was listed as a member of the Institution of Civil Engineers of Ireland until 1864.

Henry Clements' brother, Edgar, qualified at the county surveyor examinations in 1841 but never received an appointment. Another of his brothers, Edward, was a barrister, author of a number of legal works in the 1840s and 1850s and the commissioner whose report in 1856 led to the abolition of the last of the turnpike trusts; he died in January 1862 and was buried with his uncle, David Aher, in Mount Jerome cemetery.

COFFEY, Denis (1896–1955), was born on 10 August 1896 in Cork city where his father carried on a drapery business at Great George's Street. He was educated by the Christian Brothers in Cork and at University College, Cork, where he gained a BE degree in 1916 and a BSc in 1918. He taught for some years at Cork Grammar School before becoming assistant county surveyor in the Schull area of west Cork in 1922, working mainly on roads, bridges and marine works. He was appointed to one of the two chief assistant county surveyor posts in Cork in 1928 and became county surveyor in Cavan in August 1930.

In April 1947 Coffey returned to Cork to work as one of the three deputy county engineers who managed the engineering business of the county in the absence of a county engineer post from 1940 to 1958. He had responsibility for south Cork where he was regarded as a courteous, painstaking official who did particularly good work on road construction and maintenance. He died in office on 20 July 1955.

COFFEY, Jeremiah Gerard (1908–88), was born in Midleton, Co. Cork, on 27 April 1908. Having graduated with a BE degree from University College, Cork, in 1928 when he was only 20 years of age, he worked with his father, a building contractor, on a Midleton water supply scheme and on a number of building projects in the Cork and Waterford areas. He was employed as a junior engineer by Henry Ford Limited in Cork in 1929–30 and joined the staff of the South of Ireland Asphalt Company, Dublin, in 1931. He was appointed temporary engineering inspector in the Department of Local Government and Public Health in May 1937 and was assigned to work on water and sewerage schemes and roads.

In February 1939 Coffey was seconded to the Department of Industry and Commerce to supervise the construction of the Irish pavilion for the New York World's Fair – the famous 'Shamrock Building' designed in steel and glass by Michael Scott. He sailed for New York on 25 February and was highly commended by the Minister, Seán Lemass, for his services there. The pavilion was completed on schedule and was opened by Seán T. O' Kelly, Tánaiste and Minister for Local Government and Public Health, on 13 May 1939. Coffey resumed duty in the Department in July 1939 but resigned with effect from 31 March 1940 to take up duty as Kilkenny county surveyor.

Coffey was elected president of the Institution of Civil Engineers of Ireland for 1960–1 and dealt in his presidential address with the role of the local authority engineering services. He was critical of what he described as the unrealistic policy of spending twice as much on county as on main roads – an approach which he attributed to political considerations rather than to economic or social considerations – and he drew attention to 'the slaughter and maiming' which in his view was caused, in the case of a large proportion of road accidents, by inadequate roads. He argued that a central national roads authority should be established to plan and carry out all major improvement works on arterial roads, and he suggested that this could be financed by the assignment of excise and import duties to road reconstruction.

When he retired in mid-1974, a year after he had reached 65 years of age, Coffey was the last serving official in the local service whose original appointment was to the post of county surveyor. He died in Kilkenny on 21 December 1988, aged 80. His brother, Patrick, was professor of engineering at University College, Cork, until 1967, and his son, Frank, became the first county engineer for the new county of South Dublin in 1994.

COLLEN, William (*c*.1860–1932), was a son of John Collen of Portadown, a deputy lieutenant for Co. Armagh, one-time high sheriff of the county and a member of Armagh county council. Collen senior was also founder and principal of the firm of Collen Brothers Limited of Dublin and Portadown which carried out many large building and construction contracts throughout Ireland, including the Royal Dublin Society's main hall (1884) and the asylum at Portrane (1903), as well as numerous railway and water supply schemes. William Collen was awarded a BA degree at Trinity College, Dublin, in 1882, a BAI in 1884 and later the MAI degree. He was a gold medallist at Trinity and distinguished himself in classics and in mathematics. He gained experience of railway engineering and other work in Ireland, and was employed for a period under **P.C. Cowan**, then Down county surveyor, before being appointed a Dublin district surveyor on 31 December 1891 with responsibility for the northern district. Following the retirement in 1896 of the long-serving **R.A. Gray** who had charge of the southern district, the County Dublin Surveyors Act 1897 was enacted to allow one and the same person to hold the office of county or district surveyor for the two districts and to allow the grand jury to appoint as many assistants to the surveyor as they thought necessary. Thereafter, Collen was in sole charge of the engineering business of the county where the roads were in a notorious condition as the turn of the century approached, due to the cheeseparing of successive grand juries and the shortage of professional and other staff – and matters were getting worse as traffic in and around the city increased with the introduction of the motor car.

Collen developed a reputation as an energetic and effective county surveyor who made a major contribution to the improvement of Dublin's road network. He was implacably opposed to the contract system of road maintenance, describing it as a farce of a system the objective of which seemed to be to subsidize small farmers. He decided to introduce steamrolling at an early stage, the first surveyor to do so in the south of Ireland. When he failed to convince the county council of the merits of the new system he procured and maintained a steamroller at his own expense and soon converted the councillors to his views. With a force of eleven rollers at work throughout the county in 1900, Collen found himself in difficulty with the Alliance and Dublin Consumers Gas Company which claimed that the use of heavy rollers in the Kilmacud, Harold's Cross and Cabra areas had fractured gas mains which had been laid nearly forty years earlier. The company sought an injunction in the courts to restrain steam-rolling operations which could cause further damage and the case was heard over four days in March 1900. There were numerous expert witnesses, those for the company contending that their mains had been laid at an adequate depth to keep them safe from the ordinary traffic of the day, and those for the county council (including P.C. Cowan of the Local Government Board) arguing that a depth of eighteen inches was the absolute minimum acceptable for roads which carried traffic, including steamrollers, of up to thirty tons. The master of the rolls gave judgment for the gas company and his decision was upheld by the court of appeal. He took the view that, as the existence of steamrollers had not been dreamt of when the mains were legally laid, the company could not be expected, forty years later, to relay its entire network of pipes to allow for the introduction of new kinds of traffic. And while accepting that Collen had given evidence with candour and fairness, the judge told him that he was quite wrong in his understanding of the law: he had gone too far by asserting that his council had a right to work their steamrollers regardless of anything or anybody, and he firmly rejected the surveyor's claim that he alone was the judge of how the public roads were to be restored.

In spite of this setback Collen was said by 1905 to have 'worked a perfect revolution' in the roads of south Dublin so much so that he earned a glowing testimonial from Irish cyclists. Some years later, the *Irish Builder and Engineer* remarked that 'everybody knows, and every owner and driver of a vehicle appreciates, the excellent condition of the county Dublin roads'. But Collen had difficulty from time to time in persuading the county council to grant him sufficient funds to maintain the steamrolling programme. In north Dublin there were added complications because of his insistence that the local stone was 'mostly rubbish' – a view not shared by other engineers; his refusal to use the stone had obvious implications for employment in the local quarries and the county council cut his roads estimate in 1912 in an effort to persuade him to change his mind.

An action taken in the courts by Collen against a landowner who had erected a barbed wire fence along a section of the road from Balbriggan to Dublin led to a ruling by the court of exchequer that the fence should be removed, one of the judges describing it as a public nuisance and a danger to young and old, strong and feeble, and recommending that Collen's action should be followed up in every county. This led to the enactment of the Barbed Wire Act 1893 which is still on the statute book; under the act, the county surveyor (now the county council) was authorized to require a landowner to remove barbed wire from a fence adjoining a public road where the wire may be injurious to persons or animals using the road.

As county surveyor, Collen was responsible for the maintenance of small harbours at Rush and Lambay but he described these in 1902 as 'comparatively useless for purposes of trade or fishing'. However, he planned and supervised the extension and reconstruction by direct labour of Loughshinny pier at a cost of some £5,000 between 1905 and 1908 and presented a humorous paper on the project to the Institution of Civil Engineers of Ireland in 1910. With John Smith as resident engineer, he was responsible in 1904–7 for Newtown and Lissenhall bridges, Swords, and for coast protection works costing some £5,000 at Barnageeragh, near Skerries. He also undertook a considerable amount of work on public buildings in the county; his plans for reconstruction of the county offices were approved in 1901 as were plans for Rathfarnham courthouse in January 1913.

When Collen's salary came up for review by the new county council in 1899, he fared better than many of his colleagues, being allowed a rate of £850 a year and, some ten years later, he was given an increased salary to allow him to purchase a car for use in connection with his official duties. He served as president of the Institution of Civil Engineers of Ireland, 1913–15, and became district chairman of the Institution of Municipal and County Engineers in 1912. Following his retirement in June 1924, Collen went to live at Bournemouth where he died on 29 April 1932.

COTTON, Charles Philip (1832–1904), of Lower Pembroke Street, Dublin, was appointed Kildare county surveyor on 19 August 1865, a few weeks after the death in office of **John Yeats**. The appointment must have created considerable surprise in the Irish engineering world because Cotton was by then a prominent consulting engineer with offices at Westland Row, Dublin. With his substantial experience in railway and marine engineering, it is difficult to understand why he decided to apply for a post which was not particularly well paid and which offered few opportunities to practise his undoubted engineering skills. At the time of the Kildare appointment, Cotton's name had been on the list of those who had qualified at county surveyor examinations for over eight years and when he learned that his name had reached first place on the list, he asked that 'my turn be passed over for this time' while reserving the right to take up a future vacancy. However, when told that anyone who declined an offer of a surveyorship would be struck from the list, he changed his mind and decided to accept the Kildare appointment. The *Leinster Express* reported on 2 September 1865 that Cotton was to take up residence in Naas to meet the express wishes of the grand jury, and the same edition of the paper carried an advertisement from the new surveyor inviting applications for a post of assistant surveyor, the previous holder having resigned because of the low salary. The appointment was duly made by Cotton but this seems to have been one of his few official acts as county surveyor: he resigned from the position on 3 October and was replaced by **Thomas Brazill**.

Cotton was born in Dublin on 19 January 1832. His father was Henry Cotton (1789–1879), a native of Oxford who came to Ireland in 1823 as chaplain to Archbishop Richard Laurence of Cashel, married Mary, the younger daughter of the archbishop, and served as archdeacon of Cashel from 1824 to 1872. He was educated at St Columba's College, Dublin, and at

Stackallen, Co. Meath, and entered Trinity College, Dublin, in 1848. He was awarded a BA degree in 1853 and a Diploma in Civil Engineering in 1854, with special certificates in practical engineering and in chemistry and geology. In 1855 he became a pupil of W.R. Le Fanu (1816–94), then consulting engineer to the Great Southern & Western Railway Company and to numerous other railway companies, and went on to become his senior assistant. Under Le Fanu, Cotton acted as resident engineer on the construction of the Bagenalstown & Wexford railway (1858) with its spectacular sixteen-span stone viaduct at Borris; the Mallow & Fermoy line (1860); and the Roscrea & Nenagh line (1863). His *Manual of Railway Engineering in Ireland* was published in Dublin in 1861.

When Le Fanu took up an appointment in July 1863 as a commissioner of public works, Cotton carried on the consulting engineering practice in partnership with Benjamin F. Flemyng. In addition to his railway engagements, he acted as consulting engineer from 1854 to 1867 to the Corporation for Preserving and Improving the Port of Dublin which had responsibility until 1868 not only for the port but also for the lighthouses around the Irish coast. One of his assignments involved visits to the Fastnet Rock in 1865 and 1866 to inspect the cast-iron lighthouse which had been completed there by George Halpin (senior) in 1854 and to advise on how its safety and stability could be improved. After much deliberation and consultation with experts in Britain, Cotton's proposals were accepted: the diameter of the base of the tower was increased, the lower storey was filled with solid masonry and the top of the rock itself was levelled to offer less resistance to the waves which went right over the rock when gales were blowing. The work was completed in 1868 and similar strengthening of the lighthouse on the Calf Rock followed immediately. However, notwithstanding Cotton's work, the latter tower was carried away during a heavy gale in 1881 and the lighthouse at the Fasnet was replaced by the present granite structure, constructed between 1896 and 1904.

After his six weeks service as county surveyor in the autumn of 1865, Cotton continued his career in consulting engineering, working on the Wicklow to Wexford railway and on the Shillelagh branch line (1865), among others. He and Flemyng became involved in the debate which raged throughout the 1860s and 1870s about the best means of providing a central railway station in Dublin linked to all of the main provincial lines. Twenty-six separate schemes were put forward between 1860 and 1875, several of them by Cotton & Flemyng. One of these involved the construction of a railway from Ranelagh to Sydney Parade, to be called the Kingstown Connecting Branch. A later scheme of theirs, published in 1869–70, would have created a large low-level central railway station at D'Olier Street, and a subsequent modification involved a higher-level station with a diagonal link across the Liffey towards Amiens Street.

In October 1878 Cotton was appointed with two others to form a royal commission (the Exham Commission) to inquire into the boundaries and municipal areas of cities and towns in Ireland; the commission reported in 1881 by which time Cotton had been appointed chief engineering inspector of the Local Government Board. He held this position from 1879 until his retirement on 31 December 1898 but was retained temporarily in the board's service until 31 March 1900 to assist in implementing the Local Government (Ireland) Act 1898. As chief inspector, he was heavily involved in the development of better water supplies and sewerage disposal systems and he published several textbooks on these and related topics, including *Loans for Sanitary Purposes* (1886); *A Manual of Procedure by Provisional Order* (1887); *The Housing of the Working Classes Act, 1890* (1890); *The Irish Public Health Acts, 1878-90* (1891); and *The Irish Sanitary Acts* (1892). He was appointed a member of the Royal Commission on

Sewage Disposal in 1898 and was chairman of a vice-regal commission which was set up in 1900 to inquire into and report on the health of the city of Dublin.

Cotton was a member of the Institution of Civil Engineers of Ireland to whose transactions he contributed numerous papers, and was president of the institution from 1873 to 1875. He was also a member of the Institution of Civil Engineers, London. He lived in his later years at Ryecroft, Bray with his wife, Marion Louisa, eldest daughter of Sir Maltby Crofton of Longford House, Co. Sligo, whom he had married in 1878. The couple had no children. Cotton died at Bray on 10 March 1904 at the age of 72. In a tribute to him later that year, the president of the Institution of Civil Engineers of Ireland, Robert Cochrane, said that he was, perhaps, the most widely known member of the profession in Ireland and described him as an able, impartial, and painstaking officer, who was regarded with the highest esteem and respect by those with whom he came in contact.

COURTNEY, Thaddeus Cornelius (1894–1961), was born in Cork on 13 December 1894. He was one of the most prominent personalities in Irish engineering for almost a quarter of a century and had a major influence on the development of both the road and rail transport networks.

Courtney was educated at Presentation Brothers College, Cork, and at University College, Cork, from which he gained a BE in 1916 and an ME in 1932. His first post was as assistant to J.R. Kerr, general manager and engineer of the Cork, Bandon & South Coast Railway. In January 1917 he took up a position as assistant engineer with Messrs Henry Ford & Sons at Cork and was involved in the design and construction of their new factory buildings and wharves at the Marina. In July 1918 he became assistant engineer at the Harland & Wolff shipyard in Belfast, working on the construction of new slips, tower cranes and large industrial buildings, before returning to Fords in 1920. He joined the National Army in 1922, serving with the Railway Protection, Repair & Maintenance Corps which was formed in September that year and, with the rank of major, was second-in-command of the corps of engineers by 1924. In the following year he took up an appointment in the Department of Local Government and Public Health as an engineering inspector and was assigned to the roads division.

Courtney was selected in July 1930 by the Local Appointments Commission for the position of county surveyor in Tipperary north riding, but the county council initially refused to appoint him because the salary and expenses for the post had been fixed by the Department at levels above those decided by the council. The council bowed to the inevitable, as the *Nenagh Guardian* put it, in September 1930 and appointed Courtney under strong protest. Following an open competition, he returned to the department as the first holder of the new post of chief engineering adviser on 1 October 1934 and, in the course of the following fifteen years, played a major part in the reorganization of the local authority engineering services, including the introduction of unified staffing arrangements at county level. He had a major role too in the development of the main road network and in the planning of other public works programmes, including the construction of hospitals and sanatoria. He was a member of the Greater Dublin Tribunal under Mr Justice Gavan Duffy which reported in 1938 and of the Committee of Inquiry into the Housing of the Working Classes in Dublin which was set up in 1939.

When fuel supplies became scarce in 1941, county surveyors were called on to participate in a major turf production campaign for which the Department of Local Government and Public Health was given responsibility at central level. The campaign was planned and managed by a turf executive, chaired by Courtney, who carried responsibility for the administrative aspects of the scheme as well as its technical direction. At one stage, there were upwards of 20,000 men employed by local authorities on turf production and over two million tons of turf had been produced by 1944.

Courtney acted in a part-time capacity as railway inspecting officer under the Department of Industry and Commerce in the 1940s. He was appointed full-time chairman of CIE in February 1949 when the government decided to nationalize the company, replacing Percy Reynolds, a close friend of the former Minister for Industry and Commerce, Seán Lemass. Courtney served in that capacity until August 1958, presiding over the introduction of diesel and diesel-electric traction in place of steam on Irish railways to an extent which was far in advance of other European railway systems at the time.

Dr C.S. Andrews who, as a member of the turf executive, was closely associated with Courtney during the emergency years and succeeded him as chairman of CIE, described him as 'something of a hypochondriac' but 'a man of considerable charm and popularity'. He was president of the Institution of Civil Engineers of Ireland in 1943–4 and was active also in Cumann na nInnealtóirí of which he was president in 1953. He died at Waterville, Co. Kerry, on 5 August 1961.

COWAN, Peter Chalmers (1859–1930), was born in Dundee on 20 March 1859 and was educated there at the High School where he gained an engineering scholarship. He studied at the university of Edinburgh from 1878 to 1881 under Professor Fleeming Jenkin, one of the most prominent of the early British electrical engineers and the first to hold the chair of engineering at Edinburgh, and won various medals and prizes. He gained a BSc degree in 1881, with firsts in engineering, surveying and natural philosophy and was awarded an honorary DSc by the university in 1908. For a year after his graduation, Cowan was assistant to professor Jenkin. With the aid of the Vans Dunlop travelling scholarship in engineering, worth £300, he went to America in 1882 and served under the New York city surveyor for some six months before taking up a series of appointments with railway companies. These included a period as assistant engineer on the New York, Western State & Buffalo Railway, followed by a period with the Canadian Pacific Railway during which he was responsible for the construction of some thirteen miles of track. After his return to England in 1884, Cowan continued to work on railways as well as on harbours and docks, under A.C. Boothby and John Macrae.

In February 1886 Cowan was appointed county surveyor for the southern division of Co. Mayo following a ten-day examination undertaken by some fifteen candidates. In addition to the normal duties of a surveyor, he undertook a number of water and sewerage schemes for the sanitary authorities in the county and was involved in some railway work. He strongly advocated the development of the fishing industry in his area and, in evidence to the Royal Commission on Irish Public Works (the Allport commission) in June 1887, proposed that the government should aid the purchase of boats and nets and assist the construction of a railway from Westport to Mulrany, with a branch to a nearby fishing pier (a line to Mulrany and onwards to Achill Sound was, in fact, completed in 1894). In 1886 he was called on to work with the Piers and Roads Commissioners who had been allocated a fund of £20,000 to finance relief works in some of the distressed coastal unions in counties Galway and Mayo. With his colleague in west Galway, **James Perry**, he assisted in the design and supervision of eighty-six separate projects mainly involving small harbours and piers such as those at Clare Island and at Keel on Achill Island, as well as some new roads and bridges; his care and devotion to the work deserved most ample acknowledgment, according to Colonel Fraser RE who had overall responsibility for completion of the scheme in 1888.

Cowan built a number of bridges in concrete and steel throughout his area and acted as consulting engineer to his colleague, **W.P. Orchard**, on the design of the Moy Bridge in Ballina which cost £2,700. His major project in Mayo was the construction of the Achill swing bridge, designed to pivot on one central pier. Work on this began early in 1886 in accordance with plans which had been prepared by Cowan's predecessor, **Edward Glover**, but the new surveyor made some important design alterations, including the substitution of steel for iron in the superstructure. The bridge was completed in 1887 at a cost of £6,000 and was opened by Michael Davitt, founder of the Land League, and named after him. It was replaced in 1948.

In his first report to the Mayo grand jury in July 1886, Cowan declared that the roads in his area were only in fair condition and that far more regular attention would need to be paid to them by the contractors whose work, in many cases, he refused to certify for payment. 'Good roads cannot be made by spasmodic efforts before the sessions and the assizes' he declared, and the methods employed were wasteful and ineffective: loose stones were scattered in the ruts which developed on the roads, instead of using the pick-axe and rake to repair the surface, and there was insufficient attention to drainage and grass margins. At the spring assizes in 1887, with a full year's experience behind him, Cowan was able to present a more complete assessment of the condition of the road network; many of the roads were badly located and imperfectly formed, he declared, and had only 'a very tender skin of broken stone or gravel'; and a system of 'lingering starvation' was in operation, even on the main roads, with maintenance expenditure so low that 'the capital of the roads is being spent'. He spoke out on the matter again at the summer assizes that year, complaining that the amounts allowed for road maintenance were too low, and that the local farmer-contractors lacked technical ability. But all of this seemed to be too much for the foreman of the grand jury, Lord John T. Browne, who made it clear to Cowan that, while they were prepared to give him credit for taking great pains to better the condition of the roads, they were quite satisfied that local men could carry out maintenance contracts on short lengths of roads cheaply and well; besides, if they were to alter the traditional system as Cowan suggested, they could find themselves in difficulty when he left the county, as he inevitably would, since 'all good officers prove to be birds of passage'. This forecast came true in Cowan's case when he was transferred to the better paid post in Co. Down at his own request in June 1889 after little more than three years in Mayo; nevertheless, the grand jury received the news graciously and paid tribute to his zeal, activity and impartiality.

Cowan served initially as surveyor for the southern division of Co. Down (although he went to live at College Gardens, Belfast) but took on responsibility for the entire county in March 1890 when **Henry Smyth** retired. Apart from the extensive alterations and additions at the Downpatrick Asylum for which he was responsible in 1896 and a cottage hospital at Dromore, Cowan's term of office in Down was notable mainly because of the work which, with his eleven assistants, he carried out on the county's 2,800 miles of public roads. He claimed in 1891 that the system of management of county works in Ireland was twenty years ahead of the systems in England and Scotland, with all of the works, and every road, under the control and supervision of a professional officer, appointed after a competitive examination. While he was willing to concede the need for improvements in grand jury law, he himself had succeeded in operating it in a very flexible manner in Co. Down, overcoming problems of which many other surveyors complained, including the requirement to use the contract system instead of direct labour. He also managed to introduce equipment like scrapers, sweepers, and stone crushers, even though there was no express provision in law for their acquisition.

Rolling of roads had been advocated by different engineers at different time from the seventeenth century onwards but was not practised on a significant scale in Ireland or England until late in the nineteenth century. Steamrolling was introduced on London streets and in Liverpool in the 1860s and was initiated in Ireland by W.J. Robinson, the Derry city surveyor, who acquired an Aveling & Porter machine in 1874 soon after he took up office in the city. Cowan was a strong advocate of the use of steamrolling, in the interests of better road surfaces and for economy, and he secured the support of his grand jury for its introduction on the roads in his direct charge, ahead of any other county surveyor. He hired, and later purchased, an Aveling & Porter roller in 1890 and between then and 1898 spent £5,500 on steamrolling in his county, concentrating mainly on the roads nearest the city of Belfast which had a very tightly drawn boundary until 1896. In the process, as Cowan himself admitted some years later, he had driven 'not a carriage but a steamroller' through the law on grand jury road operations: 'I considered it right to go on the assumption that as no question could be raised about the use of a shovel for road repair as a necessary tool, no proper objection could be raised to the use of a more modern tool – a steamroller'.

On 23 January 1899, three months before the Local Government (Ireland) Act 1898 was to come into operation, Cowan became chief engineering inspector under the Local Government Board, based at the Custom House, Dublin, succeeding **Charles Philip Cotton** who had held the post since 1879. His term of office was a busy one, coinciding with the expansion of the role of the board following the reorganization of local government by the 1898 Act, the rapid development of road traffic, the introduction of direct labour and new road-making techniques, the expansion of public housing programmes, and greater activity in relation to water and sewerage schemes. Cowan held numerous public inquiries throughout the length and breadth of Ireland in relation to proposals by local authorities under all of these headings, as well as inquiries into hospital development proposals, the compulsory purchase of land for various purposes, and personnel and disciplinary matters, particularly those involving engineers. He was a man who 'stood for thoroughness and efficiency' and who kept himself abreast of the latest developments in engineering and science; many of the schemes constructed throughout the country during his term of office owed much to his incisive criticism, helpful suggestions, and active intervention in the public interest.

Cowan wrote and presented numerous papers on the state-of-the art of road-making in the first decade of the century at conferences in Ireland and abroad, and was instrumental in promoting better standards and practice throughout Ireland. As the only Irish member of the Road Board's six-man advisory engineering committee, he played a major part in the allocation of the state funds released for road improvement works under the Development and Road Improvement Funds Act 1909. His brief as chief engineering inspector also drew him into fields of activity in which he would have had little involvement in his years as county surveyor. Nevertheless, in areas such as housing, public health and sanitation, his public statements and the papers he presented to a wide range of interest groups, showed him to be a man who had an excellent grasp of the history and development of the different subjects, and of the most up-to-date thinking on both policy and operational issues.

Motivated, apparently, by a strong social conscience, Cowan adopted at times a campaigning role quite unlike that which might have been expected of a senior civil servant in the early years of the twentieth century. In a major address which he delivered in 1915, he spoke out about the conditions in the slums of Irish cities which, he said, ought not be tolerated in any so-called civilized society. He expressed sympathy and admiration for the patience and long-suffering of the working classes and spoke of the need for 'business methods and the division of profits to be adjusted' so as to secure for every worker a wage which would enable him and his family to live in moderate comfort and in a house of decent standard. He rejected absolutely the suggestion that the condition of the poor was due to causes within their own control and refuted the 'vain delusion that the interests of capital and labour are opposed'. This particular address ended with a call for a new crusade against the evils of urban living – 'to secure in fairer proportion for our people a reasonable measure of good health, comfort, and leisure, is a task requiring all our energies, and a marshalling of all the forces of our civilization in a manner not yet realized'. In another significant address in 1916, Cowan focussed specifically on the housing problem and how it might be solved. The housing programme, to that date, had been carried out, he believed, 'in a half-hearted, desultory manner, unworthy of our legislators and our race' leading to the 'avoidable discomfort, degradation and disease which haunt the dwellings and darken the lives of the mass of our people'. In light of this, he called for 'a well thought out and clearly defined plan of campaign against bad housing, the blackest blot on our social system, and the greatest bar to domestic peace and national efficacy'.

Against this background, the chief secretary turned to Cowan in July 1917 when a major review of Dublin housing was decided on. His report (dated January 1918 but not published until August) dealt comprehensively and competently with the full range of administrative and technical matters directly involved, but is more notable for the fact that he used the opportunity to set out again his personal views on a range of wider issues affecting social and economic conditions in the capital. 'There is a great emergency now', he wrote, 'and it must be met by emergency measures'. Influenced strongly by the garden city ideals which were being discussed at the time, Cowan advised that an expenditure of almost £9 million should be undertaken to build at least 16,500 new houses on greenfield sites served by cheap and rapid transit, and to reconstruct some 3,800 of the better-class tenement houses to cater for another 13,000 families. Dublin was the despair of many people, according to Cowan, and while the corporation was as free from corruption and jobbery as any other similar body, responsibility lay lightly on most of its eighty members. It had strayed longer and further into 'the bog of direct labour' than other authorities and, influenced by 'relief committee considerations', was a very large employer of artisans and labourers who could not find other employment. In Cowan's view, there was no satisfactory way in which the existing local government machinery in Dublin could deal fittingly with 'the unprecedented and lamentable state of affairs in Dublin'. As a result, he strongly recommended that a new expanded housing programme, with higher design standards and lower densities, should be undertaken urgently and should become the responsibility of a separate new board representing the government and all the Dublin local authorities. He concluded, however, with some rather extraordinary statements – that the rebellion of 1916 might possibly have been prevented if the people of Dublin had been better housed and that a solution to the housing problem would also be a valuable measure for the defence of the realm; this, latter suggestion, presumably, reflected the opinion he had expressed in 1905 that the state contribution to public housing 'might fairly be increased if only on account of the importance of fostering a class which furnished large numbers of valuable recruits for the navy and army'.

The *Irish Times* of 14 August 1918 commented very favourably on Cowan's report, describing it as 'an admirable document both in form and in matter, written by a man of wide views and expert knowledge who has no axe to grind and no purpose to serve but the public good'. The editorial went on to say that the report:

> deserves the careful study not only of the citizens of Dublin, but of responsible Irishmen throughout Ireland; for the ills of Dublin, though they exist in its case in an aggravated form, are not peculiar to Dublin, and, moreover, bad conditions of life in the capital city of Ireland make their influence for evil felt all over the country. The housing question, like the poor, seems to be always with us. We have had Departmental Committees and Conferences upon it from time to time, but nothing has been done. Nothing has been done despite warnings of the gravest kind that we neglect with increasing peril to the very stability of social life in the capital city of Ireland. Nothing can now be done to remove this cancer in the body politic of the capital of Ireland until the end of the war. But then at last it can, and must be grappled with seriously.

After the war, the government did indeed take action by enacting the Housing (Ireland) Act in August 1919, in part a response to Cowan's report. To promote and facilitate implementation of the act, and to secure 'the utmost efficiency and expedition in dealing with the housing problem in its manifold aspects', the Local Government Board had already established a separate housing department under the control of a four-man committee chaired by Cowan, whose salary was increased by £300 to £1,200 in recognition of his new responsibilities. Planning activity and other work under the new act built up rapidly: by the end of March 1920, almost 5,000 acres of land had been inspected and approved for acquisition and nearly 42,000 houses had been authorized to go to construction stage. Cowan seems to have devoted himself fully to housing matters from January 1920 onwards when Alfred D. Price, one of the board's assistant engineering inspectors, took over his normal duties as chief engineering Inspector, but the new programme ground to a halt later that year when the majority of the country's local authorities broke off relations with the Local Government Board. Based in separate offices at 29–30 Lower Fitzwilliam Street, Dublin, Cowan and his colleagues were not directly affected by the Custom House fire on 25 May 1921 but later that year an efficiency review conducted by the Treasury led to reduced staffing levels in the housing department in consequence of the fall-off in activity, and cuts in the additional allowances paid to Cowan and the other members of the committee.

Described as a man of brilliant intellect, shrewd, practical, clear-headed, with an infinite capacity for work and with a fund of that particular type of common sense typical of the Scotsman, Cowan had become one of the best-known public officials in the country by 1922 because of his lectures and public statements on a variety of issues and his many visits to every corner of the land on inspections and in connection with local inquiries. He was one of the very small number of senior staff of the Local Government Board who opted to transfer to the service of the provisional government on 1 April 1922; his assistant on the engineering side, Alfred D. Price, retired the previous February citing the changed circumstances, while Louis E.H. Deane, the board's senior architect, retired about the same time. Although the separate housing department and committee did not survive, Cowan continued to be engaged in the new Ministry of Local Government on housing work which, according to the Ministry, was by common consent the most urgent of the social problems facing it and 'brooked no delay'. A fund of one million pounds was set aside by the new government in March 1922 to finance an expansion of the local authority housing programme and, in the following year, the Department's inspectors set about assisting the authorities in the selection of suitable sites and the preparation of layout plans and designs. All of this, according to the secretary of the department, involved 'a considerable addition to the duties of Dr Cowan' but his association with housing work – and with Ireland – was cut short abruptly early in 1923 when the Minister for Local Government, Ernest Blythe, sent a curt hand-written note to the secretary of the department directing that Cowan should be informed that 'in consequence of the need for reorganizing the department, we must let him go'. The appointment was duly terminated with effect from 28 February 1923 leaving Cowan with no option but to apply for a pension under article 10 of the Anglo-Irish treaty. Following representations on his behalf by the chairman of the compensation committee, Judge Wylie, to the president of the Executive Council, W.T. Cosgrave, Blythe withdrew his opposition to Cowan's claim for favourable pension terms and he was finally awarded a pension of £720 in respect of his thirty-seven years service in Ireland.

Cowan's work in the housing field was the subject of many tributes, such as the following which appeared in the *Irish Builder and Engineer* in April 1923:

> To his enlightened appreciation of the needs of housing in Ireland is due much of the present advanced housing policy of Ireland in regard to the character of planning, density of acreage, etc. He was keenly alive to the pressing needs of the time, sympathetic in regard to the appalling conditions of so many of the cities and towns of Ireland, and very anxious to ameliorate them. To the movement to secure that the new houses for the working classes should be planned on modern lines, and designed and supervised by competent and qualified men, he gave every support.

The decision to discharge Cowan from the civil service in face of tributes like this, and at a time when there was a need for a rapid build-up of the housing programme, is not easily accounted for by reference to the grounds stated by Blythe ('the need for reorganizing the Department'). The decision is even more difficult to understand in the light of Cowan's pre-eminence and experience as a roads engineer, coupled with the fact that the Ministry had taken over responsibility for roads from the Ministry of Economic Affairs in October 1922 and had only one roads engineer, J.P.J. Butler, who was already contemplating retirement under article 10 of the Anglo-Irish treaty. By July 1923, only a few months after Cowan's departure, and with Butler's retirement set for 30 September, a case had to be made to the Department of Finance for sanction as a matter of urgency to the creation of a new senior post which would be filled by 'an engineer of proved administrative ability and good professional standing'; **James Quigley**, the Meath county surveyor, was subsequently appointed to the new post.

Cowan was elected president of the Institution of Civil Engineers of Ireland in May 1911 for a two-year term and was president of the Engineering and Scientific Association of Ireland from 1915 to 1918. He was elected an honorary member of the Incorporated Association of Municipal and County Engineers, an honorary member of the Institution of Water Engineers, a fellow of the Royal Sanitary Institute and an honorary fellow of the Royal Institute of the Architects of Ireland. He was a member of a variety of other scientific and professional organizations, to all of which he contributed a flow of original and well-researched lectures and papers. He was also an extern examiner at Trinity College, Dublin and at the Royal College of Science.

Cowan lived from 1899 onwards at 33 Ailesbury Road, Dublin, and later at Castlemount, Castleknock, Co. Dublin. In 1888, during his service in Mayo, he married Marion Johnston, the daughter of a doctor who had an extensive medical practice in Westport and in whose memory he erected a memorial drinking fountain at the local fair green in 1911. He lost two of his three sons in the First World War, a blow from which he was said to have never fully recovered: Captain Sidney Edward Cowan of the 29th Squadron, Royal Flying Corps, was killed in air combat over France on 17 November 1916, having been commissioned on his eighteenth birthday a year earlier and having won the Military Cross with two bars before his death; and a second son, Captain Philip Chalmers Cowan, also attached to the Royal Flying Corps, was killed in action in France on 8 November 1917 at the age of 22; both names are inscribed on the large memorial cross in the grounds of St Mary's Church of Ireland at Anglesea Road, Dublin. Within a few weeks of his forced retirement in 1923, Cowan went to live in England. He served on the council of the Institution of Civil Engineers until 1924 and died at Fleet, Hampshire, on 10 August 1930, aged 71.

COX, John (fl. 1846–84), was employed by the Board of Works as an inspector of drainage in Co. Limerick from 1846 to 1859. He took second place at an examination for county surveyorships in 1877 and was appointed to the Limerick western division post on 20 December 1877. He was allowed to have three assistant surveyors, each of whom was paid £80 a year, but the grand jury was not happy that competent and trustworthy men could be engaged at that salary level. They resolved that up to £150 a year should be paid to competent and well-educated assistants and that such men should, as far as possible, be considered for appointment to county surveyor posts when vacancies arose. These recommendations, however, could not be implemented under the law as it then stood.

Cox designed a new sewerage scheme for Rathkeale which was completed in 1882 at a cost of about £2,300; an egg-shaped concrete main interceptor sewer was constructed, 3,385 feet long, and the scheme included 500 feet of other sewers, elaborate settling tanks and filters, and an arrangement by which river water could be diverted into the system for flushing purposes. The administration of the roads works in his area was in a 'demoralized' state, according to a committee of the grand jury, when Cox was forced to resign due to failing health on 5 March 1884.

CRAWFORD, Andrew H. (c.1811–91), of Mount Pottinger, Belfast, was the third son of Arthur Crawford, a merchant, and his wife, Catherine Campbell Lundy. He was appointed county surveyor for Tipperary north riding on 16 September 1845 and, just two months later, was consulted by the Board of Works about the state of the potato crop in his county. He reported that his enquiries and observations led him to think that 'the greater part of the reports are much exaggerated' and that there was no cause for 'any well-founded alarm of a scarcity of provisions'. But while the extent of the crop damage varied from one area to another, it was clear by the end of the year that Crawford's assessment was an over-optimistic one: by August 1846 he was having to share with Board of Works engineers the responsibility for supervising a workforce of almost 19,000 men on public works schemes in his area, mainly road works. Like his colleagues in other counties, he found that labourers were unwilling to accept task work from which they could earn up to 1s. 6d. per day, but were 'perfectly satisfied to draw 10d. per day subsistence and relax their efforts'. His decision to reduce the daily rate to 8d. where labourers were not putting in sufficient effort led to strikes, protest marches and demonstrations. He had difficulties, too, with the local relief committee who demanded that he should find work for all of those to whom they had indiscriminately issued employment tickets. Overall, he reported that there had not been much violence on the works 'despite the easily excitable disposition which characterizes the inhabitants of these baronies'.

When the relief works were closed down in March 1847 by direction of the Treasury, many projects in north Tipperary were left unfinished. However, Crawford was able to report that nearly 56 miles of new roads had been completed, and almost 158 miles repaired in his area, and he was able to arrange that some of the unfinished roads were dealt with at a later stage through the normal presentment system. While admitting that some of the works would not have been carried out in normal circumstances, he told his grand jury that several of the new roads would be of the greatest importance and would much enhance the property surrounding them.

Crawford's request at the spring assizes in 1863 to have his salary of £300 increased to the maximum of £500 allowed by the act of 1861 led to a major debate. One grand juror felt strongly that such a salary would be grossly excessive amounting, according to his calculations, to half a crown an hour for every twelve-hour day in the year, including Sundays! Others felt that the surveyor should either accept less than the maximum or forgo his private practice. However, the majority view was that Crawford should be allowed the full increase, and should maintain his right to private practice, as the public had confidence in his ability and integrity, especially as an expert witness in court cases.

The unsatisfactory work of the road contractors was a constant cause for complaint by Crawford and, by the early 1860s, he was forced to maintain on a direct labour basis a significant mileage of roads for which suitable contractors had not come forward. He seems to have had little involvement in the provision of public buildings in his area, perhaps because he received very little support from his grand jury for any proposals which he brought forward in this regard. When he sought approval at the spring assizes in 1852 for an expenditure of up to £800 on a new bridewell at Newport, the response of the jury was to establish a committee to consider the matter. Nearly ten years later, when plans for a bridewell were eventually furnished 'by the authorities in Dublin', they were very deficient in Crawford's view; he complained that the plans had not been submitted to him before being produced 'for the gratification of the grand jury' and he disclaimed all responsibility for the outturn in cost terms or otherwise. The two-storey bridewell, built of sandstone in random courses, was completed in 1865, thirteen years after the previous building had been condemned by Crawford because it was no better than a dungeon; it is now one of the few remaining buildings of its kind and has been restored as a heritage centre.

Crawford had been in office for almost forty-six years when he died on 1 August 1891 at the age of 80. He is buried at Ballinaclough, near Golden, Co. Tipperary, with his wife Anne, daughter of Revd Francis Synge of Terryglass. The couple had two daughters, Catherine (Kate) and Beatrice, and a son, Cooper.

DAVISON, Henry (1819–1901), of Glenarm, Co. Antrim, qualified at the county surveyor examinations held in 1841. He declined offers of appointment in a number of counties, including the west riding of Cork, in 1845–6 before his appointment as Armagh county surveyor on 21 February 1846. Within months, he had to undertake the supervision of famine relief works in the county in addition to his normal duties and in early 1847 he was obliged to report to the Board of Works that the men in his county were so badly fed that they were not fit for any hard work. By 1856, when matters had returned to a more normal state, Davison had responsibility for almost 1,000 miles of public roads but complained that he was unable to maintain them adequately because spending per mile had been reduced by the grand jury to about 50 per cent of the 1834 level.

When the courthouse at Armagh, completed in 1809 to a design by Francis Johnston, was found to be 'very inconvenient' for the business being transacted there, Davison prepared plans in association with **Thomas Turner** for alterations, improvements and additions costing £2,500; these were approved by the grand jury and completed in 1863. Working again with Turner under the title 'Turner & Davison, Architects', he was also responsible for the design of the courthouse at Lurgan, completed in 1874.

Davison retired from his position as county surveyor in October 1886. He was 82 years of age when he died at his residence at Melbourne Terrace, Armagh, on 2 February 1901, the day of Queen Victoria's funeral. An obituary in the *Ulster Gazette* noted his great ability, judgment and tact, and lauded him as a high-minded, worthy and benevolent gentleman. He had for many years been connected with St Mark's Church of Ireland in Armagh, designed by Francis Johnston in 1811, and was buried in its churchyard.

DAVISON, Robert (1870–1951), was born on 1 October 1870 in Belfast and was educated in the city at Methodist College and the Mechanical Institute. He served a three-year pupilage under J.C. Bretland, Belfast city surveyor, from 1893 to 1897 and subsequently worked for six years as a district surveyor for Belfast Corporation. Meanwhile, he studied at the Municipal Technical Institute and at Queen's College, Belfast, where he gained a BE degree in 1906. From 1903 to 1914, he was an assistant engineer in Belfast, engaged on sewerage schemes and bridge works.

Following the retirement in 1913 of the Limerick county surveyor, the county council decided to revert to the former practice of appointing a separate surveyor for each of the two divisions into which the county had been divided. Davison was appointed on 1 April 1914 to the eastern division where he introduced improved methods in road-making and in other public works. In 1923 he reconstructed the centre arch of Annacotty Bridge using the *Expanet* reinforced concrete system, one of a number of patented systems which were being promoted at the time.

Davison favoured the development of water-power for industrial purposes and was one of a small number of county surveyors who gave evidence on the subject in December 1919 to the Commission of Inquiry into the Resources and Industries of Ireland which had been established by Dáil Éireann. He presented tables giving details of the flow of water in the river Shannon at different locations in the years 1916–18 together with costed proposal for harnessing the river at Doonass to generate the equivalent of 32,000 horsepower of electricity. In addition, he presented information on surveys of other possible sites for hydro-electric development in Limerick and adjoining counties and ideas for industrial development of various kinds in the area.

Davison retired on pension in 1929 and returned to his native Belfast where he died in April 1951.

DEANE, William Henry (1816–87), was born on 19 December 1816 into a prominent Cork family of architects and builders. His father, David Deane, was an architect and a first cousin of Sir Thomas Deane (1792–1871), builder and architect of many of the city's public and commercial buildings. He is likely to have learned his engineering in the family firm in Cork and seems to have worked there in a variety of roles until the 1850s. In 1851, when he prepared elevations of properties in Cork city which were to be sold under the Incumbered Estates Acts, he signed himself 'CE, architect and surveyor'. He was subsequently employed by the Board of Works as a clerk of works.

Deane was appointed county surveyor for Tipperary south riding on 22 March 1852 after **Samuel Jones** had been dismissed. Matters seemed to be progressing satisfactorily for him eight years later. His plans for converting the old house of industry at Clonmel into a lunatic asylum were 'highly approved' of by the Board of Works in July 1860. In the previous April, the *Dublin Builder* had commended him for summoning seven road contractors and their sureties to appear before the petty sessions court to answer charges of neglecting to repair the roads – and the journal hoped that 'the example will have a salutary effect on others nearer home'. At the 1860 summer assizes, however, Deane was heavily criticized for having failed to comply with directions of the grand jury and, when the jury voted by twelve votes to ten to dismiss him, he was forced to appeal to Judge O'Brien to allow him to have counsel heard on his behalf. The judge ruled that, while it was within the exclusive competence of the jury to dismiss their surveyor, it would be but common justice to allow Deane's counsel to present his case, bearing in mind the serious effects of a dismissal on the prospects for life of a professional man. The foreman indicated that they would act on the judge's ruling but the arguments subsequently made by Deane's counsel were not sufficient to avert his dismissal in July 1860.

Having consulted Sir Richard Griffith, then chairman of the Board of Works, the chief secretary's office took the view that Deane's offences should not debar him from continuing to serve as a county surveyor and he was appointed to the eastern division of Tyrone, with headquarters in Dungannon, in August 1860. Eight years later, he was again in difficulty when one of his assistant surveyors was obliged to take action against him in the courts for the repayment of £20 which Deane had borrowed from him at a time when he was financially embarrassed. The grand jury took the view that Deane could no longer effectively perform his supervisory and other duties following this incident and requested that he should be transferred to another county. Some months later, Deane arranged to exchange places with W.A. **Treacy** who wished to leave his post in the southern division of Mayo, and the authorities in Dublin agreed to implement this transfer in July 1868 notwithstanding the Mayo grand jury's protests about the 'inconvenience' caused by too frequent changes of surveyors. Deane moved again to Fermanagh in August 1876 and after only six weeks he was again transferred in October, this time to Kerry.

Deane was responsible for the erection of Killorglin Bridge over the river Laune, with eight thirty-five-foot masonry spans, and completed at a cost of some £10,000 in 1885. He had great difficulty, however, in maintaining the county roads because, he complained, most of the contractors whose tenders had been accepted were paupers who had neither the money nor the skill to carry out the work efficiently. He died at his residence, Castlemorris, Tralee, on 22 April 1887, aged 70. He had been ailing for some time but had apparently improved and was back in his office in Tralee courthouse on the day of his death. Two of his sons, John and David, followed him into the engineering profession and were both employed in British Columbia around the turn of the century.

DELAHUNTY, Thomas Joseph (1881–1938), was born in Dublin on 7 May 1881, the son of Michael Delahunty, a grocer, and was educated at Clongowes Wood College and at Trinity College, Dublin, from which he gained BA and BAI degrees in 1903. He was assistant to a consulting engineer, J.H. Middleton, on water supply and sewerage schemes, roads and general municipal engineering between 1903 and 1906 and worked also for Pembroke urban district council.

In September 1906 Delahunty became town surveyor in Tralee at a salary of £200 a year and served there until 1909. He then moved to Drogheda as borough surveyor and, having qualified at the examinations held in 1917, was an applicant for county surveyor posts in

Longford and Galway before returning to Kerry as surveyor for the southern division when the county council decided in 1921 to appoint two surveyors for the county. Delahunty was a conscientious public official and, at the same time, popular with his engineering staff and the workers under his control. In his spare time, he was a keen philatelist who possessed a fine collection of rare stamps. Having been ill for some time, he died at the Mater Hospital, Dublin, on 3 May 1938 and was buried at Glasnevin cemetery.

DELANEY, James (fl. 1888–1909), graduated with BA and BE degrees from the Royal University of Ireland in 1888 and gained the Royal College of Science diploma in the same year. He was subsequently a pupil of G.N. Kelly, the Midland Great Western Railway's chief engineer, and was involved in the planning of the Galway–Clifden Railway. He worked for over a year as an assistant county surveyor in Derry under **A.C. Adair**, before taking up a similar appointment in Warrington, Lancashire. Between 1894 and 1896 he was resident engineer on the construction of a waterworks scheme at Carlow and went on to become an assistant to the Kildare county surveyor, **Edward Glover**.

When the post of county surveyor in King's County was advertised in April 1900 at a salary of £400 a year (£100 a year less than was paid to the previous office-holder) a letter to the *Irish Builder* under the nom de plume *AMICE* suggested that it was obvious that 'he who has most friends at court will be the successful candidate'. Since the death of the previous surveyor, **R.B. Sanders**, in February there had been speculation in the newspapers about the filling of the post, with the *King's County Chronicle* going so far as to suggest that R.P. Gill, an assistant surveyor in north Tipperary, was a likely successor, given that he was 'one of a family every member of which has been endowed by Providence considerably beyond the gifts found among ordinary people'. In the event, there were only three candidates for the post and, with or without influential friends, Delaney was appointed in October 1900, having gained first place in the qualifying examinations held in the previous July.

There seems to have been considerable friction between Delaney and the county secretary during his term of office, often involving relatively minor matters. At the quarterly meeting of the county council in May 1909, for example, he sought the support of the chairman in a dispute which had erupted in relation to his right to have full access to certain forms and documents relating to the business of the council. Little more than a month later, while Delaney was on holiday in Harrogate, he died suddenly at the end of June 1909. At a special council meeting in August sympathy was expressed to the late surveyor's friends. He was described by members as an upright honourable man of remarkable ability who had given the utmost satisfaction to all. The *King's County Chronicle* noted that, while Delaney was 'recognized as somewhat exacting', the road contractors 'had the amplest confidence in him as a man actuated by a sense of his duty and of justice towards the council and the public'.

DEVERELL, Frederick George (c.1821–90), was the son of William Deverell, an army officer, and a native of King's County. He entered Trinity College, Dublin, in July 1838 at the age of 17 and graduated with a BA degree in 1841. He registered for the two-year diploma in civil engineering course which began at Trinity later that year but left without sitting the examination. By 1850, he was employed as Board of Works district engineer at Glencar, Co. Sligo, and was responsible for extensive drainage works and a considerable amount of bridge reconstruction and replacement in his three drainage districts. In 1851 alone, some 130,000 man-days of employment were provided and, according to Deverell 'the works kept many a family from the poorhouse'. In 1854, he was resident engineer on the Roscrea & Parsonstown Railway.

When Deverell was appointed county surveyor for the southern division of Mayo on 31 March 1856 he was the first university graduate (although not an engineering graduate) to join the ranks of the surveyors. After the grand jury of Co. Cork decided that the size of the east riding, and the nature of the duties to be performed there, warranted the establishment of two divisions with a county surveyor for each, Deverell was appointed in November 1861 to the new southern division, exchanging places with **W.A. Treacy**. The *Mayo Constitution*, in reporting Deverell's transfer, acknowledged that he had performed his duties in a firm, upright and generally satisfactory manner but regretted that he had chosen to live 'in a secluded part of the county' instead of in Castlebar; the public convenience, it said, demanded the presence of the county surveyor where the public would have continuous recourse to him.

In his first report to the Cork grand jury at the 1862 spring assizes, Deverell described the roads in his division as 'very much neglected' and requiring a major effort on his part if they were to be brought up to standard. The grand jury, however, set his salary at a level of only £350, increased to £400 in 1864, as against the maximum of £600 allowed by law for the post. Deverell objected strongly to this, declaring that he was being forced 'to do first class duty for third class pay'. He supplemented his salary, however, by working for private clients and for other public bodies in his area. In 1865 he was commissioned to advise on the provision of a sewerage system for Passage West where insanitary living conditions and open sewers were causing sickness and disease. His report noted that domestic toilet facilities in the town were primitive: out of a total of 388 houses, 76 had neither a sewer nor a backyard and 104 had yards but no sewer and had to empty refuse and waste of all kinds onto the streets. Improvements costing £794 were suggested by Deverell but were not carried out due to opposition by property owners. In the following year he designed a small water supply scheme for the town and its dockyard; this was constructed between 1868 and 1873 but never gave full satisfaction.

Because of his dissatisfaction with his rate of pay and criticism by grand jury members of the number of times his name appeared in the newspapers 'connected with private contracts', Deverell obtained a transfer in November 1871 to the less demanding post in Cavan. He was in office there until 1889 and died on 8 December 1890, aged 69. He is buried at Deansgrange cemetery in Dublin with his wife, Letitia Stapleton, whom he married in 1847; she died on 8 June 1911.

DICKINSON, James Austen (1826–1911), and his twin brother, John Abraham, born in Dublin on 29 June 1826, were sons of Dr Charles Dickinson MA DD (a native of Cork who was Bishop of Meath from 1840 until his death in 1842) and his wife, Elizabeth Russell of Limerick. In July 1844 the twin brothers entered Trinity College, Dublin, and were awarded BA degrees in 1849, but their paths diverged at that stage. John Abraham opted for a career in holy orders, as did his brothers Charles and Hercules Henry, while James Austen, who had been awarded the DCE at the newly-established engineering school in Trinity in 1848, began a career in engineering. He served his time to G.W. Hemans and Sir John MacNeill, two of the country's most prominent railway engineers and on completion of his training was employed as resident engineer on railway construction works, including the Dundalk & Enniskillen Railway.

Although he had qualified at the county surveyor examinations in December 1851, Dickinson had to wait for more than nine years for an appointment. He was assigned to the Westmeath post in January 1861 but was forced to leave the county nine years later when he believed that his life was in danger: he had received threatening letters and for several months had to be protected by the police. An address was presented to him by the Westmeath grand jury in November 1869 and he took up duty in the southern division of Tyrone in January 1870. When **A.C. Adair** left the northern division of the county in May 1874, the magistrates and cesspayers at a number of presentment sessions called on the lord lieutenant to appoint one surveyor to serve the entire county, as had been the case until 1847. However, after a short delay during which Dickinson took charge of the whole county, a new surveyor, **T.B. French**, was appointed to the northern division.

When Dickinson became unable to carry out his duties due to ill-health from the autumn of 1894 onwards a deputy was appointed in his place and he retired in July 1895 with a pension of £275 a year. Before leaving Dungannon in November of that year to live in retirement at 5 Belgrave Square North, Monkstown, Co. Dublin, he was the recipient of a presentation from friends and admirers of what the *Tyrone Constitution* described as 'a massive silver inkstand of exquisite Irish design', a purse of sovereigns and an address attesting to his good work as a public official and one who had taken a leading part in every good work in the county. He was especially noted for his 'earnest labours' in the cause of temperance, especially among the young, for his services at Sunday school, and for his assistance with technical instruction classes. Dickinson died in Dublin on 19 July 1911, aged 85, and is buried at Mount Jerome cemetery with his wife, Letitia (daughter of Revd William Lodge of Killybegs) who had died in 1909, his four sisters and his twin brother who died in 1907.

DIXON, Edward Keville (1860–1942), was born in Cork on 27 March 1860, son of Thomas Dixon of Blackrock, and was educated at Dr Knight's School in the city. He entered Queen's College, Cork, in 1876 at the age of 16 and had a distinguished academic career, gaining scholarships, an exhibition and various prizes. He was awarded a BE degree with second-class honours in 1879 and an ME in 1882. He served his articles in the firm of Lanyon, Lynn and Lanyon of Belfast before going to work in America as assistant engineer with the construction department of the Union Pacific Railway Company. From 1886 to 1893 he was on the civilian staff of the Royal Engineers.

Dixon gained first place at the county surveyor examinations held in July 1893 and was appointed to the southern division of Mayo in the following September; the post had been vacant for nearly six months – a delay which, according to the grand jury foreman, had left the contractors without payment and the county exposed to great injury. He took up duty on 18 October after he had worked out his notice with the war office and went to live with his family in what a local newspaper described as a splendid new house on The Green, Castlebar. He was to be in office in Mayo for over thirty years.

After direct labour operations had become legally possible in 1901, Dixon told the county council that he believed that a well-formulated scheme of direct labour for the upkeep of the public roads was preferable to the old contract system which, he said, was then dying out everywhere. However, the strong opposition of the councillors to both steamrolling and the introduction of direct labour prevented him from making any significant progress towards improving the roads of Mayo. After extensive discussion of the two subjects at the first Irish roads congress which Dixon attended in Dublin in April 1910, he visited Co. Down to view the condition of roads which had been rolled in the previous ten years and, later that year, made another effort to persuade his council to change their policy. In a special report, he acknowledged that it would be too expensive for the council to acquire and operate their own plant but proposed that about sixty miles of main roads should be steamrolled by a specialist contractor; the cost of £20,000 could be met by a 50 per cent grant and by borrowing. But even this was too much for the council: they were prepared to agree only to a more limited programme and then only if government grants met the full additional cost.

Dixon seems to have had some success as a bridge builder. In a paper on *Rural District Bridges of Small Span* which he contributed to the *Irish Builder and Engineer* in 1910, he described how he had set about replacing bridges affected by extraordinary rainfall in September 1908 when as much as eleven inches fell at Castlebar. Six old bridges had been carried away entirely and twelve others were badly shaken and undermined; all were replaced within four months at modest cost. In earlier years, Dixon had employed a variety of bridge types – a wooden trestle bridge of 50-foot total span, built around 1900 at a cost of £160; single 20-foot-span stone arches, costing about £250 each; and twin 16-foot-span stone arches. He also built bridges of iron girders and concrete, costing an average of £350 each; on these bridges, old rails from local railways, embedded in concrete, were used extensively for spans of up to 10 feet, and a double layer of old creosoted railway sleepers was regularly used to provide a platform in soft ground on which masonry piers and abutments could be constructed.

After the virtual suspension of road work in 1921–2, Dixon told the county council in December 1922 that, by slow degrees, conditions were returning to normal with nearly all roads of importance then being maintained in reasonable condition. But he went on to report that in large parts of the county it was still almost impossible to travel in any direction owing to the destruction of bridges and obstructions caused by the felling of trees and the digging of trenches – and matters were getting worse in this respect. Bridges wrecked before July 1921 and since rebuilt had again been destroyed and Dixon saw no point in trying to rebuild them while a civil war still raged. With more settled conditions in 1923, he set about rebuilding the seventy-one damaged bridges in the county – some of which he had built himself and many of them much older, including the attractive single-arch Delphi Bridge over the Bundorragha river, designed by William Bald in 1830. Dixon was able to tell the county council in June 1924, after an extensive tour of inspection, that with the exception of a few sections here and there, the roads were all 'wonderfully travellable' taking all adverse circumstances into account.

Dixon carried on a private practice in engineering and architecture almost from the beginning of his term in Mayo believing, as he put it, that taking private work prevented a surveyor from getting 'rusty' and took him over his district. In addition, he undertook a considerable amount of work for other public bodies, including the Congested Districts Board for whom he built some sections of the Louisburgh–Leenane road in 1894–6 through difficult terrain, and including the girder bridge at Aasleagh. He was town surveyor in Castlebar for many years and architect and engineer to the local board of guardians and rural district council. He extended the water supply system at Westport in 1896 and was responsible for the extension of the Castlebar water supply in 1916 at a cost of some £4,000. In 1902–3 he carried out major extensions and improvement works at the asylum in Castlebar (including the construction of a new chapel) and was architect also for housing schemes in Castlebar and elsewhere in the county. He undertook the extension and improvement of Ashford Castle for the earl of Iveagh in 1918 and prepared plans for a variety of smaller architectural projects including the refurbishment of the Catholic church at Glenisland, Co. Mayo, in 1927.

Dixon was regarded by the grand jury and the county council which replaced it as an excellent official and he managed to maintain good relations with all sections of the community in the difficult 1919–22 period, taking no part in any political activity. There were reports in the local press, however, in early 1922 of attacks on the homes of protestants and of the 'indignity' suffered by Dixon whose windows were smashed; the reports went on to say, however, that prompt action against the perpetrators had been taken by the Irish republican police. In May 1921, when the county council voted to suspend all road work due to 'existing conditions' in the county, they decided also to suspend all of their engineering staff without pay. Dixon protested at this and, when he refused to accept a compromise proposal that he should be paid at the rate of 50 per cent of his previous salary, the impecunious council decided in October 1921 to serve notice of dismissal on him. Dixon responded in November with a letter of resignation and a request to be placed on pension as from 1 January 1922 and, while this was agreed to by the council, they soon found it necessary not only to invite him to resume his former duties in south Mayo on a temporary basis but to act also as surveyor for north Mayo in place of **J.J. Noonan** whom they had dismissed. Dixon served as surveyor for the entire county until October 1924 when **T.P. Flanagan** was appointed to succeed him. He died in Dublin on 28 February 1942 and was buried at Mount Jerome cemetery.

DOBBIN, Peter Le Fanu Knowles (*c.*1860–1922), was born in Dublin, the third son of Revd William Peter Dobbin (1820–71), chaplain to Dr Steven's Hospital, and his wife Frances (Fanny) Le Fanu Knowles, daughter of Revd Peter Le Fanu and his wife Frances Knowles. The family was connected with Richard Brinsley Sheridan (1751–1816), playwright and politician; the novelist and newspaper proprietor, Joseph Thomas Sheridan Le Fanu (1814–73); and his younger brother, William R. Le Fanu (1816–94), consulting engineer and commissioner of public works from 1863 to 1891.

Peter Dobbin was educated in Rathmines, entered Trinity College, Dublin, in November 1877 and obtained a BA degree in 1882 and a BAI in 1884. He was subsequently employed by the Board of Works under Robert Manning, chief engineer, and was clerk of works on a project at Malin Head, Co. Donegal, in 1886–7. He became surveyor for the southern division of Mayo in October 1889 and served uneventfully there for over three years. At the spring assizes in March 1893 he told the grand jury that he had heard of two vacancies 'in the south' and had applied for a transfer to one of them. He was appointed in the following month to Co. Clare where the long-serving **John Hill** had retired. His salary was fixed at £400 as against the £600 which had been paid to Hill but, when he sought an increase some months after his appointment, he was told by the grand jury that, while they regarded themselves as the most liberal in Ireland, he would first have to prove himself and show that he deserved the higher rate; this was conceded a few years later.

The West Clare Railway had just been extended to Kilrush and Kilkee by the separate South Clare Railway Company when Dobbin took up duty in Clare and, as both companies had invoked their baronial guarantees, they were effectively under the management of the grand jury. This meant that the surveyor had general responsibility for the railways and was required to report at each assizes on their operation and on the condition of the track, buildings, locomotives and rolling stock. One result of this was that Dobbin found himself in contention in the courts with Percy French, with whom he must have been acquainted during their years together in the Trinity College engineering school, when the famous entertainer sought – and was awarded – damages in 1896 against the railway company because he and his troupe were late for a performance in Kilkee. The incident led French to compose what was to become one of his most popular ballads 'Are ye right there, Michael?'

Complaints about the deterioration in road conditions in Clare in the early 1900s led to mandamus proceedings in the courts designed to ensure that the county council properly performed its duty to maintain the roads. The council's defence – that they had entered into contracts for the repair of the roads – was rejected by the court of appeal which held that a council was obliged to see to it that the county surveyor took proper steps to make the contractors perform their duties. Subsequent allegations of mismanagement and inefficiency in the county surveyor's department led the council to request the Local Government Board to hold a sworn inquiry in February 1908 into Dobbin's performance. The inquiry lasted for over a week and heard complaints from the chairman of the council and some of his colleagues that the roads were getting worse and worse each year, despite higher spending on them. Dobbin, it was alleged, did not exercise proper supervision over the contractors, and where direct labour had been introduced, he was said to have engaged as overseers people who had been unsuccessful in various walks of life. The evidence, however, was not all one way as nine members of the council swore that the roads had improved in the previous few years. Dobbin himself made the same case and argued, perhaps unfairly, that the real deterioration had occurred in the early 1890s, during the last few years of his predecessor's long term of office; John Hill, he said, was already in his nineties before his retirement (adding ten years to Hill's age in the process) and it was impossible for him to do any work at that age. The verdict delivered by the board in June 1908 could best be described as a draw. They accepted that there had been some improvement in road conditions in the previous five years but considered that the standard was still far from satisfactory due in part to inadequate funding by the county council. They found that while the specific charges against Dobbin were not substantiated, his department could be better managed but the county council were also said to be at fault and should accept the need for 'cordial cooperation' with their officials; they should remember the difficulties the surveyor had to encounter, and should endeavour to support him in the discharge of his duty.

The 1908 inquiry failed to put an end to complaints about Dobbin's performance. In 1910, R.J. Mecredy, the influential editor of *The Irish Cyclist*, lamented that the roads of Clare, which had been among the best in Ireland in John Hill's time, had seriously deteriorated, so much so that the surfaces were 'as vile as the truly execrable surfaces to be found in north Kerry'. Conditions deteriorated still further after 1919 when, like most of his colleagues, Dobbin faced fresh difficulties in maintaining the road network. Initially, improvement work had to be postponed for reasons of economy and maintenance work was later cut to the minimum. By the end of February 1921 all road workers had been paid off leaving only Dobbin and his assistant surveyors still on the payroll and a small number of contractors at work on roads where direct labour had not been introduced. Later that year Dobbin approached the council chairman – a prominent IRA leader – with the extraordinary suggestion that, through a friend of his at the Treasury in London, he could get a grant for the relief of unemployment among ex-soldiers, many of whom were by then volunteers; the grant would be devoted to road work, but without interfering with the trenching and obstruction of roads in which the chairman himself was actively engaged! The chairman gave Dobbin permission to go ahead and it appears that a grant was later received.

In June 1922, Dobbin was asked by the county council to resign his post as they considered 'that the administrative work of his department requires the constant attendance of a younger and more energetic official'. He submitted his resignation at a special meeting of the council in the following August and was awarded a pension of £685 but continued as acting county surveyor until his successor, **Francis Dowling**, took up office in November 1923. At the time of the 1901 census, Dobbin was living in Kilkee with his wife, Clemence, and his family of three daughters and a son. With his second wife, Kathleen (née Vance), he was the father of Rachel Burrows (1912–87), actress and broadcaster and wife of Canon Jerram Burrows (d. 2003).

DORMAN, Richard Henry (1858–1949), was born in Cork on 18 May 1858, the fourth son of Revd Thomas Dorman, a member of a well-known Kinsale family, and his wife, Charlotte Isabella Hobart, daughter of a Cork medical practitioner. The family lived at Richmond House in Cork city while Thomas was serving as curate at St Ann's, Shandon. Dorman studied at the Royal Indian Engineering College, near London, and gained some early experience on the railways of west Cork which had engaged his eldest brother, J.W. Dorman, as consulting engineer, and later on the London & Northwest Railway at Preston. He was employed on irrigation schemes by the public works department at Madras between 1879 and 1882. In 1885–6 he worked in the Hornsey area of north London under T. de Courcey Meade (later to become Manchester city surveyor) and superintended the construction of the Highgate Hill cable tramway and the laying down in concrete of Highgate Hill, part of the old Great North Road out of London.

Dorman became county surveyor for the northern division of Mayo on 15 July 1886 but served there for less than six months. He expressed his regret to the grand jury for the inconvenience caused by his removal to another county so soon after his appointment but this did not prevent the foreman, Lord John T. Browne of Westport House, from publicly censuring the authorities in Dublin Castle for 'the monstrous injustice' involved in taking away a county surveyor as soon as he had any practical knowledge of the county. Dorman's appointment as Armagh county surveyor was announced in October 1886 and he took up duty there in January 1887, becoming responsible for the maintenance of over 1,600 miles of public roads, with an annual budget of about £20,500 in the 1890s, rising to more than £30,000 by 1907. As he was among the limited number of county surveyors who regularly wrote and presented papers about road maintenance and improvement in the 1890–1920 period, it is possible to trace the development of his thinking on the subject in the light of his experience and of rapidly changing circumstances. His initial approach, as he explained it at a meeting in Dublin in September 1891 of the Incorporated Association of Municipal and County Engineers, was to aim at keeping the roads in good order at the smallest possible cost and, in his view, no system could compare with the contract system for achieving this. At the same time, he was convinced of the value of steamrolling, and had estimated that four rollers would be needed in his county to deal adequately with the main roads alone, but he lamented the fact that the law did not allow the grand juries to acquire these machines. On a more personal note, he complained that the law also precluded recoupment to him of the travelling expenses involved in inspecting the contract roads even though these expenses absorbed £80 a year of his salary of £500.

By 1898 Dorman had developed radically different views on the relative merits of the different systems of road maintenance, taking account of the rapidly developing tourist traffic, the 'myriad of cyclists … inundating the country', the more regular appearance of the traction engine – 'that hydra-headed monster' as he described it – on the roads, and the likely development of motor car traffic. At that stage, he had divided his county's roads into four classes and was spending over four times as much per mile on the maintenance of the first-class roads as on the fourth class; in addition, he had introduced a limited programme of steamrolling on the more heavily trafficked routes in 1892, the third surveyor to do so, notwithstanding tremendous opposition from the ratepayers. He freely admitted that he had completely changed his opinion of the contract system in the light of his experience with 'rotten contractors' and their sureties, and the fact that, with control and ready money in his own hands, he had been able to carry out both maintenance and improvement works on main roads and on byroads for as much as 25 per cent below the contractors' prices. As a minimum, therefore, he believed that all main roads should be maintained directly by the surveyor's own staff, with materials being supplied on contract where prices and quality were reasonable. Some years later, in May 1907, Dorman was even more outspoken about the contract system which he then described as 'a corrupt and extravagant' system which gave contractors 'every opportunity of swindling the county and hoodwinking the county surveyor'.

Direct labour operations on the 500 miles of main and first-class roads in Armagh began in April 1907 and, within five years, Dorman was said to have worked wonders in producing good roads. The council had by then acquired eight steamrollers, as well as scarifiers and other equipment, and had opened a number of quarries to produce road materials; one of these operated on a very large scale, with broken stone being carried by a tramway system to a railway siding for onward transport to Armagh, at an all-in cost of 3s. 3d. per ton. While the use of binders in Ireland was only at the experimental stage, Dorman laid some road sections in tarmacadam and carried out trials of tar-grouting in other locations. In addition, he set up a plant at which up to forty tons per day of broken stone was heated and dried by a coal-fired furnace and mixed with tar and pitch in a plant actuated by two horses.

Dorman's approach to road improvement was notable for the considerable amount of attention he devoted to the camber and width of the roadway and the importance he attached to proper curbs and margins. He wrote in 1891 that, for the cross-section of a road, a 'flat elliptical curve for three-fourths of the width is preferable, and the *shoulders* well filled out'. It has been suggested that this use of the word 'shoulder', forty years before sealed, surfaced *hard shoulders* were introduced in the USA and Germany, entitles Dorman to be regarded as the pioneer in this aspect of road engineering, but this seems not to have been the case; while he certainly favoured the provision of footpaths alongside main roads, his treatment of verges or margins, as he described it in 1912, consisted of laying 'a sod curb … along the existing surface to define the proper width, and some soil packed behind to form a margin'.

When Dorman sought an increase of £200 in his salary (which had already been increased to £800 a year) to take account of the extra work thrown on him by the direct labour system, and another £200 for office expenses, the response from the councillors was not favourable. There was a particularly strong reaction from the Newry No. 2 rural district council which not only held that the existing remuneration was ample but went on to resolve unanimously that 'inasmuch as this council finds it impossible to conduct the business of the district with efficiency or economy, owing to the county surveyor's unwillingness or incompetence to discharge his duties, we call on the Local Government Board to find employment for him in some place more suited to his abilities'. Dorman appealed to the board on the salary question but they took the view that he had made no case and ruled in 1908–9 that he was not entitled to the increase.

Dorman was at the centre of controversy again in 1913 when he clashed with Armagh rural district council about road maintenance. In a report presented to a meeting of the council he complained that the councillors in their zeal had endeavoured to interfere with his management of the roads in the district and had placed such difficulties in his way that little if any progress had been made towards improving them. He claimed that the first-class roads, which were practically under his sole control, had been improved substantially in the previous ten years, and at less cost than before, while the rest of the roads, in relation to which the district council had interfered, were in poor condition with many bad and indifferent contractors still being tolerated. Important roads were being left derelict and the provision made annually for bridges was quite insufficient, with the result that these structures were rapidly deteriorating. Dorman was critical also of the refusal of the council to adopt proposals which he had brought forward for the improvement of the Armagh–Portadown road and claimed that a Road Board grant which had been on offer had gone elsewhere; if the local authority did not do its duty in the improvement of the main roads, he warned, the management of these roads would be taken away from them altogether. Councillors were unlikely to take such criticism from one of their officials without rejoinder. One councillor said that Dorman's report was a very insulting document coming from a paid official who was dictating to the representatives of the ratepayers. Another accused the surveyor of making all the trouble he could, and hindering progress in every way. After a long and heated discussion in which the report was labelled inaccurate and impertinent, rotten, dirty, and only a piece of mud-slinging, it was agreed unanimously that it should be struck from the minutes. Dorman had the consolation, however, of an editorial in the *Motor News* which suggested that a surveyor who bore an exceptional reputation for his ability should not be lectured to by 'ignorant laymen who have only the slightest smattering of road knowledge'.

While Dorman's reputation is based primarily on road development and bridge building, he was associated also with other engineering and construction projects during his long service in Armagh. In 1903, for example, he provided a substantial new building (The Hill Building) at Armagh lunatic asylum (now St Luke's Hospital), the first of the series of asylums built to designs by Francis Johnston and completed in 1824. With Henry Shillington CE, he was called in by Portadown urban district council to advise on a water supply scheme but their proposal to take a supply from the river Bann was rejected by the council who feared that the river was polluted by sewerage from Banbridge, ten miles upstream. Some years later, Dorman and J.H. Swiney, a Belfast consulting engineer who specialized in the design of water and sewerage schemes, devised a large new scheme to serve both towns, drawing a supply from the headwaters of a river in the Mourne mountains; this involved the construction of an embankment nearly 800 feet long and up to 48 feet high to create a storage reservoir, 1,000 feet above sea level, holding 75 million gallons, and service reservoirs to serve the two towns. Tenders for the first section, which included the laying of some twenty-four miles of 9-inch and 10-inch pipes, were invited in 1904 and the entire scheme was completed in 1909.

Dorman presented three papers to the first Irish Road Congress in Dublin in 1910 and also addressed the 1911 congress. He delivered a series of papers to the Institution of Municipal and County Engineers, mainly on roads matters, between 1891 and 1912, acted as the first Irish district secretary of the institution from 1891 onwards and later represented the Irish district on the council of the institution. He was also, for a time, president of the Association of County Surveyors. He gave notice in 1918 of his intention to retire from his county surveyor post but continued in office until 1921. Some years later, he moved from his Armagh residence at Tullymore Park and acquired Ballea Castle, originally a sixteenth-century building, near Carrigaline, Co. Cork, and lived there for the rest of his life.

When he died on 8 September 1949 at the age of 91, Dorman was the last survivor of the corps of surveyors who had served before 1899 under the grand juries. His wife, Beatrice Nora Jane Crooke, daughter of an army chaplain, whom he had married in 1889, predeceased him in 1942. The couple had one son and two daughters. Dorman's eldest brother, John William (1850–1911), was engineer to the Cork & Bandon Railway from 1873 to 1884 and contractor on the Clonakilty Extension Railway (1884–6), before becoming chief engineer and general manager of the British Guiana State Railway. Another of his brothers, Edward Hobart (1852–93), was secretary of the Cork, Bandon & South Coast Railway Company until his death.

DOWLING, Francis (1884–1944), was born on 19 August 1884, son of Richard Dowling, an assistant county surveyor of Ballynacally, Co. Clare. He was educated privately and at University College, Galway, where he was awarded a BE degree in 1911. He spent his entire engineering career in Co. Clare, working initially as an assistant county surveyor and as engineer to the rural district councils of Ennis and Ennistymon in connection with their labourers' cottage building programmes. He took up an additional appointment as Ennis town surveyor in 1916 and continued to hold this post when he became Clare county surveyor in November 1923. His brother, Richard, was one of his assistant surveyors.

In a report presented to his county council in 1925, Dowling delivered a scathing indictment of the direct labour system of road maintenance. He had been driven to the conclusion that the system was a failure because the attitude of the men was to do as little work as they possibly could and the overseers were not doing much to improve matters. There was slackness and general laxity among the labour force and, where workers were disciplined or replaced, there was intimidation, sometimes by armed men. Farmers were adding to the difficulties by charging exorbitant prices for access to land and by forming 'rings' to maintain high prices for quarry materials. In sum, direct labour had become 'a sheltered occupation', not exposed to competition; the approach was to demand the highest possible payment for the least possible amount of work.

Dowling's difficulties were added to in the late 1920s when heavy lorries used by contractors on the Shannon Scheme did serious damage to many miles of roads. In some areas, he reported, the passage of eight-ton trucks had virtually isolated villages from the rest of the world. Ten years later he continued to be frustrated by the reluctance of the county council to allocate the necessary funds to improve the roads. In his estimates for 1939–40 he sought a provision of £119,000 for road works, but this was cut to £70,000 by the council, the same level as in the previous year. Dowling told the councillors that this was farcical; it was a waste of time for him to prepare estimates if they were to be rejected by the council, and he proposed to report the situation to the Department of Local Government and Public Health. However, the outbreak of war later in 1939 put an end to Dowling's hopes of making major progress with his road improvement programme; instead, his energies and the resources of the county council had to be diverted into turf production to alleviate the wartime fuel shortage. Anxious that Clare, the constituency of the then Taoiseach Éamon de Valera, should make the maximum contribution to the national effort, Dowling was said to have worked night and day to make a success of the scheme.

After a long period of ill-health, Dowling died on 5 February 1944 at his home in Ennis and was buried, after an impressive funeral procession, at Drumcliff cemetery just outside the town. An obituary in the *Clare Champion* described him as a model public servant who was held in the highest repute for his professional ability and who enjoyed the regard and esteem of all ranks and classes in the community he served so loyally. He was said to have been an industrious and zealous official, with an amazing capacity for work and tireless energy, a man who aimed at the highest standards, and was always considerate and courteous to his men and ready to defend them when he felt they were unjustifiably censured.

DOYLE, Valentine Denis (1888–1958), was born on 14 February 1888, probably in Scotland where his father, J.J. Doyle, worked with the customs and excise service. Doyle senior (1855–1927) was a native of Kerry and was better known in later life as Séamus Ó Dúbhghaill. He was one of the pioneers of the Gaelic League and, under the nom de plume, *Beirt Fhear*, was author of numerous works in Irish, including *Beirt Fhear ón dTuaith, Caint* *na Cathrach, Tadhg Gabha, Prátaí Mhichíl Thaidhg, Cathair Chonroí agus sgéalta eile, Cléibhín Móna, Muintear na Tuatha,* and *Beartín Luachra.*

After an early education at St Columb's College and the technical school in Derry, where his father was then stationed, Doyle gained a first-class honours BE at Queen's College, Galway, in 1908 and an ME in 1912. In 1908–9, he was assistant to W.J. Robinson, Derry city surveyor, and later to **Christopher Mulvany**, Roscommon county surveyor. He was appointed in 1910 as town surveyor in Athlone where he carried out a substantial public housing programme and was a partner in an extensive private practice under the style Doyle & Loughlin, civil engineers.

When the Longford county surveyor gave notice in December 1913 of his intention to retire in the following month, the county council decided that the post should be advertised at a salary of only £250, with £50 for expenses. As they had done in other cases, the Local Government Board protested that this would not be sufficient to induce men of experience and ability to apply and suggested a minimum of £300. However, the council refused to alter their decision, one member pointing out that fully qualified men were already canvassing actively for the post at the advertised salary. Another remarked that a salary of £25 a month was a very good one and wondered 'how would a man spend that at all'? By February 1914 six applications were on hands but the only applicant who was fully qualified came from Cork and did not seem to be acceptable to the council. Another of the six was a native of Longford who was said to have been working as an engineer in Chicago for twenty-five years, and even though he failed to submit evidence of his qualifications, the council decided to appoint him. What happened after that is unclear. However, having sat the qualifying examination held in January 1915, Doyle was appointed to the post in the following June. According to the report of the council of the Institution of Engineers of Ireland for 1915–16, he was one of twenty-five members of the institution who had been 'accepted for service in his majesty's forces' but his name did not subsequently appear in the roll of honour listing members who had actually served in the army or navy.

When **Singleton Goodwin**, Kerry county surveyor, gave notice in 1919 of his intention to retire, the county council decided that the county should be divided into two divisions with a county surveyor for each, as had been the case between 1887 and 1891. At a meeting in May 1920 – the last before the June elections which swept Sinn Féin into power in twenty-nine of the thirty-three county councils – the two positions were filled after an acrimonious debate in which there were allegations of jobbery and references to demands by the Sinn Féin clubs that the business be left to the incoming council. Leo Casey, described as 'an old Tralee man and a member of a respectable family' was proposed to fill one of the vacancies as was **T.J. Delahunty**, a former Tralee town surveyor. The third nominee was Doyle whose proposer and seconder spoke in glowing terms of his father's services to the Irish language movement and argued that the council should pay him the compliment of electing his son to office in his native county. All three candidates had been examined in Irish which the council had declared to be essential for the post; Casey was awarded 34%, Delahunty got 35%, and Doyle 100%. However, when the votes were counted, it was Casey and Delahunty who were declared elected. But this was not the end of the matter: in July 1920, the new council under the chairmanship of Pádraig Ó Siochradha (An Seabhach) decided that the posts should be advertised again, and in the following September, Doyle was appointed county surveyor for north Kerry at a salary of £400 a year. Because of what they described as 'the economic orders of Dáil Éireann', some members of the council felt at that stage that only one surveyor should be appointed and Doyle was therefore assigned to act temporarily for the southern division also while the issue was being reviewed. In the event, the members generally agreed to adhere to the earlier decision to have two surveyors and they appointed T.J. Delahunty to the south Kerry post in 1921.

At the time of Doyle's appointment, road conditions in Kerry, as in many other parts of the country, had seriously deteriorated due to a variety of factors – additional motorized traffic, including military traffic; reduced funding from government grants; the low levels of rate collection; and damage caused directly to roads and bridges as part of the war of independence. Kerry ratepayers argued for the restoration of the contract system of road maintenance but Doyle claimed that the real issue was not the principle of direct labour but the poor quality of the supervisors and the inadequate organization of the work. There were far too many idle gangers, in his view, some illiterate, some not physically fit, and some with conflicting interests. But matters got worse before they got better: by the end of 1921, less than ten per cent of the county's roadworkers were still at work and the unemployed majority was threatening to close down the limited works that were still in progress. When more settled conditions developed Doyle and his colleague began a drive to improve the standards of Kerry's roads. By the early 1930s, a major works programme was underway and a large stock of steam-powered plant and machinery was employed, including thirteen traction engines and stone crushers, twelve steamrollers and a steam lorry.

Doyle became responsible for the whole county after Delahunty, the south Kerry surveyor, died in May 1938. He resigned in October 1943 on grounds of ill-health, but was subsequently able to carry on private practice in Tralee. He died in April 1958, aged 70. With his wife, Elizabeth Connell of Athlone, he had one daughter and four sons, one of whom, Dermot Joseph Doyle BE, was Westmeath county engineer from 1955 to 1977.

DRAPER, Carter (1845–1902), was born at Hereford on 28 February 1845, the son of Carter Ebenezer Draper. Two years later, he came to live in Dublin where his father set up in business as a druggist. He was educated privately before entering Trinity College, Dublin, in 1863. He was awarded the LCE in 1867, having qualified in engineering, practical mechanics and physics, and obtained a BA in spring 1868. He went on to become a pupil at the Dublin mechanical engineering firm of Courtney, Stephens and Bailey, and was subsequently employed on various projects by Messrs Bewley, Webb and Co. and by Messrs Ross, Stephens and Walpole. He then turned to railway work and was employed for three years as an assistant engineer by the Great Southern & Western Railway Company, engaged mainly on the extension of the line to the North Wall in Dublin which was completed in 1877.

Draper gained first place at an examination held by the Civil Service Commissioners for a county surveyor vacancy in March 1878. He was awarded a very high overall mark and won particularly good marks in the papers on marine, railway and canal engineering. He received his certificate from the commissioners in May 1878 and was appointed county surveyor for the western division of Galway shortly afterwards. He was transferred to Wicklow in July 1882 in place of **Henry Brett** who had died in office, and he served the county for over twenty years.

Draper was described as a keen lover of the profession of engineering who gave close attention to the smallest details of every work which he undertook. He was said to be unflagging in his attention to duty and was held, professionally as well as personally, in great esteem throughout his county. He rebuilt Boleyhorrigan Bridge, near Lough Tay in 1896 at a cost of £400 and designed some additional buildings at the registrar's house at Deansgrange cemetery in 1897. An unusual assignment fell to him in the 1890s when a committee was established to erect a memorial in Wicklow town to Captain Robert Halpin (1836–94), commander of Brunel's *Great Eastern* during its years of service as a cable-laying ship and, in later life, chairman of the local harbour commissioners; an obelisk in polished granite was designed by Draper, carved near Belfast and brought with difficulty by road to Wicklow where it was unveiled in October 1897.

In 1900, Draper prepared plans for a new water supply system for Greystones, then described as a rising watering-place south of Dublin with a summer population of more than 2,000. The scheme involved laying a five-inch main to bring a supply of Vartry water to the town at a cost of about £2,000; it was designed to give a supply of forty gallons a day for up to 3,000 people. From 1897 onwards, he had also been working on plans for a main drainage scheme for Greystones at an estimated cost of £4,000 – an interesting comparison with the expenditure of more than IR£20 million on the Greystones sewerage scheme constructed between 1993 and 1996. Draper's scheme was the subject of a local inquiry in December 1897; it involved the provision of two sea outfalls which, he claimed, on the basis of flotation tests, would ensure that the sewerage was carried out to sea. But environmentalists were at work even in those days: there was opposition to the discharge of raw sewerage to the sea, complaints that the scheme would pollute the beach and foreshore; and demands for the provision of a treatment works. The controversy delayed commencement of the scheme and Draper did not live to see his plans implemented; a scheme designed by consultants, J.H. Ryan and R.M. Butler, was eventually constructed in 1905–6 at a cost of about £10,000.

After the harbour at Greystones, originally constructed in 1846, had been redeveloped by the Board of Works as a direct labour scheme at a cost of more than £21,000, with an extended pier, 200 feet long, on the southern side and a new pier on the north, it was proposed in the late 1880s that the works should be handed over to the county council for maintenance. On Draper's advice, the grand jury (and subsequently, the county council) refused to have anything to do with the harbour although an order was made in 1896 transferring the works to them – a wise decision as things transpired for the foundations of the piers were defective and the harbour was said to be in ruins and absolutely useless by 1910.

The Dublin & Blessington Steam Tramway, constructed along the public road from Terenure to Blessington between 1887 and 1888, caused considerable problems for Draper. The company had planned to open the line for traffic in June 1888 but Draper insisted that drains and culverts which had been damaged should first be repaired, that additional ballast should be laid between the sleepers, and that a variety of other road restoration work should be undertaken. When these matters had been attended to, the line was allowed to open on 1 August 1888. In subsequent years, Draper was obliged to monitor closely the operation of the tramway and had to take action regularly to ensure that minor works were carried out along the route in the interests of the ordinary road users. The line was extended by 4½ miles to Poulaphouca in 1895.

Ireland was said to have lost one of her most capable and faithful officials when Carter Draper died at his home at Grosvenor Road, Rathmines, Dublin, on 7 December 1902, having been ailing for a considerable time. He is buried in the grounds of the Church of Ireland in Enniskerry with his wife, Sophy, who died in 1926. His son, Charles Frederick Draper (1876–1952) who graduated from Trinity College, Dublin, with a BAI degree in 1899, served as chief engineer, water and sewerage, in the Department of Public Works at Montreal and later as assistant engineer on the Canadian Pacific Railway. A younger son, Francis Edwin (1880–1935), who also qualified as an engineer, was employed by the Board of Works until his death.

DUBOURDIEU, Francis Rawdon Hastings (*c*.1784–1861), was a member of a notable Huguenot family both in south-west France and, after the revocation of the Edict of Nantes in 1685, in England and Ireland. One member of the family settled at Lisburn, Co. Down, in 1689 and served as chaplain to Frederick, duke of Schomberg, King William's commander-in-chief in Ireland who was killed at the Boyne in the following year. A later member of the family – Revd John Dubourdieu, Rector of Annahilt, near Hillsborough, Co. Down – compiled *A Statistical Survey of County Down* (1802) and *A Statistical Survey of County Antrim* (1812).

The four sons of Revd John Dubourdieu and his wife, Margaret Sampson, joined the army and two were killed in action. One of the survivors was Francis Rawdon Hastings who served as a captain in the Royal Hanoverian Engineers. He became a Knight Grand Cross of the Order of Guelphs of Hanover and in 1830 published, in translation from the German, *Instructions for the Choice, Fortifying, Occupying, Defence, and Attack of Military Fortifications.* On his return to Ireland Dubourdieu worked initially as an engineer for the Newry Navigation Company on the enlargement and extension of the ship canal which, when completed in 1742 was Ireland's first canal. Writing from Fathom Lodge, Newry, in 1831, he suggested to the lord lieutenant that a grant of £3,000 should be made available to enable the refurbishment of the canal to be undertaken under his direction. The works he had in mind were designed primarily to provide employment which would relieve the 'urgent distress' in the area and they included clearing out the locks, repairing the banks and scouring the blocked feeder streams and drains. Nothing seems to have come of this proposal.

Dubourdieu was employed as an engineer by the Board of Works on the improvement of the Kells-Bailieboro road in 1831–2 and became the first county surveyor for Sligo on 17 May 1834. He had been an applicant for the position of engineer to the Corporation for the Improvement of Belfast Harbour and protested that **Thomas Woodhouse**, the Antrim surveyor, who got the appointment could not possibly carry out both duties effectively. In addition, he told the chief secretary's office that he 'sincerely lamented' the fact that he had not been placed in a northern county and alleged that he had been promised that, if he could not be appointed to his native Co. Down, he would be assigned to Monaghan, Armagh or Antrim. He wrote on five different occasions to the chief secretary during the following two years, seeking a transfer and was eventually appointed to Co. Louth on 14 October 1836, taking on also the separate post of surveyor for the county of the town of Drogheda. However, he seems to have continued to be responsible until sometime in 1837 for work on the Sligo-Ballyshannon post road which was being carried out under the general direction of the Board of Works.

In Louth, the parsimonious grand jury fixed Dubourdieu's salary at £250 a year, refusing to allow him the legal maximum of £300 which his colleagues in most other counties received. At the summer assizes in 1838, however, they allowed him a special extra payment of £25 'in consequence of the additional expenses he has been put to in mapping the new lines of road from Dundalk to Ardee, from Dundalk to Louth, and from Ardee to Aclint'. Like his predecessor, **John Yeats**, Dubourdieu's differences with the grand jury related to more than salary matters. In February 1837, he sent two candidates for appointment as assistant surveyors to be examined by the Board of Works and received certificates of qualification for them, but the grand jury refused initially to make the appointments. He complained to the board about this and they took the matter up with the chairman of the grand jury, Sir Patrick Bellew MP, asking why the jury objected to 'an arrangement so very desirable in the interests of the county'. The grand jury relented but fixed the assistants' salaries at only £30 a year as against the legal maximum of £50. There was more bad feeling when the surveyor refused to send a particular man for interview by the Board of Works but this difficulty seems to have been overcome and two further assistants were appointed in 1840.

Dubourdieu was forced to resign his post in August 1840 because of continuing differences with the grand jury and was succeeded in September of that year by **John Neville**. He continued for some time to press the chief secretary's office to carry out an investigation into the operations of his former employers and sought in 1842 and again in 1852 to be assigned to another county surveyor post; his applications were not successful, even though others who had been forced to resign were later reappointed on the recommendation of the Board of Works. After his retirement, he published a book of verse *Wild Flowers from Germany* comprising translations of German folk-songs, ballads and marching songs. He was living near Moy, Co. Tyrone, in the 1850s and operating a charcoal manufacturing business, but subsequently lived with one of his four sisters at Broommount, Aghalee, Co. Antrim. Captain Dubourdieu, who never married, died on 1 December 1861 and was buried at Soldierstown.

DUFFIN, William Edmund L'Estrange (*c.*1843–1925), was born in Co. Down. His father, Revd William Duffin (1802–82), was vicar of Maghera and his mother was Lucy L'Estrange, a native of Co. Offaly. William entered Trinity College, Dublin, in July 1860 at the age of 17 and was awarded a BA degree in 1865 and the LCE in 1866. He was subsequently a pupil of Sir John MacNeill and, for three years, was an assistant with William Lewis, a Dublin civil engineer, working on railways, sea walls and other projects. This was followed by a period of employment on the planning of the Central Ireland Railway. In May 1871, Duffin became a temporary assistant geologist on the staff of the Geological Survey of Ireland which then had its headquarters at Hume Street, Dublin, later transferring to a permanent post of assistant geologist following an examination conducted by the Civil Service Commissioners. He gained first place in the county surveyor examinations in December 1873 and was appointed to the western division of Co. Limerick in February 1874. He transferred, at his own request, to Co. Waterford in October 1877.

Duffin was always a severe critic of the contract system which he described in 1891 as 'one of the great blots on the Irish system of road maintenance'. He claimed to be able to maintain those roads for which he had direct responsibility at a price at which no contractor would take them and, as well, to keep them in better order. After the introduction of a direct labour scheme in the county in 1905 and the gradual introduction of new machinery and equipment, he made good progress on the improvement of the county's roads. He was an innovator in bridge-building and was able to claim responsibility for the construction before 1910 of the first Hennebique ferro-concrete road bridge in Ireland; this was a relatively simple and economical structure at Knockmahon, Co. Waterford and was replaced in the 1950s. Ballyduff Bridge, constructed over the river Blackwater six miles west of Lismore in 1887, was also designed by Duffin; the bridge consists of two wrought-iron double lattice-girder spans, each roughly 100 feet long, separated by a masonry river pier. A somewhat similar but much smaller bridge with a steel superstructure on masonry and concrete abutments was built at Ballyneety, near Dungarvan, in 1904–5 to replace a bridge swept away by floods in the river Colligan two years earlier; the superstructure in this case has since been replaced by reinforced concrete. Around the same period, Burtchaell's Bridge over the river Nire was rebuilt with steel girders.

Proposals were made by Duffin and others to the Royal Commission on Irish Public Works in September 1887 for the construction of a railway from Waterford city to Dunmore East with a view to the development of the herring fishery there. Duffin carried out preliminary design work on the line himself, and estimated that 11½ miles of narrow-gauge track could be built for £45,000, as against about £70,000 for a standard-gauge line. The scheme, which would have provided Waterford with a fourth separate railway terminus, was never carried through.

Soon after his appointment to Waterford, Duffin realized from a study of the six-inch ordnance survey maps prepared in 1841 that considerable erosion had taken place in the intervening forty years all along the coast. In the 1880s he began a programme of groyne construction, smaller ones in timber but also some more substantial structures using rocks and concrete. There were particularly severe problems at Tramore where the average annual rate of erosion was as much as three feet per year. In 1867, the grand jury had piled portion of the beach-front but despite further expenditure in the following years, the piling had to be abandoned in 1879 as the cost of repairs was too great. In 1881, an experiment was made with timber groynes extending about 100 feet seawards in the same area and planting of sea buckthorn to stabilize the shingle bank was tried at a later stage. In 1896, Duffin built a sea wall across the frontage of the bog marsh at a cost of £5,000; this was badly damaged in the following year during a heavy gale and repairs cost £300. By 1913, further damage had been done to the wall, and it was badly outflanked by the sea at its eastern end. Another sea wall and promenade was then commenced with the aid of a grant of £4,000 from the Board of Works and was completed in 1915; however, the basic problem remained to be grappled with by his successors long after Duffin had retired.

A new pier at Tramore was completed by direct labour under Duffin's supervision at a cost of about £5,000 in 1907 and he constructed small waterworks schemes at Passage East and Tramore in 1889 and 1894, respectively. In 1908, he prepared plans for the improvement of the navigation of the river Blackwater as far as Cappoquin and for similar improvements in the river Bride. The plan for the Blackwater, estimated to cost £4,000, involved dredging the river channel and building walls of dry stone at particular points so as to allow steamers of nine or ten feet draught to reach Cappoquin at high water on any tide. On the Bride, he recommended works costing £2,300 so as to allow navigation from its junction with the Blackwater to Tallow Bridge, a distance of about 7½ miles. The Royal Commission on Canals and Waterways reviewed Duffin's plans and reported in 1911 that 'the estimated cost … is small indeed in comparison with the results which are anticipated … and we consider the carrying out of schemes such as those which have been laid before us to be necessary and urgent'. Nothing was done, however, to implement Duffin's plans.

Until the end of its days in 1899, the Waterford grand jury met at the county courthouse in Waterford despite efforts from time to time since the 1830s to transfer all business to the county town of Dungarvan. The new county council took up the matter at its very first meeting in April 1899 and a committee was appointed to work with Duffin on the adaptation of the courthouse in Dungarvan to allow meetings to be held there. When, in the following year, the council decided that the offices of the county secretary and county surveyor should also move to Dungarvan, Duffin built new offices beside the courthouse to accommodate the staff.

Duffin tendered his resignation to the county council in October 1914 after more than forty years service as a county surveyor. By then, according to the *Irish Builder and Engineer*, he was 'one of the dwindling survivors of the old competitive system which in Ireland produced such a splendid service'. During his years in Waterford, he lived at Whitechurch House, Cappoquin, and at Larkfield, Waterford, with his wife, Alice Susan Eager, daughter of the rector of Drumgooland, whom he had married at Newcastle, Co. Down in 1880. The couple had two daughters, Maeve and Pearl, who never married, and a son, Guy, who died on 7 September 1884 when only 2 years old. The family moved to Dublin following his retirement and he died at Berea Terrace, Rathfarnham, on 25 July 1925, aged 82. He is buried at Mount Jerome cemetery with his wife who died in 1953 and their two daughters.

DUFFY, Edward J. (1893–1947), was born at Castlepollard, Co. Westmeath, on 22 June 1893. He was educated at St Finian's College, Mullingar, and entered the engineering school at University College, Dublin, in 1912. He graduated with first-class honours BE and BSc degrees in 1915 and gained an ME in 1922. He was a member of the teaching staff of St Mary's College, Dundalk, from 1915 until July 1919 when he became a research assistant under professor Pierce Purcell at the Fuel Research Board premises at Turraun, Co. Offaly, working on the properties of peat and the development of electrical and mechanical systems for harvesting and drying it. He worked in London in the early 1920s for the Anglo-Danish Concrete Construction Company, specializing in reinforced concrete design. While there, he was instrumental in promoting an investigation into the possibility of establishing a cement industry in Ireland, using peat as a fuel.

Duffy returned to Ireland on his appointment as Meath county surveyor in October 1923 at a salary of £600 a year. Soon after his appointment, when experiments were being conducted on different types of road construction, he constructed about ten miles of cement-bound macadam roads in Meath, with a sand-cement mixture laid on a bed of three inches of broken stone and a further layer of broken stone laid on top before rolling. In addition to his roads duties, he was responsible for carrying out a number of water and sewerage schemes in the county and for extensive rural housing schemes. He was a progressive engineer, always seeking to improve methods and practice in road construction and design. He acted in the 1940s as a special lecturer in University College, Dublin, on the design and construction of roads and on aspects of public administration and he presented papers to the Institution of Civil Engineers of Ireland on reinforced concrete beams, cement-bound macadam roads and turf production.

Continuing his interest in peat development, Duffy visited Denmark in 1933 to investigate the possibility of establishing a briquetting plant in Ireland. In 1935 he was a member of a Turf Development Board delegation, led by Dr C.S. Andrews, which visited Germany and Russia to study new and more effective methods of large-scale turf production and the use of turf for electricity generation. The trip is described in some detail in Dr Andrews' autobiography *Man of No Property* (Cork, 1982). According to Andrews, Duffy was basically a mathematician and a man of great intellectual gifts who had no interest in, or knowledge of, political affairs; he was an essentially gentle, pious person who was not very happy about being in Russia. Nonetheless, Andrews clearly valued Duffy's participation in the study visit and recounts how, with P.G. Murphy, the Electricity Supply Board's chief engineer, Duffy subsequently prepared 'a really fine report on the engineering aspects of turf production and its use in power stations'. Duffy filled his county surveyor post with distinction until his death at his home near Navan in February 1947. His daughter, Joyce, who had been Andrews' secretary, married him following the death of his first wife in 1967.

DUGGAN, Thomas Stanislaus (1892–1961), was born in Belfast on 9 March 1892 and was educated at St Vincent's College, Castleknock, Co. Dublin. He served a pupilage in Belfast with J.H. Swiney, consulting engineer, and worked for some time as resident engineer on a variety of water and sewerage schemes in the north of Ireland before taking up an appointment as assistant county surveyor in Cavan in 1914 under **Henry J. O'Reilly**.

Early in 1920, Duggan was one of four applicants for the post of county surveyor in Tipperary south riding and was appointed in March on the casting vote of the council chairman. He took up duty soon afterwards on an acting basis pending the approval of the Local Government Board which, in due course, advised the council that **T.P. Meade** was in fact the only qualified candidate, having obtained first place in the examination conducted by the Civil Service Commissioners. On a number of occasions during the following months, the newly-elected council, dominated by Sinn Féin members, discussed how the situation might be resolved. There were suggestions that the post should be re-advertised and that some time should be allowed for Duggan to sit the qualifying examination again; the matter was complicated by the fact that Meade, with other members of the Dublin brigade of the IRA, had been convicted by a court martial of unlawful assembly and was serving a two-year sentence in Cardiff gaol. Eventually, in March 1921, it was decided to appoint Meade to the permanent post and Duggan, who had in the meantime passed the qualifying examination, continued as acting county surveyor. When Meade was released from prison in January 1922 on completion of his sentence, Duggan was reluctant to step aside to allow him to take up his duties. Some members of the county council took it on themselves to resolve the impasse and there were reports that a revolver had been laid on Duggan's desk with a suggestion that a Belfast man was not welcome in the county. At the March meeting of the council, Duggan protested about his treatment and called for a public inquiry: he had given up his previous employment and carried out the duties assigned to him by the council for two years but was then 'practically thrown out by the back of the neck'. However, the council stood by their decision to appoint Meade, forcing Duggan to seek employment elsewhere.

During his two-year term in Tipperary, Duggan had been unable to carry out any road work due to the disturbed conditions of the times and the poor financial position of the county council which had lost both rates and grant income; in June 1921, up to 200 road workers were laid off and steamrollers had to be laid up. As a result, the roads were a mass of potholes by the end of the year, according to Duggan.

Duggan was engineering manager with Messrs Charles Tennant & Co. of Dublin for a time before re-entering the local service as special assistant to the Dublin county surveyor, **J.A. Ryan**. He worked from 1924 to 1928 on the reconstruction of the Naas, Bray, Howth and Swords roads as part of the government's two million pounds scheme of main road improvement. He was among the unsuccessful applicants for a county surveyor appointment in Mayo in 1924 but was appointed Offaly county surveyor in August 1928. In the early 1930s he designed a number of reinforced concrete beam and slab bridges to replace older structures such as those at Ferbane and Clonbullogue, and he designed a concrete arch at Kilbeggan to replace an eighteenth-century hump-backed bridge. He was actively associated with the activities of Cumann na nInnealtóiri from a very early date and was its president in 1948–9. He retired on 31 March 1954 and died at his home in Blackrock, Co. Dublin, on 5 January 1961, survived by his wife, Eileen, two daughters and a son, Thomas, who was also an engineer. He was buried at Deansgrange cemetery.

FARRELL, James Barry (1810–93), was born at Bristol on 3 December 1810. He served a pupilage to Joseph Burke, a civil engineer, and was employed from 1833 to 1839 by the Board of Works as superintendent of the works – effectively resident engineer – under Richard Griffith on the construction of the road (including bridges and tunnels) between Glengarriff and Kenmare. Robert Graham, a much-travelled Scottish whig who visited the south of Ireland in 1835, inspected the new road which, he wrote, 'will certainly be the finest work of its kind in Ireland and in examination will not be surpassed by many works of the kind even on the continent' (*A Scottish Whig in Ireland: the Irish Journals of Robert Graham of Redgorton* [Dublin, 1999]). He noted that the road was 'engineered with great science' through very difficult terrain and that local people had been trained by Farrell to carry out admirably the masonry work required to form long retaining walls. At the same time, Farrell told him that 'the difficulties of the engineering are nothing to overcome in comparison with the management of the people' of whom between 800 and 1,600 were employed at different times on the works.

The construction of the tunnels at the Glengarriff end of the route, using substantial quantities of gunpowder, attracted Graham's particular attention and he noted that it was Farrell's intention that the longest of these (183 yards) on the Cork-Kerry border would have two separate carriageways, separated by a four-foot-wide footpath, raised three feet above road level. (As built, the tunnel had only a single undivided carriageway). He was particularly interested in Farrell's ideas on the design of the suspension bridge, with a central pier, which was to carry the new road over the Kenmare River and into the town. The road had been laid out in 1832–3 by William Bald and Farrell worked with him on early plans for the bridge after Bald had been asked by the Board of Works in October 1835 to submit plans and estimates for different kinds of bridges – 'stone, iron arches, iron suspension, or mixed'. After a suspension bridge was decided on, Farrell was credited with the design of the massive central masonry tower from which the chains were to be suspended. In September 1837 he submitted a bid to carry out all of the masonry work himself, as contractor, but this was rejected by the board and having failed to get an acceptable tender from any other established contractor, they decided in July 1838 to construct the abutments and pier by day labour and to invite Captain Sir Samuel Brown RN (who had pioneered the design of suspension bridges in the 1820s) to tender for the supply and erection of the iron and wood work. Farrell then organized the construction by direct labour of the tower and the abutments of the bridge but had left the project team before the bridge – the first of its kind to be built on a public road in Ireland – was completed in 1841. The final design of the superstructure was provided by Captain Brown and the cast-iron girders and chains were fabricated at Cardiff and erected at Kenmare under his supervision.

Farrell was appointed county surveyor in Tyrone on 7 March 1839 but moved to Wexford on 12 August 1840 because he considered that Tyrone (which was subsequently divided into two divisions) was too large a county for one surveyor. By 1868, he had built 135 miles of new roads in Wexford and was associated with several major bridge projects. Having completed the wooden bridge at the Deeps (Killurin) on which work had begun under his predecessor, **William Mackenzie**, Farrell's first significant bridge was a massive stone structure at Carigmenan, about six miles from Wexford on the Killurin road; this bridge, completed in 1844, has 3 arches, each of 50-foot span, and a height at the centre of 120 feet over a ravine and small stream. Farrell designed and built another large stone bridge at Cooladine, near Enniscorthy, around the same time and numerous others in subsequent years. In 1847 the approval of parliament was obtained for the construction of a new bridge at Wexford, half a mile upstream of the timber toll bridge, originally 1,571 feet long, which had been built by Lemuel Cox in 1795; a cheap timber bridge was initially proposed but when the price of iron fell Farrell came up with another design, retaining wooden piles and substructure in the interests of economy but substituting a superstructure of iron girders and decking to bridge the twenty spans, and with a bascule opening span of fifty feet at the centre. In January 1851 he outlined his plans in a paper presented to the Institution of Civil Engineers of Ireland and exhibited a model of the bridge which he had tested by putting a ton weight on it. Due to shortage of funds, the new bridge was not completed and opened to traffic until March 1866. It cost about £25,000 and lasted until 1959 when it was replaced by a prestressed concrete structure.

In January 1867 the bridge at New Ross – a wooden trestle structure also constructed by Lemuel Cox – was swept away by ice in the river Barrow, which was about 650 feet wide and up to 38 feet deep at that point. Farrell worked with **Peter Burtchaell**, the Kilkenny surveyor, and **Samuel Ussher Roberts** to design a new iron bridge consisting of 4 fixed spans of lattice girders, each 88 feet wide with a manually operated swinging section near the centre, providing two 50-foot openings. The bridge piers consisted of pairs of cast-iron cylinders, up to nine feet in diameter, which were sunk with difficulty through sandy clay and gravel to the rock and filled with concrete. The ironwork in the piers and abutments weighed 1,182 tons and the iron superstructure weighed another 650 tons. The entire work was carried out between April 1868 and July 1869 and cost £36,250. The bridge was described by a contemporary commentator as 'not a very strikingly grand or imposing looking structure, with piers too dumpy in appearance' but it served its purpose for nearly one hundred years (although weight restrictions and a three mph speed limit had to be applied in 1927) until the present precast concrete bridge – O'Hanrahan Bridge – was opened to traffic in February 1967.

During the 1846–7 famine period, Farrell was engaged under the Board of Works on the supervision of relief works on which nearly 1,300 men were employed in Wexford. In early 1847 one project alone – the construction of a road from Enniscorthy to Kiltealy – involved the employment of an average of 1,100 men and 100 horses each day. Under the Drainage (Ireland) Act, he was also employed until 1851 by the board as district engineer for the Cahore drainage district and submitted a report on the proposed drainage of flooded lands near Cahore Point in March 1844. Some years later, he superintended the construction works at Cahore pier for the board.

Farrell had a long association with efforts to improve Wexford harbour which was ideally located for shipping to and from Britain and close to busy shipping lanes, but failed to provide a safe haven for ships or adequate docking facilities, due to continuous silting up of the channels through which the river Slaney reaches the sea. The *Dublin Builder* noted in 1862 that the number of 'engineering doctors' called in to advise on the problem in the years up to 1846 was no less than twelve; these included Sir John Rennie who reported in 1831, Sir John MacNeill, Sir John Coode, Sir William Cubitt (1838), Captain Vetch RE (1841) and James Rendel (1844). Farrell's first report on the improvement of the harbour (published in the Tidal Harbour Commissioners Report of 1845) was said by Captain Washington RN to have established him as second to none in Ireland in the theory of hydraulic engineering; some of the recommended works were carried out by Farrell in the late 1840s as engineer for the Harbour Improvement Company which had been established by legislation enacted in 1846. In addition to major dredging operations, these included the construction of some 3,430 yards of embankments, over thirteen feet above low water, extending to Raven Point on the northern side of the harbour, to permit the reclamation of nearly 2,500 acres of slobland. Embankments 2,580 yards long built on the southern side near Drinagh in the 1850s brought the total area of reclaimed land to some 4,600 acres. There were sluices, tidal gates, wrought-iron siphons, and steam-powered pumping stations to maintain the reclaimed areas.

During the following twenty years, further proposals were put forward by Farrell at regular intervals to improve the harbour by widening and dredging the channels and by the construction of two long training walls through the main channel, and a long wall outside the harbour to promote the growth of the Dogger Bank into a permanent island which would improve the scouring action of the tides; while some of these works were not carried out due to lack of funds and questions about their cost-effectiveness, the Admiralty funded a major dredging project which was carried out under Farrell's supervision between 1863 and 1867 at a cost of nearly £18,500. In 1873, harbour improvements and drainage and reclamation works were carried out by Farrell at Courtown harbour (built to the design of Alexander Nimmo between 1824 and 1827) and he was engaged regularly by the Admiralty to report on harbour projects at other locations around the coast. One of the most significant assignments of this kind arose in 1847 when Farrell was appointed to hold a public inquiry in Belfast on a proposal to constitute a new organization – the Belfast Harbour Commissioners – with wide powers to improve the harbour; legislation on the subject was enacted in June 1847 after a favourable report from Farrell which included the prophetic statement that 'your trade will prosper to a degree that will astonish yourselves'.

In June 1864 Farrell submitted proposals to Wexford Corporation for a water supply scheme to serve the borough; there was to be a reservoir with a capacity of just over eight million gallons, providing a supply of about 130,000 gallons a day, and the whole scheme was to cost just over £8,200. However, the corporation were not happy with Farrell's scheme and called in Thomas Hawksley (1807–93), a London-based consulting engineer who was then reaching the pinnacle of his career in municipal engineering. Hawksley told the water committee that his role, as he saw it, was to examine the scheme proposed by Farrell and to suggest any improvements 'which my greater professional experience may enable me to point out, and which gentlemen only locally employed cannot be expected to attain to'. He went on to advise that Farrell's storage reservoir would be much too small and located at too low a level, and that trunk mains of at least ten inches in diameter would be required as against the seven-inch cast-iron mains proposed by Farrell. Hawksley's alternative proposals were based on the assumption that a daily supply of 450,000 gallons would be required for a growing population and he proposed a storage reservoir holding forty-nine million gallons to ensure this; he advised that no filtration would be necessary, as the source was so pure, and that the entire works could be constructed at a cost of £14,000, much less according to Hawksley than the cost in other towns of similar size. Farrell vainly protested that his scheme would deliver 30,000 gallons more than had been sought in the corporation's advertisements but his views carried less weight than those of the urbane London consultant who had already been responsible for water supply schemes in major British cities such as Nottingham and Liverpool. The matter was settled when the water committee, congratulating themselves on their good fortune in having obtained the services of 'so able, so talented, and so experienced a gentleman', voted to give Hawksley their best thanks and to recommend his scheme to the corporation. The waterworks at Wexford were duly completed in accordance with Hawksley's plans in the early 1880s at a cost of some £30,000 – one of more than 150 such schemes for which he was responsible during his long career; with output augmented in 1927 and again in the 1940s, the scheme served the town until the 1980s.

Farrell seems to have undertaken relatively few architectural projects throughout his career. He was responsible in 1862–3 for the alteration and enlargement of the county courthouse at Wexford which had been designed in 1803 by Richard Morrison; the building was burned down in November 1922. In partnership with **James Bell**, one of the Dublin surveyors, he designed the Wexford lunatic asylum at Enniscorthy, now St Senan's Hospital, situated on a hillside overlooking the river Slaney; it is probable that Bell, with his greater experience and training in architecture, contributed most to the design and may even have worked on it during his employment by the Board of Works in the late 1850s. The 630-foot-long granite and brick three-storey building in an Italianate style, with five lofty water towers, cost about £40,000 and was completed in 1868.

Having served with distinction in Wexford for over fifty years, Farrell retired on a pension of £400 a year in July 1891 when he was over 80 years of age. He was said to be a generous and kind-hearted man and a friend to many and his death on 3 January 1893 was widely mourned. He was buried in a small cemetery beside the Franciscan church in Wexford but his grave cannot now be located due to the removal and destruction of headstones during cleaning operations in the 1960s. He was elected to membership of the Institution of Civil Engineers (London) in 1870 and was awarded the rare distinction of honorary membership of the Institution of Civil Engineers of Ireland in December 1892, shortly before his death, having been an ordinary member since the establishment of the organization in 1835. His son, Martin John, served a pupilage to his father between 1859 and 1864 and subsequently worked with him as J.B. Farrell & Son on the Wexford harbour improvements, on the design of the Wexford and New Ross bridges, and as contractor on sections of the Wexford-Rosslare railway and at Rosslare harbour. Farrell, however, seems to have had no connection with Isaac Farrell, the leading Methodist church architect in Ireland in the first half of the nineteenth century, or with the four Farrell brothers who were so prominent in architecture and sculpture in Dublin during his lifetime.

FEEHAN, Malachi Andrew (1897–1963), was born on 4 February 1897 in Co. Laois, the youngest son of James Feehan of Ballickmoyler who had been an assistant county surveyor since 1874 and had acted as county surveyor for a few months in 1914–15 while the post was vacant. He was educated at Rockwell College and at University College, Cork, where he gained a BE degree in 1917. In 1917–18, he worked as assistant to R.F. Waller on the construction of the ten-mile-long Athy to Wolfhill Railway and the Castlecomer Collieries Railway which was seven miles long; both projects were promoted by the government because of the wartime scarcity of coal and they were to be the last railways built in Ireland. Feehan remained in Castlecomer throughout 1919, working as assistant engineer to the British Ropeway Engineering Company on the construction of an aerial ropeway at the collieries.

Feehan became an assistant county surveyor in his native county in 1920 and was appointed county surveyor in 1924. On taking up duty, his first task was to complete the repair and reconstruction of roads and bridges after the neglect and damage caused by the war of independence and civil war. He was responsible for the construction of a number of reinforced concrete bridges and for the provision in 1931 of a reinforced concrete road between the Heath and Portlaoise. He planned and implemented water and sewerage schemes in several towns around the county and was responsible for the design and construction of one of the country's first cattle marts at Rathdowney.

During the Second World War Feehan had to turn his hand to turf production. In 1941–2 twenty-one bogs were in use around the county and the number of council employees engaged in turf production reached 766 in June 1942. Even as late as April 1947, with continuing coal shortages, Feehan was reporting to the council that turf production must be given absolute priority in the allocation of labour, with the inevitable result that not even urgent road works could be carried out.

Arising from adverse reports by a local government auditor, a local inquiry was held over a number of days in July 1945 by an inspector appointed by the Minister for Local Government into the manner in which Feehan and his county manager had discharged their duties. There were extensive reports of the inquiry in *The Nationalist and Leinster Times* going into minute detail about matters such as store keeping, stocktaking, record keeping and the acquisition of machinery, materials and small amounts of land. The only complaint of substance came from Deputy Oliver J. Flanagan who contended that there were no worse roads in Ireland than in Co. Laois. At the conclusion of the inquiry, Feehan's counsel submitted that there had been no allegation of dishonesty or of loss to the council through the neglect of the surveyor, and no allegation of other negligence on his part, except by Deputy Flanagan.

Father Frank Browne SJ, whose photographs captured so many aspects of Irish life in the first half of the twentieth century, recorded the scene on the Cork–Dublin road in 1932 when the crew of a breakdown truck were attempting to recover Feehan's overturned Ford motor car. Feehan survived the incident and was in office until 1947 when he resigned. Described as a practical engineer and a kindly, unassuming man, he died on 8 February 1963 and was buried in the new cemetery, Portlaoise.

FERGUSON, James R.A. (fl. 1875–1912), was awarded a BE with second class honours at Queen's College, Belfast, in 1875 and an ME in 1882. He was an assistant county surveyor in Antrim before taking up duty as county surveyor for the southern division of Donegal in January 1892. In reporting the appointment, the *Donegal Independent* expressed the hope that Ferguson would 'do his utmost to bring the bad roads of the district into passable order', many of them having been badly cut up by the construction work which was in progress on the Donegal–Killybegs narrow gauge railway, completed in 1893, and the branch line to Glenties which opened in 1895.

In 1895 Ferguson was transferred to the northern division of the county where the maximum salary was £500 as against £400 in the south. When he transferred again to the service of the new county council in 1899, he sought an increase in his salary, as the law allowed, to take account of the additional duties devolving on him under the new local government system. The council were willing to allow an increase of only £50, but Ferguson maintained that an extra £160 was justified and appealed to the Local Government Board to determine the matter. The board's decision to fix a new salary of £680 brought a strong reaction from the council; the decision was condemned as an arbitrary one made without hearing the council's views, and the council refused to implement it. They took the view that if, as he claimed, Ferguson had given his whole time to the job before 1899, his new situation could not logically involve any extra work. Moreover, his claim that his new situation involved extra travelling was met by a demand that he should transfer his office and his residence from Derry to Letterkenny. The dispute between the council, the surveyor and the Local Government Board went on for months until eventually in May 1901 a bargain was made: Ferguson would be paid a salary of £550 but could continue to live in Derry.

At the end of 1902 Ferguson and his council decided to raise a loan to purchase a steamroller for £400, with a scarifier costing £80, a sleeping van, £30, and a watering cart, £30; a second set was decided on for the southern division shortly afterwards. There were strong protests from the Inishowen area because of the extra taxation involved and claims that use of the new machinery would deprive the working men of the area of their means of support. But the council went ahead with the purchase of the new machines and the appointment of drivers and flagmen.

Ferguson told the grand jury at its last meeting in March 1899 that the foundations of the lattice girder bridge over the Gweebarra estuary, designed by **Samuel Ussher Roberts** and completed only three years earlier, had failed due to defective construction work but neither the Congested Districts Board nor the county council was willing to provide the funds required to remedy the defects; the bridge was not replaced until 1953. Ferguson resigned in July 1912 'on the peremptory advice of his medical adviser' and lived in retirement at The Rock, Londonderry.

FLANAGAN, Thomas Patrick (1895–1980), was born near Claremorris, Co. Mayo, and was educated at St Jarlath's College, Tuam, and at University College, Galway, where he graduated with distinction in engineering in 1917. He had joined the volunteers in Galway in 1914 and, on his return to Mayo after graduation, was active in the various strands of the independence movement. He was, according to himself, often employed in several capacities by the local active service unit while holding the position of assistant county surveyor in the Westport area from 1919 onwards. After the truce of July 1921, Flanagan was appointed as engineer officer with the west Mayo brigade of the IRA on a full-time basis, but returned to his post with the county council early in 1922. On the outbreak of civil war he was back in action on the anti-treaty side. Government forces soon took over the main towns in Mayo, having landed by sea in Westport, leaving the irregulars to fight on in the mountains and more remote areas. In the autumn, Flanagan was arrested and held in the military barracks at Athlone but was released towards the end of the year and was allowed to resume his duties as assistant county surveyor.

When **E.K. Dixon**, the long-serving Mayo county surveyor, who had been carrying on his duties in a temporary capacity since the beginning of 1922, gave notice in the autumn of 1923 of his wish to retire, the county council requested him to continue in office until all of their assistant surveyors had an opportunity of sitting the qualifying examination for county surveyor posts. Examinations were duly held in April 1924 and when the results became available in June the council set about finding a replacement for Dixon at a salary of £450 a year. Flanagan was the only one of the candidates who met all of the statutory qualifications for the post, but the council voted in August by fourteen votes to twelve to appoint a Mr M. Martin of Ballinasloe whose supporters on the council claimed that in addition to his engineering qualifications he had an impeccable national record: he was said to have been imprisoned during the 'Tan War', had carried on a hunger-strike for thirty-eight days and was interned during the civil war. The Department of Local Government and Public Health refused to sanction Martin's appointment and forced the council to advertise the post again. Before the matter came up for consideration by the council at the end of September, Flanagan withdrew his application, leaving **T.S. Duggan**, who had been acting county surveyor in Tipperary south riding from 1920 to 1922, as the only qualified candidate. The council decided, however, to invite Dixon to carry on until the end of March 1925 and he reluctantly agreed to this, provided he was allowed to take a month's holiday 'to make me more fit for the winter'; the council agreed and appointed Flanagan to act as his deputy while Dixon was away. Once again, the Department objected to the council's plans, holding that Dixon's continuation in office was indefensible, given that he had been declared almost a year earlier to be incapable by reason of old age of efficiently discharging his duties. The matter was listed for consideration once again at a special meeting of the council in October 1924 but, at the appointed time, only eleven members (described as *Free Staters* by the *Connaught Telegraph*) were present and they voted to appoint Flanagan. The majority of the members (described as the *Republicans*) had been attending a caucus meeting at which it had been decided to support the candidature of a Mr Glynn of Cong, and when they arrived at the council meeting a 'piquant and incisive discussion' ensued, according to the *Telegraph*. The meeting broke up in disorder, but Flanagan's appointment stood.

Flanagan's first task in Mayo was to complete his predecessor's programme of replacing bridges which had been destroyed during the war of independence and the civil war, a task he had completed by 1926 at a cost of about £15,000. Reconstruction work on some bridges, including those at Pontoon and Ashleagh, was let out on contract. Other bridges were rebuilt in reinforced concrete using direct labour – a lasting monument to Flanagan's efficiency, according to one account; one of these, the bridge at Castlemagarrett was said to be the best example of reinforced concrete in the county. Some of the older stone arch bridges were replaced by concrete and steel girder structures while others, including the bridge at Shrule, were rebuilt. Like his contemporaries, Flanagan had to contend with the growth in freight and bus traffic on the roads, and called in 1928 for restrictions on vehicle weights and speeds. He protested in particular about 'the increasing menace of bus traffic to the roads, the road user and the road workers'; buses were, he said being driven at terrific speed over newly laid surfaces – and one had been timed at 38mph!

Roads, housing, hospitals and sanitary services were all developed very considerably in Mayo during Flanagan's long term of office and he was heavily involved in turf production during the 1941-45 period. He retired in March 1960. In his youth, he was active in athletics, particularly sprinting, and throughout his life, maintained an interest and active involvement in sport, including athletics, gaelic games, coursing and fishing at which he represented Ireland internationally. He was President of Cumann na nInnealtóirí in 1954–5. He was elected to Seanad Éireann from the Industrial and Commercial Panel in 1961 and took the Fianna Fáil whip. On that occasion he was described in the *Engineers Journal* as 'a hard hitter, straight-forward and determined in argument – yet kindly, courteous and considerate, a fine grámhar type'. He was re-elected at the 1965 general election and was one of the eleven senators nominated by the Taoiseach, Jack Lynch, to the 1969–73 Seanad. He died on 7 October 1980.

FOGARTY, Gerard Peter (1900–65), was born on 4 June 1900 at Ninemilehouse, Co. Tipperary. He was educated at Rockwell College and gained a BE degree at University College, Dublin, in 1922. He joined the army's Railway Protection, Repair and Maintenance Corps as a second lieutenant in the autumn of that year and served in the Clonmel area until July 1923 when the corps was disbanded. After six months further service in the army, he was demobilized in December 1923 and took up a temporary appointment as a draftsman in the Department of Local Government and Public Health where he served until June 1925.

After a period with Dublin Corporation, Fogarty was appointed borough surveyor in Clonmel in 1928. He moved to one of two new posts of chief assistant county surveyor in Cork in 1931 and immediately became involved, under **R.F.M. O'Connor**, in the construction of the Healy Pass, a spectacular mountain route across the Beara peninsula. In 1940, after O'Connor's sudden death in office, Fogarty became one of two acting county surveyors. He was appointed Carlow county engineer on 11 December 1944 and transferred to Louth as county engineer in October 1946.

Fogarty introduced a high degree of mechanization as well as modern scientific methods into the road engineering operations of Louth county council and established a fully equipped soils and materials testing laboratory in 1957. He was a strong advocate of the direct economic benefits of main road improvement work and, as early as the mid-1950s, was producing data to demonstrate the value of the savings in time, fuel consumption, and wear and tear achieved

by particular road projects. The reconstruction and tasteful landscaping of the Dundalk–
Newry road in the 1950s and of the Dublin–Dundalk road at Monasterboice were among his
noteworthy achievements; the former was described at the time as 'a magnificent highway
through the beautiful tree-clad mountains of north Louth, a most fitting entry for visitors to
the Republic'.

Forced to retire prematurely in 1960 due to ill-health, Fogarty died at Sandycove, Co.
Dublin, on 18 February 1965 and was buried at Deansgrange cemetery.

FORREST, Edward V. (fl. 1828–41), took up a position as accountant in the Boundary
Survey Office, Dublin, headed by Richard Griffith, in January 1828 at a salary of one guinea
a week. He was earning £1 10s. in 1830 and a few years later, when he had begun to practise
as an engineer, he prepared plans for sections of a new road from Castleisland to Cork city.

Forrest became the first county surveyor for Monaghan on 17 May 1834. He has been
credited with the design of the courthouse at Clones although it follows a standard plan
attributed to William Caldbeck and may not have been completed until after his transfer to
Queen's County in May 1839 in an exchange with Alexander Harrison. At the spring assizes
in March 1840, the *Leinster Express* reported that several members of the grand jury of
Forrest's new county 'bore testimony to his efficiency', and his maps and plans, which were
exhibited at the assizes, were said to have 'gained universal commendation'. Little over a year
later in July 1841, he was removed from office by the grand jury for general neglect of his
duties and for granting certificates to contractors in many cases before works had been
completed. His summary dismissal created difficulties for other contractors like Thomas
Millie who was working on the construction of over four miles of a new line of road in the
Abbeyleix area at the time and who told the agent for the de Vesci estate that receipt of an
advance which he had been expecting for the work would be delayed because Forrest 'had a
squabble with the grand jury at the assizes' and had been superseded. According to a return
presented to parliament, Forrest had died by July 1856.

FORTH, Charles Gerard (c.1808–45), was one of the earliest – and possibly the first – of
the assistants engaged by Charles Vignoles when the latter began his career as a consulting
engineer in England. He worked with Vignoles on the Cheltenham Waterworks Survey in
1825 and subsequently on a variety of English railway schemes, including the Liverpool &
Chorlton Railway (1829–30). On the Dublin & Kingstown Railway survey in April 1832 Forth
was Vignoles' senior assistant. Later, he and **Thomas Jackson Woodhouse** were left in
charge of detailed planning of the line while Vignoles was engaged on his numerous other
assignments in England.

Forth was appointed Carlow county surveyor on 17 May 1834 when construction of the
Kingstown line was nearing completion. His major project in the county was the
reconstruction of the old bridge over the river Slaney at Tullow which was felt to be very
dangerous because of its steepness, its roadway width of only 18 feet and the awkward
alignment of the approaches to it. As no diminution of the waterway would be acceptable for
fear of flooding, and as it would have been inconvenient to raise the approaches, the original
spans, varying from 17 to 28 feet, were retained, but with flat arches in place of the existing
ones. By adding to the abutments and piers on the upstream side, the width of the roadway
over the bridge was increased to 28 feet, and with the new arches, the gradient of the roadway
was reduced from one in seven to one in forty. The entire operation cost only £485 and a
short account of the work written by Forth was submitted to the Institution of Civil

Engineers by Charles Vignoles in 1842 to illustrate what he considered a successful work. The bridge, which is still in service, was reconstructed and strengthened by the county council in the 1980s. A bridge at Moatabower over the Slaney, north of Tullow, and another over the river Direen on the Hacketstown–Rathvilly road, have similar features to the bridge at Tullow and seem likely to have been designed by Forth.

Although he told an official inquiry in 1841 that he carried out only a trifling amount of work for private individuals – 'I do not suppose it exceeds £10 to £15 a year' – Forth carried out some assignments for the Board of Works (such as the supervision of the Tullow–Carnew road scheme costing £2,000 in 1835) and maintained his interest in railway engineering. When Vignoles agreed in June 1836 to undertake an immediate survey of a route for a Cork–Limerick railway for a fee of £500, he arranged with Forth 'who was brought up in my office' to do the work. As a first step, Forth was to meet the Cork committee so as to get a loan of, and permission to take notes from, the plans and sections made by Richard Griffith for an abortive 1825–6 scheme; these had been shown to Vignoles in Cork and he thought they would 'greatly facilitate our operations'. However, the 1836 scheme, like its predecessor, came to nought.

Forth was transferred to the post of Waterford county surveyor on 31 July 1844 in an exchange with **John Walker**, and was appointed also as surveyor for the county of the city of Waterford. There were complaints during the following ten months that great inconvenience was arising because Forth had not actually taken up his new duties and Walker was pressed by the chief secretary's office to stand in for him. Forth's absence was presumably due to the fact that in 1844–5 he was again head of a team under Vignoles, this time planning the Limerick & Waterford Railway. He died suddenly in July 1845 just as the Bill to authorize the line was completing its passage through parliament. When his wife, Catherine Jane McCartney, died soon afterwards, Vignoles was concerned to set up a fund for the support of their children.

FOSBERY, Thomas (1820–93), was born on 1 May 1820, the third son of George Fosbery of Curraghbridge, Co. Limerick. He qualified for appointment as a county surveyor in December 1851 but had to wait for an appointment until June 1862 when he was assigned to the eastern division of Limerick following the death of **Thomas Kearney**. One year later, his salary was increased to the maximum of £400 fixed by law for the post. While he was allowed by his grand jury to have two assistants, his personal workload still seems to have been a heavy one: at the summer assizes in 1879, seventeen years after his appointment, he notified the grand jury that he proposed to go away for a month, trusting that there would be no inconvenience to anybody and pointing out that 'it is the first time I have done so since I was appointed county surveyor in 1862'.

Fosbery had difficulties in maintaining the Limerick road network which had increased by 60 per cent in the twenty-five years before his appointment. The normal annual maintenance cost was about £13 per mile but flooding due to the exceptional rains of several winters caused major damage to the roads and bridges; many of the replacement bridges designed and built by Fosbery are likely to be still in service. Flooding in the Mulcair drainage district was a particular problem in the 1860s and led to the appointment of Fosbery by the Board of Works in 1869 as one of the engineers for the district. Between then and 1875 he supervised the construction of embankments and the deepening and straightening of the river channel at a cost of about £28,000. However, after only a few years, there were complaints by the landowners that the scheme had been a failure and some even suggested that the last state was worse than the first. Well over one hundred years later, flooding in the Mulcair catchment is still a recurring problem.

When the surveyor for the western division of Co. Limerick, **John Cox**, resigned for health reasons in March 1884, the grand jury decided that it would be best if one surveyor carried responsibility for the entire county and invited Fosbery to take on this new role. A committee which had examined the question advised that he would be ideally suited to the task of 'bringing into proper order' the affairs of the western division which were said to be in a 'demoralized state' and that it would be impossible for a new surveyor – likely to be a young man of little experience – to manage the division. Fosbery agreed to the arrangement and was allowed to appoint one of his sons as an additional assistant surveyor to help him. Three months later, the foreman of the grand jury spoke of the extraordinary and almost miraculous changes which Fosbery had already brought about in the western division but noted with regret that the law would have to be amended to allow the arrangement to be continued. He feared that the vacancy in the western division would be filled by the appointment of a man who would be a 'distinguished engineer and a scholastic gentleman' but who would be entirely unfit to be a county surveyor. In the event, **John Horan**, the person appointed in the following October to succeed Cox, was a capable engineer who was to serve in Limerick for almost thirty years.

In 1887 Fosbery was allowed an additional assistant to cope with an increasing volume of work, including activity arising from the rush to promote light railways and tramways which was a feature of the 1880s. His health began to fail towards the end of the 1880s and he retired on pension in March 1893 with the thanks of the grand jury for more than thirty-one years of faithful service. He died less than a month later on 1 April 1893 at his home at Kilgobbin, near Patrickswell, which he had inherited from his brother, Francis, on the latter's death in 1859; he had also inherited the family estate at Clorane on the death of his eldest brother, George, in 1875. The *Limerick Reporter and Tipperary Vindicator* described him as a popular figure, active in hunting and sporting circles, especially hare coursing, and with many friends among the country gentlemen of Limerick. Fosbery was survived by his wife Georgina, daughter of St George Smith of Greenhills, Drogheda, whom he had married in 1852 and with whom he had two sons and four daughters.

FRASER, John (fl. 1834–74), was described by the *Irish Builder*, when reporting his death in 1874, as a man who rendered the country some service, borrowing from Shakespeare's *Othello* as did a Taoiseach in his resignation speech well over a century later; his biography, if written, would form a valuable record of the advancement of several of the northern counties, according to the journal.

Fraser was appointed on 17 May 1834 to be the first county surveyor for Co. Down and went to live at Downpatrick, the assizes town, on taking up his official duties. From the outset he carried on a substantial private practice in engineering, surveying and valuation and by 1841 was earning up to £300 a year from this, almost as much again as his official salary. He became heavily involved in railway engineering at a later stage; his lines included the Belfast & Hollywood, the Newry & Warrenpoint, the line from Newry to Enniskillen, and some of the Co. Down lines. With a local architect, he was chosen to design and construct a memorial to Major-General Sir Robert Rollo Gillespie in the town of Comber, Co. Down where he was born in 1766 (Gillespie had died in action leading an attack on a fort in the Himalayas in 1814). In 1852, he was commissioned by Henry Spencer Chichester, Baron Templemore, to draw up a detailed development plan for his property in Ballymacarret, then an industrial suburb of Belfast; the plan proposed a new street pattern and designated sites for villa development and for industry, but little of it was implemented.

Fraser's obituary recorded that he laid out miles upon miles of excellent roads in Co. Down. He was a strong advocate of the benefits of improved roads and regularly produced cost-benefit analyses of his schemes, showing the economies that would result from the reduction in the number of horses required to haul heavy loads on the improved roads. There are indications, however, that his grand jury was not impressed by his efforts at road engineering and that his relationship generally with them was an unhappy one. In 1847, he was admonished because the Board of Works had complained about the condition of some of the post roads in the county and the jury, obviously suspecting a lack of attention by Fraser to his public duties, directed that he must keep a diary of his inspections of the roads. Two years later they were dissatisfied with his explanation for the employment of his younger brother, Leslie, to superintend, simultaneously, three separate road contracts with separate payments for each. William Fraser, another brother, had been certified as an assistant county surveyor in May 1837 but left some years later to work in India where he died in April 1874. In apparent explanation of this rather fraught relationship with the grand jury, Fraser noted in 1852, in his final report to them, that 'during eighteen years and upwards, various circumstances must necessarily have arisen in which different views have been taken by some of the grand jury and their surveyor'.

Although he had worked with **Thomas J. Woodhouse**, the Antrim surveyor, in the mid-1830s on the design of Queen's Bridge in Belfast, to replace the old Long Bridge, and had overseen the construction of the bridge with **Charles Lanyon** in the 1840s, Fraser was passed over on several occasions by the grand jury when designs for major structures were required; Thomas Duff, a Newry architect, was given the commission to design Newry courthouse (built between 1838 and 1841) and William Caldbeck was brought in to design the courthouse at Newtownards in 1850. In each case Fraser was critical of the designs, especially as regards the internal arrangements.

Fraser left Co. Down in September 1852 to exchange places with **Henry Smyth** from whom he took over the northern division of Co. Donegal. While there, he surveyed a possible Derry–Letterkenny railway route in 1855. He agreed in March 1864 to resign from his post in Donegal 'at the instance of the grand jury' but on the understanding that he could arrange an exchange with another surveyor. In the following July, he was appointed by the lord lieutenant to the Cavan post in an exchange with **Frederick Gahan** who moved to Donegal. His term of office in Cavan was a short one – his resignation was reported by the grand jury to the chief secretary's office on 8 March 1867 and he was replaced in May by **Alfred Gahan**. In March 1870, he presented a petition to the House of Commons praying for a retiring allowance but there was no statutory provision for this at the time. On his death in August 1874, the *Irish Builder* wrote that 'during his long and successful career, he gave evidence of high powers and corresponding industry, and the scenes of his labours present memorials of him of which any public official might be proud'.

FRENCH, Theophilus Bolton (*c*.1820–98), was the son of George French, a barrister, of Eccles Street Dublin. He attended the Feinaglian school and was awarded a BA degree at Trinity College, Dublin, in 1842. He was well established in practice as an engineer by 1856 when he was admitted to membership of the Institution of Civil Engineers of Ireland. Having qualified at examinations held in 1857 he became county surveyor for King's County in May 1867 – the last man to be appointed on the basis of the original, but by then discredited, system of examinations which was replaced in 1869 by a system based on examinations conducted by the Civil Service Commissioners. His son, Theophilus junior, was appointed as one of his assistants in 1873.

In what the lord chief justice, presiding at the summer assizes in Tullamore, described as an exceptional case, a road contractor was prosecuted in June 1873 for attempting to bribe French. According to a report in *The Times*, the contractor, Patrick Duffy, a respectable-looking man, had called on French at his home to ask him to certify that road work, priced at £10 5*s*., had been properly completed and left a one pound note on the hall table. French flung the note after him, but it was returned later 'as a present to the surveyor's child'. Duffy was sentenced to three months' imprisonment with hard labour for half of the time as he was over 63 years of age.

French was transferred to the northern division of Tyrone in June 1874 and brought his son with him as an assistant surveyor. According to one of the local newspapers, he continued to be sorely tried in the course of his new duties by the 'dishonest and the unscrupulous'. He retired on a pension of £217 in July 1895 on grounds of old age and died at his residence at Mullaghmore, near Omagh, on 1 June 1898, aged 78. The *Tyrone Constitution* described him as 'a gentleman in the truest sense of the word – upright, honourable, large-hearted and generous to a fault'. His funeral a few days later to the churchyard at Cappagh attracted a 'very large and representative cortege' with blinds drawn in all houses and business premises along the way.

FRITH, Richard Hastings (*c.*1815–73), of Islandview, Enniskillen, became district surveyor for the No. 1 (northern) district of Co. Dublin in early 1845 and took up residence at Balbriggan House. Members of his family – originally from Devon – had practised for some years as builders and architects in Co. Fermanagh and a local newspaper noted that 'from what we know of Mr Frith's engineering attainments, we think the Board [of Examiners] could not have selected a more qualified person to fill the important office' for which eighteen candidates had been examined.

By the autumn of 1846 Frith was involved in the organization of relief works in the Fingal area under the general supervision of the Board of Works. He reported that after considerable perseverance he had got acceptance of piece work which was then in place on 'the cutting at Swords'; under this system, men could earn up to 1*s*. 4*d*. per day and, by what Frith described as 'active exertion', could even make two shillings a day. The labourers who were employed in cutting the hill at Garristown were not, however, earning adequate wages, and he found it necessary to bring in a new overseer to arrange matters so that earnings would be increased. He took pleasure in reporting that the workers were being paid regularly on Fridays and Saturdays and reported that his assistants and clerks worked on the accounts through the night each Thursday and Friday so that the pay clerks would be able to pay the men the following day.

Frith transferred to the No. 3 (southern) district of Co. Dublin after the death in August 1855 of **Robert Hampton** and served there for ten years. He was responsible for improvement works at Main Street, Blackrock and designed a new bridge over the Dodder linking Londonbridge Road and Bath Avenue, Sandymount; this three-span masonry-arch structure was completed in 1856–7 and replaced an earlier wooden bridge. He carried out improvement works for the commissioners of Bray commons in 1861, laying out building sites and a ten-acre park, and prepared plans for a new harbour at Bray which was not proceeded with. He was involved also in laying out the cemetery at Deansgrange, Co. Dublin, where the first interments took place in 1865.

A pamphlet by Frith on *Macadamized Streets compared with Paved Streets*, published in 1857, was essentially an elaboration of a paper which he had presented to the Institution of

Civil Engineers of Ireland a year earlier. In it, he strongly advocated the provision of adequate surface and sub-surface drains and described the system he had installed on the road from Milltown Bridge to Sandyford. Frith's adage 'without efficient drainage, perfection in road-making is unattainable' is as true today as it was a century and a half ago. So too are the remarks with which he prefaced his paper: 'Road-making, being intended for the accommodation and comfort of the public, when inefficiently executed, is a fruitful source of discontent and murmuring'. A further pamphlet by Frith on *The Drainage and Water Supply of Dublin*, published in 1867, dealt with 'the Liffey Nuisance' and criticized on engineering grounds the suggestion that the river might be purified by the construction of intercepting sewers on each side. Details and drawings of Frith's Patent Double Action Valve (for preventing loss of water or damage in high pressure water supplies) were published in the *Irish Builder* in November 1867 – 'a perfect and comparatively inexpensive system well worthy of attention by builders and others having to do with house property', according to the journal.

Frith seems to have had an interest in railway development although no record has emerged of any direct involvement. With **John Bower** he attended the opening of the new street tramway system in Birkenhead in August 1860. Three years later, in October 1863, in a long letter to the *Dublin Builder*, he advocated the building of a central railway terminus in Dublin and was one of the many engineers who put forward proposals about that time for a system of connecting lines to link the different railway termini; it was to be another thirty years, however, before the Loop Line was built to provide in part for such a system.

Frith was a member of the Royal Irish Academy and lived in the 1860s at Leinster Road, Rathmines. He retired in 1865 when the number of surveyors in Dublin was reduced to two, and died on 17 December 1873, aged 58. He is buried at Mount Jerome cemetery, Dublin, with his wife, Lavinia, née Lambert, whom he had married at Athenry in 1863.

GAFFNEY, Stafford (fl. 1869–1920), was born in Wexford on 7 November 1869, son of a local industrialist, and was educated in England at Beaumont and Stoneyhurst colleges. He entered Trinity College, Dublin, in April 1890 and gained a BA degree in 1893, a BAI in 1894 and MA and MAI degrees in 1903. In 1894–5 he trained with Sir Douglas Fox & Partners while working on railways in London and worked as assistant engineer on the Banbury branch of the Great Central Railway and on the Great Northern & City Railway (an underground line) until 1899. He was resident engineer for Fox & Partners on the Cork, Blackrock & Passage Railway extension to Crosshaven between 1902 and 1904 and lived and worked in Dublin for some years after that.

Following the enforced resignation in March 1909 of **Henry Webster** from his £500 a year position as Wexford county surveyor, the county council received twenty-three applications for the post in response to their advertisements. The Local Government Board advised that eleven of the applicants were qualified but when the council met on 7 June it soon became clear that only two men were really in the race, **W.F. Barry**, the Monaghan county surveyor, whose work some years earlier on water schemes in Enniscorthy and Ferns had brought him to the attention of some of the councillors, and Stafford Gaffney, whose supporters on the council pointed out that, as well as being a native of the county and the son of an old and respected citizen, he was a man of independent means – but 'it is no crime ... to be well-off'. The council voted by thirteen votes to eleven to appoint Gaffney with effect from 1 July 1909 even though he had received lower marks than any of the other seven candidates at the April qualifying examinations.

In view of the circumstances of his predecessor's resignation, it was clear that Gaffney would have to set about a major programme of bridge works on taking up duty in Wexford. By early 1910 planning of a new bridge to replace the wooden bridge built by Lemuel Cox in 1794–5 at Ferrycarrig was well advanced. The bridge was to be a ferro-concrete twelve-span structure, with an opening span, designed by Messrs L.G. Mouchel and Partners of London on the Hennebique system – the third major structure of its kind in Ireland, following St John's Bridge in Kilkenny and Waterford Bridge. A construction contract with Messrs Colhoun of Derry was signed in July at a price of about £7,000 and the bridge was due for completion one year later. However, difficulties were experienced with the harbour authorities in relation to the width of the opening span and the headroom and there were delays in the delivery of steel. By April 1911 the contractors were at work on the new bridge under Gaffney's direction and it was opened in November 1912. The bridge was subject to weight restrictions from 1972 onwards but survived until 1980 when it was replaced by the present eight-span precast concrete structure.

In 1911 when Gaffney reported to the county council on the condition of the bridge at Wexford (which was still being described as 'the new bridge' although it had been completed by **J.B. Farrell** in 1866), the council decided to seek a report from an expert on the exact condition of the underwater sub-structure, and an estimate of the cost of the works needed to prolong the life of the bridge; they also sought an estimate of the cost of a completely new ferro-concrete structure capable of carrying a load of up to twenty tons. Some months later, Gaffney told the council that the bridge required immediate repairs at an estimated cost of £2,600, with further work costing over £7,000 likely to be needed in the following few years to put the bridge in good order for up to twenty years. English consultants put the cost of the necessary repairs at £4,540 and James Price, a Cork consulting engineer, submitted three different reconstruction schemes at costs varying from £9,300 to £21,700. The council set up a committee to consider the question of reconstructing the bridge, or building a new one, and some urgent repairs costing £725 were carried out on the bascule section under Gaffney's direction. It was left to his successors, however, to grapple with the problem of maintaining the bridge for nearly fifty years more until it was replaced in 1959.

Gaffney's term of office in Wexford was a short one. In a letter of 9 November 1911, he told the county council that it had been evident to him for some time that he was unable to satisfy their requirements and he submitted his resignation with effect from the end of January 1912. By 1914, he had moved to Greystones and he was living at Chestnut Lodge, Enniscorthy, in 1920.

GAHAN, Alfred (*c*.1821–71), a native of Co. Kilkenny, was one of two brothers who became county surveyors. He was employed by the Board of Works in 1846–7 as one of the engineers in charge of the relief works in Co. Tipperary and part of King's County and was living at Inistioge, Co. Kilkenny, in 1847 when he qualified as a county surveyor at the examinations held in June that year. There were about forty candidates in all and Gahan's name was one of four on the initial list of qualified candidates sent by the examiners to the lord lieutenant in the following September.

Gahan was appointed surveyor for the western division of Tyrone, with headquarters at Omagh, on 14 October 1847 when the county was first divided and seems to have served satisfactorily and without incident there until 1864. At the spring assizes that year, Gahan reported that he had detected a road contractor blatantly claiming payment for work which had not been done but the grand jury failed to support his proposal that the man should be prosecuted. The contractor, who was 'a class leader amongst the Methodists and was supported by a strong Methodist party', according to Gahan, then sued him for libel and to make matters worse arranged that the trial should be held in Derry 'where he could secure a jury from amongst the merchants' with whom he had family connections.

In light of this incident, Gahan told the chief secretary's office in March 1864 that he would be 'a heavy loser' and could no longer work with the Tyrone grand jury, and he applied for an immediate transfer to Donegal on foot of an exchange arrangement with John Fraser, one of that county's surveyors. In the event, it was his brother, **Frederick Gahan**, who was transferred to Donegal to replace Fraser in July 1864 and Alfred had to wait until May 1867 for a transfer to Cavan. He died on 23 October 1871 at Drumbar, Co. Cavan, at the age of fifty. His widow, Alice, née Herbert, died in 1917, aged 87, and is buried at Deansgrange cemetery.

GAHAN, Frederick (*c*.1822–1904), was the youngest son of Major Beresford Gahan of Bishop's Hall, Co. Kilkenny, and a brother of **Alfred Gahan** who was also a county surveyor. He was one of over seventy candidates who sat the county surveyor examinations in 1851 and was one of the sixteen qualifiers whose names were sent to the lord lieutenant in the following December. He was appointed Cavan county surveyor on 5 March 1853, having already served for a year to the full satisfaction of the grand jury as deputy county surveyor in the absence of **Thomas Turner** who had been appointed to the post in February 1852 but did not, apparently, take up duty. When Turner resigned, Gahan's appointment as county surveyor was called for by unanimous vote of the grand jury and, even though his name was not next on the list of qualified candidates, the appointment was agreed to by the lord lieutenant.

Gahan sought unsuccessfully in 1863 to move to a southern county because, he wrote, his wife's and his own relations were nearly all resident in Co. Cork, but one year later, in July 1864, he was transferred to the southern division of Donegal, in an exchange with **John Fraser**. On taking up duty there, he found that his predecessors had left him no reports on the condition of the county works nor any records of work or expenditure undertaken in previous years. He complained, too, of the arduous duties of the Donegal post, especially in the severe winters when he found difficulty in travelling on inspections through his 1,000-square-mile area with its wild mountains and indented coast. His workload was increased periodically by the allocation of Government funds for relief works in his area; in 1890–1, for instance, he was responsible for the construction of substantial lengths of new roads with the aid of these funds in the Killybegs, Ardara and Glencolumbkille areas, including long sections of the Killybegs–Kilcar coast road. He was responsible also for the maintenance of sixteen piers and landing places on the coast. The only architectural project with which he was associated seems to have been the enlargement of the courthouse in Donegal town which he completed in 1892; an unusual feature of the scheme was the construction of the dock which, when not in use, could be let down into the floor in what was described as 'an ingenious arrangement'.

After twenty-seven years of service in Donegal, Gahan submitted his resignation to the grand jury at the summer assizes in July 1891, citing his advancing years and his inability due to ill-health to carry out his duties. His valuable service to the county to which, it was said, he had given the best years of his life was duly recognized in a resolution of the jury, members of which described him as an upright and faithful official. His salary at retirement was £450 a year and he was awarded a pension of £240. Gahan agreed to continue to carry out his duties pending a new appointment, but his successor, **J.R.A. Ferguson**, was not named until

December 1891 and was unable to take up duty in Donegal due to illness until late January 1892. Gahan's additional period of service was acknowledged by the grand jury foreman at the assizes in March 1892: 'he has never done anything but fulfil his duty with the greatest care and industry and in every way to our satisfaction'. Gahan lived in retirement at Kennilworth Square, Dublin, where he died of heart failure on 26 July 1904, aged 82. He is buried in Mount Jerome cemetery with his second wife, Catherine Jane, who died in 1920 and who was a close relative of one of his county surveyor colleagues, **H.U. Townsend**, whose grave adjoins theirs in Mount Jerome. His first wife, Henrietta Byrne of Carlow, whom he had married in 1853, died suddenly in Cavan in 1856.

Gahan's son, Frederick George Townsend Gahan (1866–1955), a civil engineer who had worked for a short time with the Cavan & Leitrim Railway, became the first member of the outdoor staff of the Congested Districts Board after its establishment in August 1891 and served initially as a temporary inspector with responsibility for the preparation of detailed baseline reports on parts of Co. Donegal. He went on to serve as senior inspector with the board in Donegal and Mayo until its dissolution in 1923 and then transferred to the staff of the Land Commission from which he retired in 1926. He subsequently took up residence in Dublin and carried out valuation work for the ESB in connection with the Poulaphouca hydro-electric scheme and the development of the peat-fired power station at Portarlington. His daughter, Muriel (1897–1995) was the founder of The Country Shop in Dublin, Country Markets Limited, and the Crafts Council of Ireland, as well as being a major figure in the development of the Irish Countrywomen's Association.

GALLAGHER, Stephen Gerald (*c.*1871–1959), was born in Waterford and was educated at Synge Street Schools, Blackrock College and Queen's College, Galway, where he was awarded a BE degree in 1896. He was engaged as contractor's engineer on the Blackrock and Kingstown main drainage scheme (1895–6), on the Dublin main drainage scheme (1896–9) and on the Clontarf and Howth Tramway (1899–1900). He later undertook a variety of projects for Wicklow county council as assistant county surveyor and as engineer to Rathdrum board of guardians and Rathdrum rural district council. He also carried on a private practice.

In February 1903 Wicklow county council invited applications for the post of county surveyor. This was one of the first posts to be filled under the new rules for the selection of county surveyors which had been drawn up by the Local Government Board in 1899; under the new system, it was left to the county council to vote into office the candidate of their choice, subject only to the person having passed, or subsequently passing, the qualifying examinations held by the Civil Service Commissioners. The closing date for applications had not long passed when the *Irish Builder* reported that 'very strong canvassing is going on for the position ... candidates who have not yet begun their canvass may take it from us that they will have little chance of securing the post'. The canvassing went on, apparently, for six months and the chances of A.T. McDermott, a nephew of Charles Stewart Parnell who had strong links with Co. Wicklow, were much fancied. However, when the council eventually voted on the matter at a meeting in August 1903, Gallagher was the winner by twelve votes as against nine for McDermott, even though he had received lower marks than any of the candidates who sat the qualifying examination in May 1903; the other four candidates received no votes.

With growing motor-car traffic and regular damage being caused to roads by traction engines and their wagons and by heavy steam lorries, Gallagher attempted in 1907 to persuade his council to allow him to acquire steamrollers and to introduce direct labour schemes to strengthen and improve the Wicklow roads but the council was slow to do so, due to fears about the cost and pressure from the existing road contractors. Motorists and cyclists complained bitterly about the state of the roads and the lack of steamrolling but had to wait another ten years for action to be taken, beginning with a major improvement scheme on the Bray–Wicklow road. A direct labour scheme applying to the main roads in the county was finally approved by the council in November 1920 and a substantial fleet of road machinery had been built up five years later. Due to pressure from farmers and small contractors, however, direct labour was not relied on for all road maintenance in Wicklow until 1935–6.

In July 1907 Gallagher reported that Aughavanagh Bridge – a single-span masonry arch over the river Ow – was in parts 'shaky, distorted, tilted, sagged, bulged, with stones missing and others split'. The bridge was the more westerly of a pair of almost identical structures, only about half a mile apart, completed in 1809 at the extremity of Major Taylor's famous military road through the Wicklow mountains. Initially, Gallagher proposed a scheme of repairs costing £300 but, after all the available options had been reviewed by a council committee, the bridge was substantially reconstructed a few years later: the arch was removed and the ashlar masonry abutments were taken down to the arch springing line and rebuilt in concrete to carry a new ferro-concrete deck slab, supported on two deep longitudinal beams and a series of smaller transverse beams. The reconstructed bridge had a span of 35 feet but was only 16 feet wide, 8 feet less than the bridge it replaced. The concrete beams were constructed on the Hennebique system, making Aughavanagh one of the earliest examples in Ireland of the use of this patented system. By the 1990s, spalling of the concrete had exposed the steel rods and flat stirrups which were a feature of the system and these, in turn, had suffered serious corrosion. This led to the imposition of a ten-ton weight limit and the construction of a replacement bridge upstream.

Gallagher served in Wicklow until his retirement in May 1938. He was president of the Institution of Civil Engineers of Ireland in 1931–2 and was the oldest member of the institution when he died at his home at Novara Road, Bray, on 7 April 1959, aged 88. He was survived by his wife, Catherine Elizabeth.

GARRETT, Robert (1823–57), died in the early stages of the siege of the city of Cawnpore on the river Ganges during the Indian mutiny of 1857. The siege itself, and the carnage that followed it, was one of the epic events in the mythology and folklore of Queen Victoria's empire. Men, women and children died during the siege, or were slaughtered indiscriminately when the city fell, including some who had become caught up inadvertently in the fighting. One of these was Robert Garrett who had left his post as Donegal county surveyor a few years earlier to deploy his talents as an engineer in more challenging and more remunerative fields.

Garrett was born on 8 June 1823 at Cromac, near Belfast, and was educated at the city's Academical Institution. He became a pupil of **Charles Lanyon**, the Antrim county surveyor, in 1839 and worked under him on the construction of the Belfast & Ballymena Railway in 1846–7. His appointment as Donegal county surveyor in November 1847, before the county was divided into two divisions, followed an extraordinary mix-up in the selection and appointments system operated by the Board of Works. In September of that year, after the board had advised the lord lieutenant that the name of William L. Perrott of Fermoy, an assistant surveyor in Cork east riding, was on a preliminary list of candidates who had qualified at the county surveyor examinations held in June 1847, Perrott was notified of his appointment to the Donegal post. Shortly afterwards, however, the board told the lord lieutenant that Perrott's name had been included on the list because of a clerical error (presumably bad handwriting) and they advised that, as he had not yet taken up the appointment, it was not too late to cancel it. The chief secretary's office took the view that, as he had been notified, it would 'hardly be profitable to cancel the appointment unless his incompetence is established'. When the Board of Works confirmed that Perrott would not be 'among the qualified class when the examination shall have been completed' the way was open for Garrett to be appointed.

Garrett served for more than two years to the entire satisfaction of the Donegal grand jury. When he gave notice of his proposed resignation in July 1850 the jury passed a resolution expressing their great regret at the prospect of losing his valuable services and recording the 'efficient manner in which he had carried out the various duties of his office which was evident from the great improvement in the roads, bridges, etc. since they were placed under his charge'.

In 1851 Garrett went to work in London where he was engaged as resident engineer under Messrs Walker & Burges, one of the most prominent firms of consulting engineers of the day, on construction works at the London docks and at Dover. He joined the Army Works Corps on its formation and, as second-in-command of a team of some 2,400 navvies and artificers, set off for the Crimea in 1855 to assist in the construction of roads between the port of Balaclava and the British army's positions on the heights above the besieged city of Sebastopol. He returned to London in the summer of 1856 with the intention of practising there as a consulting engineer in partnership with W.T. Doyne, a native of Co. Carlow who had considerable experience of railway engineering in England and who had been superintendent general of the Works Corps in the Crimea. Garrett was said at that stage to be a man of great knowledge, energy and determination, with a wide circle of friends, who could obviously have gone far in his chosen profession. However, the fates ordained otherwise.

In January 1857 Garrett took on an assignment in India involving surveys for a projected railway between Cawnpore (now Kanpur) and Lucknow, the capital of Oude, a distance of some sixty miles which took eight hours by mail coach. Oude had been annexed by the East India Company in the previous year and the railway was to form part of a major programme of administrative reorganization and infrastructural development. Garrett had completed his site investigations by early May and had returned from Lucknow to Cawnpore when large sections of the Bengal army mutinied and overran much of the province. Unable to travel onwards by the grand trunk road to Calcutta, almost 600 miles away, Garrett and his team joined the small number of British troops in Cawnpore, under the command of the elderly Major-General Sir Hugh Massy Wheeler who had gathered the British community into a makeshift entrenchment built around two old barracks on relatively open ground on the south of the city. In all, nearly a thousand British subjects, including about 300 women and children, crowded into these poorly fortified positions, surrounded only by low earthworks and sandbags, and with limited supplies of food, water and ammunition. Garrett and more than twenty of the railway survey staff were formed into the Railway Rifle Corps, the very mention of which, he wryly told his brother in one of his letters 'no doubt strikes terror into the hearts of the disaffected'. They were, according to Garrett, so 'honoured by the confidence of the General as to be placed as an advance or outer guard in one of the most responsible positions'. They were, in fact assigned to man a line of nine unfinished barracks that came within about 100 yards of the entrenchment at the nearest point. Garrett and his colleagues manned these pathetically inadequate forward positions day and night through the last weeks of May 1857 and the early days of June, confident with their shotguns and hunting rifles that 'we will be able to pick a lot of the devils off' when the attack came.

The mutinous Indian regiments opened fire on the fortifications at Cawnpore with artillery and muskets at noon on 6 June 1857 and a siege which was to last for eighteen days began. Garrett must have been among the first to die in the dust and heat of Cawnpore – he and his railway colleagues came under sustained attack in the early stages and died, one by one, victims of rebel shells or of snipers' bullets. Terms of a sort were offered and accepted by General Wheeler on 26 June, when some 300 hundred of the 1,000 souls who had occupied the entrenchment were already dead. The survivors were promised a safe passage to Allahabad, about 100 miles away, but most of the men were massacred on the banks of the Ganges. The surviving women and children, some 200 in all, were butchered a few weeks later and only four of the military garrison lived to tell the tale of how Cawnpore fell. For almost a year, there were hopes among his family and friends that Garrett might also have survived but his fate was confirmed in 1858 when coordinated lists of the dead finally reached army headquarters in England. His name was subsequently inscribed on the Garrett family memorial in the city cemetery in Belfast, with an indication that he was 'killed at Cawnpore, East Indies, in the year 1857, about the 8th June'.

GIBBONS, Barry Duncan (1798–1862), began his engineering career in 1828 when he superintended for the commissioners of fisheries the construction of an extension to the pier at Kilrush which had been designed by Alexander Nimmo. He also supervised extension works at Ardglass harbour in Co. Down before becoming an inspector under the newly established Board of Works in 1832 at a salary of £167 a year. He was immediately assigned to act as resident engineer at the royal harbour at Dunmore East which had just become part of the new board's responsibilities; the harbour (also designed by Nimmo) had suffered severe storm damage and repair and improvement works costing some £15,000 were initiated under Gibbons' supervision. He also carried out some other engineering assignments for the board in the early 1830s, including inspections of various road and bridge works which were being assisted by loans and grants.

Gibbons was appointed to be the first county surveyor for Wexford on 17 May 1834. He planned and constructed a number of important sections of new roads in the Wexford and New Ross areas and seems to have carried out his duties conscientiously, travelling in his gig over the entire 1,500 miles of roads in his county twice each year in the company of whichever of his four assistant surveyors had day-to-day responsibility for the district. Each circuit occupied six to eight weeks and a change of horses was required three or four times each day.

When Richard Thomas, who had been resident engineer under the Board of Works at Kingstown (Dun Laoghaire) harbour since 1832, died in April 1838, Gibbons opted to resign his post in Wexford and to return to a career in marine engineering. By then, well over £500,000 had been spent on the construction of the harbour under the general supervision of Thomas Rhodes who had succeeded John Rennie as directing engineer. As resident engineer, Gibbons took responsibility for the remaining construction work, and in particular for the harbour entrance and landing quays which took another five years to complete. Simultaneously, he acted as engineer to the Dublin & Kingstown Railway Company and supervised the building of the company's atmospheric line extension to Dalkey which opened for public traffic in March 1844. He told a meeting of the Institution of Civil Engineers in London later that year that the 'arrangements had given perfect satisfaction' and that he was 'satisfied of the applicability of the plan to all such positions as that under consideration'; however, the atmospheric line turned out to be a costly failure and was converted to the normal system of traction and operation in 1854. Gibbons worked under Brunel on surveys for the Dalkey–Wicklow line and served as engineer to the Dublin & Wicklow Railway Company (which had taken over the Kingstown company) until the line to Wicklow was completed in 1856.

Gibbons became harbour engineer under the Board of Works in 1846 just as a new programme of marine works was being initiated under the Fisheries (Ireland) Act 1846 with the aid of a government grant fund of £50,000. He was responsible for vetting the 125 applications for grants received in 1846 for piers, small harbours, quays, landing places and approach roads, for much of the design work on some of the projects, and subsequently for the supervision of over forty small projects completed in the period up to 1853. There were also some larger projects such as the Belmullet canal and swivel bridge, and the building of dry stone walls to enclose and maintain the navigable channel of the river Boyne at Drogheda. Gibbons continued to have overall responsibility during this busy period for the royal harbours at Ardglass, Donaghadee, Dunmore East and Howth and for the harbour at Kingstown where he designed and constructed Traders Wharf, a new pier for colliers and merchant ships, which was completed in 1852. He was responsible also for Carlisle Pier, opened in December 1859 to accommodate the paddle steamers of the City of Dublin Steam Packet Company which then held the contract to carry the mails to and from Holyhead; the ships had been using the east pier since 1827 when the mail service transferred from Howth.

In the 1850s Gibbons was called on to advise on a variety of harbour improvement proposals advanced by local interests, mainly in the west of Ireland. He was one of a three-man expert group commissioned by the Admiralty in 1858 to report on 'the capabilities and requirements of the port and harbour of Galway' which was then being promoted by local groups as a suitable packet station for trans-Atlantic shipping. In 1862 he was consulted by the Ballast Board about a proposal to construct, at an estimated cost of £70,000, a fifty-foot-high lighthouse on the Great Foze Rock, about sixteen miles off Valencia and the most westerly land in Europe. Gibbons advised that the proposal was quite feasible but the Ballast Board, the Board of Trade and Trafalgar House did not agree that the building and maintenance of a lighthouse on such a low-lying rock, exposed to the huge Atlantic seas, was the right solution; the construction of a light on Inistearaght, one of the Blasket islands, was decided on instead and this was completed by 1870.

Gibbons lived at Connaught Place, Kingstown. He was elected an honorary associate of the Royal Institute of the Architects of Ireland in February 1850 and was vice-president of the Institution of Civil Engineers of Ireland in 1861. He became principal engineer in the Board of Works in 1857 and filled the post with distinction for a number of years. According to the *Dublin Builder*, he appeared to be a man of robust constitution and 'might naturally have been expected to number many more summers before he should have shuffled off this mortal coil'. But he died unexpectedly in Dublin on 24 October 1862 when he was 64 years of age.

GLASGOW, Benjamin George Little (*c*.1898–1989), was born in Cookstown, Co. Tyrone, the elder son of Harry L. Glasgow, co-founder and first editor of the *Mid-Ulster Mail*. He entered Trinity College, Dublin, in October 1915 and, having served in the latter stages of the First World War as a private in the Royal Dublin Fusiliers, graduated with BA and BAI degrees in summer 1919. He worked for several years in England, including a period with the Great Eastern Railway at Ipswich, before taking up an appointment as deputy county surveyor in his native Tyrone in October 1924 when **J.W. Leebody** took on the duties of surveyor for both divisions of the county.

When Leebody gave notice in September 1934 of his intention to retire at the end of the year, a county council committee recommended a reorganization of the engineering services in an effort to absorb the retiring surveyor's pension of over £700 within the existing budget. Under their proposals, Glasgow was to be appointed county surveyor at a salary of £700 a year, rising by annual increments to £900 (as against Leebody's salary of £1,060), with Leebody's son, Ian, moving up from the grade of mechanical supervisor to become deputy county surveyor. After an acrimonious debate, the proposals were agreed by the council in November 1934. 'Poison me and that will remedy the whole matter' was Leebody's jocose advice to the council.

During his term of office Glasgow made good use of government grants to reconstruct many miles of main roads in his county and improved a large mileage of third-class roads in the years after 1949 to benefit the farming community. With his brother, he inherited his father's printing and publishing business, and was a director of the company from 1920 to 1979. He also inherited from his father a sharp and sometimes waspish wit, which often enlivened county council meetings. He was awarded the OBE in the New Year honours in 1958, when he retired, in recognition of his services to Co. Tyrone. He lived in Omagh for most of his life and took a keen interest in the affairs of the Royal British Legion. He died at the age of ninety at a nursing home at Aughnacloy on 15 January 1989.

GLOVER, Edward (1847–1931), was born on 9 December 1847 at Churchtown, Buttevant, Co. Cork, son of James Glover, and was educated at Bandon Grammar School and at Trinity College, Dublin, which he entered in January 1870. Having graduated in 1874 with MA and BAI degrees, he served a pupilage in 1875–6 with **Peter Burtchaell**, Kilkenny county surveyor, and travelled in France, Germany and Belgium before being appointed county surveyor for the southern division of Mayo in November 1876. He had taken third place (behind **E.T. Quilton** and **A.E. Joyce**) at the examinations for three vacant surveyorships held by the Civil Service Commissioners in the previous September.

Glover designed the viaduct and swing bridge at Achill Sound which is 240 yards wide at its narrowest point, with a tidal range of fifteen feet. In the early 1880s a fund for building the bridge was initiated and in the summer of 1884, the grand jury approved Glover's estimate of £4,500 for the work. This was subsequently increased to £6,000, necessitating a further appeal by the local clergy for funds. Work on the bridge was just beginning in 1886, with a Scottish firm as main contractors and Glover's brother-in-law, Edward Brewster, as resident engineer, when Glover left Mayo; the bridge was completed with some modifications by his successor, **P.C. Cowan**, and was opened by Michael Davitt in 1887. The work consisted of two causeways, one extending 140 yards from the mainland and one projecting 60 yards from the island, joined by a light girder bridge of 120-foot span, pivoting on a central pier and providing a roadway 8 feet wide. By 1939, the bridge was in poor condition, it was deemed to be inadequate for modern traffic and its replacement was decided on. A contract for the construction of a new opening bridge was awarded in 1941 but due to wartime scarcity of materials the work was not completed until 1948.

Glover was engineer and architect to the board of Mayo lunatic asylum from 1877 to 1893 and provided the plans for a substantial enlargement of the facilities in Castlebar, begun in 1888 at a cost of some £9,500. He designed and carried out sewerage works for Ballinrobe, Castlebar and Westport and a waterworks scheme at Westport, reconstructed the pier at Clare Island and the quay walls at Newport, and had some involvement also in proposals for the development of light railways in Mayo. He designed a wrought-iron lattice-girder bridge to span the river Moy at Clounannana, near Swinford, in 1883 but the bridge had not been completed when the time came for him to leave Mayo; it seems likely that Glover was also responsible for the fine single-arch masonry bridge over the tidal inlet at Newport and the associated flood arches and long embankment.

On 29 June 1880, when Glover was returning to his home in Castlebar from the presentment sessions in Ballinrobe, he came upon David Ferrick, a land steward, lying injured in the roadway, having been shot a short time earlier by three men. Glover returned to the town to notify the priest, the doctor and the police all of whom attended the scene, but the unfortunate man died of ten bullet wounds some days later. At the subsequent inquest a verdict of wilful murder by some person or persons unknown was recorded.

Following his transfer in January 1886 to Kildare, Glover received fulsome tributes from the Mayo grand jury at the March assizes for his ability, impartiality, attention to duty and valuable services which had given entire satisfaction to the public. In his new county, his energetic approach to his duties soon impressed the grand jury and they unanimously agreed to increase his salary to the statutory maximum of £500. He took on responsibility for the maintenance of some 1,150 miles of public roads, with growing traffic volumes, and built a number of bridges of stone, steel and concrete, including in 1897 the bridge which carries the N78 over the canal at Athy. In addition, he was engineer for a number of drainage districts in Kildare, Carlow and Meath. In 1888, he was one of a number of engineers who put forward proposals for a new water supply scheme for Athy.

In a humorous paper which he presented to the Institution of Civil Engineers of Ireland in 1897, Glover claimed that county surveyors were 'at their wits end to know how they are to meet the peremptory demands of cyclists who will soon be the majority'. In his view, the bicycle had but one advantage for an engineer – it was a 'sanitary animal' which did not leave dirt on the roads. However, to bring roads fully up to the standards required by cyclists would lead to bankruptcy for many ratepayers and to allow cycling on the footpaths would be an option only if 'scorchers and cads' could be induced to give the same degree of respect to old people, women and children as they gave to the 'strong man with the blackthorn'. He took the view, however, that 'the bicycle has a long lease of the ground it now occupies, after a hard up-hill fight of many years, and in spite of the sneers and contumely heaped upon it by many well-off, well-dressed indolent people, who until lately, looked upon cyclists as cads on castors'. He therefore proposed that special cycle tracks should be constructed on roadsides, according to a detailed specification which he presented – making him almost certainly the first engineer in Ireland to put forward this solution and, perhaps, one of the first to advance it in any country.

Glover's ideas on the treatment and disposal of town sewerage were also ahead of their time. In the discussion on a paper read to the Institution of Civil Engineers of Ireland in April 1894 he noted that, for Irish engineers, the real difficulty was not the location and construction of outfalls to deliver crude or quasi-treated sewage into tidal waters – for that was child's play – but the problem of 'manufacturing potable water (practically) out of complex and compound filth of varying strength' which an engineer was legally required to do by virtue of the Rivers Pollution Act where it was proposed to discharge sewage from an inland town to a river or stream. In the absence, therefore, of a solution to the problem of sewage treatment or purification, thousands of people in inland towns and villages would be condemned to live in dirt and filth, because their local sanitary authorities were not allowed to construct sewers discharging to a river, and whoever succeeded in resolving the problem was sure to earn both money and fame. For Glover, existing methods were unsatisfactory:

purely chemical methods were false in principle although the brute forces of money and ignorance would give them a certain lease of life, while land treatment also had many drawbacks, chiefly the amount of land required. Success was likely to be found, according to Glover, by some method based on the principle of sticking as close to nature as possible while providing artificial means of speeding up digestion by 'land life' and oxygen.

At the spring assizes in 1888 Glover presented a report to the grand jury on the possible impact on the county of a bill then before parliament to authorize a major drainage scheme in the river Barrow catchment. He was particularly concerned at the extent to which roads and bridges in the county would be interfered with in the guise of 'improvement', and at the amounts which would be levied on the grand jury to recoup the cost of these works. Given the long series of attempts to obtain government support for a Barrow drainage scheme, the grand jury were reluctant to oppose the bill, but decided, nevertheless, on Glover's advice, to petition parliament in an effort to secure modifications which would protect their interests. In the event, the estimated cost of the proposed works was so great (£360,000) that a Barrow drainage scheme was not commenced until the late 1920s.

In 1893 Glover was concerned about a proposal by the War Department to close the public road at Donnelly's Hollow, which crossed the Curragh rifle grounds, during firing exercises as this would lead to very great public inconvenience. In that year too, he was engaged in some controversy with the Dublin & Blessington Steam Tramway Company which demanded that he should relax his requirements if a proposed 4½ mile extension of their Blessington line to Poulaphouca was to go ahead; the differences with the company were ultimately resolved and the extension opened in May 1895.

When it was decided by the British and Irish Automobile Club that the 1903 Gordon Bennett Cup race would be held in Ireland, Glover and his colleague in Queen's County, H.V. White, became heavily involved with the organizers in the selection and preparation of the route. The Dublin correspondent of *The Times* reported on 23 March 1903 that 'these gentlemen, and everybody in the two counties, down to the humblest peasant, are co-operating in the preparations with delightful zeal and enthusiasm'. The course finally chosen comprised two circuits, one from Kilcullen to Carlow and back to Kilcullen via Athy and a western circuit running from Kilcullen to Kildare, Portlaoise, Stradbally, and back to Kilcullen via Athy. On each of these routes, the surveyors agreed to have the roads steamrolled to give the best possible surfaces, to have roads widened and corners rounded off, trees felled and hedges trimmed, and surfaces levelled near culverts and other elevated spots where fast-travelling cars might 'jump-off', as the organizers put it, and leap thirty or forty feet through the air. Work of this kind went on urgently between March 1903 and race day on 2 July when an international field raced around the prepared circuits watched by thousands of spectators. Overall the event was hailed as a triumph of organization and Glover was warmly commended by the local authorities for the part he had played in making it possible.

Following his transfer to the service of the newly-elected Kildare county council in April 1899, Glover seems to have maintained an excellent relationship with his new masters. He designed a seal for the council, combining the arms of the towns of Naas and Athy with the harp of Leinster to form a design which was in use until 1991 when arms were formally granted to the county by the Chief Herald of Ireland. He became a fellow of the Royal Institute of British Architects in June 1893, one of the few county surveyors to achieve this status, and was involved in private engineering and architectural work in partnership with his brother-in-law, Charles Astley Owen FRIBA from 1887 onwards, with offices at Molesworth

Street, Dublin. He was president of the Institution of Civil Engineers of Ireland in 1900–2, the first serving county surveyor to attain that distinction.

Glover became ill towards the end of 1913 and was granted an extended period of sick leave during which his duties were assigned to Francis Bergin who was in practice in the area as a consulting engineer. On his retirement at the end of June 1914, he was granted a pension of £400 a year, based on his years of service in Mayo and in Kildare. The Local Government Board argued that his Mayo service could not be reckoned for pension purposes but the county council adhered to its original award, with the chairman publicly declaring that the whole county wanted the surveyor to have the maximum rate of pension. Glover lived with his family at Prince Patrick Terrace, North Circular Road, Dublin, during most of his term of office in Kildare. He married, firstly, Sarah Mary Brewster, daughter of **Henry Brewster**, surveyor for north Mayo, and secondly, Margaret Young, a widow from Co. Cavan who also predeceased him. He died at a Dublin nursing home on 20 March 1931, aged 83.

GOODWIN, Singleton (b. 1852), son of Revd William Goodwin, was born at Maryborough (Portlaoise) and educated at Kilkenny College. He entered Trinity College, Dublin, in November 1870 and obtained a BA in winter 1874 and a BAI degree in winter 1875. He was employed as resident engineer by John W. Dorman, contractor for the nine-mile-long Clonakilty Extension Railway, between 1884 and its completion in August 1886.

On the death of the Kerry county surveyor, **W.H. Deane**, in April 1887 the lord lieutenant agreed to a request by the grand jury that two surveyors should be appointed to replace him, one to be based in Tralee and the other in Killarney. Goodwin was appointed to take charge of the new southern division on 14 September 1887 but transferred to the northern division in October 1889 when the surveyor who was serving there, **E.A. Hackett**, moved to south Tipperary. Less than two years later, in July 1891, the division of the county was terminated, leaving Goodwin to carry responsibility for the whole county for the next twenty-nine years.

Goodwin always had problems in maintaining the extensive Kerry road network. These were exacerbated in 1910 when the horse-drawn coaches which had operated for many years on the famous Prince of Wales route between Glengarriff and Killarney, and on the Ring of Kerry, were replaced by what he described as 'motor-driven charabancs of a juggernaut type'. Goodwin claimed that the weights of these vehicles (which he estimated to be as much as six tons) and their speeds of up to twenty miles an hour were altogether unsuitable for the roads involved which were narrow and tortuous, with a thin crust and poor foundations in many places. 'For a few weeks', he told the Irish Road Congress in April 1911, 'everything went well'. But soon 'the strain became too great for the thin crust … the wheels went right through … the bog subsoil was squeezed up … long stretches of road were pot-holed … and what had formerly been a delight, became an abomination and a danger to the tourist and to the traveller'. And matters were aggravated by what was, even by Kerry standards, a wet season! The tunnel road above Glengarriff was particularly badly affected: it was reduced to a quagmire of slush and stones through which coaches had to take what Goodwin described as 'a flying leap at twelve or fifteen miles an hour'. How were the ratepayers of a poor and heavily-taxed county like Kerry to face the complete destruction of their roads, Goodwin asked the congress? His solution was new legislation which would enable the county council to recover damages from parties who unduly injured the roads – and, he went on, 'if that were to mean that the Irish Tourist Association would no longer run tourist coaches you may be very sure somebody else will, and be very glad too to pay a percentage of their profits towards the upkeep of the roads'. Government grants which were becoming available for improvement works on tourist roads were not the answer for Goodwin: roads of higher standard would attract higher traffic volumes, leading in turn to higher maintenance costs and higher rates for the farmers and cottiers.

Goodwin's appeals for user charges were, of course, strongly criticized by those involved in the growing tourism and motor-coach businesses but, almost 100 years later, his plea for some financial contribution from those responsible for the introduction to minor county roads 'of a class of traffic which they were never intended to carry' would probably still find a receptive audience among road engineers. His protests, however, brought some results: in 1910, an axle-weight limit of three tons was imposed in relation to the Kenmare suspension bridge; in the following year, the Local Government Board banned heavy motor traffic on the Kenmare–Killarney road and imposed a speed limit of six mph on other traffic; and in 1912 the Road Board allocated grants of £18,000 for steamrolling and other improvement works between Killarney and Glengarriff.

Before Goodwin took up duty in Kerry construction of the thirty-eight-mile-long Tralee & Dingle Light Railway had been approved by the grand jury following consideration of a report from his predecessor, **W.H. Deane**, in 1884. Construction began in 1888, the line opened for traffic on 31 March 1891 and it made headlines two years later when it was the location of the most serious accident on any Irish narrow-gauge line. On 22 May 1893, the driver of a special train carrying livestock and passengers from Dingle pig fair to Tralee lost control of his train on a steep bank near Camp and, having reached a speed of about 40 mph, the engine and seven wagons were derailed at a viaduct on a sharp curve and plunged into the stream below, killing the driver and two others and many of the pigs. Goodwin, who was a frequent traveller on the line and was required to report annually to the grand jury on its condition, happened to be walking towards the nearby Castlegregory Junction at the time of the accident, and was one of those who gave evidence at the subsequent Board of Trade inquiry to the effect that the speed of the train was greater than normal before the derailment. In 1896 continuing deficits, which had to be made good from the local rates, led to the transfer of the management of the line to the grand jury, operating through a committee of management; this, effectively, meant that Goodwin had overall responsibility for the operation of the line from then on.

Goodwin gave notice of his intention to retire in May 1919 but continued in an acting capacity until May 1920. County councillors paid tribute to him at that stage for the careful, fair and honest way in which he had served the county and because 'he had guarded the council against any surcharge'. The fact that the county was again divided into two divisions following his retirement, with a county surveyor for each, may be taken as a more positive acknowledgement of his work.

GORDON, Stewart (fl. 1828–60), worked as a civil assistant with the Ordnance Survey between 1828 and 1833. He was appointed Londonderry county surveyor on 17 May 1834 and, according to the *Dublin Builder*, held the office with much credit and the esteem of a wide circle for twenty-six years. He acted also as surveyor to the Honourable the Irish Society and while he told an inquiry in 1841 that he did private work 'more by deputy than in person', he seems to have succeeded in getting most of the best architectural commissions in Derry during the quarter-century he spent there. He certainly needed the income from a private practice if his evidence to the Select Committee on County Cess in 1836 is to be believed; he told the committee that travelling expenses in his first two years as county surveyor had cost him £400, and there were many other incidental expenses, as a result of which he was 'out of pocket a considerable sum'.

Gordon's buildings have been said to be rather solid in appearance, reflecting his engineering training, and stylistically old-fashioned. The best of them – and the best known – is probably the Scots Church at Great James Street, Derry (1837) which has an enormous Ionic portico at the top of a flight of steps and an attractive galleried interior carried on slender cast-iron columns. Another of his Presbyterian churches at Strand Road, Derry (1847) employed the Gothic idiom. In the late 1850s, he acted as overseer ('resident architect') of works at Magee College designed by E.P. Gribbon of Dublin, and reconstruction works at Dungiven Castle have also been attributed to him. As county surveyor, he was responsible for the design of a bridewell erected at Magherafelt in 1839 at a cost of £515, a courthouse built at Coleraine in 1852, and additions and modifications to the gaol at Derry, now demolished.

Gordon was actively engaged in new road construction work between 1834 and 1844 and designed a significant number of bridges. In the early 1840s, he arranged for special surveys of the old bridge over the river Bann at Coleraine which had been built by the Irish Society in 1748, and was responsible for the construction of a new bridge on the same site in 1844, with **Alexander Tate**, his assistant surveyor, in direct charge of the work. The three-arch Brickhill Bridge over the river Agivey near Coleraine (1855) was another of his projects. Gordon lived at Aberfoyle House, Derry, a large Victorian house in its own grounds on the north of the city, and died there on 17 June 1860. He was buried in the cemetery at Derry.

GRANTHAM, Richard Boxall (1805–91), was born in London on 13 December 1805, the eldest son of John Grantham, a surveyor and civil engineer. He was educated in Kent and when little more than 16 years of age entered the London office of Augustus Charles Pugin (1762–1832), a distinguished draftsman, drawing-master, architect and architectural illustrator and father of A.W.N. Pugin, architect of the English Gothic Revival. Pugin had established a flourishing school of architectural drawing and some of the measured drawings prepared by Grantham during his long hours of attendance there (from 6.00 a.m. to 6.00 p.m.) were later to appear in Pugin and Britton's *Illustrations of the Public Buildings of London* (2 vols, London, 1825, 1828); one of them is a plan and elevation of the novel Diorama building designed in part by Pugin which opened in Regent's Park, London, in 1823.

Grantham's father was commissioned in April 1821 by the prominent consulting engineer, John Rennie, acting for the government, to survey the river Shannon from Rooskey to Killaloe with a view to reducing the level of the river so that bogs in the midlands and other land liable to flooding could more readily drain into it. After Rennie's death in October 1821, the elder Grantham continued the work and submitted a detailed report in 1822 incorporating an accurate chart of the river and proposals for the drainage of some two million acres at an estimated cost of about £310,000. Richard Grantham joined his father in Ireland in 1823 and worked with him on rebuilding the bridges over the Shannon at Limerick, Killaloe and Portumna, with arches of increased span to alleviate flooding.

With his father and younger brother John who was also a civil engineer (and was to become surveyor of iron ships for the Admiralty, an advisor to Brunel on the design of his ill-fated Great Britain and author of a textbook on iron shipbuilding), Richard Grantham introduced steam navigation to the lower Shannon. In the mid-1820s the family's Shannon Steam Navigation Company brought the *Marquis Wellesley*, a twin-hulled iron-built steamer with a paddle wheel between the hulls, from Dublin to the river via the Grand Canal. The ship had a registered tonnage of 120 and a speed of three knots and was the first iron steamer to ply within Ireland. She was in passenger service on the Shannon until 1858 but the Granthams' direct association with the service ended in 1829 when their company was taken over by the

Irish Inland Steam Navigation Company set up by Charles Wye Williams and which itself was absorbed shortly afterwards by the rapidly growing City of Dublin Steam Packet Company set up by Williams in 1823.

Grantham's father took responsibility as resident engineer for the completion of Wellesley Bridge in Limerick (now Sarsfield Bridge) following the death in 1831 of Alexander Nimmo whose design of the attractive five-arch structure was based on plans which had been submitted by the elder Grantham himself in May 1822. After a period working on English railway surveys with his father who died in February 1833, and his brother, Richard was appointed county surveyor for King's County on 17 May 1834. He was transferred, probably at his own request, to the Clare surveyorship on 9 December 1834. Some months later at the spring assizes in March 1835 he submitted a very detailed report on the state of the Clare roads and put forward a major programme of improvement works. Roads were said to be 'in indifferent repair' in some baronies because of the 'badness of the material' and were too narrow and winding in places for the use made of them, especially in the vicinity of Ennis. In the Burren, roads were few and far between but were little used; flooding of roads was caused in some places by eel-weirs on adjoining rivers; and many roads in coastal areas were in need of improvement to cater for the heavy traffic in sea sand for use as manure. In relation to the main roads, Grantham proposed improvements to the Limerick–Ennis road and suggested that a new mail coach road should be constructed from Ennis to Dublin, running via Tulla, Scarriff, Mountshannon and Portumna.

Grantham could have made little impact on the roads of Clare by September 1836 when he resigned his county surveyor post and returned to England. He was engaged by Isambard Kingdom Brunel as an assistant engineer on the construction of the Great Western Railway; he worked on the Brent Viaduct alongside Joseph Colthurst, a Cork-born engineer, and one of his juniors was Christopher Bagot Lane (1814–77), who was later to become assistant to the professor of engineering at Trinity College, Dublin, and first occupant of the chair of civil engineering at University College, Cork, in 1849. Grantham went on to serve as resident engineer on a branch line from Gloucester to Cheltenham but left the service of the Great Western in 1844. For the next two years he was employed principally by Sir John Rennie in making surveys for railways in the English midlands. He also carried out survey work for the Newry and Enniskillen line.

After the railway mania had abated, Grantham continued to work as a consulting engineer, turning his attention initially to sanitary matters. In 1847 he went to Paris to examine the city's abattoirs and subsequently, in the hope of establishing a similar system in London, presented a paper on the subject to the Institution of Civil Engineers and published *A Treatise on Public Slaughter Houses*. Some years later he returned to railway work, surveying sections of the lines from London to Portsmouth and to Bury St Edmonds. With his brother John he was appointed in 1860 as engineer to the Northern Railway of Buenos Aires and for some years he also acted as engineer to the Quebrada Railway Company in Venezuela and supervised construction of part of the line.

In the early 1860s, Grantham became involved in a campaign which led to the passing of the English Land Drainage Act 1861 and he was subsequently appointed as an inspector of drainage under the Inclosure Commissioners for England and Wales who had been set up under the act. He held numerous public inquiries during the following twenty years into arterial drainage and flood relief proposals for districts throughout England, including the Thames Valley, and he superintended a major reclamation project at Brading Harbour in the Isle of Wight. In later life, he interested himself in questions relating to the treatment and disposal of sewerage and chaired a committee of inquiry which reported in favour of disposal on land as a means of purifying sewerage, rather than sea disposal. He went on to design and carry out a considerable number of sewerage disposal works.

Grantham lived to be one of the oldest members of the Institution of Civil Engineers of which he had become a member in 1844 and to which contributed a substantial number of papers on a variety of subjects. A memoir in the proceedings of the institution noted that from the time he entered Pugin's office until 1891, almost seventy years in all, his life was one of continuous hard work and that, to the end, the desire to be at work was strong within him. 'Conscientious and ardent in his work, and of a sanguine temperament, difficulties never caused him to despond or ruffled his prevailing good temper. Genial, bright and courteous in manner, he was fond of society, particularly that of men of science'. Grantham attended his office on Friday, 27 November 1891, but the effort overtaxed his strength; he collapsed some days later and passed peacefully away on Saturday, 5 December 1891, shortly before his 86th birthday.

GRAY, Richard Armstrong (1819–99), was the contractor who built the five-arch masonry bridge over the river Shannon at Carrick-on-Shannon in 1846 for the Shannon commissioners. He was subsequently employed by the Board of Works and became district engineer for the Upper Boyne drainage district in 1849 and later for the Boyne and Deel districts. His salary as a full-time employee was £230 a year until 1853 but, with the winding down of the Board's drainage organization, his subsequent employment was on the basis of one guinea a day, when required. Gray became one of the three Dublin district surveyors in 1856, with responsibility for the northern division, but continued to have responsibility for the maintenance of the Boyne drainage works into the 1870s. His workload as district surveyor was increased substantially in 1866 when, following the retirement of **R.H. Frith**, the grand jury reduced the number of district surveyors to two, leaving Gray with responsibility for an enlarged southern division of the county but without the services of a clerk or of assistant surveyors whose employment was not permitted by the grand jury laws relating to Co. Dublin.

In this situation, it was not surprising that the condition of the roads and bridges in Dublin was extremely unsatisfactory and gave rise to regular criticism. In the 1860s and 1870s the state of the Rock Road, from Merrion to Blackrock, was the subject of particular controversy and, in view of the importance of the road as the link between Dublin and its fashionable new residential suburbs, as well as the mail boat harbour at Kingstown, special legislation was enacted to allow the Board of Works to arrange for works costing £7,200 to be carried out in 1876–8, at the expense of the city and county grand juries, to restore the road to good order. Gray personally came in for severe criticism on account of the condition of the road which was said by the board to be in 'utter disrepair' but he defended himself stoutly in letters to the newspapers: he could neither raise money nor spend money of his own will for any purpose; he could not make, prevent nor break any contract; he could pay no contractor nor prevent him being paid; in short 'it is out of my power to do anything to the Rock Road, except to look at it, and to report to others the state in which it is'. He claimed that his specifications for the repair of the road had been reduced by the grand jury and that the contract had been let for five years, against his advice, at a price some fifty per cent less than in earlier years.

The fact that Gray was allowed no assistants either for the growing volume of office work or to supervise the increasing amount of road work he was being obliged to carry out directly, led the *Irish Builder* to comment that 'he is more an object for our pity or sympathy than for our resentment'. But the journal went on, quite legitimately, to point out that while Gray might have been unusually unlucky with the contractors and others whom he had to deal with, there was legal machinery available which he could have used to remedy the problems. Gray, however, was able to advance other excuses, attributing the condition of the Rock Road partly to the fact that it had been dug up repeatedly to lay water, gas and sewer pipes. He had difficulty also with the tramways company, as did his colleague **James Bell** in the northern division, following the introduction of horse trams in 1872. Twenty years later he was still having difficulties with the transport companies: he had to sue the Dublin, Wicklow & Wexford Railway Company to force them to maintain the bridge over the railway and its approach roads at Shankill and he had regular difficulties with the Dublin and Blessington Steam Tramway Company which left the roads in some areas in a dangerous and objectionable condition.

Gray struggled on, without assistants or clerk, into the 1890s, and with growing traffic volumes the roads of Co. Dublin continued to be in a notoriously bad state. In 1891, the grand jury finally decided to take some action to improve the situation and as the law did not allow them to appoint assistant surveyors (as every other grand jury had power to do), they opted to revert to the situation which had obtained before 1866 by dividing the county into three districts and appointing a third surveyor. However, to make up the pay of the third man, the pay of the two existing men was to be reduced. Gray objected to this and the matter came before the privy council, one of whose members remarked that it was a strange suggestion to cut down the pay of old men (Gray and his colleague, **Thomas Turner**, who had succeeded Bell, were then in their seventies) who had spent much of their lives in the service of the county, because their work had increased. To add insult to injury, counsel for the grand jury suggested that Gray might retire on pension if he objected to the reduction of his pay. But Gray had no wish to retire and the privy council had no hesitation in refusing the application to reduce his salary.

Evidence given by Gray at the privy council enquiry confirmed that a reorganization of engineering services in Co. Dublin was long overdue. For instance, when asked by the lord chancellor who measured the stones and road materials on contract roads, Gray replied, to the obvious surprise of the council, that he carried out this work himself, with his own hands, generally in the first three weeks of September, and when 'hard-run' he had to employ a man at his own expense to help him, as the quantity involved ran to over 52,000 tons per year. There appeared to be a general impression, according to the *Irish Builder*, that it was 'derogatory to the office and lowering of the surveyor's influence in the eyes of the people ... and bad policy, for a grand jury to force such a service upon a man whose qualifications were unnecessarily high for such an occupation'. Yet, when the County Surveyors (Ireland) Bill 1893 was before parliament, an effort to include provision for the appointment of up to four assistant surveyors in Dublin was defeated. It was to take another four years to have the law amended by the County Dublin Surveyors Act 1897 so as to bring the engineering organization in Dublin into line with the rest of the country, with a single county surveyor, and as many assistants as the grand jury thought necessary.

In spite of his problems with the roads, Gray seems to have found time to engage in a certain amount of private practice in engineering and architecture. He prepared a plan for a Catholic chapel at Ballymun in 1864 and in 1875 he was engaged by Blackrock town commissioners in connection with the reclamation of slob lands, drainage works, and the development of the Peoples' Park. Two years later, the works were dragging on and there were complaints that the project had been badly managed. An example of Gray's architectural work – a gate lodge 'of an ornate character' at Johnstown Kennedy, Rathcoole, Co. Dublin – was illustrated in the *Irish Builder* of 15 August 1879. In 1886, when criticized for undertaking the design of dwellings at Finglas for the North Dublin union, contrary to his employment contract, he defended himself by arguing that this was not private practice as it only occupied his leisure time!

Gray was in office until the spring of 1896. He lived at Fortfield House, Cowper Road, Rathmines, where he died on 3 April 1899, aged 80. He is buried at Mount Jerome cemetery with his daughter, Sarah Maria Gray, who died in 1916. His wife, Susanna, lived in British Columbia after his death. His son, Robert de Witt Gray (b. 1861) was apprenticed to his father and was employed by the Board of Works as superintendent of works at Howth, Dun Laoghaire and Dunmore East harbours in the final decades of the nineteenth century.

GRAY, Roderick (fl. 1825–76), was one of the large team of engineers, overseers and clerks assembled by Alexander Nimmo in the 1820s to work on the planning and construction of roads at government expense in the western district of Ireland. Gray was employed as a pay clerk at Nimmo's Dublin office from 1825 onwards and was earning £1 5s. per week in 1831. In the following year he sought employment with the Ordnance Survey as a civil assistant for which the pay was 4s. a day.

Gray became Fermanagh county surveyor on 17 May 1834. Initially at any rate, he devoted his entire time to his public duties which, according to himself, were so onerous as to prevent him from earning even one farthing from private work up to 1841. He had constructed 275 miles of new roads by 1868, many of them in the western part of the county between Beleek and Enniskillen. One of the new roads in the barony of Glenawley gave rise to considerable controversy and Gray himself received a threatening letter through the post in May 1845 reminding him of the fate of his late neighbour, Captain McLeod, should he attempt to construct the road. The threat was taken seriously by the authorities and a reward of £50 was offered for information leading to the apprehension and bringing to justice of the writer.

Like the other surveyors of his era, Gray shared responsibility with Board of Works engineers for the supervision of famine relief works in his county. In 1846, before it had become official policy to do so, he was involved in the organization of soup kitchens in Enniskillen and Lisbellaw, but the fact that a payment of one penny was sought for the service gave rise to anger and tension among those seeking to avail of it. In 1853, when work on the Ballinamore and Ballyconnell Canal was well under way, opening up the prospect of commercial navigation from the Shannon into Upper Lough Erne, Gray put forward extensive and costly plans for drainage, flood-relief and navigation works in the upper and lower loughs. These were heavily criticized by the Board of Works inspector, **Barry D. Gibbons,** himself a former county surveyor. Gibbons doubted the need for costly navigation works centred on Enniskillen at a time when a railway link between the town and Belfast was rapidly nearing completion, and when there were prospects also of railway links to Londonderry and Sligo. He was severely critical of Gray's plans, costings and estimated rates of return: 'with every respect for Mr Gray', he wrote, 'I am bound to declare my conviction that if his plan of operations were carried out to the full extent, the flooded lands would still remain unrelieved and unprotected to an appreciable extent … If his principle of contracting the lake surface [by heavy embankments] were acted upon at all extensively, the damage to the low lands would be materially increased instead of prevented'. This seems to have marked the end of Gray's foray into the field of drainage and navigation which had already become a very controversial one and the subject of a House of Lords enquiry the previous year.

A house at Derrygore, Enniskillen, built in 1850 but long since demolished, and a small Church of Ireland church at Drumlone, Co. Cavan (1854), are the only new building projects attributed to Gray but he was also responsible in the 1860s for some remodelling of the courthouse at Enniskillen, completed to a design by William Farrell in the early 1820s. He resigned on account of ill-health in 1876 with a pension of £333 a year.

GREEN, Charles Frederick (1845–86), was born on 9 December 1845, the second son of Abraham Green of Essendon, Hertfordshire, and was educated at Bedford Public School. He showed a remarkable early talent for science and mathematics, prompting his headmaster to write that he had 'rarely seen good natural talent combined with such indomitable energy and perseverance'. On leaving school, Green became a pupil under C.H. Gough, then resident engineer on the Mid-Sussex Railway, and after three years went to work as assistant engineer on harbour works and as resident engineer on the Lewes & Uckfield Railway. He then spent four years in private practice in Hertford, and carried out sea defence works at Newhaven and at Bognor.

At the age of 26, Green was one of two Englishmen who took the first places at the county surveyor examinations held by the Civil Service Commissioners in June 1872; with 717 marks out of a possible maximum of 1,000, he was 120 marks ahead of the next-placed competitor, **S.A. Kirkby**. He became surveyor for the southern division of Cork east riding in August 1872 but served there for less than two years. A new pier at Crosshaven, and associated sea walls, seems to be the only project with which he can be associated.

Green was a candidate in April 1874 at the first-ever open competition for the position of assistant engineer in the Board of Works; there were seven candidates in all and the examinations in fourteen subjects extended over fourteen days. Green won the appointment in June 1874. The salary of £400 a year for the post, with increments of £20 a year and a maximum of £600, was not much better than the rates paid to county surveyors at the time but the post was a more prestigious one and there were prospects of internal promotion without further examination to the principal engineer position. Green, however, died at his home, Eldorado, Stillorgan, Co. Dublin, at the early age of 41 on 26 September 1886 before any further promotion came his way, and was buried at Mount Jerome cemetery. He was survived by his wife, Sarah Rose Green and a young family. Cecil, one of his sons, an officer in the 19th Hussars, died in Natal in March 1900 during the Boer War.

During his twelve years in the Board of Works, Green worked under Robert Manning (1816–97) whose promotion to principal engineer in 1874 had created the vacancy to which Green was appointed. With no other engineering staff and the assistance of only one draftsman, the two men had to undertake a wide variety of projects all over Ireland, including piers, harbours, inland navigation schemes, arterial drainage works, bridges and roads. In the early 1880s Green had overall charge of a new phase of Shannon flood regulation works including the Meelick Cut, about half a mile long, which occupied up to 600 men at times.

In spite of the pressure of work, Green seems to have maintained his interest in railway engineering and contributed two detailed papers on aspects of the subject to the Institution of Civil Engineers of Ireland in 1878 and in 1881. He was said to be a man of singular piety, and although possessed of great intellectual powers was simple and gentle in character. On his death, Robert Manning wrote:

It is my melancholy duty to express my grateful acknowledgement of the able and willing assistance which I received from Mr Green on all occasions, and to bear my testimony to the invariable kindness and consideration with which he treated everyone with whom he came in contact, and which earned for him the respect, esteem, and friendship of all who knew him. His early death has been a great loss to the public service, and especially to me, and I deeply feel the sudden termination of so many years of intimacy with him.

Manning himself, author of the famous formula for open channel flow, retained his post as principal engineer for another five years, retiring in 1891 when he was 75 years old; he died in Dublin in December 1897.

GRIGOR, William (fl. 1935–1962), was Antrim county surveyor from 1935 to 1962. Described as 'a canny Scot', his chief interest was the economical maintenance of roads. He was chairman of the Northern Ireland association of the Institution of Civil Engineers in 1951–2.

GUNNIS, John William (c.1862–1914), was born in Windsor, Berkshire, the eldest son of James W. Gunnis (described in the 1871 census as 'Musician to the Queen') and his wife, Ellen. He was articled to the architect, Alfred Drew of Margate, between 1879 and 1882 and later served as assistant in the offices of Hubert Bensted and of Ruck, Son & Smith, both of Maidstone, before returning to the office of Alfred Drew. He was elected an associate of the Royal Institute of British Architects in 1889 and was living at Dover in 1891 when he competed successfully for a county surveyor appointment. He was assigned to the Longford post on 19 December 1891 to replace **J.O. Moynan** who had transferred to Tipperary north riding.

According to a report presented to the grand jury by its courthouse committee at the 1899 spring assizes, the courthouse in Longford had just been refurbished and remodelled substantially 'due ... above all to the county surveyor, Mr Gunnis, for his invaluable gratuitous assistance and services'. As a result, the county now had 'one of the most commodious and comfortable courthouses in the country, at a cost of about as many hundreds of pounds as the proposed new courthouse would have been thousands'. The warmest thanks of the grand jury were given to Mr Gunnis 'a most deserving and hard-worked county official'.

Gunnis had two assistant surveyors, one for the north of the county, and one for the south, but experienced the same difficulties with small local road contractors that had influenced his predecessor to leave the county. His efforts to improve matters were resisted by the county councillors who voted unanimously in 1905 not to allow him to introduce steamrolling on the grounds that it would deprive these contractors of a source of income.

Having been ill for most of the year, it came as no surprise to the county council that Gunnis gave notice of his retirement at a special meeting in December 1913 when he had completed twenty-two years service as county surveyor. According to the *Longford Independent* reporter, he told the council that he had been finding it 'impossible for years past to get the job done with the road contractors'. On top of that, he complained that he was forced 'to knock a half-crown's worth of work out of men for a shilling, and it has certainly knocked me out'! He left the service of the council on 10 January 1914, with a pension of £216 a year.

HACKETT, Edward Augustus (1859–1945), was born on 1 December 1859, the fifth and youngest son of a well-known Co. Offaly gentleman, Thomas Hackett JP of Castletown Park, Ballycumber, and his wife, Henrietta Clementina Fawcett. He was educated privately, entered Queen's College, Galway, in 1877 and obtained the BE degree in 1880 with first-class honours, a gold medal and first place in Ireland; in 1882, he was awarded an ME degree without examination. He served a pupilage to James Price of Dublin in 1880–1 and in the following year was employed, as assistant to B.F. Flemying of Dublin, on the planning of a branch of the Great Southern & Western Railway to serve Sallins and Baltinglass. Between 1882 and 1886 he worked for the government of Natal as assistant resident engineer on the Ladysmith Extension Railway and on his return from South Africa was engaged as contractor's engineer on the construction of the Craigmaddie reservoir which was to serve Glasgow. In 1886–7 he was one of ten engineers engaged by the Piers and Roads Commissioners to supervise construction works on a variety of projects in distressed coastal areas in Galway and Mayo; his main assignment involved the building of a concrete pier and breakwater, 155 feet long, at Clare Island.

Hackett took first place at the county surveyor examinations held in 1887 and was appointed to the newly-established northern division of Kerry in September 1887. He moved to the better-paid post in Tipperary south riding in September 1889 and took on the separate position of engineer to Carrick-on-Suir urban district council in 1899, earning an additional salary of £50. He developed a reputation at an early stage for excellence and the use of new techniques in the construction, maintenance and management of the roads of Tipperary. He was a strong advocate of the direct labour system provided, as he put it, that the labourers would not strike for a big wage and that one could engage honest and skilled gangers and stewards. Hackett was also one of the pioneers in the introduction of steamrolling and mechanized stone crushing plant; a pamphlet of his entitled *Economical Steam Rolling of Irish Country Roads* was published in 1904 on the initiative of the Irish Roads Improvement Association. By then, over 500 miles of roads in south Tipperary were being maintained by direct labour and some forty miles were being steamrolled each year. All of this, according to Hackett, brought important benefits to the county generally and, as well, would lead to a reduction in roads expenditure in the medium-term; it provided absolute conclusive proof, according to the *Irish Cyclist*, that good roads and economy go hand in hand. In 1906, he was critical of existing methods of applying tar to roads to make them dust-free: he contended that in wet weather the tar became diluted with water, producing an offensive black mud which was carried on people's boots into their dwelling houses, while in dry weather the tar adhered to particles of dust and flew about. Hackett announced at that stage that he was experimenting with a new procedure which involved the application of a coating of tar and lime over a dry-rolled stone base, and then applying a coating of chips or fine gravel, rolled smooth – this was, in effect, the system which was commonly used in Ireland in the succeeding decades.

Hackett and two other county surveyors, **W.E.L. Duffin** of Waterford and **A.M. Burden** of Kilkenny, were members of a bridge commission established in 1907 to consider the need for a new crossing of the river Suir at Waterford in place of the wooden toll bridge which had been completed by the American, Lemuel Cox, in January 1794. In October 1908 the commission recommended the erection of an elaborate steel girder bridge with stone abutments and four concrete piers at a total cost of £114,000, but their design was denounced for reasons of cost and on aesthetic grounds. The recommendation was formally challenged by a number of the local authorities who would have had to share the cost, with Waterford corporation, in particular, arguing strongly for a ferro-concrete bridge, estimated to cost about £40,000, for which designs had been put forward by Mouchel & Partners of London and Sir William Arrol as early as 1903.

In April 1909 the judicial committee of the privy council rejected the report of the commission and held that a new bridge could be provided with less expense either by the erection of a different form of iron bridge or by constructing a ferro-concrete structure, instead of what Mr Justice Kenny described as 'the extreme ugliness of the bridge designed by the commission'. But that was not the end of the matter for Hackett who maintained his trenchant opposition to the ferro-concrete option. There were different views on the subject generally. Reinforced concrete, incorporating wrought-iron or steel bars to enhance the tensile strength of the material, was quite well known in Ireland by 1900 and its use in buildings and structures of different kinds had been endorsed in 1907 by the Royal Institute of British Architects which reported that 'it was at least as durable as brick or stone'; one year later, however, the president of the English Local Government Board told the House of Commons that there was still need for caution in dealing with the material as it was doubtful whether it was suitable for structural works under all conditions. Hackett, essentially, took the latter view, claiming that ferro-concrete was untrustworthy in water, that it was still at the experimental stage, and that it was being pushed and advertised to quite an extraordinary extent: 'not even our famous ointments and soaps have their virtues so highly extolled'.

Hackett's view was challenged by engineers and others who argued that ferro-concrete had long since passed the experimental stage. It was pointed out that it had been used to construct substantial bridges in the United States with clear spans of well over 100 feet and that reinforced concrete, of a kind, had been in use in Ireland since the 1880s. By the early 1900s, the system devised by the specialist French designer, François Hennebique (1843–1921), and promoted by his British agent, L.G. Mouchel, had been employed at a number of places around the country. In Waterford itself, where the extension of the Great Southern & Western Railway Company's station had necessitated diversion of the main road into the city from Dublin and Kilkenny, a 720-foot-long viaduct was completed parallel to the bank of the Suir in 1906, with ferro-concrete piles and decking; this structure, built by the Yorkshire Hennebique Contracting Company, was Ireland's first major reinforced concrete structure. At Passage East, only a few miles from Waterford, ferro-concrete had been used around the same time in the construction of a breakwater, 120 feet long, and in 1909 a footbridge in precast ferro-concrete, 172 feet long, was being completed to link the new Mizen Head signal station to the Co. Cork mainland; this was the longest single span of its kind in Britain or Ireland at the time.

The Waterford Bridge controversy – one of the greatest debates in Irish engineering circles since the argument about the Dublin loop-line in the 1880s – continued through 1909, with Waterford corporation reversing its position on the ferro-concrete option, the privy council refusing to go back on its decision, and the estimated cost of a ferro-concrete bridge creeping upwards to £71,000. Finally, in December 1909, the privy council formally ordered a ferro-concrete bridge to be built and designs for this on the Hennebique system were finalized by Mouchel and Partners. The bridge was built between 1911 and 1913 by a Glasgow firm, Kinnear Moodie & Co., but it was Burden, the Kilkenny surveyor, rather than Hackett, who carried primary responsibility for supervising the work. The bridge was formally opened by John Redmond after whom it was named. Before its construction, it had been written-off by another county surveyor, **J.O. Moynan**, as at best, a costly experiment. It survived until 1984, a much shorter life span than that of the masterpiece built by the Boston carpenter all those years before; it was replaced by Brother Edmund Ignatius Rice Bridge which was completed on the same site in November 1986.

After the coming into operation of the Local Government (Ireland) Act 1898 and his transfer to the service of the new county council, the Local Government Board directed that Hackett's salary as county surveyor should be increased from £600 to £800 a year but the council refused to pay the higher amount. Hackett sought redress in the courts and was successful in the first instance. However, on appeal by the county council, the court of appeal held against him on a technicality and he was directed to pay all of the council's costs. Hackett was granted the salary increase by his council some years later when he claimed that his official duties had become so demanding that he could not attend to private business. But when he sought a further increase in October 1907 his request brought a strong response from the council. John Cullinan, the nationalist MP for Tipperary South, complained that the surveyor's duties now appeared to be so light that he was able to leave his work in Tipperary, devote his time to activities elsewhere, and mulct the ratepayers. Cullinan was referring to Hackett's involvement in what was being referred to as the Glenaheiry Lodge affair – an early-morning explosion on 14 August 1907 at a shooting lodge owned by Lord Ashtown near Ballymacarbery, Co. Waterford, following which a substantial claim for compensation for malicious damage was made by Lord Ashtown, supported by evidence given by Hackett.

During his years in Tipperary, Hackett carried on a substantial private practice, much of it relating to small water and sewerage schemes, and regularly appeared as an expert witness in law suits. He was engaged in 1898 in the carrying out of alterations and additions at country houses, including Birdhill House, Clonmel, and Johnstown Castle, near Fethard; he was appointed to design new sanitary arrangements at the workhouses in Tipperary (1898) and Clonmel (1902); and, with J.F. Fuller of Dublin, he prepared plans for substantial additions at the lunatic asylum in Clonmel in 1901. This latter assignment prompted the Royal Institute of the Architects of Ireland to send the county council a copy of their resolution suggesting that county surveyors should be required to devote all of their time to their official duties but this intervention was rejected by the council in what the *Irish Builder* described as an ignorant and contemptuous fashion.

Hackett was responsible for the establishment of Clonmel golf club in 1912 with a nine-hole course three miles from the town. He was prepared also to become involved directly in economic enterprises of different kinds both within and outside his county. In December 1919 he told the Dáil Éireann Commission of Inquiry into the Resources and Industries of Ireland that he had carried out a survey for English interests of possible sites for a large carbide production plant and that, when this did not go ahead, he had himself established an undertaking of this kind at Askeaton, Co. Limerick. He told the same commission that with some others he had been 'foolish enough to run a coal mining property' in the Slieveardagh area some years before and while a seam of good quality coal, thirty-six inches thick, had been located, he and his partners found that the game was not worth the candle because of difficulties with the employees; the chief problem, according to Hackett, was that the men, having been paid on a Saturday night, bought a barrel of porter and 'sat around drinking it until the following Wednesday'. He advocated, nevertheless, that the establishment in the area of a briquette-making plant should be considered because of the high proportion of slack produced by the local mines.

In 1883 Hackett married Emilie Elliott Henry, daughter of Captain J.W. Henry RN of Ballyshannon, and they lived at Greenville, Clonmel, during his service in Tipperary. The couple had three daughters and three sons. Eric, aged 21, a lieutenant in the Royal Irish

Regiment, was killed in action at Ginchy during the battle of the Somme in September 1916, while Learo, the eldest, a captain in the Royal Irish Rifles and the holder of the Military Cross, was killed in action at Ypres in April 1918; they are both commemorated on the Men of Thomond war memorial in St Mary's cathedral, Limerick. The family's losses were tragically increased by the death from pneumonia in London on 13 October 1918 of one of their three daughters, Venice, on her return from service in France with the Voluntary Aid Detachment. Hackett himself, travelling to London to visit his ailing daughter, was one of the survivors of the sinking of the RMS *Leinster* by German torpedoes off Dun Laoghaire on 10 October, in which over 500 lives were lost.

For a man of Hackett's background and whose family had suffered so much in the service of the crown, the events of 1919–20 must have been the last straw: IRA violence began in south Tipperary in 1919, the county council transferred their allegiance from the Local Government Board to the Dáil Department of Local Government soon afterwards and it was virtually certain that the elections due in June 1920 would give Sinn Féin candidates an overwhelming majority on the council. All of this led Hackett to resign his county surveyor post in May 1920 when one of the councillors stated that while 'they might disagree with him politically, that did not prevent them from recognizing his courtesy and sterling worth'.

Hackett seems to have severed his formal links with engineering in November 1922 when his name was removed from the list of members of the Institution of Engineers of Ireland although he was nominated in 1925 as the Farmers' Union representative on the Roads Advisory Committee which operated under the Ministry of Transport Act 1919. He had by then retired to Castletown Park, Ballycumber, Co. Offaly, where he took over the farm of his older brother, Billy, who had died, but his farming was not very successful. The farm was acquired by the Land Commission in the 1930s, the house demolished, the stone used to pave the local roads, and the bodies in the family grave on the estate exhumed and reinterred in the graveyard at Liss, Ballycumber, near Clara. E.A. Hackett was buried there after his death at Woodenbridge, Co. Wicklow, on 2 February 1945, the gravestone noting that all six of his children had predeceased him.

HALL, Thomas Dugall (fl. 1834–57), worked as a surveyor for some years in the Sligo-Roscommon area before qualifying for appointment as county surveyor at the first examinations held in April 1834. He was not among the first group of surveyors appointed in the following month but became Leitrim county surveyor on 14 October 1836 when **N.R. St Leger** was transferred to Sligo.

At the spring assizes in 1838 Hall submitted a detailed twenty-eight-page printed *Report on the State and Progress of Public Buildings, Bridges, Roads and Other Works* in the county. In general, he found the roads and bridges to be very poorly maintained by inefficient and incompetent contractors and begged the grand jury to adopt some decisive measures with respect to contractors in the future; as a minimum, he suggested that they should examine the sufficiency and ability of each and every contractor and surety before putting works into their hands. Legal proceedings which he had initiated in a number of cases against defaulting contractors were no solution, in his view: the legal costs involved were excessive, the county still had to meet the cost of the road work and the delay was such that the public could be inconvenienced for up to two years. As regards bridges, a major programme of work was required in Hall's view. He drew particular attention to the state of the large bridge over the Shannon at Battlebridge; the foundations were very shallow, the masonry was composed of 'wretchedly bad materials' badly put together, and the arches were cracked from the springing to the crown. Money would be better spent on a new bridge than on repairs, according to Hall and this view was shared by W.T. Mulvany, the Board of Works engineer, who surveyed the bridge about the same time and described it as ruinous.

Hall's 1838 report provides some valuable comments on the impact of the Ordnance Survey on the work of civil engineers in Ireland. The first six-inch maps of Co. Leitrim had been published in 1836 and Hall had been provided with a set of these by the kindness, as he put it, of Colonel Thomas Colby, the survey's director. He found the maps to be 'of the greatest and most minute exactness, combined with the closest detail', and he attached particular importance to the data they provided on levels and heights; had these been available earlier, he said, many expensive bridges would have been built at different locations, or not built at all, and many roads would not have taken their present course.

Hall served in Leitrim for over twenty years. He died on 12 January 1857 at 108 Dorset Street, Dublin, leaving a wife and large family. He was, according to the *Leitrim Journal*, 'one of those good-natured men who in all the transactions of life, aimed at doing good; as a father, friend and neighbour, his only fault was to be too open-hearted'.

HAMPTON, Robert John (*c*.1821–55), was the eldest son of **William Hampton**, Wicklow county surveyor from 1834 to 1845, who arranged to have him certified in February 1837, when he was only about 16 years old, to serve as one of his assistant surveyors. He continued in that position until January 1845 when he became one of the first three Dublin district surveyors and was assigned to the No. 3 (southern) district. After the sudden death of his father in November 1845 Robert sought to be appointed to replace him as Wicklow county surveyor, but while the Board of Works reported that they had considered his application sympathetically in the sad circumstances, they felt bound to advise the lord lieutenant that the younger man had not sat the examination for county surveyors (but only the separate examination held in 1844 for positions in Dublin) and could not therefore be appointed without departing from the 'the principle which has been laid down and which has been universally followed'.

Hampton lived at 2 Grantham Street, Dublin, and was honorary secretary and treasurer of the Institution of Civil Engineers of Ireland from 1853 until his death from fever on 22 August 1855 at the early age of 34. He was buried with his parents and other members of the family at Mount Jerome cemetery in Dublin in a grave which has neither a kerb nor a headstone; however, near the entrance to the cemetery, a granite obelisk, carrying the insignia of a freemason of worshipful master rank, was subsequently erected to his memory by the road contractors 'as a mark of their esteem for him and as a testimony of their confidence in his integrity and impartiality in his dealings with them'.

A fine single-arch bridge over the river Dodder at Oldbawn with a span of fifty feet, which was erected in the 1840s to replace an older three-arch structure, can probably be attributed to Hampton. Erosion of the bed of the river, coupled with the haulage from it of sand and gravel, had seriously undermined and exposed the foundations by 1899 leaving them standing some twenty feet above the water level. Major works to underpin the structure were undertaken by the county council at that stage and the bridge is still in use and carrying heavy suburban traffic of a kind that its designer could never have imagined.

HAMPTON, William (*c*.1786–1845), was the son of a Ballinasloe merchant but little is known of his training and early career as a land surveyor. His older brother, John (1782–1846), served an apprenticeship in the office of Richard Morrison, the well-known architect, and was employed by the military authorities between 1803 and 1808 on planning, survey and land acquisition work when signal towers and fortifications were being constructed all around the coast, and along the river Shannon, in prospect of a possible French invasion. John went to become one of nearly sixty engineers and surveyors engaged on surveys for the Bogs Commissioners between 1809 and 1813; he worked mainly in Co. Longford under Richard Lovell Edgeworth, one of the district engineers engaged by the commissioners. He subsequently served under John Rennie in London and under Alexander Nimmo in Ireland in the early 1820s and was in practice as engineer and architect in Ballinasloe by 1827.

William Hampton, like his brother, worked as a surveyor for the Bogs Commissioners in 1812–13 and seems to have been well established in practice as a land surveyor at that stage. In 1813, he carried out a topographical survey of Co. Roscommon and produced a small book of attractive manuscript maps of the county's roads and villages. He was subsequently responsible for a number of county maps and he worked also with William Edgeworth on mapping and road surveys. The Hampton brothers attempted in the 1820s, after they had both married, to establish themselves in Ballinasloe as engineers, architects, and teachers of surveying and levelling, describing the business in their advertisements as 'an artists' and mechanics' warehouse'. They were engaged also in the making of 'accurate and elegant' estate maps'. However, the establishment of the Ordnance Survey in 1824, directed by military officers and staffed by salaried employees, created uncertainty as to the future prospects of existing and future land surveyors and forced the Hamptons to seek other employment. John, having rejected an offer of employment at four shillings a day under the Ordnance Survey, took up a post in the General Valuation Office in 1830 where he worked until his death on 27 December 1846 at the age of 64. William was described as a merchant on the baptismal certificates of his children baptized in the parish of Creagh, near Ballinasloe, between 1828 and 1832, but he turned to a career in engineering in the early 1830s when he was employed by the Board of Works on a number of road projects, including the building of a new Killaloe–Tuamgraney road along the western shore of Lough Derg a cost of nearly £3,000.

On 17 May 1834 William Hampton became the first county surveyor for Wicklow and established his offices at Flannel Hall in Rathdrum. He was simultaneously surveyor to the trustees of a section of turnpike road from Kilcullen to Carlow from which assignment he earned £50 a year. When the question of appointing an assistant county surveyor arose early in 1837, Michael Coogan of Baltinglass was strongly recommended for the post by James Grattan, MP for Wicklow, but he complained to the Board of Works that he had been prevented by Hampton from going forward for examination. The board, in reply, told Coogan that they could not believe that Hampton 'as a professional man, and in a public employment, would act so imprudently and improperly as to take into consideration the religion or politics of persons to be employed by him, or in any matters connected with his public duties'. The fact that Hampton's 16-year-old son, **Robert John Hampton,** was certified some weeks later to be qualified to serve as assistant surveyor may throw some light on Hampton's real motives in frustrating Coogan's application for the post.

Hampton completed new roads from Rathdrum towards Carnew, Co. Wexford, and Tullow and Hacketstown, Co. Carlow, in 1838, including a number of bridges but appears to have built a relatively small mileage of other new roads. On the other hand, he was responsible for the construction of a significant number of bridges throughout his county including the bridge at Greenane in 1844. In addition, there were presentments each year for the reconstruction or repair of bridges, such as those at Ashford, Annagowlan, Glencullen and Mucklagh, which were regularly damaged by floods.

Few public buildings in Wicklow can be attributed to Hampton. He was responsible for the provision of courthouses at Arklow and at Tinahely in 1842–3 at a cost of about £1,000 in each case, although it seems likely that an outside architect (either W.D. Butler or William Caldbeck) may have provided the basic designs. He had general responsibility also for improvement works carried out at Wicklow gaol in the early 1840s at a cost of some £6,000.

In response to an enquiry in 1841 from commissioners who had been appointed to review the grand jury laws, Hampton reported that he resided and kept an office in his county, and was 'always attending at it when not travelling or inspecting the works of the county'; in all, he thought that he spent up to fifty days each year in the office 'drawing plans, writing letters, instructing my assistants in the discharge of their several duties'. He was responsible at that stage for the maintenance of 870 miles of road and complained that, because he had only three assistants, there was a lack of control over the operations of the road contractors.

Hampton married Clementina St. Clair around 1820 and they had eleven children. On taking up his Wicklow appointment he and his family lived initially at Grove Hall near Rathdrum but moved to Enniskerry some years later and finally to 3 Harrington Street, Dublin. Having applied unsuccessfully for a Dublin district surveyor post in November 1844, he died suddenly at his home on 23 November 1845 at the age of 59, due to the rupture of a blood vessel. Less than two years later, in June 1847, Clementina joined him in his unmarked grave in Mount Jerome cemetery. The second youngest of their children, Emma Margaret (b. 1837), emigrated with her husband, John Fleming of Co. Longford, and their five young children to New Zealand in 1865 and died in Auckland in 1912; one of their great-grandchildren was Sir Edmund Hillary who, with Sherpa Tensing Norgay, was first to reach the summit of Mount Everest in May 1953 and the first man after the ill-fated Captain Robert Scott to reach the south pole by an overland route.

HANNIGAN, James Joseph (1881–1964), was a native of Ballybofey, Co. Donegal, where he was born on 24 August 1881. He was educated at St Columb's College, Derry, and at Queen's College, Galway, from which he graduated with honours in arts in 1904 and in engineering in 1905. He worked for a short time on the Central Railway in New York and was employed in London when he qualified at the county surveyor examinations held in December 1910.

Hannigan had competed unsuccessfully for vacancies in Offaly in 1909 (where it was held that his Irish was deficient) and in Mayo in 1910 before being elected Monaghan county surveyor at a special meeting of the county council in February 1912, with thirteen votes in his favour as against eleven divided between the only other two candidates who were in serious contention. The Local Government Board had reservations about the appointment because of Hannigan's lack of his experience in 'responsible engineering positions' but they agreed to approve the appointment for a one-year probationary period; in the event, Hannigan was to serve in Monaghan for over thirty-five years, winning the trust and respect of his council and the esteem of his staff. It was said of him that he remained urbane and courteous even in the most heated scenes in the council chamber or when his estimates of the expenditure required to maintain various services were under prolonged attack. He was particularly praised for his efforts to increase the production of turf during the emergency years from almost inaccessible bogs on the north Monaghan hillsides.

The Farney Development Company was set up by Hannigan and, though it ultimately failed, it led to significant economic development in and around Kingscourt in the 1920s. Later on, he discovered the large deposit of gypsum in the area and founded the Gypsum and

Bricks Company to exploit it because, according to the *Northern Standard*, 'it did not excite anyone but himself'. The company later developed into the successful Gypsum Industries Ltd of which Hannigan was a director until his death. After his retirement on 5 April 1947 he attempted to develop a marble industry in his native Donegal, as one of the Donegal surveyors, **William Harte**, had sought to do eighty years earlier. Despite Hannigan's claims that the product was equal to the best Italian marbles, he never succeeded in attracting sufficient interest and support to put his proposals into effect.

Hannigan, who was unmarried, lived at Tirkeenan, Monaghan, and died on 26 May 1964; he was buried at Latlurcan cemetery, Monaghan. 'Few men of his time made a greater impact on the people of Co. Monaghan', wrote the *Northern Standard* after his death … not only will he be remembered as an able and devoted engineer … but for his foresight, energy and determination in establishing one of its most noted industries'.

HARRISON, Alexander (fl. 1834–51), served an apprenticeship to Messrs Sherrard, Brassington & Gale, the leading Dublin firm of land surveyors and valuators, before becoming the first county surveyor for Queen's County on 17 May 1834. He seems to have gone about his business in a methodical manner, dividing the roads in his county into four classes for maintenance purposes. The mail coach roads and the other main roads were to be kept in constant repair under his scheme, while the cross roads, linking the first two categories, were to receive occasional maintenance on a three-year cycle. Finally, the byroads, or green roads, were to be inspected and repaired, where necessary, every seven years.

Harrison designed a small sessions house at Mountmellick with an imposing neo-classical facade, completed in sandstone ashlar in 1841, and he carried out some improvement works at the courthouse in Portlaoise around the same time although, in this case, the plans may have been prepared by an outside architect. He may also have designed the courthouses at Mountrath (demolished in the early 1960s) and at Abbeyleix for which presentments were approved in 1838.

In 1839 Harrison applied to the chief secretary for a transfer to a smaller county because, he said, his health had been 'much impaired' and he could not continue to function without any assistants. Shortly after an exchange between him and **Edward V. Forrest** of Monaghan had come into effect in May 1839 the *Leinster Express* noted that 'a very unfounded report' had been circulated – and had gained credence in 'high quarters' – suggesting that Harrison had been obliged to leave the county 'from a feeling of a political or sectarian character against him'. But while implying that there may indeed have been differences between Harrison and the grand jury, the journal was happy to assure its readers that it had never considered the surveyor to have 'evinced any political or sectarian feeling' and that the alleged reasons for his transfer were completely inconsistent with the facts; as evidence of this, it recalled that both radical and conservative gentlemen had 'assailed Mr Harrison together' on different occasions, while he had been defended by similar cross-party groupings.

In Monaghan Harrison had to undertake substantial programmes of famine relief works in 1846–7 and seems to have been successful in introducing a system of task work under which, he reported, the men could see that they earned something more than the fixed rate of ten pence per day and were well satisfied. He was transferred to Carlow on 7 August 1850, in an exchange with **John Walker**, but his term of office there was a very brief one. He died at his residence at Athy Street, Carlow, on 25 March 1851 leaving a wife and two children. The *Carlow Sentinel* noted that he had been universally respected for his kindness of heart and social qualities and that he had discharged his duties with unremitting zeal and ability and to the public advantage. The cause of his death, after a short illness, was said to be typhus fever contracted in the discharge of his duties, a fact which suggests that, like his predecessor, he may have been involved in efforts to redesign and replace the malfunctioning sewers in Carlow town; indeed, one of the first projects submitted to the grand jury by the unfortunate man's successor, **Peter Burtchaell**, was the replacement of the John Street sewer.

HARTE, William (*c.*1825–95), was employed by the Board of Works as one of the engineers in charge of the relief works in Co. Cork in 1846–7, based at Bandon and later at Mallow. He seems to have acted on a temporary basis as county surveyor in Monaghan in 1849 before becoming county surveyor for the southern division of Donegal on 15 September 1851. He was transferred to the northern division in July 1864, effectively on promotion, as that division was in the second class for salary purposes and carried a salary of £500; the southern division was in the third class and paid only £400. Apart from this the Donegal grand jury who had forced **John Fraser** to leave his post in the northern division and transfer to Cavan, wanted 'to avail themselves of his tried services in bringing the roads, and the county surveyor's department, then in very bad order, into good condition'.

Harte described himself in 1887 as an experienced railway engineer and architect as well as a civil engineer but, apart from his involvement as county surveyor in the assessment of the many tramway and light railway schemes which came forward in the 1880s, it is difficult to find any evidence of his railway engineering activity. In the field of architecture the picture is somewhat clearer. He was reported to be responsible for the construction of a 300-seat concert hall at Convoy, Co. Donegal in 1867; for extensive alterations, additions and new offices at the courthouse and gaol in Lifford, completed in the same year at a cost of about £1,500; and for the design and construction of Carndonagh courthouse in 1873. In 1886 he was one of those who took part in a competition for the design of a new town hall in Derry; when the results were announced, he protested bitterly, pointing out that the plans submitted by the winner were incomplete and defective and did not comply with the conditions; the Guildhall, designed by John Guy Ferguson, was completed in 1887.

The *Irish Builder* described Harte in 1895 as a man 'whose tastes and gifts ran oft-times outside the dry routine of duty into geological and kindred research'. He had developed a substantial interest in archaeology and in 1866 was reported to have found shell middens on the shores of Lough Swilly, attributable, according to a contemporary account, to 'the aboriginal inhabitants of Ireland' and similar to those found in the Danish islands. In the course of his work in Donegal he noted the ready availability of granite and set about promoting its use for building work in Ireland and England. He claimed that the richer colour and greater durability of the Donegal granites made them suitable for a wide range of uses and in 1870 maintained that quarries which he was then operating near Dungloe and Bunbeg could compete on price, as well as quality, with best Scottish granite. The *Irish Builder* agreed at the time that Donegal was yielding excellent granites of different colours, quite as good as those of Dalkey or Ballyknockan or the much acclaimed Aberdeen red granite; however, the journal was forced to note, with regret, some twenty-five years later, that 'the tide of time and circumstance' was against Harte and that he had failed to profit from his enterprise and investment.

Harte became involved in some controversial drainage schemes in the early 1880s one of which included the construction of embankments 12½ feet high and extending for some 2½ miles along one bank of the Foyle below Lifford to protect the property of the earl of Erne and others from flooding. When it was put to him by the Royal Commission on Irish Public Works in March 1887 that the scheme merely diverted flood waters to the lands of the duke of Abercorn on the other side of the river, he readily agreed but insisted that 'people should look out for themselves' and anyway 'my embankment is twice as solid as theirs'. Another of his schemes in the Swilly catchment also ended up in dissension. In this case, Harte told the commission that he had done all of the preparatory work, legal as well as engineering at his own expense, only to have the plans rejected by a Board of Works inspector 'who had already blundered in his engineering with railways, a very extraordinary sort of man altogether'. A few years later, when Harte resubmitted the plans, they were reviewed on behalf of the board 'by a gentleman of eminence in his profession and passed with éclat'. But then, the local drainage board which was made up, according to Harte, of farmers who thought they were engineers, became incensed with him because he would not incorporate further changes which they thought necessary; this led to his dismissal as engineer to the project and proceedings in the courts to recover the fees due to him.

Criticism of another kind emanated from cyclists, one of whom wrote to the *Irish Cyclist and Athlete* in March 1886 stating that Harte's roads had always been the worst in the neighbourhood and that any criticism of them was met by Harte in a 'plausible and sarcastic style'. In the light of comments like these, and the fact that he had been the subject of various outrages at different times, including an incident in 1871 when a grenade was thrown into his house, Harte told the Royal Commission on Irish Public Works in 1887 that the position of the county surveyor was a difficult one: he was 'at the mercy of a set of jobbing ignorant local people' and had to perform unpleasant duties from time to time. Surveyors, he went on, must knock their heads against local parties every day, contractors and cesspayers, and even dukes and earls and men of high position; and, he added, 'the higher grade on the social scale does not exclude human nature'. Harte assured the commission that he had been able to take care of himself but, with other surveyors, he sought to have the law amended so as reduce, if not eliminate, the control grand juries had over the tenure and conditions of service of surveyors; these efforts had come to nought, Harte claimed, because witnesses before a committee of the house of commons 'were told to hold their tongues'.

It might have been expected that public comments like these about the conduct and competence of his employers would not have made life easy for Harte in his remaining years in Donegal but there was to be no question of early retirement. In fact, when **Frederick Gahan** decided to retire from his post in the southern division of the county in July 1891 the grand jury advised the lord lieutenant that, instead of filling the vacancy, they wished to appoint Harte as surveyor for the entire county, a course which had already been adopted in counties Down and Kerry. Harte indicated that he would be pleased to accept this 'flattering offer' but the Dublin authorities refused to sanction the appointment (with the increased salary of £700 which Harte and the grand jury had agreed upon) and insisted that the vacancy should be filled in the normal way. Harte made no secret of his annoyance with this: if the decision is to stand, he told the grand jury in March 1892, the county would forever be saddled with an unnecessary charge but, by then, a new surveyor, **J.R.A. Ferguson**, had already taken up duty in the southern division. Harte continued to serve in the northern division until 15 March 1895 when he died in office at the age of 70, having spent nearly forty-four years in the service of the Donegal grand jury. He was buried in Derry where he had lived. Some months later, a monument was erected over his grave – a simple granite block taken from one of the quarries which he had worked near Dungloe many years before.

HAUGH, Patrick John (1892–1959), was born on 21 July 1892 in Doonbeg, Co. Clare, and after a Blackrock College education, gained a position as officer of customs and excise by competitive examination in 1912. He left Ireland a few years later to study and gain experience of engineering in America and went on to qualify for associate membership of the American Society of Civil Engineers as well as membership, by examination, of the Irish and English engineers' institutions. From October 1918 to February 1920 he was a field assistant on the preliminary survey for a large irrigation project in Wyoming. He later worked on highway and bridge construction and maintenance for the Wyoming State Highway Department and was employed for five years by the Standard Oil Company on design and construction work at oil refineries in Indiana.

On his return to Ireland in 1926 Haugh took up a position as assistant engineer with Messrs Siemens Bauunion on the Shannon hydroelectric scheme. In 1928 he was clerk of works on the construction of a reinforced concrete water-tower at Portrane mental hospital and from 1929 to 1931 was employed by the Board of Works as a temporary engineer on the river Barrow drainage scheme. He joined the local authority service in April 1931 as assistant county surveyor in Offaly and was appointed county surveyor in Leitrim in April 1936.

Haugh moved to Sligo as county surveyor on 1 January 1943. As all subsequent vacancies were filled by the appointment of county engineers, Haugh thus became the last man to be appointed to a county surveyor post in the state. However, **J.T. O' Byrne** who was appointed to the Wicklow post in 1940 was the last man to attain the rank of county surveyor excluding, of course, those appointed in Northern Ireland.

Haugh was responsible for a considerable post-war programme of road reconstruction and main road realignment in Sligo involving a significant degree of mechanization and new quarry plant. He was a popular and prominent figure in engineering circles and earned the respect, admiration and loyalty of his immediate staff and of all those with whom he came into contact. He reached retiring age in July 1957 but served for some time after that. He died in September 1959.

HERON, James (fl. 1885–1924), was a son of James Heron, professor of ecclesiastical history and pastoral theology at the Assembly's college in Belfast and author of *The Celtic Church in Ireland: the story of Ireland and Irish Christianity from the time of St Patrick to the Reformation* (London, 1898). He was educated at the Royal Belfast Academical Institution and had a distinguished career at Queen's College, Belfast, where he graduated with second-class honours in engineering in 1885 and in arts in 1891. He worked for a year as assistant to B.D. Wise, engineer to the Belfast & County Down Railway and subsequently, while a member of the staff of Belfast Corporation under J.C. Bretland, city surveyor, he was involved in the design of a new Albert Bridge to replace an earlier bridge which had collapsed in 1886.

Heron became Monaghan county surveyor in January 1892, having taken second place at the Civil Service Commissioners examinations in the previous year. He was selected in August 1899 from among twenty-eight applicants for appointment to the surveyorship of Co. Down, with an initial salary of £800 which was conditional on his giving his entire time to the post; the appointment was not confirmed, however, until March 1900 and he then resigned from his post in Monaghan. A letter under the pseudonym *AMICE* to the editor of the *Irish Builder* in May 1900 was highly critical of the appointment, claiming that Heron had succeeded only 'because he can exert most influence through his father who is a professor of divinity in the Presbyterian College, Belfast'.

Unlike many of his contemporaries, Heron was not an advocate of the direct labour system. As late as 1906, when the system was being introduced in many other counties, he took the view that as there were many good contractors available it would be inadvisable for the county council to go over to direct labour at that stage; it would be better to work at reorganizing and perfecting the contract system and, when the results had been evaluated, it would be time enough to consider alternatives. Heron's views may have been influenced by the fact that he had an unusually large team of assistant surveyors who could exercise close supervision of the contractors – ten in 1899 and twelve in 1915. At that stage, however, the county council proposed to lay off the six junior men and to pay the remainder at the rate of £200 a year, with increments which would bring the total to £250, but this scheme was not, apparently, implemented.

Heron reconstructed the harbour at Newcastle in 1902 and carried out repairs to Kilkeel harbour in 1904 and again in 1912, following storm damage. He built a reinforced concrete water tower at Down lunatic asylum around 1910. He was obliged to retire in 1920 due to ill-health and died on 16 February 1924 at his residence in Belfast.

HILL, John (1812–94), was born on 18 December 1812 in Co. Meath. At an early age he was apprenticed to William Armstrong, a Dublin engineer and, on completing his professional education, joined the staff of Sir John MacNeill with whom he was engaged for several years on preliminary surveys for railways, including an abortive Dundalk & Western Railway. He was a resident of Grey Abbey, Co. Down, when admitted to membership of the Civil Engineers' Society of Ireland in 1836 and, around 1838, started work on his own account as a consulting engineer. He also worked intermittently in partnership with James Thompson of Glasgow and gained wide experience in Scotland in surveying, drainage works, building and road-making. He was one of those who tendered unsuccessfully for the construction of the workhouse at Dundalk in August 1839.

In 1834 when Hill was living in Co. Meath, he was recommended for appointment as county surveyor by Sir Charles Bellingham who considered him to be 'a most deserving, well-conducted young man, attentive, active, sober and honest' but it seems that his application may not have been received in time. He passed the qualifying examination for county surveyors in 1841 but did not receive an appointment for some years. In the meantime, he took up a position under the Board of Works at a salary of £156 a year as superintendent of the central roads district, with responsibility for the maintenance of bridges and some 220 miles of roads which had been constructed by government engineers in counties Galway, Clare and Limerick in the 1820s and early 1830s. He built up and managed a large direct labour organization, with fifteen overseers, gangers, and large teams of labourers, and his average annual budget was over £4,000. In April 1850 and again in April 1856 he described the techniques and organization he employed, and the costs involved, in papers which he read at meetings of the Institution of Civil Engineers of Ireland. He continued to discharge this responsibility, concurrently with county surveyor duties, until 1854 when the roads were handed over by the Board of Works to the grand juries. Within two years after the transfer, he claimed that maintenance costs had risen under the grand jury contract system while the condition of the roads had seriously deteriorated.

Hill became Clare county surveyor in December 1845 and almost immediately became involved in planning and supervising major programmes of famine relief works in the county, initially under the aegis of the grand jury and from the autumn of 1846 onwards under the general supervision of the Board of Works. In conjunction with the board's inspector, Captain Edmond Wynne, and in an atmosphere strained by threats, protests, disturbances and violence, Hill shared responsibility for the conduct of hundreds of relief work schemes in Clare in 1846–7; the county was allocated a much greater sum for public works than any other between August 1846 and January 1847 when some 53,000 people were employed on the works there, mainly on inessential road schemes. Two years later, when shortage of funds resulted in the virtual collapse of the new system of outdoor relief in parts of Co. Clare, Hill took it on himself to advise the authorities in Dublin that all sectors of society were 'sinking beneath the pressure of distress', with even the better class of landholders 'diminishing fast' and some of the largest farmers ruined. He anticipated that the numbers seeking relief in Ennis union alone would increase to 40,000 by April 1849, involving a level of expenditure which the poor rate could not be expected to finance without disastrous effects on farming and industry.

In the years following the famine, Hill went on to discharge the more normal duties of a county surveyor with great distinction and constructed a substantial mileage of new roads. However, he found the duties of the post too demanding and transferred to the smaller King's County on 4 June 1855, having received 'a handsome testimonial of plate' from the Clare grand jury. Although he had been in office for less than two months, Hill's report to his new grand jury at the 1855 summer assizes provided a comprehensive assessment of road conditions in the county: roads were generally in a very poor condition as were many bridges and retaining walls, maintenance contracts related to road lengths of less than half a mile and improvement works were poorly specified. Hill proposed to begin immediately to put matters on a more organized footing; roads were to be divided into three classes, with maintenance contracts applying to sections of between three and five miles and running for up to five years. Classes 1 and 2, the more heavily trafficked roads, were to be actively dealt with, while it would be left to the local inhabitants to propose the initiation of maintenance or other works on the class 3 roads. Standards for maintenance works were to be laid down for all contractors and more assistant surveyors were to be employed to ensure better supervision of the work. All of this seems to have been well received by the grand jury and Hill went on to serve the county 'with great credit to himself and advantage to the public service' for twelve years. When the law was changed in 1857 to allow the grand jury to entrust road projects to the county surveyor in cases where an acceptable tender had not been received, he took on direct responsibility for some 400 miles of roads in the county which had become almost impassable. His salary was increased from £300 to £400 a year – the new statutory maximum for the county – soon after the law was amended in 1861 to allow for increases but, when a vacancy arose in Clare in May 1867, he returned to the county at the invitation of the grand jury as that post then carried a maximum salary of £600.

One of his colleagues, **P.C. Cowan**, described Hill as a man of great force of character and ability, who was most conscientious and energetic in the discharge of his various duties. The public works for which he was responsible were always said to be in good condition and under his supervision the main roads in Clare were reputed to be among the best in Ireland – a view repeated in 1905 by Cowan, who by then was the Local Government Board's chief inspector. The unusual 'spectacle bridge', as it came to be known, still stands as a monument to Hill's ability to innovate. The bridge was built in 1875 to carry the Ennistymon to Lisdoonvarna road across a narrow gorge and 46 feet above the level of the river Aille. The novelty of Hill's design arises from the fact that, instead of deep spandrel walls above the semi-circular masonry arch, the bridge incorporates a circular opening, 18 feet in diameter, thus lessening considerably the weight of the structure itself.

Hill was consulting engineer to the Limerick, Ennis & Killaloe Railway Company in the mid-1840s. That project did not proceed because of financial difficulties and, although a new

Limerick & Ennis company was formed to construct the line in the 1850s, Hill does not seem to have had any involvement at that stage. He was engineer in the early 1860s to the abortive Clara & Portumna Railway and to its successor, the Midland Counties & Shannon Junction Railway, which sought to build a line on the same route some ten years later; the line was eventually constructed, but only as far as Banagher, and opened in 1884. He also undertook a number of arterial drainage schemes, two of which he described in a paper presented to the Institution of Civil Engineers, London, in 1880; one of the schemes was completed in 1867 in the Rathdowney area of Co. Laois while the other and more substantial one – affecting over 2,000 acres – was carried out in the Sixmilebridge district of Clare between 1865 and 1872. While this latter was in progress in 1867, a dam which had been erected as part of the scheme burst, releasing a large volume of water which gushed down the river and badly damaged a bridge at Annagore on the road from Kilkishen to Sixmilebridge; Hill's successor in Clare, A.C. Adair, reported the damage to the grand jury at the spring assizes but it was Hill himself, who was back in the Clare surveyorship later that year, who had to undertake the rebuilding work.

The Board of Works regularly engaged Hill to act as an inspector on proposed drainage schemes throughout the country. In later life, his vast experience and impartiality led to his being in frequent demand as an arbitrator in contract and railway disputes and to inspect and report for the government on the condition of various public works. One such assignment, undertaken when he was already 79 years of age, seems to have hastened his death. In the autumn of 1891, on a tour of inspection of the Balfour relief works in the north-west, he had to undertake long drives on outside jaunting cars through wild, open country, fully exposed to inclement weather. Though apparently in good health on his return to Ennis, it was not long before his strength began to fail and he was never again restored to good health. He was the doyen of county surveyors when forced to retire on pension in March 1893 at the age of 81. Notice of his retirement was received with great regret by his grand jury who expressed their 'deep sense of his merit as a public official, his courteous bearing and impartial conduct' and wished him well in his honourable retirement.

Hill was an elected member, and sometime chairman, of Ennis town commissioners from 1870 onwards, a position which led to a certain conflict of interests when discussions arose about responsibility for the unsatisfactory and insanitary condition of the town's streets. He was also, for a time, vice-chairman of Clarecastle harbour board. He was a member of the Institution of Civil Engineers of Ireland and of its sister institution in London and he was elected to membership of the Royal Irish Academy in 1867 and to fellowship of the Royal Society of Antiquaries of Ireland in 1871. He was an enthusiastic freemason and held high rank in the masonic order. He died on 24 January 1894 at his residence in Bindon Street, Ennis, and is buried with his wife, Maria (d. 1896) in Drumcliff cemetery, a few miles north of the town.

HORAN, John (1853–1919), son of an engineer, was born on 10 June 1853. He was educated at Cookstown Academy and Queen's College, Belfast, where he gained a BE degree with first-class honours in 1874 and an ME in 1882. He worked until 1879 as assistant engineer with Messrs Farrell Patten & Co. on the construction of Rosslare harbour and as assistant engineer with T.J. Dwyer, a contractor. In 1879–81 he was assistant to Messrs Fowler & Bacon, engineers, and between 1881 and 1884 was chief assistant engineer at the Alexandra Dock, Hull. He gained first place out of eleven candidates in a competition for the post of Cork city surveyor in 1880 but a local engineer, M.J. McMullen, was elected to the post by the

1 The three-arch masonry bridge over the river Erne at Belturbet, Co. Cavan, designed by **Alexander Armstrong** in 1836–7 and which still carries the main road from Cavan to the Northern Ireland border.

2 The bridge over the river Bann at Coleraine, Co. Derry, constructed by **Stewart Gordon** in 1844 with **Alexander Tate** as resident engineer.

3 The bridge at Bandon, Co. Cork, reconstructed in 1838 by **Edmund Leahy**.

4 One of the tunnels on the Glengarriff–Kenmare road constructed under the supervision of **J.B. Farrell** between 1833 and 1839 and photographed over 100 years later.

5 The wooden bridge over the river Slaney at the Deeps, Killurin, Co. Wexford, completed by **J.B. Farrell** in 1844; the bridge was replaced in 1915 by the ferro-concrete structure shown in Fig. 34.

6 The bridge over the river Dargle at Enniskerry, Co. Wicklow,
designed by **Henry Brett** in 1865.

7 The old seven-arch bridge at Wicklow, reconstructed and widened on the upstream side by **Henry Brett** in 1862–3.

8 The attractive granite three-arch bridge at Avoca, Co. Wicklow, completed at a cost of £3,000 by **Henry Brett** in 1869; the bridge was seriously damaged by the floods which followed 'hurricane Charlie' in 1986 but was subsequently reconstructed.

9 The unique 'spectacle bridge' in Co. Clare, designed by **John Hill** in 1875 to carry the Ennistymon–Lisdoonvarna road 46 feet above the level of the river Aille.

10 Killorglin Bridge, Co. Kerry, completed by **William Henry Deane** in 1885 at a cost of £10,000, with eight thirty-five-foot masonry spans.

11 Bridges at Ballyvoyle, near Dungarvan, Co. Waterford: the road bridge (foreground) was completed by **Charles Tarrant** in 1862–3 and restored by **J.K. Bowen** in 1923 after damage by explosives during the civil war; **Tarrant** was also responsible for the construction in the mid-1870s of the viaduct (background) which carried the Mallow–Waterford railway; this was destroyed during the civil war and replaced by an iron structure.

12 The 120-foot-span wrought-iron lattice-girder Obelisk Bridge over the river Boyne
at Oldbridge, three miles above Drogheda, designed by **John Neville** and completed
in 1869; the obelisk, erected in 1736 near the place where William of Orange was
believed to have crossed the river in 1690, was blown up in 1921.

13 The lattice-girder bridge constructed in the 1860s by **Samuel Ussher Roberts**
on the abutments of Alexander Nimmo's masonry-arch bridge at Maam,
Co. Galway, which was destroyed by a flood.

14 St Mary's Bridge, Drogheda, Co. Louth, a two-span masonry-arch structure, designed by **John Neville** was completed in 1868 and replaced by a reinforced concrete bridge in 1984.

15 St Dominick's Bridge, Drogheda, a light steel structure completed in 1894 by **Patrick J. Lynam** but incorporating the massive stone abutments of an earlier wooden bridge built by his predecessor, **John Neville**, in 1863; use of the bridge is now confined to pedestrian traffic.

16 Ballyduff Bridge, a wrought-iron lattice-girder bridge over the river Blackwater, west of Lismore, Co. Waterford, completed to a design by **W.E.L. Duffin** in 1887.

17 Another lattice-girder bridge on masonry and concrete abutments, at Ballyneety, near Dungarvan, Co. Waterford, constructed by **W.E.L. Duffin** in 1904–5 to replace a bridge which had been destroyed by floods in the river Colligan.

18 The 1,300-foot-long wrought-iron Youghal Bridge over the river Blackwater, completed by **Samuel A. Kirkby** in 1883 and replaced in 1963, after years of controversy, by a reinforced concrete bridge 580 yards upstream.

19 **Samuel A. Kirkby** also provided a number of small bridges on slim cast-iron screw-piles in Co. Cork such as this bridge at Ringnanean, near Kinsale, completed in 1876 and replaced in 1995.

20 The large iron bridge over the river Barrow at New Ross, Co. Wexford, completed by **J.B. Farrell** in 1868–9; the bridge had 4 fixed spans of lattice girders, each 88 feet wide, and a central swinging section, providing two 50-foot openings.

21 Another view of New Ross Bridge, Co. Wexford; the bridge was replaced by a precast concrete bridge on the same site in 1967.

22 The 660-foot-long Poulgorm Bridge between Glandore and Union Hall, Co. Cork, completed by **Nat Jackson** in 1886.

23 The swing bridge at Achill Sound, Co. Mayo, designed by **Edward Glover** and completed by his successor as Mayo county surveyor, **Peter C. Cowan**, in 1887; the bridge, which was only eight feet wide, was replaced in 1948.

24 A view of the Achill swing bridge in the open position – the light girder bridge pivoted on a central pier to allow shipping to pass through the sound.

25 North Gate Bridge, Cork, designed by **Sir John Benson** and completed in 1864, was one of the longest single-span cast and wrought-iron bridges constructed in Ireland; it was in use until the late 1950s.

26 Dillon Bridge, Carrick-on-Suir, Co. Tipperary, a lattice-girder structure with three ninety-two-foot spans, completed by **J.L. Worral** in 1882.

27 Two of the earliest road bridges constructed in Ireland using the Hennebique ferro-concrete system: the bridge at Knockmahon, Co. Waterford (top) provided by **W.E.L. Duffin** *c*.1909 and the single-span bridge over the river Tempo at Drumlone, Co. Fermanagh (also 1909) for which **J.P. Burkitt** was responsible; the latter is still in service.

28 St John's Bridge, Kilkenny, designed by **A.M. Burden** in association with L.G. Mouchel & Partners and completed in 1910 – a ferro-concrete arch which was then the longest of its kind in Britain or Ireland.

29 The long wooden toll-bridge built by the American engineer, Lemuel Cox, at Waterford in 1794 and which was in service until 1913.

30 The controversial reinforced-concrete Redmond Bridge which was completed at Waterford in 1913 under the supervision of **A.M. Burden** to replace the 1794 bridge; Redmond Bridge was itself replaced in 1986.

31 One of the main steel plate-girders in course of erection at Portumna Bridge over the river Shannon in 1913 under the supervision of **J.O.B. Moynan**.

32 Construction of the road surface on the iron deck of the new Portumna Bridge in 1914.

33 Hartley Bridge over the river Shannon, north of Carrick-on-Shannon, Co. Leitrim, an early reinforced concrete structure, completed by direct labour under the supervision of **Eugene O'Neill Clarke** in 1912–13.

34 The 360-foot-long reinforced concrete bridge completed under the supervision of **W.F. Barry** in 1915 at the Deeps, Killurin, Co. Wexford to replace an earlier wooden bridge (see Fig. 5); the bridge is now the earliest surviving example of its kind in Ireland.

35 Lady Brooke Bridge, with fourteen spans constructed on the Hennebique reinforced concrete system in Upper Lough Erne by **J.P. Burkitt** in 1935 to link Lisnaskea via Trasna Island to the south of Co. Fermanagh.

36 View southwards from the summit of the Healy Pass, Co. Cork, constructed under the supervision of **R.F.M. O'Connor** in 1931.

37 An Aveling & Porter steamroller of the kind which was brought into use by many county surveyors in the early years of the twentieth century.

38 Machinery, such as this 'Acme' stone breaker, powered by a steam-driven traction engine, was introduced on road works in some counties at the beginning of the twentieth century notwithstanding demands by local councillors for the maintenance of labour-intensive methods.

39 Work in progress in Co. Antrim in 1915 under the supervision of **David Megaw** on the laying of a stretch of reinforced concrete road.

40 The Dublin–Cork road near New Inn, Co. Laois, in 1932, showing an overturned car believed to be that of the county surveyor, **Malachi A. Feehan**.

41 A steamroller and a diesel roller in action on the Dublin–Cork
road near Portlaoise in 1939.

42 Road-making in 1940 was still a labour-intensive business as illustrated in this
photograph of work in progress on the Dublin–Cork road near Portlaoise.

43 The ninety-foot-wide screen of Glasgow sandstone across the forecourt of St Patrick's pro-cathedral, Dundalk, was designed by **Thomas Turner** in 1851 in an open competition.

44 The 185-foot-high tower and spire at the Mariner's church, Dun Laoghaire, was designed by **Thomas Turner** in 1866.

45 Plans by **Thomas Turner** for rebuilding Stormont Castle in Scots Baronial style, as illustrated in *The Builder* in December 1858; the rebuilt castle and its demesne were acquired in 1921 as the site for the Northern Ireland parliament buildings.

46 Elevation of the new Ulster Bank, Longford, by **James Bell (junior)**,
as illustrated in the *Irish Builder* in February 1863.

47 Elevation of the new National Bank at Waterford, designed by
Henry Brett & Sons in 1878.

48 New wine stores at Clanbrassil Street, Dundalk, designed by **John Neville** in 1868.

49 The Penrose Quay terminus in Cork city, designed by **Sir John Benson** in 1858 for the Great Southern & Western Railway.

50 Woodlawn House, Co. Galway, redesigned and extended by **James Forth Kempster** in the 1860s for the second Lord Ashtown.

51 An 1874 photograph of the 1,000-foot-long main frontage of the Co. Down
lunatic asylum (later known as the Downshire Hospital) designed
by **Henry Smyth** and completed in 1869.

52 An early twentieth-century view of the entrance and frontage of the
Downshire Hospital as extended by **Henry Smyth**.

53 A new school at Nenagh, Co. Tipperary, designed by **J.O.B. Moynan** and his partner, R.P. Gill, in 1906.

54 A gate lodge near Rathcoole, Co. Dublin, designed by **R.A. Gray** in 1879.

55 The central portico of the county courthouse at Nenagh, Co. Tipperary (left), as rebuilt by **J.O.B. Moynan** in the 1890s in conjunction with Walter G. Doolin, and the town hall at Nenagh, completed in 1889 by Moynan and his partner, R.P. Gill.

56 The Wellington monument at Trim, Co. Meath, designed by **James Bell** in 1817.

57 The memorial of Major-General Sir Robert Rollo Gillespie in Comber, Co. Down, designed and constructed by **John Fraser** in conjunction with a local architect.

58 The clock tower on the quay at Waterford, designed by **Charles Tarrant** and completed in 1861.

59 The mausoleum of **Noblett Rogers St Leger** (who died in 1872) near the main entrance to the cemetery at Cleveragh Road, Sligo.

60 The granite obelisk erected by the road contractors at Mount Jerome cemetery, Dublin, in memory of **Robert Hampton**, Dublin district surveyor, after his death in 1855.

61 The family grave of **James Bell (junior)** at Mount Jerome cemetery; the attractive carved headstone was probably designed by Bell himself who died in 1883.

62 Charles Philip Cotton,
appointed as Kildare county
surveyor in 1865 and afterwards
chief engineering inspector
with the Local Government
Board for Ireland.

63 Bernard B. Murray,
Down county surveyor
from 1862 to 1889.

64 **Edward Glover**, county surveyor, Mayo 1876–86 and Kildare 1886–1914.   65 **Joseph Henry Moore**, county surveyor, Westmeath 1870–4 and Meath 1874–1907.   66 **Peter Chalmers Cowan**, county surveyor, Mayo 1886–9 and Down 1889–99; subsequently chief engineering inspector with the Local Government Board.   67 **John Ouseley Bonsall Moynan**, county surveyor, Longford 1883–91 and Tipperary north riding 1891–1930.

68

69

70

71

68 Charles Littleboy Boddie, Derry county surveyor, 1900–24. **69 John Henry Brett**, county surveyor, Limerick 1863–9, Kildare 1869–1885 and Antrim 1885–1914. **70 Richard Henry Dorman**, county surveyor, Mayo 1886–7 and Armagh 1887–1921. **71 Edward Augustus Hackett**, county surveyor, Kerry 1887–9 and Tipperary south riding 1889–1920.

72 Henry Webster, county surveyor, Kerry 1889–91 and Wexford 1891–1909.
73 Samuel Alexander Kirkby, county surveyor, Longford 1872–4 and Cork east riding 1874–1909.   **74** Richard William Frederick Longfield, county surveyor, Donegal 1895–8 and Cork west riding 1898–1925.   **75** Patrick John Lynam, county surveyor, Mayo 1883–6 and Louth 1886–1912.

76 **Joseph Albert Ryan**, county surveyor, Queen's County (Laois) 1915–24 and Dublin 1924–53.   77 **Stephen Gerald Gallagher**, Wicklow county surveyor 1903–38.
78 **Thaddeus Cornelius Courtney**, county surveyor for Tipperary north riding 1930–4 and subsequently chief engineering adviser in the Department of Local Government and Public Health (1934–49) and executive chairman of CIE (1949–58).   79 **Jeremiah Gerard Coffey**, Kilkenny county surveyor 1940–74, the last serving official whose original appointment was to the grade of county surveyor.

80 One of the first group of county engineer appointments: **Thomas Kelly**, Wexford 1944–5, Kildare 1945–59 and Dublin 1959–62.

81 **Thomas Aloysius Simington**, Clare county engineer 1945–9, another of the early county engineers.

corporation even though he had gained fewer marks than any of the seven qualified candidates.

Horan was appointed county surveyor for the western division of Limerick in October 1884. Following the retirement in March 1893 of the surveyor for the eastern division, **Thomas Fosbery**, the grand jury renewed a demand which they had originally made before Horan's appointment that a single surveyor should be responsible for the entire county; this was agreed by the government but special legislation had to be enacted in September 1893 to permit it. Horan then became county surveyor for the whole county at a salary of £600, an increase of £200. He moved both his public office and his private practice from Rathkeale, where he had lived since 1884, and took up residence at Templemungret, near Limerick. Even with his wider public responsibilities, he managed to continue a substantial private practice, generally dealing with water supply, drainage and harbour works. He became a member of Foynes harbour trustees when that body was established in 1890 and some years later designed a wooden extension to the pier to provide berthing in twenty-six feet of water at low tide. He also reconstructed and extended the pier at Glin in 1894–5 at a cost of £2,600.

Reports from Horan to his grand jury in the early 1890s demonstrated a pioneering approach to the organization of the road maintenance work of the county. As early as 1893, the contracts which he drew up provided for the appointment by the county surveyor of part of the labour force on such roads as he thought fit – generally the main roads and the streets of towns and villages. These labourers were known as *surfacemen*; their wages were from nine to twelve shillings a week and they were required to be paid by the contractors at intervals of not more than two weeks. To ensure the employment of these men by contractors all the year round for cleaning, draining and spreading stones on roads, without significant extra cost, Horan succeeded in forming the streets of towns and villages into single contracts and in doing away with contracts for very short road lengths elsewhere. In addition, he had introduced a limited steamrolling programme by 1898. While the grand jury was not enthusiastic about these modifications of the traditional contract system, Horan claimed that the effects were very beneficial on the sections of roads involved and on the performance of neighbouring contracts, by providing a corps of skilled and experienced men for road work.

Horan made no secret of the fact that he viewed with apprehension the introduction in 1899 of 'a popular form of government'; while making it clear that 'as loyal men' the surveyors would do their best to work under the new county councils, he expressed the hope 'that the new bodies will rise to a sense of the increased responsibilities placed on them'. In the event, Horan's county council had been in office for only a few months when it came under pressure from unemployed labourers and their supporters, and from several of the district councils, to terminate the contract system entirely and to introduce direct labour for all road works. In face of this agitation, the council decided in January 1900 to hand over some 600 miles of road to the county surveyor for maintenance on a direct labour basis, even though such an arrangement was not authorized by law at the time. After the law had been amended, the council formally adopted direct labour proposals in October 1902 and these were approved by the Local Government Board in March 1903. From then on, 1830 of the 1835 miles of roads in Limerick were under Horan's direct charge, making Limerick the first county to go over almost fully to direct labour. Labourers were employed and dismissed by the county surveyor, surfacemen appointed by the surveyor acted as gangers, there were up to eighteen foremen to supervise generally, and there were seven assistant surveyors. According to Horan, the new arrangements were 'far and away more efficient' and more economical, especially when coupled with an expanded programme of steamrolling; in 1904–5, three rollers were at work in the county, making up the streets of towns and villages and weak and worn sections of the main roads, and the purchase of two additional rollers was being considered.

Horan continued to achieve good results in road maintenance during his remaining years as county surveyor. The new working arrangements obviously added considerably to his workload but the county council were unwilling to concede the salary increase which he sought to take account of this. In June 1900, when he suggested an increase from £600 to £1,500, the council would agree to no more than £800. After a local inquiry in 1905, the Local Government Board granted him an extra £300 a year bringing his salary to a new total of £900. The state of his health forced Horan to resign his surveyorship with effect from 31 March 1913 after which the county council decided to revert to the practice of appointing two county surveyors (but at reduced salaries of only £300 each). He died at his home at Templemungret on 26 April 1919.

HORNE, William (fl. 1836–7), of Shannonbridge, Co. Offaly, was appointed Limerick county surveyor on 26 November 1836. There were complaints by the grand jury about his competence after the spring assizes in 1837 and the foreman sought to have him dismissed by the lord lieutenant. The Board of Works advised that Horne had been a 'bona fide appointment' and they were 'not aware of any reason for the grand jury thinking otherwise'; no specific charge of misconduct had been brought against him, according to the board, and if the jury thought that he was unequal to the duties, then they, rather than the lord lieutenant, should dismiss him. Horne had been replaced by the summer of 1837 and had died by 1856.

HUMPHREYS, Henry Temple (1839–91), was born in London on 20 December 1839, the youngest child of Griffith Humphreys. He was educated mainly at private schools near the family home in Bayswater and, with a view to a career in engineering, was placed for six months under tutors in mathematics and technical drawing. In 1854 he was apprenticed for five years to Messrs John Rennie and Co. at their London workshops, keeping workmen's hours. However, because of his delicate constitution and the distance he had to travel, morning and night, he was transferred to the firm's drawing offices, where attendance was not required until 9.00 a.m. On completion of his apprenticeship in 1859 he was retained as an assistant for two years and then worked for a year as chief draftsman at an ironworks in London.

In December 1861 Humphreys went to Karachi to make preliminary surveys for the Oriental Canal and Irrigation Company. He was back in England in 1863, working as an assistant engineer on the Thames Valley Railway. His next assignment was as chief district engineer on the Buenos Aires Great Southern Railway but he returned again to England late in 1864 and was engaged for some years on a variety of railway, water supply and other projects. He took up an appointment as executive engineer under the Indian public works department in March 1868 and was placed in charge of irrigation works in the Bangalore area. But the climate in India did not suit Humphreys, his health suffered, and he gave up his post in 1869 and returned to London.

The first county surveyor examinations arranged by the Civil Service Commissioners after they had been given responsibility for the certification of candidates for appointment to these posts were held in November 1869. Humphreys took second place to **Leveson Francis Vernon-Harcourt** and was appointed county surveyor for the western division of Limerick in January 1870. Within a year he was engaged in litigation with an important road contractor and talked of leaving Limerick. However, having won a libel action against the contractor, he decided to remain. In 1873 he advised the grand jury that it was no longer possible to get men to break stones for macadamizing the roads, due to the general improvement in the conditions and pay of labourers in the county. He proposed, therefore, that stone-breaking should in future be done by machinery rather than by hand, an operation which was costing one shilling per ton at the time. He designed a what he described as a 'travelling steam stone-breaker' which he claimed could be built for £500, would crush 100 tons of stone per day and would bring considerable savings after a few years, but the grand jury declined to act on his suggestion that the machine should be built. Another idea put forward by Humphreys was that prizes should be offered for the best road contractors, to encourage good work, but this scheme did not find favour either.

In 1873 Humphreys contracted rheumatic fever in the discharge of his duties and he was transferred to the western division of Galway with effect from 1 December of that year; he was appointed also as surveyor for the county of the town of Galway for which he received an extra £100. The *Galway Vindicator*, in announcing the appointment of the new surveyor, described him as 'a man of high standing and special merit in his profession' but complaints about his work were already being made when the grand jury met only a few months later for the 1874 spring assizes and they refused to approve the presentment for his half-year's salary. As time passed, Humphreys' health became much worse due, he claimed, to continuous exposure to the cold damp air of west Galway and fatigue brought on by long hours of travel around his extensive district. He resigned from his post in January 1878 and, although he was still less than 40 years of age, his engineering career was practically at an end at that stage.

Humphreys had some involvement in the debate which had been going on since the 1860s about the best route for a Galway to Clifden railway. He was a strong supporter of the southern route, via Barna, Spiddal and Roundstone, rather than a line through Oughterard and Maam. He believed that the mail-coach road had developed the northern districts as far as was possible whereas there was a greater population density nearer the coast where a railway could lead to the development of the sea fisheries and better marketing of the catch. Even on this route, however, Humphreys suggested that a railway would have to be constructed economically, with costs kept to less than £4,000 per mile, and very carefully managed, if it was to pay a dividend of 4 per cent. In the event, the railway was built on the northern route, but not until 1895, and it was to last for only forty years.

After his retirement, Humphreys lived in London for two years and spent the rest of his life in various parts of Scotland. He remained an inveterate inventor. His activity in this field had begun when he was less than nineteen years old with a design for a new kind of projectile. His other inventions included an elevated railway which could be quickly and cheaply constructed in the colonies without disturbance of land; the *Clamor Oestis*, described as a self-regulating sea-screech to warn ships during fogs; and an instrument called a *Cycloscope* for setting out railway and other curves, which he exhibited and explained at a meeting in London of the Institution of Civil Engineers in May 1866. Humphreys also maintained his interest in economic and social questions and despite continuing poor health he was active to the end. He never married but, according to an obituary, his genial disposition and musical ability made him very popular in society and he had many friends. He died at Clady House, Cairnryan, near Stranraer, on 27 October 1891, aged 51.

JACKSON, Nathaniel (*c.*1825–1900), was the youngest son of Erasmus Jackson of Portsmouth. He was employed as a draftsman by the Board of Works from 1850 to 1853 when he became a fellow of the Royal Institute of the Architects of Ireland. In August 1850, *The*

Builder reported that the Society of Antient Concerts was spending almost £1,600 on the construction of additions to the rear of their premises at Great Brunswick Street (now Pearse Street), Dublin, in accordance with plans prepared by Jackson; new dining, drawing and practice rooms were being provided. Three years later, the same journal recorded that additional works designed by Jackson had been carried out; the roof of the music hall had been raised by nine feet and the room lengthened by ten feet, bringing seating capacity to upwards of 1,000; the accommodation for the orchestra was much enlarged; a new ventilation system was provided; and the entire building had been decorated 'in the Corinthian style'. The premises had been occupied by the Dublin Oil Gas Company in the 1820s and, before Jackson became involved, adaptations had been made by the prominent Dublin architect, Patrick Byrne, to fit it for use as a concert hall. The first performance staged by the National Theatre Society (later the Abbey Theatre) took place there in 1902 and both James Joyce and John McCormack sang in a concert there in 1904. Before its most recent refurbishment for use as modern offices the building housed the Academy Cinema.

Jackson was appointed county surveyor for the west riding of Co. Cork on 31 March 1855 and served there for forty-three years. Although his duties in this far-flung district were obviously quite demanding, he was never well paid. In 1862, a year after the law had been changed to allow a maximum salary of £600 for the post, the grand jury granted him an increase of only £100, bringing his annual salary to £400 which, in the view of the *Cork Constitution*, was not an extravagant rate of pay for a gentleman who did his work so satisfactorily. In face of threats by the surveyor to move to a smaller county a further increase of £50 was conceded in 1864 but, while the salary was increased again to £500 in 1873, Jackson was not allowed the legal maximum of £600 until March 1898 when he was on the point of retirement.

Road contractors caused problems for Jackson throughout most of his long career. On several occasions in the 1860s when there were complaints about the condition of the roads, he sought unsuccessfully to get approval from the grand jury for the employment of stewards on the main roads to compel the contractors to keep the water-tables clear and to remove scrapings. In the 1880s when some of the roads in the barony of West Carbery were said to be in a disgraceful condition, Jackson complained that the rule of law had broken down in the area, making it impossible to prosecute defaulting contractors who had taken on work at very low prices due to intense competition. Following some improvement in local economic conditions in the 1890s the situation seemed to improve – contracts were taken at more realistic prices and were more often completed satisfactorily. In all, Jackson claimed to have built about 1,000 miles of new roads by 1898.

Due to penny-pinching on the part of the grand jury and the presentment sessions in some of the baronies, Jackson was forced to build a number of important bridges in wood rather than iron or stone. At Glandore, for example, a wooden bridge was decided on against his advice and completed in 1863; the bridge was about 700 feet long and 16 feet wide, with a swivel opening section to allow ships used in the slate export trade to go right up to the village of Leap. The *Cork Constitution* reported that the bridge reflected 'great credit on the county surveyor who planned it' and that 'from the beauty of its construction, it adds greatly to the appearance of the splendid harbour'. But despite heavy expenditure on maintenance, the bridge was in a bad state by 1878 and was 'in a shaky condition' when tenders for an iron replacement bridge costing nearly £11,000 were received in 1884. The new bridge – 'light and graceful in appearance' according to a contemporary newspaper report – was completed in 1886 by the Stockton Forge Company which had built the new bridge at Youghal three years earlier. It was supported on screw-piles driven to an average depth of 50 feet below the water level, and was 660 feet long with a swinging section which gave two openings of 34 feet each for shipping. The bridge is still in use but with a new steel and concrete deck resting on the original piles and piers.

Jackson was one of the pioneers in the use of mass concrete on bridge works in Ireland, finding this a particularly economical solution especially in the case of skew bridges. In a paper read to the Institution of Civil Engineers of Ireland in February 1879 he reported that he had been using concrete in the construction of bridges on county roads for some years and described his methods in detail. The span of the arches he had constructed varied from 20 feet to 35 feet but, in one instance where foundations for a central pier could not be obtained without going to considerable expense, Jackson constructed an arch with a span of 45 feet and a rise of 7 feet 6 inches; the concrete was 3 feet thick at the crown and 4 feet at the springing. The great advantage of concrete, according to Jackson, was that after a flood, any damage that might arise to the foundations of a bridge could be made good at a cost of a few pounds, whereas in a stone-built bridge, one stone loosened at the foot of an abutment or pier would often cause the destruction of the whole structure or a large part of it. He was also reported to have built a 'concrete tank' at Bandon, possibly the large domed tank provided in the mid-1880s on a hill above the town as part of the town's first water supply scheme and which served the town well into the twentieth century.

During his long tenure as county surveyor, Jackson made some further significant contributions to the infrastructure of west Cork. He designed and constructed numerous piers, including those at Schull, Dursey Island and Castletownbere, and he had a substantial involvement in the development of the tramway and light railway network in the 1880s and 1890s. At the spring assizes in March 1884 no less than six separate tramway proposals in his area came before the grand jury, all of them requiring detailed examination by the surveyor before the jury could decide whether to approve them and to provide baronial guarantees on the capital. One of those which went ahead – the Schull & Skibbereen Tramway & Light Railway – engaged one of Jackson's east riding colleagues, **S.A. Kirkby**, as one of its joint engineers and opened to traffic in 1886. One year later, following local complaints, Jackson was asked by the Board of Trade to report on the condition of the line; he found significant faults in the construction and in the locomotives and rolling stock which required a further investment of £3,000. By 1892 the line was in serious financial difficulty and its operation had to be taken over by a committee of management appointed by the grand jury, leaving Jackson to undertake most of the responsibility for the engineering aspects from then on. The same situation applied in relation to the Ballinascarthy, Timoleague & Courtmacsherry Light Railway.

Jackson's official position in west Cork was not a very demanding one in architectural terms, involving no more than the maintenance and minor modification of bridewells and small courthouses. He seems, however, to have maintained a private architectural practice with an office at the South Mall, Cork, during his years as county surveyor although few of his projects are recorded. It was reported in 1873 that a marine residence designed by him was being erected at Coolmain, near Kilbrittain, for William Shaw MP and in later life he acted as architect to the Rushbrooke estate in which capacity he supervised substantial rebuilding work in Cobh (then Queenstown) in the last decade of the century. Dominick J. Coakley – architect of several churches in the Cork area and of the national monument unveiled in 1906 at the Grand Parade in Cork city – spent two years of his training in Jackson's office before gaining associate membership of the Royal Institute of British Architects in 1881.

In 1863, after the English architect, William Burges (1827–81) was declared the winner of an architectural competition for the design of a new St Fin Barre's cathedral in Cork, there was considerable controversy as to whether his plans were in conformity with the competition's rules and could be implemented within the cost limit of £15,000 which had been stipulated. Jackson – who seems to have been one of the sixty-eight competitors – took up the issue in the letter columns of the *Dublin Builder*, suggesting that it would have been more honest if competitors had been advised that 'any architect conscientious enough to follow these instructions will only lose his labour, health and money'. However, Burges defended his scheme, arguing that the body of the church could be completed for £15,000 and that it would be better to leave the spires and other features until later rather than erect a building unworthy of the established church. The committee responsible for the management of the project accepted this view and confirmed Burges as the winner. The foundation stone was laid in January 1865 and the building was consecrated, although incomplete, in 1870; the project had cost about £100,000 by 1881 and the final figure was greater still.

In his early years in west Cork, Jackson lived at Gurteen, near Bandon, but took up the tenancy from the Devonshire estate of a large Georgian townhouse at Devonshire Square, Bandon in 1890. With his wife, Elizabeth, daughter of Daniel Porter of Dublin, whom he had married in April 1859, Jackson had one daughter, Helen, who married Thomas Meade of Rushbrooke, a civil engineer, and a son Alfred, born in 1868. The latter had an adventurous spirit, enlisting in the French foreign legion at only 16 years of age, later serving in the British army during the Boer war and ending his days in South Africa where he worked as a builder's labourer and as a blacksmith's assistant.

Early in 1898, when he had reached his seventy-third year, Jackson gave notice of his intention to retire and when he addressed the summer assizes in the following July, he noted that this was his 87th attendance at an assizes without a break. The grand jury awarded him the maximum pension of £400 a year – a handsome sum, according to the *Cork Examiner*. Jackson subsequently went to live at Gordon Villas, Monkstown, Co. Cork, but his health failed rapidly and he died on 21 February 1900. His life had been a long, laborious, useful and gentlemanly one, according to the *Examiner* and the monuments of his engineering skill which he had erected all over the west riding would last, as his footprints in the sands of time, long after his memory had begun to fade. For all that, his estate was a modest one, with little over £1,000 passing to his widow under his will.

JOHNSTON, William (d. 1840), was one of the many surveyors who were employed under the Bogs Commissioners between 1809 and 1813. He was engaged as an assistant engineer by Alexander Nimmo on road planning and construction work in Connemara in the late 1820s, earning three guineas per week with a daily allowance for his horse. When Nimmo's assignment was terminated in 1831 Johnston went to live in Dublin where he advertized both his engineering and land surveying services; an 1832 advertisement suggested that he was willing to undertake a wide variety of activities: 'surveying, laying out roads, bridges and engineering generally' as well as estate surveys, preparation of maps and plans, copying of maps and reducing them for attachment to leases, and designing and executing improvement works.

Johnston was appointed to be Waterford's first county surveyor on 17 May 1834. He seemed to be in some difficulty with his grand jury six years later when he was forced to appeal to the judge at the 1840 summer assizes to order payment of his salary which was eight months in arrears. Less than ten days later he was dead, having collapsed on 27 July 1840 as he was entering the courthouse in Kilkenny.

JONES, Charles Booth (1843–1919), was born at Enniskillen on 11 June 1843, son of Charles Jones, and was educated by Revd Mr Steele before entering Trinity College, Dublin, in October 1861. He obtained the LCE in 1864 and a BA degree in winter 1865 and was employed in the office of **J.H. Brett**, Kildare county surveyor, when he took first place at the December 1871 county surveyor examinations. Jones was appointed as surveyor for the southern division of Cork east riding on 1 February 1872 at a salary of £400 a year but moved to Sligo in the following May. He was heavily involved in public works programmes under the Relief of Distress Act between 1880 and 1882 and was paid an additional £300 by the grand jury in recognition of the additional work involved. He was responsible for the construction of a significant number of new bridges in Sligo, including Balla Bridge, an attractive 36-foot-span single-arch masonry structure constructed over the river Owenbeg near Ballysadare in 1887.

In March 1906, when he had served in Sligo for almost thirty-four years, Jones submitted his resignation to the county council citing medical grounds and the fact that he did not feel equal to the volume of additional work which had arisen since the Local Government (Ireland) Act 1898 had come into operation. His salary had continued to be £500 a year since grand jury days and he qualified for a pension of £325. In reporting a good-humoured discussion of his entitlement to a pension, the *Sligo Champion* noted that the council was a strong nationalist body while Jones 'had not the slightest pretence to nationality' and was known to wear an orange tie on the 12th July. Councillors commended him for his zeal, efficiency, courtesy and integrity, with some of them remarking that the colour of his tie, or where he said his prayers – or if he never said them at all – was no concern of theirs! Jones' resignation came into effect on 4 May 1906 but he agreed to continue to carry out the surveyor's duties until a successor was appointed. Following his retirement, he moved from his residence at The Mall, Sligo, to Maxwell Road, Rathgar, Dublin, where he died on 30 November 1919. He is buried at Mount Jerome cemetery in Dublin with his wife, Jincie Pollock, who died in 1945.

JONES, Samuel (1797–1859), was born in Edgeworthstown, Co. Longford, the son of a carpenter who was a tenant of the Edgeworth family. He was educated and trained as a surveyor by William Edgeworth before going to work with Alexander Nimmo who, as government engineer in the western district of Ireland between 1822 and 1831, carried out numerous road, bridge and harbour construction projects. He married Elizabeth Campbell in Clifden in 1825, built Ardbear House near the town, leased 400 acres of land and took on the trappings of respectability. Meanwhile, according to Maria Edgeworth's account of her tour in Connemara in 1833 'his brothers were in rags and whiskey at Edgeworthstown' and begging for work, and she wondered where his money came from: 'I fancy he has just job-job-jobbed and I don't suppose his hands are very clean as to public money, road money, etc'.

In 1828 Jones was contractor for an important section of the new central Connemara road from Galway to Clifden, one of the many projects planned by Nimmo. He was found to be unfit for the job and work came to a standstill in 1828 'after terrible quarrels with Nimmo', according to Maria Edgeworth, leaving the road in an impassable condition and the labourers unpaid and half-starving. The dispute about the contract was still not resolved when Nimmo died in January 1832 and the last section of the road west of Oughterard had still to be constructed in 1833 when Maria and her companions visited the area. In that year the authorities reluctantly agreed, on the recommendation of Hamilton Killaly, the engineer who succeeded Nimmo, to pay £700 to Jones on foot of the work he had done and so as to enable him to discharge the debts he had incurred with suppliers, subcontractors and labourers. He was contractor also in the early 1830s for the repair of the pier at Tarmon on Blacksod Bay, Co. Mayo, built as part of Nimmo's programme of fishery piers some years earlier.

Jones was the first county surveyor for Co. Tipperary, appointed on 17 May 1834. When the law was altered in 1836 to allow surveyors to appoint assistants at salaries of up to £50 a year, he was fortunate to receive approval from his grand jury to the appointment of as many as six assistants. His next step was to enquire from the Board of Works if it would be in order to send on his apprentices for the qualifying examination; he was told that this would be in order except in the case of one sixteen-year-old. By 1836 complaints about the poor state of the roads in north Tipperary led to calls for the appointment of a second county surveyor but Jones told the grand jury in lengthy reports at the spring and summer assizes in 1837 that matters were improving; he claimed to have done much to prevent the system of jobbing which had eliminated competition for road contracts and led to higher prices in earlier years. He complained, however, that he still had to contend with negligent contractors who, through ignorance, inability or wilful neglect, had left works unattended in some areas for up to eighteen months, causing serious inconvenience to the public. Contractors like these, he claimed, had received indulgence at the hands of former grand juries and, as a result, nearly one-third of the works presented for in previous years now had to be re-presented. In December 1838 Tipperary was divided into two ridings, with Jones initially acting as surveyor for both. However, at the first assizes for the north riding in March 1839, the grand jury decided that it should have its own surveyor and thereafter Jones served as surveyor for the south riding only.

In March 1837 Jones presented plans, specifications and cost estimates to the grand jury for new sessions houses at Clogheen, Carrick-on-Suir, Tipperary and Roscrea, having been directed by the authorities in Dublin to do so. In March 1841 he reported that the building at Tipperary would soon be completed but little progress was being made at Carrick-on-Suir. As constructed, the two courthouses appear to be identical and follow a standard William Caldbeck design, while the building completed around the same time at Roscrea appears to have been designed by Jacob Owen. A fever hospital was completed in Tipperary in 1842 under Jones' supervision and a bridewell in 1852.

In his early years, at least, Jones seems to have been active in developing the road network in south Tipperary. In 1841, for example, he was able to report that a number of new bridges had been completed, including a bridge at Arraglin, and he presented proposals for a new line of road from Clonmel to Cashel. He consistently urged the grand jury to allocate funds for the development of the Suir navigation from Clonmel to Carrick-on-Suir by removing shoals and deepening the waterway. Although he argued that substantial economic benefits would flow from his proposals through the reduction of freight costs, the grand jury was reluctant to do more than meet the cost of occasional repairs to the quay walls in Clonmel and to the trackway used by the horses which towed barges on the waterway.

At the spring assizes in March 1852, when the state of the roads attracted severe criticism, Jones' defence was that the roads were in a poor state in many places, and in some cases unsafe, because competent persons would not undertake the work at the approved prices. In addition, he complained that the Waterford & Limerick Railway Company had done great damage to the roads and had not met its promises to restore them. Jones, however, was heavily criticized for refusing to certify payment to a contractor who claimed to have done all that was required of him and who suggested that payment was being withheld because of the 'injustice and caprice' of the surveyor. Immediately after the assizes, the grand jury held a special investigation into the complaint at which the contractor was represented by counsel and, after a long debate, Jones was dismissed.

Jones died at his residence, Ardbear House, Clifden, in March 1859 when he was 63 years old. He had purchased additional land in the Bunowen area and in 1852 Griffith's Valuation showed him to be the owner of a substantial acreage. According to an obituary in the *Galway Vindicator*, he was a liberal and kind landlord and 'for many years since his retirement from active public service, he had discharged the duties of magistrate and country gentleman in his neighbourhood with benefit to the poor and a liberal hospitality. In every work of charity and religion, his name was prominent and his generosity munificent. Towards the erection of Catholic institutions in his vicinity, he was a bountiful contributor, and no man sought his aid in vain when good works were to be performed'.

JOYCE, Arthur Edward (1851–1927), was a member of a well-known west of Ireland family, son of Thomas Appleyard Joyce of Rahasane Park, Co. Galway, who was high sheriff of the county in 1882. He was born on 10 May 1851 and gained a diploma in engineering at the Royal College of Science for Ireland in June 1875. He was subsequently a pupil of James Price, chief engineer of the Midland Great Western Railway Company for about eighteen months. He was placed second at the county surveyor examinations held in September 1876 and in the following December was appointed Westmeath county surveyor, becoming the first graduate of the Royal College of Science for Ireland to enter the ranks of the surveyors. He was to hold this position for over fifty years, successfully making the transition from the grand jury system to the service of the new county council in 1899 and to the service of the new state in 1922. When he applied after 1899 to have his salary increased to £550 – a modest figure relative to the size of county and its road mileage – the council were reluctant to agree but, as one member pointed out, Joyce had treated them 'very honourably and kindly' and he was ultimately granted the increase.

By 1910 Joyce had managed to persuade his county council to introduce steamrolling and three years later the change in the condition of the roads in the Mullingar area, in particular, was said to be scarcely credible – road surfaces which had been of the worst description were now said to be a credit to the council and its engineers and there had been a large increase in the number of motor cars in use in the area. However, the council then began to have second thoughts about steamrolling and decided to suspend all operations when the existing contracts expired. Like many of his colleagues, Joyce had difficulty in attracting and retaining assistant surveyors of good calibre. Because the salaries were very small, the assistants were forced to look for private work and this, of course, adversely affected the carrying out of their official duties. Besides, the fact that no travelling expenses were allowed meant that a conscientious man who did his work well was punished far more than the man who did nothing. Joyce sought to resolve the situation in 1913 by proposing that his assistants should be supplied with motorcycles at the expense of the rates but the council unanimously rejected this novel scheme.

Joyce himself was able to supplement his official earnings by undertaking a variety of projects of an engineering nature for bodies other than the grand jury and the county council. He was engineer to the nine-mile-long Loughrea to Attymon branch railway which opened in 1890 and he carried out the survey work for the twelve-mile-long Ballinrobe to Claremorris branch, opened in 1892. In the 1890s he worked with Edward Townsend, professor of civil engineering at Queen's College, Galway, on planning and promoting the Galway and Clifden (via Oughterard) Light Railway, based on plans prepared years earlier by **S.U. Roberts**, with a branch from Recess via the Lough Inagh valley to Killary Harbour which was then an important base for the British navy. Some years later he was involved in the planning and promotion of the Mullingar, Kells & Drogheda Railway which was intended to allow shipments of cattle from the midlands to be exported via Drogheda instead of Dublin but, although the project was supported by the Great Northern Railway and was authorized officially in 1903, it failed to go ahead because of funding difficulties and the opposition of the Midland Great Western Railway. A similar fate befell proposals with which he was involved for a railway from Mullingar to Mountmellick.

Joyce had some competence in architecture and undertook a significant amount of work in this field both for his county council and for private clients. In the 1890s, he was responsible for the remodelling and refurbishment of the asylum at Mullingar – now St Loman's Hospital – and he submitted plans for a new water supply scheme for the complex ten years later. In Mullingar, new county buildings which he designed and had built at a cost of about £5,000 on the site of the old county gaol were opened in 1913; they have been described as 'elegant and restrained Italianate, with a shallow, bow-fronted portico of contrasting delicacy'. His rebuilding of the courthouse at Athlone, completed in 1916, cost about £4,000; at the first quarter sessions held there, the judge remarked that it was probably the best sessions courthouse in Ireland and paid great compliments to Joyce on what he described as his second triumph – his first being the new county buildings. Technical schools in Mullingar, completed in 1920, and extensions, including a new operating suite, constructed in 1909–10 at the county infirmary were among his other projects.

When the Dáil Éireann Department of Local Government called on all local authorities and their staff, after the June 1920 local elections, to sever all connection with the Local Government Board, Joyce voluntarily resigned his position and accepted reappointment by the county council on the basis of a verbal assurance that his pension position would be protected. He lived initially at Prospect, Mullingar; he moved later to Belmont, Ballinagore, Co. Westmeath – an impressive 1740s house eight miles from Mullingar and now used as an equestrian centre; and finally to Earl Street, Mullingar. He was an ardent sportsman, a keen supporter of the Westmeath hunt for fifty years and one of the founders of the Westmeath polo club which went out of existence on the outbreak of war in 1914. Joyce was said to be the oldest man hunting in Westmeath when, on 3 January 1927, he was thrown from his horse on the hunting field. He died at his home in Mullingar as a result of his injuries on 11 January and was buried two days later at Walshestown cemetery; High Mass in the cathedral at Mullingar was followed by an impressive funeral in which large numbers of public representatives, council officials and road workers marched in formal processional order, with all shops closed and blinds drawn.

By the time of his death Joyce had come to be regarded as a local institution and was one of the most outstanding characters in Westmeath. He had been held in high esteem by the long succession of grand juries and county councils which he served and by all of those who worked under him, and eloquent tributes were paid to him at a specially-convened public meeting in Mullingar and at a special meeting of the county council. He was an energetic and highly competent county surveyor and it was regularly claimed that the roads of the county were among the best in Ireland during his long term of office. He was described by a local Dáil deputy as a devoted officer, a slave to duty, socially irresistible and a great sportsman. 'He died as he had lived' was the comment of another Deputy, P.W. Shaw, 'facing at 76 years of age the biggest double-ditch in the county's hunting fields, namely the Bryanstown double'.

KEARNEY, Thomas (fl. 1831–62), was described in the annual report of the Board of Works for 1831–2 as a man who had much experience in the execution of road works. In that year, he had submitted plans and estimates for some twenty miles of new roads through the Knockmealdown mountains, from Clogheen to Lismore and Cappoquin, and the newly established board allocated a loan to finance the work which was proceeding satisfactorily by 1833. Thus did the renowned scenic route incorporating 'The Vee' come into existence. Other projects with which Kearney was involved in the period 1826–33 included roads from Fermoy to Tallow, Gorey to Wexford, Waterford to Tramore and Kilkenny to Pilltown; his drawings of these schemes are preserved in the National Archives. Kearney also acted as contractor on some road projects, including part of the new Kanturk–Cork road through the Boggera Mountains which was completed in 1830.

Kearney, according to a Board of Works report in 1837, was 'an unexceptionable man' and had sat the county surveyor examinations purely 'for form's sake'. He had, presumably, not accepted a county surveyor appointment in 1834 because of the extent of his road construction commitments but was appointed to the eastern division of Co. Limerick on 29 June 1837 when the county was first divided into two divisions. He took up residence near Pallas Grean, carried on a steady programme of road and bridge building in the following years and seems to have had a good working relationship with his grand jury. At the same time he continued to advise and assist a number of his brothers who operated as road contractors in other counties, and he carried out some work for the Board of Works, inspecting and reporting on proposals from other counties for loans and grants for road and bridge projects.

When railway development began in Limerick in 1845, Kearney found it necessary to monitor the activities of the railway companies very closely and sought through the courts to force them to comply with their obligations in relation to roads as he understood them to be. In winning a landmark judgment against the Limerick & Waterford Railway Company, he succeeded in establishing that the company was liable to repair the approaches and surfaces of bridges over the railway, but he failed to enforce a similar obligation in relation to roads under railway bridges, even where the level had been lowered considerably by the railway company to give sufficient headroom.

During the famine period Kearney, like many of the other county surveyors, had to undertake major additional responsibilities arising from the extensive programmes of road and drainage works which were initiated to provide some relief for the destitute poor. The role of the surveyors in selecting the projects to be financed and the persons to be employed was always a difficult one. Special presentment sessions at Kilfinnane on 28 September 1846 developed into a riot both inside and outside the courthouse, according to a report from Lieutenant Inglis, the Board of Works inspecting officer, and as the riot became more and more violent, Kearney and himself were fortunate to be able to make their escape to the local hotel. At Hospital on the following day, thousands assembled outside the sessions and, according to another report from Inglis, 'all their passion was directed at the county surveyor' who was subjected to 'the most vehement expressions of uncontrolled indignation' as he attempted to leave. The fury of the mob 'amounted to nothing short of madness' and 'poor Mr Kearney was actually hunted like a mad dog by the whole country population'; the police had to intervene with loaded carbines before he could drive off in his gig 'under awful groaning and pelting of stones' and he was saved from certain death only by the fact that he 'so placed himself on the car that the mob could not conveniently get at him without first killing a very favourite parish priest'.

The reports from Lieutenant Inglis were passed on to the Treasury in London by Colonel Jones, chairman of the Board of Works, who added that while Kearney had stood his ground manfully during the earlier programme of relief works, it would be impossible for any public official to hold his ground if mobs were to be incited to similar acts of outrage. Jones feared that Kearney would resign as a result of the attacks on him, but when the unfortunate surveyor came under attack from a different quarter some weeks later, Jones was less supportive. Lord Monteagle complained, as the relief works were getting under way, that Kearney was 'wholly incompetent to take charge of executive relief works' and Jones, instead of attempting to defend the surveyor, responded that his board 'may upon fair grounds, complain of the grand jury sanctioning the continuance in office of a person who is incompetent'; he added, gratuitously, that 'difficulties of a similar nature, I am sorry to say, exist in other counties'.

Kearney survived the difficult famine years and his duties returned to a more normal pattern in the 1850s, with responsibility for the maintenance of over 800 miles of roads in the eastern division and for the Limerick city grand jury roads which he took over from **John Hill**, the Clare surveyor, in May 1851. He was in bad health by the early 1860s and found it necessary to apologize to the grand jury at the summer assizes in July 1861 because of his inability to give the roads the same degree of personal attention as in the previous twenty-four years. Close supervision was obviously necessary because, he reported, the road contractors, with few exceptions, scarcely ever used the necessary quantities of stone on the roads; as a result the surfaces were being cut into ruts in bad weather and then badly repaired by scattering 'half-broken stone' in the ruts, making travel very dangerous. He was concerned, too, that few contractors kept the water-tables clear and that many of them were failing to cut the weeds on the roadsides. The grand jury allowed Kearney to recruit a special assistant during his illness and this arrangement continued until he died, still in office, on 3 June 1862 at Sunville, Co. Limerick.

KELLY, John (*c*.1808–78), served an apprenticeship to his father who was in practice as a land surveyor in Co. Down. He was employed as an inspector by the Board of Works in 1832 in connection with works costing over £2,500 on the Drumsna–Bundoran road and became the first Roscommon county surveyor on 17 May 1834. He continued to work for the board on road construction projects for some years, including schemes arising from the improvement of the Shannon navigation, and claimed to have built a total of 239 miles of new roads by 1868. He also carried on a private practice as a land surveyor and employed a large number of men on land reclamation and improvement works. An unusual assignment came his way in 1840 when, for a fee of £25, he acted as returning officer at the first elections of guardians of the newly constituted Roscommon union.

During his service of more than forty years in Roscommon Kelly was said to have discharged his onerous duties 'in a perfectly honourable and independent manner' and to have been held in high esteem by successive grand juries and by all classes in the county. Like the other early surveyors, he had to undertake the supervision of famine relief works in his county where there was severe distress and above-average mortality; demand for relief work was relatively high and over 20 per cent of the total workforce were engaged at one stage. In August 1846 he had to contend with striking labourers who threatened to plunder the town of Roscommon if their demands were not met; they were averse to task work, according to Kelly, and raised every objection to it.

Kelly had a significant role in land matters in Roscommon in the 1830s and 1840s, both as a land surveyor and as land agent. He was closely involved with the Mahon family of Strokestown House and its 30,000-acre estate, made famous by the sensational murder on 2

November 1847 of Major Denis Mahon. After the murder Kelly, who held a small tenancy on the estate, took over the potentially dangerous role of agent for the estate and acted as such for the new owner for some years. At nearby Ballykilcline, where the Commissioners of Woods and Forests succeeded in 1848 in clearing the 600-acre crown estate of its 500 inhabitants by eviction and assisted emigration to America, Kelly, who had been among those engaged to advise the commissioners about the management of the lands, was one of the first to show interest in purchasing them. 'From the disturbed state of the district' he told the commissioners 'many would hesitate to purchase those lands but as I am well known in that part of the county and have always treated the people with kindness, I do not apprehend any annoyance from them'. He went on to suggest that with his 'professional exertions' and some investment in drainage, the estate might become for him 'a tolerable reserve for the decline of life'. In the event, Kelly did not acquire the land when it was sold in 1849 but he continued his involvement in land reclamation and development in the area and, according to one of his obituaries, continued to give a good deal of employment.

Following exceptional floods on the river Shannon, Kelly travelled to London to give evidence in 1865 before a House of Lords select committee on the drainage and navigation of the river, qualifying for a payment of seven guineas in travelling expenses and just over £40 in subsistence allowances for his thirteen-day absence from home. His evidence was very critical of the activities of the Shannon commissioners who, in his view, had achieved nothing and had done injury instead of advantage to the drainage in some places. He retired from his county surveyor post due to ill-health at the summer assizes in July 1876 and was granted the maximum pension of £333, making him one of the first of the surveyors to benefit under legislation which had been enacted in the previous year. Although he had taken a house in Kingstown for the benefit of his health, he was instructed by the under-secretary, T.H. Burke – murdered in the Phoenix Park a few years later – to return to Roscommon and to continue to carry out his former duties until his successor was appointed in October; he insisted at the following assizes on his right to receive a salary for this period, in addition to his pension, and succeeded in persuading the grand jury to concede this.

Kelly became a justice of the peace and a poor law guardian after his retirement and was a member of the grand jury at the summer assizes in 1877. He was in his seventieth year when he died on 8 May 1878 at his residence, Essex Lawn, Roscommon; two days later, according to the *Roscommon Journal and Western Reporter*, a large and highly respectable cortege accompanied his remains to the family vault at the Abbey cemetery, the hearse being preceded, according to the *Roscommon Messenger*, by his tenants and employees in scarf and hat bands.

KELLY, Thomas J. (fl. 1908–65), was born on 8 November 1908 in Co. Kilkenny, educated at Patrician College, Ballyfin, and awarded BE and BSc degrees at University College, Dublin, in 1929. He was a student engineer with Messrs Siemens Bauunion on the Shannon Scheme in 1928 and in 1929–30 undertook a post-graduate course in geology. As assistant to D.S. Doyle, consulting engineer, he was employed in 1930–1 as clerk of works on the construction of the new Clonmel mental hospital. In 1931–2 he was engineering assistant to W.J. Doherty of Derry, working mainly on water and sewerage schemes, and was subsequently engaged as clerk of works on another hospital project.

In July 1932 Kelly joined the local service as temporary assistant county surveyor with Kilkenny county council and became a permanent assistant in September 1935. He served also as engineer to the Kilkenny board of health. In 1943, with thirty-five other candidates, he was interviewed for the new post of county engineer in Wexford and was placed first on the list of qualifiers. He took up the position on 1 June 1944 and transferred to Co. Kildare in November 1945. In the 1950s, he was responsible for the commencement of major realignment and improvement works on the Dublin–Naas road, leading eventually to the opening in June 1963, at a cost of about £500,000, of the first three-mile section of dual-carriageway near Kill. Roads of this standard were still something of a novelty in Ireland at that time, with little more than ten miles of dual-carriageway in the entire state.

Following the retirement of **J.A. Ryan**, the post of Dublin county engineer was advertised in August 1953 at a basic salary of up to £1,200 per annum. This was regarded as grossly inadequate by Cumann na nInnealtóirí which requested candidates to withdraw their applications. The resulting impasse was not resolved for six years during which the post was filled on an acting basis by Laurence Gorman. Eventually, following resolution of the dispute, Kelly was appointed in 1959. While continuing to advance the council's own road improvement programme as actively as resources allowed, he put forward a well-argued case for the establishment of a central main road authority in 1961, noting the economic importance of the main roads, the inability of a multiplicity of local authorities to carry out reconstruction programmes effectively and expeditiously, and the priority given by local councillors to catering for purely local needs.

Kelly was president of Cumann na nInnealtóirí in 1960–1 and of the Institution of Civil Engineers of Ireland in 1964–5; his presidential address to the Institution provides a useful historical survey of the changing role of the county surveyor/county engineer. He retired due to ill-health in 1962.

KEMPSTER, James Forth (1816–93), son of William Henry Kempster and his wife Frances, née Greetham, was born in Piccadilly, London on 3 August 1916. He was living at Camden Street, Dublin, when he became county surveyor for the eastern division of Galway on 28 May 1838 at the age of 22. He based himself initially at Portumna but later set up offices at Ballinasloe.

As county surveyor, Kempster seems to have carried out his official duties satisfactorily but unspectacularly and his biennial reports to his grand jury rarely dealt with items other than those of a routine nature. At the 1867 spring assizes, for example, he noted that the unusually severe winter had caused a great deal of damage to the roads and that repairs had to be delayed until the substratum had dried out; that the new road from Portumna to Woodford was nearing completion; that the wooden piles on the bridge at Portumna needed to be tarred during the summer; that improvements were highly desirable at the courthouse at Gort; and that he was planning improvements at the Market Square in Ballinasloe and on the approaches to the Catholic church. In all, he had built a total of more than 200 miles of new roads by 1868.

Kempster was one of the early surveyors who achieved some status as an architect although his private practice was mainly confined to the area near his base at Ballinasloe. He was commissioned by the Board of Works to design and supervise extensive additions and repairs to Ballinasloe lunatic asylum (later St Brigid's hospital) between 1847 and 1851 and was called on again when further additions were required in 1880. For the second Lord Ashtown, he redesigned Woodlawn, Kilconnell, Co. Galway – a large seven-bay three-storey house refaced and embellished in an Italianate style and enlarged by the addition of two two-storey wings in the 1860s, but now semi-derelict. At Lismanny, also in Co. Galway, he was responsible for additions to Mr Pollok's residence in 1863, and in the following year his commissions included a new gate lodge at Garbally Court for the earl of Clancarty and a new four-storey house in rock-faced ashlar at Main Street, Ballinasloe. Another commission in the 1860s involved the design of a small hotel near the railway station in Ballinasloe, described at the time as Tudor Gothic in character and bearing the arms of the earl of Clancarty on the pediment; the *Dublin Builder*, noting the construction of the building and the regular flooding which occurred in and around the town, doubted that 'under the present prospects of the town – which are by no means cheering – this hotel will be in much demand except during fair time'. The former hotel still stands but is now in use as an FCA hall. Also in Ballinasloe, Kempster designed a new convent chapel in Gothic style, with a tower and spire, for the Sisters of Mercy (1864). He carried out a number of refurbishment and alteration projects at country houses in Galway and Roscommon in the 1860s and 1870s and in Co. Tipperary he remodelled Ballyquirke castle in 1872. A new rectory at Vicarstown in Co. Laois (1877) was another of his projects.

After fifty-three years in office, Kempster retired in 1891 and died at his home, Mount Pleasant, Ballinasloe, on 14 August 1893, aged 77; he was buried at Glenlahane graveyard near the town. According to the *Tuam Herald*, he was 'deeply regretted by everyone who knew him and admired his efficiency combined with courtesy and conscientious discharge of the duties of his office'. A brass plaque was subsequently erected in his memory in St John's Church of Ireland in Ballinasloe where he had worshipped for over fifty years. With his wife, Catherine Elizabeth Maher who died in 1858, Kempster had nine children. His son, William Henry (1843–1920), practised as an architect in Ballinasloe and, among other things, designed the cross which commemorates the battle of Aughrim. Another son, John (1846–1918), who was awarded a BA and LCE at Trinity College, Dublin, in 1866, wrote to the chief secretary in 1867 expressing interest in obtaining a county surveyor post but was told to await notice of the first of the new-style examinations which were to be held by the Civil Service Commissioners. In 1871 his father paid £1,000 to Benjamin Patterson (1837–1907), a Trinity College engineering graduate who had been providing quantity surveying and costing services since 1860, on foot of which the younger Kempster was taken into partnership; together the two men developed a substantial quantity surveying and costing business in Dublin under the title Patterson & Kempster, later Patterson Kempster & Shorthall, and still thriving as Davis Langdon PKS.

KENNEDY, Michael Joseph (1887–1939), was born on 4 July 1887 near Nenagh, Co. Tipperary, where his father was a farmer. He was educated at the local Christian Brothers' school and subsequently at Rockwell and Blackrock Colleges and studied at Queen's College, Galway, where he gained a BA degree in 1909 and a BE in 1910. His first appointment was as assistant in the office of the Galway borough surveyor and he was clerk of works between 1911 and 1913 on the construction of a major new bridge over the river Shannon at Portumna, for which **J.O. Moynan**, the north Tipperary surveyor, had overall responsibility.

In November 1913, when Galway county council decided to revert to the practice of having two county surveyors, Kennedy was appointed surveyor for the eastern division with headquarters at Ballinasloe. He was appointed to the better-paid post in the western division in June 1916 after the death of **John Moran**, and he took on responsibility for the entire county, on a temporary basis, a year later. On the appointment of **George Lee** to east Galway in August 1919, Kennedy's remit was again confined to the western division.

Kennedy's work on the county's roads was widely acclaimed and he was said to have built or reconstructed hundreds of bridges including a large number which had been blown up or badly damaged between 1919 and 1923. He was responsible for the reconstruction of the central hospital in Galway and for the design of the new county buildings, the city's Wolfe Tone Bridge and several vocational schools. He was actively involved in the planning of development works at the harbour in Galway and was associated with a number of industrial projects.

Described as 'leonine in mien' and a strong personality, Kennedy played rugby for Connacht in his younger days and earned a trial for the Irish team. He was an active member of the committee who planned the relocation of Galway golf club from Barna to a new site at Salthill on the outskirts of the city, and he personally supervised the work of draining, reclaiming and laying out a large area of land to form the club's first nine-hole course which was completed in 1925. Kennedy became seriously ill in mid-1937 and died on 16 December 1938, aged 52. Survived by his wife, the former Chrissie McCormack and five children, his funeral to the new cemetery was a large and impressive tribute to a man who, according to an obituary in the *Connacht Tribune*, was 'one of the most remarkable men to have served Galway and its people' in modern times.

KILBRIDE, Bernard J. (fl. 1890–1955), was born on 15 August 1890. He gained BA and BE degrees at University College, Dublin, and was an assistant county surveyor in Cavan and later in Longford between 1919 and 1922. He was acting county surveyor in Longford between 1922 and 1924 and was elected unanimously to the permanent post in August 1924. He was allowed to appoint two assistant surveyors and was granted a loan of £400 by the council to enable him to purchase a car.

Soon after his appointment Kilbride was under some pressure arising from intermittent complaints from the ratepayers and letters to the local press about the poor condition of the roads in some parts of the county. Matters came to a head towards the end of 1926 when **James Quigley**, the chief roads inspector of the Department of Local Government and Public Health, advised the council that the standard of road improvement work being carried out was so bad that he would no longer be able to authorize grant payments. He pointed in particular to the Longford section of the main Dublin–Sligo road which was in an absolutely disgraceful condition; it had been improved and steamrolled at great cost but had since been neglected and poorly maintained and was beginning to disintegrate. A committee of the council which met Quigley to discuss the situation reported to a special meeting on 11 December that he had referred to the need for reform 'in no uncertain terms'; they went on to advise that 'our survey staff will have to seek to turn out better work' and that the councillors themselves would have 'to shoulder the responsibilities which we must carry'. The outcome was a decision by Kilbride to resign his post with effect from 31 January 1927.

In August 1927, Kilbride was appointed acting county surveyor in Offaly in place of **Ignatius J. O'Sullivan** who was terminally ill and continued in the post until a permanent appointment was made in September 1928. He became borough surveyor in Clonmel on 1 October 1931 and served there until his retirement at the end of August 1955.

KIRKBY, Samuel Alexander (1845–1927), son of Edward Kirkby of Sheffield, was born at sea near Saint Helena on 24 June 1845. He was educated in England at Haversham Grammar School and began his engineering career by working for a year with George Underwood at the Sefton foundry in Liverpool and for four years with Messrs George Forrester & Co. at the Vauxhall foundry in the same city. In February 1868 he entered St Catharine's College, Cambridge, but migrated six months later to Christ's College where he held a Whitworth scholarship and gained a BA degree in 1871 and an honours MA in mathematics in 1874. He subsequently worked as assistant engineer with Sir John Fowler on major construction works, including the Metropolitan District Railway in London, and the Egyptian railways.

Kirkby took second place in the county surveyor examinations held in June 1872 and was appointed Longford county surveyor in the following August. He transferred to the southern division of Cork east riding in July 1874, replacing **C.F. Green** who had taken first place in the 1872 examinations but had vacated the Cork post in favour of an appointment as assistant engineer with the Board of Works. Kirkby took on responsibility for the entire east riding when **A. Oliver Lyons**, who had served the northern division for some thirty-five years, retired in October 1898. The grand jury had resolved at the assizes in the previous March that it would be more economical and more efficient to have two wholetime county surveyors in Cork, with higher salaries, rather than three, as had been the case since 1861 but, because of concerns about the scale of his private practice, Kirkby was initially denied any increase in salary to reflect his extended duties.

Perhaps because of his early training, Kirkby seems to have had a strong preference for the use of iron in bridge works and considerable skill in the design and carrying out of such works. In 1876 he designed Kinsale Eastern Bridge and Ringnanean Bridge, both in iron, to carry the Cork–Kinsale road over an estuary near Belgooly. These apparently fragile bridges, with slim cast-iron columns rising from screw-piles, metal decking, and ornamental hand railings, were constructed by Messrs J.O. & C.E. Brettell of Worcester with ironwork supplied by Messrs John Steel & Sons of Cork. The bridges had become badly corroded by 1990 and were subject to weight limits; they now serve pedestrian traffic only having been replaced by reinforced concrete beam and slab structures which were completed in 1995. At a later stage Kirkby built new bridges in concrete and steel at Blackstone, Ballincurrig and the Gearagh.

In the late 1870s Kirkby designed a new Kinsale Western Bridge, some 1,300 feet long, to replace Heard's Bridge, a wooden structure which had been erected across the estuary of the river Bandon three miles west of Kinsale in 1860–1; the wooden bridge was said to be dangerously defective by 1869 and it collapsed completely in 1881. Kirkby's new bridge which had a wooden deck supported on iron screw- piles and a manually-operated revolving opening span on the southern side, was completed in 1882 and was one of the longest in the country. The contractors (again, Messrs Brettell) had taken four years to complete the work instead of the three years allowed by the contract, and had to sink some of the steel piles more than ninety feet (instead of the expected maximum of sixty-five feet) to gain a good foundation. When the delay in completing the bridge was raised in the House of Commons on 9 June 1881 by Eugene Collins, MP for Kinsale, the attorney general for Ireland said that he had been advised by the county surveyor that the delay was caused by the severity of the weather in the two previous winters and by the fact that some of the screw-piles had been carried away by a passing schooner. However, while Kirkby was willing to concede that an excellent bridge had been constructed, he was not prepared to recommend any payment over the contract price of nearly £18,000 to meet the contractors' claim for payment of an additional amount of £2,658; instead he was reported to have suggested, laughingly, at the presentment sessions that 'perhaps a good testimonial may satisfy Messrs Brettell'. Kirkby's Western Bridge was itself nearing the end of its useful life by 1967 when weight restrictions had to be imposed due to corrosion of the structure. By 1971 there was serious concern for the safety of the bridge and the weight limit was reduced to three tons. The bridge survived, with some repairs, until 1977 when it was replaced at a cost of more than one million pounds by the present prestressed concrete structure, named Archdeacon T.F. Duggan Bridge, over two miles downstream; this new fixed bridge is 1,120 feet long and rises by some twenty feet as it crosses the estuary.

At the other end of his area, Kirkby was appointed by the lord lieutenant in 1876, together with **Charles Tarrant**, the Waterford surveyor and William Forsyth of the Board of Works, to report on the need for a new bridge to replace the wooden toll bridge over the river Blackwater at Youghal which had been designed by Alexander Nimmo and erected by contactors George Nimmo and John Bennett between 1829 and 1832. After Tarrant's sudden death in July 1877, it fell to Kirkby to see the new wrought-iron lattice-girder bridge, 1,300 feet long, through to completion in 1883 on the site of the earlier bridge at a cost of almost £39,500. The bridge was built by the Stockton Forge Company and had a plated deck which was covered by concrete; it was 21 feet wide, with 5 spans of about 100 feet, 20 spans of 30 feet each and 2 opening spans each 50 feet wide. It was supported on screw-piles and cast-iron cylinders filled with concrete and sunk to an average depth of some 100 feet below road level; in some places, according to Kirkby, where there was a stratum of quicksand, the heavy cylinders sank up to 18 feet at a run.

Kirkby's Youghal Bridge was not a success: it had seriously deteriorated even before his own retirement and he had been obliged to erect notices limiting traffic to vehicles with axle weights of less than seven tons. The saga of the bridge and its possible replacement was to run for more than fifty years after that – probably the longest ever controversy of its kind in Ireland. By 1927 the continuing deterioration of the bridge forced the county surveyors of Cork and Waterford to introduce further weight restrictions, and still more severe restrictions were imposed in 1939 – a maximum gross weight of 3½ tons and a speed limit of 5 miles an hour. Weighbridges were erected on each side of the bridge at that stage to facilitate enforcement and traffic on the bridge had to weave in and out through a system of gravel-filled tar barrels and planks, forcing it to slow down to a snail's pace. This decisive action followed a report from the Cork county surveyor, **R.F.M. O'Connor**, who told his council that 'for modern traffic, the structure is not only useless but a positive danger, without drastic limitations'. While the material used in the construction of the bridge appeared to have been very good, O'Connor's inspections established that the bridge had been badly damaged by the vibration caused by traffic and by the corrosive action of sea water which had caused a loss of one-eight of an inch on all exposed surfaces. The floor concrete had been reduced to debris and the bracing of the piles was very seriously corroded, causing a side sway in the bridge under the weight of traffic. According to O'Connor, the bridge was too light and too weak and, in the light of present knowledge, it was 'of a most unsuitable type ... and, even on its erection, appeared to have had very little margin of safety for the traffic of that time'. The serious inconvenience arising from the 1939 limitations on the use of the bridge and the fact that there was no other bridge for more than ten miles upriver, led the two county councils to set up a joint committee to decide what action should be taken and, following a report from the two county surveyors, the Minister was requested to allocate a grant to meet the cost of a new bridge. In 1947 the councils agreed to seek a bridge order with the objective of having a new bridge completed in about four years. Nearly ten years later, with the bridge being referred to as a long-derelict eyesore, the need for a replacement was still the subject of controversy and of an intense lobbying campaign; there were arguments about whether a replacement should be in steel or in concrete; whether it should be a high-level bridge or a low-level structure with an opening span; whether the same site or a new one should be used; and, at the last minute, claims that the old bridge could be reconstructed economically, retaining the old piers and piles, and without any interruption of traffic. In the end, Kirkby's bridge was replaced by an entirely new reinforced concrete structure on a new site at Ardsallagh, 580 yards upriver; this cost almost £600,000 and was formally opened in January 1963.

For ordinary road work, Kirkby was an unrepentant supporter of the contract system not only because he found it more economical but also because, 'from the selfish point of view, it is more satisfactory to the county surveyor'. Nevertheless, in his first report to the county council in April 1899 he put forward a range of novel suggestions to the new councillors as to how they might exercise the additional powers and functions they had been given in relation to roads; he suggested, for example, that the council itself should open quarries in central locations with suitable stone-crushing machinery powered by a steam traction engine which would visit the different quarries 'just as a threshing engine goes about in harvest time to the different farms'. He invited them also to consider the question of providing cycle paths, surfaced with cinders, on certain roads and paid for, possibly, by a small tax on each bicycle and he was anxious that new 'enlightened' policies should be adopted in relation to the cleansing of streets, refuse collection and sanitary matters generally. After he had been obliged to go along with the introduction of the direct labour system in 1904 on 159 miles of roads in his area, he reported that their condition was better than it had been under the contract system and that the cost was just below what the average cost had been. He conceded at that stage that direct labour gave better results on important roads but still believed that the contract system was more economical on lightly-trafficked country roads. Like many other surveyors of the day, he found it difficult to cope with the damage being caused to the public roads by steam locomotives and wagons. This was a particular problem in the Ballinhassig area where a seventeen-ton traction engine was being used daily to haul three loaded wagons of bricks to Cork city from a brickworks which had been developed by John Sisk at Ballinphellic. In July 1898, the grand jury refused to support a scheme for a 4½ mile-long railway to link the brickworks to the Cork, Blackrock & Passage Railway but the problem was resolved in 1901 by the construction of an aerial ropeway linking the brickworks with the Cork, Bandon & South Coast Railway at Ballinhassig, some three miles away.

Kirkby found time to carry on a substantial private practice during his years in Cork and to undertake work for a variety of public authorities as well as the grand jury. The short Kanturk & Newmarket Railway was completed under his direction in 1889 and he had some involvement also in the planning of the Clonakilty Extension Railway, completed in 1886. He was engaged as engineer, jointly with J.W. Dorman, brother of the Armagh county surveyor, **R.H. Dorman**, in planning and supervising the construction in 1885–6 of the narrow gauge Schull & Skibbereen Tramway and Light Railway; his colleague in the west riding of Cork, **Nat Jackson**, found himself responsible for the maintenance of this line from 1892 onwards when, after continued working deficits, it had to be taken over by a committee of management appointed by the grand jury. Kirkby was engineer for new water schemes for Midleton, Youghal and Cobh and for the laying of a watermain from Cobh at a depth of eighty-five feet under the waters of Cork harbour to the naval base at Haulbowline.

Following his transfer to the service of the newly-established Cork county council in April 1899, Kirkby told the finance committee in his first report that he looked forward to working with the councillors who would be able to 'advise and direct' him whereas, with a grand jury which met only twice a year, a degree of responsibility was thrown on the county surveyor which was 'often found very burdensome'. He appears to have developed a good working relationship with the council in the following years and was one of the few surveyors whose salary was increased by his new employers to a reasonable level – £750 – without recourse to the Local Government Board. Having introduced steamrolling in the east riding, he was able to advise the council in 1908 that the Aveling & Porter 15-ton machine which he had been allowed to acquire had been in constant use and that he had been able to improve the condition of the roads and absorb all of the attendant costs without increasing the road maintenance budget.

When he gave notice in March 1908 of his intention to retire, having served for a total of thirty-six years as county surveyor, the county council decided to revert to the arrangement which had been in operation before 1998 and to appoint two surveyors for the east riding. Kirkby finally left the service of the council in January 1909 having continued in office on a temporary basis while arrangements were being made to replace him. On taking up duty in Cork in the mid-1870s, he had lived at Castle Jane, Glanmire, with his widowed mother until she died in 1881 leaving a personal estate of about £10,000 – a very considerable sum in those days. After his marriage in 1886 to a widow, Agnes Augusta Woodley, he lived with her and their daughter, Agnes Muriel, at Miramar, Queenstown. A year after the death of his wife in 1896, he married Eleanor Rubie Young with whom he had a son and another daughter. He continued to live at York House, Cork, for some years after his retirement but later moved to London where he died on 29 January 1927, aged 82.

KIRWAN, Robert Joseph (1869–1949), was born on 21 August 1869, the eldest son of Henry Kirwan JP, a member of Tuam rural district council, and his wife Jean. He was educated at the Royal College of Science in Dublin and at Queen's College, Galway, where he took a BA degree in 1892 and a BE degree in 1893 and was subsequently an articled pupil of J.H. Ryan and of Edward Townsend, both consulting engineers. He worked on parliamentary and contract surveys for the Galway–Clifden railway and subsequently as resident engineer on a section of the line which was completed in 1895. He then went into private practice, carrying out projects such as the Tuam waterworks and sewerage schemes, various small water supply schemes in Galway and Mayo, improvements to the workhouse at Tuam and construction of a new post office in the town. He joined the staff of the Congested Districts Board around 1905 as assistant land inspector for Mayo and worked from a base at Claremorris.

Kirwan became Sligo county surveyor in unusual circumstances. Having gained the highest marks in the qualifying examinations held in May 1903 in connection with a vacancy in Wicklow for which he competed unsuccessfully, he was one of nineteen applicants when the Sligo post was advertised three years later at a salary of £400 a year which was to cover all travel and subsistence expenses. At a meeting of the county council in December 1906 he received the same number of votes as one of the other candidates, **W.F. Barry**, the Monaghan county surveyor, but the latter won the appointment on the casting vote of the council chairman. It subsequently emerged, however, that one of the councillors who voted for Barry was an undischarged bankrupt – and therefore ineligible to act as a member of the council – and Kirwan thereupon gained the appointment in February 1907.

Arising from his engagement by one of the rural district councils in Sligo as architect for a small scheme of labourers' cottages – an appointment of a kind held by many of his colleagues at the time – Kirwan became involved in a long and sometimes bitter controversy about the private practice rights of county surveyors. In a letter to the editor of the *Irish Builder and Engineer* in January 1911 he strongly refuted the journal's 'contemptible insinuation' that surveyors might be influenced in their official dealings with contractors by reason of the fact that the contractors might also be potential private clients. The surveyor's work, he pointed out, is carried on under the eyes of the public, any dereliction of duty is apparent to all, and the ratepayers could not reasonably complain if a surveyor, having done his official duty, devoted his spare time to earning sufficient money to enable him to live comfortably.

In a debate at the Institution of Civil Engineers of Ireland in December 1918 on post-war transport needs, Kirwan argued strongly for concentrating investment on roads, rather than railways or canals. He suggested that there should be a state-supported network of about 1,500 miles of main roads and a secondary system of about 1,500 miles, all of which would need to be improved at an average cost of about £600 a mile, giving a total reconstruction cost of up to two million pounds. His costings were based on his own work in Sligo where fifty miles of roads had been reconstructed at a total cost of £25,000 in the previous few years, and the entire scheme could be financed over ten years, according to his calculations, from existing rates of motor taxation and duties on petrol. These proposals were remarkably similar to the trunk road improvement programmes introduced and financed in the 1920s by the new government.

As far as the minor roads were concerned, Kirwan had to grapple, without state assistance, with the problems caused by rapidly increasing motor traffic and the greater demands made by the new owners of motor cars. In defending the spending estimates for 1932 which he submitted to the county council, he pointed not only to the number of heavy lorries coming onto the roads but also to the fact that 'every man up the mountains who buys a third-hand Ford now wants a steamrolled road made up to his door'. His pleas for extra funds to meet the growing cost of county road maintenance must have had some effect as his estimates and works scheme were accepted without amendment by the county council in the following year, leading the *Irish Builder and Engineer* to remark that this was 'as rare as it is refreshing'.

Kirwan retired in November 1942 and died in 1949. He was a keen geologist and in 1917 led a group which leased disused lead and zinc mines at Abbeytown, on Ballysodare Bay, from the Congested Districts Board. The mines had been worked sporadically in the eighteenth and nineteenth centuries and a survey carried out by the war department in 1916 concluded that the mines' potential was 'not unfavourable'. Kirwan's group had some initial success and exported some high-yielding ore. The mines continued to be worked by a succession of owners until final closure in 1961.

LANAUZE, Richard (*c*.1809–71), was one of the early surveyors who, by his own account, had trained and worked in the offices of Jacob Owen at the Board of Works 'for many years'. He had served for some months as acting county surveyor in Antrim, while **Charles Lanyon** was in London, before his appointment as surveyor for the western division of Limerick on 12 August 1840. He lived at Glin and had his office at Newcastlewest.

At the summer assizes in July 1862, Lanauze sought to have his salary increased by £100, as the act of 1861 allowed, pointing that he then had over 900 miles of road to maintain and that the cost of living had doubled since his appointment twenty-two years earlier. His claim was rejected by the grand jury because the harvest was a bad one and the cess-payers could not afford to pay more. At the following assizes in March 1863, the jury heard complaints about the poor state of the roads especially in the Newcastle and Abbeyfeale areas; some witnesses attributed this to a lack of attention by the surveyor to his duties, while others argued that different factors, including bad weather, were involved. The grand jury resolved unanimously that Lanauze 'from incapacity and neglect of his official duties, had forfeited the confidence of the grand jury and is no longer fit to hold his present office'. Only one member of the jury expressed reservations about dealing so harshly with a long-serving official who, he pointed out, had served with perfect honesty and without any charge of corruption or want of integrity. Lanauze felt that the grand jury's resolution left him little option but to submit a letter of resignation effective from the close of the assizes in July 1863. He sought, through his solicitor, a resolution from the jury attesting to his moral character and wishing him success in his application for a post in another county but even this was denied him, because it was felt that he had not been firm enough in his resistance to 'interested pressure'.

By agreement with the Cavan surveyor, **Frederick Gahan**, who was anxious to move to a southern county, Lanauze then sought to be appointed to Cavan, with Gahan moving to Limerick. He sent several memorials to Sir Thomas Larcom, the under-secretary, seeking approval for the transfer or for an appointment in another county, pleading that he had never engaged in private practice and that, apart from his salary, he had no other means of supporting his wife and family. However, notwithstanding representations on his behalf by over 170 prominent residents of west Limerick, including magistrates, clergy and the local conservative MP, Lt-Col. Dickson, the lord lieutenant refused to sanction the exchange or to offer him another appointment. Lanauze died at Omagh on 10 May 1871, aged 62, and is buried at Mount Jerome cemetery in Dublin with his wife, Sarah Hannah, who died in 1891.

LANYON, Sir Charles (1813–89), who served as Antrim county surveyor from 1836 to 1861, has been the subject of numerous epithets: the paragon among county surveyors; one of the most important and prolific Irish victorian architects; the most important architect of his generation in Ireland; the master of the Italianate style; the prince of railway architects; a prominent figure not only in Belfast and Co. Antrim, but throughout the whole of Ireland – these are only a few of the expressions used to describe him.

Born in Eastbourne on 6 January 1813, the son of John Jenkinson Lanyon, a naval purser, and his wife, Catherine Anne, Lanyon had begun his training as an apprentice with Jacob Owen at the Royal Engineers' establishment at Portsmouth before the latter came to Ireland in May 1832 to take up the position of engineer and architect to the Board of Works which had been established the previous year. Lanyon followed Owen to Ireland and worked with him in the board's offices in the Custom House, Dublin, first as a pupil and later as one of his assistants. His links with Owen were strengthened in 1837 when he married Elizabeth Helen Owen, one of the seventeen children of his master.

Lanyon sat the examination for county surveyorships in April 1834 at which Owen was one of the examiners and, while official records are no longer available, the weight of evidence suggests that he took second place behind **Thomas Jackson Woodhouse**. He was appointed county surveyor for Kildare on 17 May 1834 when he was only 21 years of age. In this capacity he seems likely to have been the designer of a three-arch stone bridge over the Liffey near Naas – later named Victoria Bridge to celebrate the young queen's accession to the throne. When Woodhouse resigned from the Antrim surveyorship to return to his railway engineering career, Lanyon was appointed to replace him with effect from 14 October 1836. The new assignment was clearly an attractive one for an able and ambitious young man as Belfast had prospered under the Union and by the 1830s was the fastest growing town in Ireland, with a busy port and expanding shipbuilding, textile and engineering industries.

Lanyon undertook an enormous workload as Antrim county surveyor but soon had the assistance of no less than twelve assistants – unlike his predecessor who had none. His road maintenance techniques were innovative and ahead of his time and his achievements in road-building were formidable, both as county surveyor and as surveyor to the trustees of four of the six turnpike roads leading out of Belfast – the roads to Antrim, Crumlin, Carrickfergus and Lisburn – from which assignments he earned an additional £180 a year. His major projects included the completion by 1842 of the Antrim coast road from Larne to Ballycastle – a road designed by William Bald and begun by him in 1832 – including the massive three-span Glendun Viaduct (1839) and a whole network of roads opening up the Glens of Antrim. In the early 1840s he supervised the building of Queen's Bridge in Belfast, which had been designed by Woodhouse in co-operation with **John Fraser**, the Down county surveyor; and in the early 1860s he was responsible with **Henry Smyth**, who had succeeded Fraser in Co. Down, for the design of the city's Ormeau Bridge.

In his capacity as county surveyor, Lanyon's buildings include Crumlin Road gaol, Belfast, built of black basalt between 1843 and 1845 and in continuous service until 1996, and Crumlin Road courthouse (1850), which was connected to the gaol by an arched passage under the roadway; both of these were necessitated by the decision some years earlier to transfer all grand jury business from Carrigfergus to Belfast. Lanyon also designed and built courthouses at Larne and Glenarm (1838), Ballymoney (1837–8) and Ballymena (1846). His fame as an architect, however, arose largely from the fact that, with the full consent of his grand jury, he was able to carry on a vast and lucrative private practice from his early days in Belfast. His first significant country house commission seems to have been Drenagh House, near Limavady, completed in 1837. Other major assignments of this kind in the following years included Laurel Hill, Coleraine (1843); Ballywalter, Co. Down, and Dundarave, Co. Antrim (both 1847); and Ballyedmond and Craigdarragh, both in Co. Down (1850). Many of these houses had distinctive *porte cocheres* and attractive Italianate gate lodges. Working with the famous Dublin ironmaster, Richard Turner, Lanyon was involved in the design and construction in 1839–40 of the wings of the palm house at the Belfast botanic gardens, one of the finest and earliest surviving examples of curvilinear glass and cast-iron work; the central dome was added in 1852 to complete the design.

From 1854 onwards, Lanyon was in partnership with William Henry Lynn (1829–1915) who had joined him as an apprentice in 1845. Subsequently when his son, John (1839–1900), joined the practice, the style became Lanyon, Lynn & Lanyon, Civil Engineers & Architects, with offices in Dublin as well as in Belfast from 1860 onwards. As an architect, Lanyon himself was said to have been professionally employed in every county in Ireland, except one. In Belfast, he or his firm were responsible for the Institute for the Deaf and the Dumb and the Blind on the Lisburn Road (1843–5); the Queen's College (1847–9); and the Presbyterian Assembly's College (1852–3) where the northern parliament sat before the new buildings at Stormont were completed in 1932. The Custom House in Belfast, regarded by some as his finest work, followed in 1859. Other buildings of his include the Belfast Bank headquarters in Belfast (1845); the Northern Bank head office in Victoria Street, Belfast (1852); and the Belfast Bank buildings at Armagh (1850) and Derry (1853). He designed Killyleagh Castle, Co. Down, reconstructed in 1849–50; Castle Leslie, Co. Monaghan; Castle View House, Cahir; Old Conna House, Bray; Scrabo Tower in Co. Down, a memorial to the Marquis of Londonderry (1856); and Shane's Castle and Moneyglass, both in Co. Antrim. There were numerous other country houses, as well as schoolhouses and warehouses. In addition, Lanyon designed a group of sixteen churches, without payment, for the Down and Connor Church Accommodation Society in the 1830s and 1840s, as well as a number of churches for Presbyterian congregations. In the Dublin area, there were urban housing schemes such as Prince of Wales Terrace and Florence Terrace in Bray; the Unitarian Church on St Stephen's

Green; the Church of Ireland church on St Andrew Street; the 100-foot-high campanile at
Trinity College (1854); and individual houses in the more affluent southern suburbs. Lanyon
was deprived, however, of the opportunity of contributing a major public building to Dublin
when his plans for the proposed national gallery were set aside in 1858, for cost reasons, in
favour of a scheme prepared by Captain Francis Fowke RE; not surprisingly, perhaps, the
latter attracted serious criticism from the Irish architectural community.

Lanyon had a substantial involvement in the development of Irish railways and in later life
was to become chairman or vice-chairman of the boards of several railway companies. He was
responsible for the Londonderry & Enniskillen and Londonderry & Coleraine railway surveys
in 1844 and assisted the companies in having their schemes approved by parliament in 1845;
the former was not, however, carried through. He was chief engineer to the Belfast &
Ballymena Railway, carrying out the preliminary surveys in 1844, overseeing construction
between 1845 and 1848, and maintaining his connection with the company until the 1860s.
He served also as engineer to the Belfast, Ballymena, Ballymoney, Coleraine & Portrush
Junction line, completed in 1855; the Carrickfergus & Larne line, completed in 1862; the
Belfast, Hollywood & Bangor on which work began in 1862; and several other minor lines. He
designed the impressive York Road terminus in Belfast (1849), the stations at Coleraine,
Cookstown, Larne and Bangor, and many others.

When Lanyon resigned his county surveyor post at the end of 1861 he immediately became
mayor of Belfast and a member of the grand jury which he had served for twenty-five years; he
continued as a grand juror and member of Belfast corporation for many years and became high
sheriff of Co. Antrim in 1878, a justice of the peace and deputy lieutenant of the county. As a
conservative, he was elected as one of Belfast's two MPs at a by-election in November 1866 but
he and the other Belfast Conservative Association candidate were defeated at the general election
of 1868 by a liberal and a leading orangeman, the candidate of the protestant working men. He
was knighted in 1868 in recognition of his professional achievements, bringing to three – with
John Benson and Thomas Deane – the number of Irish architects who held that honour
simultaneously. He retired from practice as an engineer and architect in 1872.

Lanyon had been the first secretary of the Civil Engineers Society of Ireland when it was
established in 1835 but, presumably because of his transfer to Belfast and the pressure of work
there, he played no part in the affairs of the society after 1836 or in the business of the
Institution of Civil Engineers of Ireland into which the society transformed itself in 1844. He
became a member of the Institution of Civil Engineers (London) in 1840, and a Royal
Hibernian Academician in 1860, the only county surveyor ever to gain this distinction. He
was president of the Royal Institute of the Architects of Ireland from 1863 to 1868 – again
the only county surveyor to fill this role – and a fellow, and for a period, vice-president, of the
Royal Institute of British Architects.

Lanyon was obviously an excellent businessman as well as a gifted architect and engineer. He
was a lifelong member of the masonic order and a long-time grand master of the province of
Antrim. He was at one time or another a member of the board of virtually every public and
charitable body in Belfast. He died after a protracted illness on 31 May 1889 at his residence, The
Abbey, Whiteabbey, Co. Antrim, a large Italianate house which he had designed about 1850 for
Richard Davison, MP for Belfast from 1852 to 1860, and which he bought after Davison's death.
He was buried in Newtownbreda churchyard.

LEAHY, Edmund (1814–88) was only twenty years old when he and his father, **Patrick
Leahy**, became the first county surveyors for Cork in May 1834, Edmund in the west riding

and Patrick in the east. Leahy had learned his engineering by apprenticeship to his father and claimed also to have been a pupil of Alexander Nimmo. He worked with his father in Ireland on various surveying, mapping and engineering projects and assisted him in England on railway surveys in 1830–3. He was employed for a short time as a civilian assistant on the staff of the Ordnance Survey in Co. Donegal before taking up his appointment in Cork.

Leahy claimed considerable success as a road-builder in Cork and while the figures he quoted were clearly exaggerated, it does appear that he was one of the more active of the early surveyors and was probably responsible for the construction of about 200 miles of new roads, mainly in the coastal and more remote western parts of his extensive area. His claim to have built over 200 bridges can also be discounted but some significant structures which are still in use can clearly be attributed to him, including the bridge at Bandon (1838) and the three-arch Donemark Bridge (1839–40) on the Bantry–Glengarriff road. In 1844 his *Practical Treatise on Making and Repairing Roads* was published in London; this was one of a very small number of publications by Irish engineers of the period and a second edition was published in 1847, presumably in response to demand.

In his evidence to an official enquiry in 1841 Leahy admitted 'I do all the private business I can get, which is very little'. In reality, with his father and his brothers, Denis and Matthew, Leahy conducted a thriving civil engineering practice during the whole of his years of service as county surveyor, operating from offices at South Mall, Cork. His income from private business must have far exceeded his official salary, a fact which led inevitably to neglect of his public duties. This regularly gave rise to criticism from grand jurors who claimed that the family relationship between the two surveyors allowed them to channel private and public business between them so as to maximize their income and that a serious conflict of interest arose when a scheme prepared by one of them in his private capacity came before the other acting as county surveyor.

Edmund Leahy took the initiative in promoting and designing the Cork & Bandon Railway which was authorized by parliament in July 1845 and had himself appointed as engineer-in chief to the company. However, instead of concentrating on the construction of the line which began in September 1845, he embarked on a headlong rush to promote additional railway schemes in 1845–6 in Co. Cork and throughout Munster. Based on hasty and inadequate surveys, and with the support of his father and his brothers, he had promoted no less than ten separate companies by the end of 1845, involving a planned railway mileage of more than 500. Having failed to obtain leave of absence and to be allowed to appoint a deputy to carry out his county surveyor duties, he was forced to resign his position in January 1846 in order to advance these schemes and to attend meetings in Westminster of the parliamentary committees which were responsible for the examination of railway bills. Several of his schemes were rejected by these committees because of defective plans while those who had agreed to invest in the remaining schemes withdrew their support when the climate for railway investment changed in the early months of 1846. Some months later Leahy was dismissed from his position with the Cork & Bandon company, having virtually abandoned his duty to the company and left the works in the hands of inexperienced engineers and contractors who were forced to work from inadequate plans; after a hearing which lasted thirteen days in the high court in Dublin, he was awarded £325 in damages against the company in November 1847 but, with only a derisory six pence towards his enormous legal and other costs, this can have amounted to no more than a pyrrhic victory, gained at great cost.

In March 1849 Leahy and the extended family arrived at the Cape of Good Hope and he was soon engaged in railway promotion and other civil engineering and contracting activity.

Six months later, when serious agitation developed at the Cape because of plans to develop a penal colony there, Leahy supported the governor against the colonists and was apparently ostracized and punished for doing what he considered to be the duty of every loyal British subject: according to one of his letters, he was 'irretrievably injured', his machinery was maliciously damaged, all business intercourse with him was broken off and he was 'completely ruined and forced out of existence'. His subsequent efforts to obtain compensation, directly or indirectly, from the British government came to nought.

Having returned to London in 1851, Leahy promoted a scheme to develop a railway network between Vienna and Constantinople, capital of the vast Ottoman empire, and secured some support for this from the Turkish ambassador in London and from Sir Charles Fox, the eminent civil engineer and contractor and Thomas Brassey, the greatest railway contractor of the day. Having arrived at Constantinople in February 1852 with his younger brother Matthew, he conducted surveys in the Balkans in the following eighteen months and prepared outline plans for a 700-mile-long railway. Whatever agreements existed with Fox, Brassey and the Ottoman government broke down in 1855–6 and Leahy failed to receive the sum of £7,000 which he claimed was due to him by the government, notwithstanding intervention on his behalf by the British ambassador. Leahy and his brother, with a Greek partner, did however secure a mining concession in Thessaly and Macedonia and quickly developed mines which were reported to be producing considerable quantities of gold and silver and yielding about £1,000 a month in 1856. For reasons which are far from clear, the Leahys appear to have abandoned their apparently profitable mining operations in early 1858 and returned to London.

Efforts by Leahy in 1858 to secure British government support for a number of new steamship services and submarine cable-laying schemes failed but he managed to secure appointment by the Colonial Secretary in September of that year to a position as roads engineer in Jamaica and was promoted to the better-paid position of colonial engineer and architect in July 1859. He was alleged by the governor to have conducted his public business in a loose and careless manner but the immediate cause of his downfall in Jamaica was his promotion of a scheme to provide a tramway along one of the island's main roads; by falsifying the cost estimates, he set out to ensure that completion of the tramway would be financed entirely by government funds leaving the private company in which he was the major shareholder in possession of the line and entitled to all revenues and profits. When the fraud was exposed, Leahy was dismissed in July 1863, notwithstanding intervention on his behalf by his eldest brother, Patrick, who by then was Catholic archbishop of Cashel and Emly.

Back in London in the mid-1860s, Leahy attempted without much success to resume his engineering career. At the general election in 1868 he persuaded a friend of his – a wealthy English barrister named Henry Munster – to stand for election for the borough of Cashel where there were only 203 electors, and personally planned and managed the campaign. Although Munster failed to be elected, subsequent public inquiries found that bribery and corruption, in which Leahy played a prominent part, had prevailed extensively at the election and, as a result, Cashel was disfranchised in 1870. After attempting to obtain a number of public positions in Ireland, Leahy went to America in 1871 and early the following year became president and chief engineer of the Gilpin County Tram Railway Company with a capital of $100,000 and plans to build up to thirty miles of tram lines along the steep hillsides above Central City and Black Hawk, west of Denver, where a gold rush had been in progress for ten years. The company bought over 3,000 acres of forested land with a view to hauling wood to the growing towns for building and for use as fuel. The line was in partial operation by August 1873 but was crude and inefficient; round timbers were used as sleepers and pine poles about sixteen feet long and set three feet apart formed the rails. Wooden flat wagons, running on broad wheels with deep flanges, operated by gravity on the downhill journey and were returned by horse-power to the hillsides. The line was never a financial success; Leahy borrowed heavily and mortgaged all of his possessions in an effort to maintain it in operation but was forced to surrender everything to the bank in January 1875 and leave Colorado.

Leahy attempted in the early 1880s to re-establish himself as a consulting engineer in London but eventually took up a position as manager of turkish baths in Finsbury Park. He died on 4 March 1888 when he was run over by a train at Edgware Road Station on the Metropolitan Railway. He was buried with one of his sisters at St Mary's Roman Catholic Cemetery on the Harrow Road, London, immediately adjoining the better-known Kensal Green cemetery. His first wife, Catherine King-Fitzgerald, whom he had married in London in 1848, seems to have died in 1851, shortly after giving birth to a son, Gerald, at sea presumably on the journey back from the Cape. He had married again by the mid-1850s but appears to have abandoned his second wife, Juliet, who was much younger than he, and his son when he left Colorado in 1875. They both lived until the early years of the twentieth century, Gerald working as a day labourer but only, according to one account, 'when faced with starvation or drought'. He spent his final years in a home for the aged and was buried at Mount Oliver cemetery, Denver. Juliet died some years earlier; local folklore recorded that her body was taken back to Ireland for burial but this seems unlikely to be true.

LEAHY, Patrick (1780–1850), was born in Co. Limerick, probably on the estate of Sir Vere Hunt at Curraghchase. By his own account, he was engaged in land surveying from 1797 onwards but nothing is known of his training or early experience. He appears to have worked intermittently on projects relating to Vere Hunt's other estate in the Killenaule area of north Tipperary between 1805 and 1815 and was one of the large team of surveyors who were engaged to support the district engineers employed by the Bogs Commissioners between 1810 and 1813; working with an English engineer, Thomas Townshend, Leahy was assigned initially to survey and map bogs in counties Offaly and Westmeath and subsequently to survey 65,000 acres of bogland in the counties surrounding Lough Neagh. He worked also with Townshend on surveys commissioned by the Directors General of Inland Navigation between 1809 and 1812 with a view to establishing possible canal routes in counties Laois, Tipperary and Kilkenny.

Patrick Leahy might be described as a competent all-rounder in the first decades of the nineteenth century when the demarcation between professions was not as sharp as it was to become as the century progressed. On his return to Tipperary in 1813, he based himself in Thurles and advertised his services as land surveyor, engineer and valuer. He prepared estate surveys and maps for various clients and was willing and able to offer advice on drainage, planting, land valuation, geology and mineralogical development, and to undertake infrastructure projects. He was a skilled cartographer; in 1818 he completed a large-scale map of the barony of Slieveardagh for the grand jury and other mapping assignments followed, including a map of the coal district near Killenaule, a map of the Thurles estate of the earl of Llandaff and finally a substantial mapping programme which was carried out for Waterford Corporation between 1831 and 1834. Leahy was one of the 95 candidates at the abortive competition for the appointment of county surveyors in 1817 and applied unsuccessfully some years later for employment by the Ordnance Survey and on the government-funded programme of relief works in the western and southern districts. In Tipperary, where most of his early work was carried out, he appears to have been well thought of by the titled families and landed gentry – an essential requirement for any surveyor hoping to make a living at the time. When seeking a government appointment in 1823 he was able to offer references from the earl of Llandaff, the earl of Clonmel, Lord Waterpark and Lord Norbury, the notorious hanging judge who had estates in the county. However, in spite of his good connections, he consistently failed in his attempts in the 1820s to gain recognition at national level, to win important commissions from the public authorities, or to establish himself as a significant figure in the emerging civil engineering profession.

In 1826, when the currencies of Britain and Ireland were being assimilated and imperial standards were being introduced, Leahy published his *New and General Tables of Weights and Measures* but failed in his efforts to have these purchased in large quantities for use in the public service. He assisted Alexander Nimmo in 1825–6 in the preparation of plans for the Limerick & Waterford Railway – the first Irish railway scheme authorized by law – and worked again with Nimmo on railway surveys in the Liverpool and Leeds areas in 1830–33. As three of his sons, Denis, Edmund and Matthew, grew to manhood, Leahy instructed them 'in his own line of business'; they went on to become apprentices and assistants to their father and, after the family had moved to Clonmel, worked with him in the late 1820s as Messrs Leahy and Sons on various projects.

Leahy became an associate member of the Institution of Civil Engineers (London) in 1834, one of a very small number of Irish engineers who were admitted to membership of this prestigious body in its early years. He was appointed to be the first county surveyor for the east riding of Cork and for the county of the city of Cork on 17 May 1834 and his son, **Edmund Leahy**, was appointed at the same time as surveyor for the west riding. The mileage of roads maintained by the grand jury increased considerably during Leahy's term of office and some significant new roads were constructed, including the Lower Glanmire Road which carried traffic to the city from the Dublin and Waterford directions and the Carrigrohane straight on the western approach to the city. Leahy also built or reconstructed a substantial number of bridges, including a novel fly-over bridge carrying a new road to Carrigaline over an existing local road.

The most significant feature of Leahy's twelve-year tenure of the post of county surveyor in Cork was the extensive private practice in which he and his three sons engaged, blurring the distinction which should have been maintained between private practice and public duty. Under the title Messrs P. & E. Leahy, Patrick and Edmund, assisted by Matthew Leahy and occasionally by Denis, undertook civil engineering work of all kinds throughout Munster and occasionally took on construction contracts such as those for the provision of workhouses at Mallow, Lismore and Listowel in the early 1840s. Having designed and promoted an abortive scheme for a Cork-Passage railway in 1836, Leahy and his sons became heavily involved in railway promotion in the mid-1840s, beginning with the Cork & Bandon railway in 1844–5 and building up to a portfolio of ten separate hastily-designed schemes and a route mileage of more than 500 by 1846. All of these schemes, except the Cork & Bandon, crashed out of existence in the early part of 1846 when the investment climate changed. To make matters worse, Leahy was dismissed by the grand jury in March 1846; his relationship with them had deteriorated seriously in the previous few years, largely because of the extent of his private practice activity, but the immediate reason for his removal from office was his involvement in a questionable transaction involving payment for the construction of a sea wall at Youghal.

Leahy and his family emigrated to the Cape of Good Hope colony in 1849 and attempted to re-establish themselves there as civil engineers and contractors. Having supported the governor during the convict agitation of 1849–50, Leahy sustained an injury in the course of some disturbances and died on 5 October 1850. He was buried at the Cape in an unmarked grave. His wife, Margaret Cormack, whom he had married at Gortnahoe, Co. Tipperary, in 1805 and with whom he had four sons and five daughters, died in Cork in 1845. Their eldest son, Patrick (1806–75), who was ordained at Maynooth in 1833, became vice-rector of the Catholic University in Dublin in 1854 and Catholic archbishop of Cashel and Emly in 1857. The other three sons, mentioned above, all followed their father's profession and worked with him in Ireland until 1846. Matthew and Denis served in succession as Superintendent of Public Works in Trinidad and died there in 1860 and 1862, respectively. Only one of the five Leahy sisters married; the others lived mainly in the London area from 1850 onwards and were dependent on their brothers for support until the end of their lives.

LEE, George (1891–1961), was born in Gort, Co. Galway, and educated at St Joseph's College, Ballinasloe, Mount St Joseph's, Roscrea, and University College, Galway, from which he graduated with a BE degree in 1912. Having worked initially in the Galway shell factory he was employed as an assistant county surveyor in west Galway when the surveyor for the division, **John Moran**, died in April 1916. Having advertised the post in the normal way and received twelve applications, the council decided in June to appoint **M.J. Kennedy**, who had been serving in the less-well-paid post in the eastern division; Kennedy received 17 votes at the council meeting as against 13 for Lee whose proposers had strongly urged that he should be appointed as he came from 'one of the oldest and most respectable Galway families and was himself 'a Galwayman by birth, adoption, sentiment and patriotism'. Kennedy's appointment left a vacancy in the eastern division which, following public advertisement, was filled in August by the election of Lee. The Local Government Board, however, objected to the appointment, because Lee had not passed the qualifying examination and had insufficient experience of major works. Lee was allowed to continue in the post on an acting basis until February 1917 when, after further objections by the board, the council re-advertised the vacancy. In May, before an appointment could be made, the council was told that Lee was going to the front for six months and rather than appoint one of the other thirteen candidates, most of whom were well qualified, they decided to ask Kennedy, now the western division surveyor, to take general responsibility for the eastern division also with the assistance of a supervising assistant surveyor.

When Lee was demobilized in August 1919 he sought reinstatement in his former position in the east riding and by the end of the year had succeeded in gaining the appointment as county surveyor for which he had originally been proposed three years earlier. It appeared that his term of office was to be cut short when the Department of Local Government and Public Health notified the county council in May 1927 that he had been suspended following a report from **T.C. Courtney**, then one of the Department's roads inspectors, about the lack of progress of the work on some of the main roads near Ballinasloe. The council passed a resolution expressing amazement at this action, asserting that there was no justification for the suspension of an official who had been in their service for fourteen years without the slightest complaint and demanding that the suspension should be withdrawn.

Following the death of Kennedy who had been county surveyor for the western division since 1916, the county council agreed in February 1939 to a proposal from the Department of Local Government and Public Health that Lee should take responsibility for all of the county's engineering services. His salary was increased to £1,000 a year to reflect his new responsibilities but with a proviso that he should not undertake private practice of any kind.

Lee went on to make significant progress in road development and in hospital construction in Galway in the period of more than forty years during which he served the county. He devoted most of his energies to the improvement of the main roads and of the tourist roads in Connemara. The new promenade at Salthill was constructed by direct labour under his supervision in the 1940s and he was responsible for the management of a major turf-cutting programme during the emergency years, contributing an annual total of some 90,000 tons to the national pool. A tall, spare man with rugged features, he retired in June 1956 and died at his home in Galway on 11 March 1961. An obituary in the *Engineers Journal* noted his great organizing ability and his special personality and boyish charm which endeared him both to young engineers and to his older colleagues and friends. He was president of Cumann na nInnealtóirí in 1951–2.

LEEBODY, John Welsh (fl. 1868–1934), was born on 8 January 1868 in Derry. His father, J.R.A. Leebody MA DSc, was president and professor of mathematics at Magee College, Derry, and county analyst under the grand jury, while his mother, Mary Isabella, was a keen amateur botanist. He was educated at the Academical Institution in Derry and at Queen's College, Belfast, where he was awarded a BA degree in 1888 and a BE in 1890. He was assistant to Abraham McCausland Stewart of Derry in 1890 and worked with him on plans for the Glenties railway. Subsequently, as assistant to James Otway of Waterford, he was engaged on railway work, harbour engineering and building construction.

Having gained first place at the county surveyor examinations in 1895, Leebody was appointed to the southern division of Co. Tyrone on 23 November that year and set up his headquarters in Dungannon. After the retirement at the end of September 1924 of **Francis Lynam**, surveyor for the northern division, the county council decided that one surveyor could more efficiently and more economically deal with all roadwork in the county, provided he had adequate full-time support staff, properly organized. Leebody then took on responsibility for the entire county, supported by a deputy county surveyor, a mechanical supervisor (his son) and eight assistants, each responsible for about 400 miles of roads.

In his first report to the grand jury in March 1896 Leebody had expressed concern about the administration generally of the contract system of road maintenance in his area. He was disturbed by the performance of the contractors and, in particular, by the fact that many of them seemed to have been allowed to use pit gravel rather than broken stone for maintenance work. He set out to change this by withholding certificates from defaulters and prosecuting them in the courts. In addition, he introduced hired steamrollers in 1896 in the face of tremendous opposition from the ratepayers but, while he reported that the results spoke for themselves, he soon had to revert to traditional methods of road-making when funds to continue the experiment were denied him by the grand jury. To make matters worse, he had to cope with what he described as 'an epidemic of footpath-making', with proposals for new footpaths absorbing some 23 per cent of all the funds available for roads in 1898. In 1899, Leebody persuaded the new county council to authorize the acquisition of steamrollers but, ten years later, he was still complaining of the difficulty of getting support for the kind of investment in road improvement which he believed to be needed; he found that an antipathy to motorists still existed in the rural districts, and that 'county surveyors have to keep the motorist in the background altogether when advocating road improvement'. He was a firm believer in the operation of the direct labour system, for reasons of economy and efficiency in the maintenance of first and second class roads, while retaining the old contract system for roads of lesser importance.

Leebody claimed to have pioneered the direct production of road metal by county surveyors in a planned and organized way. In a paper which he presented to the first Irish Roads Congress in 1910, he described how, on becoming a county surveyor, he had been struck by the extreme laxity of the system under which the quantity and the quality of the materials used by road contractors was, in effect, left to the contractors themselves. Taking advantage of the changes in the law which had been made in 1898, Leebody was able to persuade his county council in 1904 to allow him to become his own 'quarry master' by leasing a three-acre whinstone quarry near Cookstown; road contractors were then supplied directly with the 'splendidly durable road metal' from this quarry, thus converting the contracts into labour only contracts, and allowing the 'small' man, with limited capital and without a quarry, to compete successfully and bring about an overall reduction in maintenance costs. By 1910 annual output from the quarry had grown to more than 8,000 tons each year at a cost of less than twenty pence per ton, and an Aveling & Porter steam tractor with tipping wagons was being used to transport the stone to the road works sites, or to the loading bank at Cookstown railway station for onward transportation.

By the 1930s Leebody was one of the best-paid surveyors in Ireland, having been awarded a salary of £1,060 as early as 1924. When he gave notice in September 1934 of his intention to retire at the end of the year, an editorial in the *Tyrone Constitution* declared that 'no other public official rendered better or more consistent service to the public'. It noted that when this 'faithful official' took up duty thirty-nine years earlier, road-making and road maintenance were carried out on very primitive lines but, with the introduction of modern machinery and methods, Leebody had done everything possible to keep the county abreast of the times; vast progress had been made on the resurfacing of the principal roads, on the rebuilding of bridges and in many other ways, all of which was eloquent testimony to the surveyor's work. Leebody's son, Ian, served as Tyrone county surveyor from 1959 to 1965.

LINDSAY, Henry L. (fl. 1800–47), appears to have been born in Dublin in October 1800, the youngest son of Charles Lindsay, bishop of Kildare and dean of Christ Church cathedral from 1804 to 1846, and grandson of the Scottish earl of Balcarres in Fifeshire. He had been engaged for more than ten years as engineer on various grand jury works, mainly in the south of Ireland, including the improvement of the Limerick–Croom road, before his appointment as Armagh county surveyor on 17 May 1834.

Lindsay must have considered himself to be something of an expert on road-making: he presented a paper on *Laying Down New Roads in a Thickly Undulating Country* at the third meeting of the Civil Engineers Society of Ireland in September 1836 but the text does not appear to have survived; in the following year, he planned to publish *An Introduction to Road Making, Simplified* but it seems that the book never appeared in print. His grand jury, however, were not impressed by Lindsay's performance as a roads engineer: by 1837 they were openly proclaiming their dissatisfaction with his services and resolved that 'we feel ourselves called on to declare, both from general reports and our own personal observation, that the roads in the county are in a worse condition now than when the changes in the grand jury laws were introduced'. The jury had received a petition 'very numerously signed' complaining that Lindsay had been occupied in private business, to the neglect of his public duties, but Lindsay insisted that he had a legal right to private practice and was determined to continue it; his practice must have involved engineering and, perhaps, surveying, for there is no evidence that he had any influence on the architecture of Armagh. He reported to the grand jury in 1835 on the condition of the courthouse at Armagh (built by Francis Johnston in 1809) advising them that, as hardly any repairs had been carried out since the building was completed, it was falling into a state of dilapidation, but it does not appear that he was authorized to carry out any substantial works to the building. Ten years later, when it was decided to construct additions to Armagh gaol, it was William Murray, the Dublin architect, who was engaged to plan the work.

Lindsay was very critical of grand jury law and practice on which he set out his views and suggestions extensively in *The Present State of the Irish Grand Jury Law Considered as it respects the Promotion and Execution of Public Works; and an Improved Plan of Jurisprudence Recommended* (Armagh, 1837). He believed that there was no point in attempting to reform the system of presentment sessions and assizes and that the whole apparatus should be swept away and replaced by a system which would be directed and regulated by scientific men, capable of knowing how to proceed systematically and judiciously. There should be a central board made up of three professional men, with district boards similarly constituted for groups of up to four counties, and an engineer for each county, to take over full responsibility for all county and public works. Views like this, widely disseminated, were unlikely to assist in improving the already stained relations between the surveyor and his grand jury.

In a thirty-six-page pamphlet published in Armagh in 1836, Lindsay had set our equally trenchant comments on the farming community of his county. In *An Essay on the Agriculture of County Armagh* he recorded his view that the 'manners of the people are rude and uncultivated, and particularly so on the borders of the county of Monaghan, where the people seem to be of a fiercer race than those in the interior of the district'. He went on to offer the view that if the people were employed they would be contented and 'saved from the contaminating influence and misrepresentations of designing and ill-disposed men, who wander about for the express purpose of spreading sedition and of disorganizing the social state of the country'. All of this was followed by some practical comments on soil types, methods of cultivation and crop rotation, land reclamation and other local improvement works. The pamphlet was strongly recommended to the gentlemen and farmers of the entire province by the *Newry Telegraph* as 'a well-written and exceedingly interesting pamphlet'.

When grand jury law was altered in 1836 to allow assistant surveyors to be appointed, Lindsay was the first of the county surveyors to send a candidate forward to the Board of Works to have his qualifications for the new post examined; he received a certificate for this first assistant in November 1836, and certificates for several others followed in January 1837. But because of their dissatisfaction with his private practice, the grand jury refused at first to allow Lindsay to employ these assistants; they relented in the following year and agreed that he might have three, but only on condition that he undertook no private business whatever. Lindsay seems to have won this particular contest, having six assistants in 1841 while continuing 'to do business for private individuals, whenever I get it which is very seldom', as he told an official inquiry that year; he claimed to have devoted no more than ten days each year to private work in the previous three years.

The conflict between Lindsay and his grand jury flared up again in 1845 when it was resolved 'that, in consequence of the authenticated reports laid before the committee of the neglected state of the roads ... we are of the opinion that increasing exertion on the part of Mr Lindsay will be absolutely necessary; and we therefore wish to call his attention to a more active discharge of his duties for the future, by which alone, we conceive the evils so loudly complained of can be remedied'. Lindsay, however, blamed the road contractors and claimed that he was already doing his best to 'awake them from their slumbers' by withholding payments from many of them. He relied also on the fact that he had left the management of what he called 'the parish roads' to his assistants because many of them were so narrow that 'I could not go over them even if I were inclined', but promised to take a greater interest in these roads in the future. In practice, however, Lindsay seems to have taken little notice of the grand jury's warnings and went on, later in 1845, to become heavily involved in survey work for a proposed Dublin & Armagh Inland Railway which was to run via Carrickmacross and Castleblaney. The patience of the grand jurymen was exhausted at this stage and Lindsay was forced to resign in February 1846.

Lindsay was elected to membership of the Royal Irish Academy in 1843. He was employed by the Board of Works as one of the engineers in charge of relief works in Co. Meath in 1846–7 but no record has emerged of any subsequent activities in Ireland. He appears to have been living in Melbourne from 1851 onwards and, while based there, applied in 1863 to take up an agency for the Sun Fire Office – a long-established London-based fire insurance company. He seems to have returned to live at Blessington Street, Dublin, in the 1870s and he died there, unmarried, on 17 February 1879.

LONGFIELD, Richard William Frederick (1861–1926), was born in Tallow, Co. Waterford, on 17 January 1861 into a family with a landed background and strong links with conservative politics in Cork city and county. He was the second of five sons of Revd Richard W.F. Longfield, rector of Mogeely, and his wife, Wilhelmina Golock, daughter of Revd James Gollock, rector of Desertserges, near Bandon, and a substantial landowner. His paternal grandfather, Revd Mountiford Longfield, had been an earlier rector of Desertserges from 1798 to 1850. Other well-known members of the family included George, who held the chair of hebrew at Trinity College, Dublin, for many years; Mountiford who was professor of political economy and feudal law at Trinity; and Robert, a QC who was MP for Mallow from 1859 to 1865.

Richard Longfield was educated at home before entering Trinity College, Dublin, in June 1879. Having been senior moderator and gold medallist in experimental physics, he graduated in arts in 1883 and in engineering in 1885. He served a pupilage under the Waterford county surveyor, **W.E.L. Duffin** and was engaged from 1887 to 1892 on railway work in the Argentine republic and in Uruguay, including surveys for the Bahia Blanca and North-Western Railway. He was employed by John G. Meiggs Son & Co. on surveys for a proposed Chumbicha and San Jeran Railway and worked also on the construction of the northern extension of the Central Uruguay Railway under Robert Crawford who had been professor of engineering at Trinity from 1882 to 1887.

On his return to Ireland Longfield took first place at the county surveyor examinations held in June 1895 and was appointed to the southern division of Co. Donegal shortly afterwards; he had served for some nine months before that as deputy county surveyor in the southern division of Tyrone where the surveyor, **James A. Dickinson**, was ill. At the end of 1898 he moved to the west riding of Cork to replace the long-serving **Nat Jackson** who had retired. Some months before his transfer in April 1899 to the service of the new county council he spoke of his 'apprehension and curiosity as to what these new bodies will do' and expressed the view that 'there will be a considerable amount of trouble on the whole, and a considerable amount of odium on the county surveyors'. Nevertheless, Longfield seemed to make the transition without too much difficulty and went on to serve the county council for twenty-five years. He was said to have been one of the most energetic and capable officers on the staff and to have invariably discharged his duties with thoroughness and attention. His service covered the entire period of the Anglo-Irish war and the civil war during which, as the

Cork Examiner put it, his position was an arduous and exceedingly difficult one due 'to the condition of things which prevailed in the country'.

Longfield told the county council's finance committee at their first meeting in April 1899 that many of the main roads in west Cork were in a most unsatisfactory condition and he urged the councillors to give the matter serious consideration. However, even though he continued in the following years to argue for a substantial improvement in the condition of the road network to cater for bicycles and the growth in the number of touring cars in west Cork, he had great difficulty in persuading some of the councillors that steamrolling should be introduced as had already been done in the neighbouring counties of Waterford, Limerick and Tipperary. When direct labour was introduced in west Cork on a limited basis in 1904 he reported that he was able to repair the roads with a better quality of stone than the contractors had been using and at a lower cost in many cases, but he was frustrated by the refusal of councillors to allow him to introduce steam-driven stone crushers instead of having stones broken by hand by labourers on the roadside. When he again sought funds for steamrolling at a meeting of Macroom rural district council in October 1906 he met strong opposition: one councillor thought that the proposed expenditure would be extravagant; another felt that the money should be spent on local labour; yet another questioned whether steamrolling would improve the roads at all; and a fourth was worried for the safety of the gullets if heavy new machinery was introduced. And so, as the *Irish Builder and Engineer* put it, the roads around Macroom (and in other parts of west Cork) were left for some years more to be darned at the leisurely convenience of a local road contractor, the road users were allowed to go on grumbling, and the tourists were left to anathematise Irish roads and all those responsible for them.

The situation changed in 1908–9 when the county council finally recognized the urgency of improving their roads and provided substantial funds for the purchase of steamrollers and other plant. In 1910, in anticipation of grants from the new Road Board, Longfield put forward proposals for a major programme of works costing £17,000 on the tourist routes in his area, including the roads from Bantry to Glengarriff and Kenmare and from Macroom to Ballylickey, the latter including the Pass of Keimaneigh; the surfaces on these roads had been badly cut up by what were then described as 'touring motor cars' each weighing up to five tons and which occasionally sank to their axles and had to be dug out. Longfield went on to rebuild and improve many of these roads in the following years. By the mid-1920s, the local councillors had come to recognize him as an efficient road engineer, noting that some of the roads he had reconstructed showed no serious deterioration notwithstanding the unexpected increase in heavy traffic which had taken place.

Longfield was responsible also for a significant amount of work on piers and harbours on the coast of his extensive area, work which for the most part was aided by the Congested Districts Board. In 1905 he enlarged and improved the pier and quay at Courtmacsherry, built in the 1870s, and piers at Adrigole and Castletownbere were completed in 1907. In 1911 the county council decided to go ahead with the building of a new pier which he had planned at Castletownshend, subject to a cost limit of £600, and plans for harbour improvement works at Baltimore were approved in 1913 at an estimated cost of £10,000 and completed in 1917. In 1914 plans for a causeway and sluices across the estuary at Ringabella were completed and work was planned to start that year at a cost of about £4,000. In 1919 planning for pier and harbour improvements at Schull was well advanced.

In September 1925 Longfield retired due to ill-health which he attributed to the risks and hardship he had to endure during the previous few years. He had married Maria Louisa, daughter of Revd Thomas Henry Gollock of Forest House, Co. Cork, in 1899 and the couple lived initially at Devonshire Square, Bandon, taking over the tenancy of a house on the Devonshire estate from **Nat Jackson**, Longfield's predecessor. They later lived at Millbrook House, Bandon – now the Bandon nursing home; Longfield died there on 2 July 1926, a day when the temperature in Cork reached a record 98 degrees. He was buried in the churchyard at Desertserges where both his grandfathers had ministered.

LYNAM, Francis Joseph (1863–1933), was born on 8 October 1863, the youngest of the ten children of James Lynam CE, Ballinasloe, and one of five brothers who pursued careers in engineering; an older brother, **Patrick Lynam**, became a county surveyor in 1883. Lynam gained a first-class honours BE degree at Queen's College, Galway, in 1884 and worked initially as assistant engineer for the river Suck drainage board. In 1886-7 he was assistant county surveyor in Co. Louth under his brother, and in 1890 he assisted Edward Townsend, consulting engineer, on surveys for a proposed Loughrea & Woodford Railway. He was appointed county surveyor for the northern division of Co. Tyrone, based in Omagh, on 25 November 1895.

In his first report to the grand jury in March 1896 Lynam expressed serious concern about the performance of many of the road contractors and sought the assistance of the county solicitor in bringing prosecutions against the defaulters in the local courts. An editorial in the *Tyrone Constitution* commented favourably on the new surveyor's warnings to the contractors, noting that the roads of the county 'should no longer be permitted to remain in their present unsatisfactory condition'. In the years that followed, however, Lynam's performance as road engineer does not seem to have impressed the new county council. In 1915, they proposed to withdraw Strabane rural district entirely from his area of responsibility and assign it to an assistant surveyor, with Lynam's salary being reduced by £75 and that of the assistant increased accordingly. The Local Government Board refused to sanction this 'change of a very unusual and drastic nature' but the council gave effect to it at a later stage, leaving Lynam with responsibility only for the Omagh and Castlederg rural districts.

Lynam was responsible for the construction of a reinforced concrete bridge with two thirty-foot spans at Cranagh in 1914 and for a concrete bridge over the river Strule at Omagh with a span of more than 100 feet. In 1906 he undertook alterations at the county courthouse in Omagh, designed by John Hargarve in 1814, including the replacement of two wood and glass domes.

After twenty-nine years service in Tyrone Lynam resigned from his post with effect from 30 September 1924 when he was 60 years of age, indicating that his health did not allow him to continue active work. In retirement, he lived in Dublin and died at Rathfarnham on 10 September 1933, survived by his wife, formerly Frances Todd, and two daughters, Maeve and Nora. The *Tyrone Constitution*, in a brief but sympathetic obituary, noted that Lynam had been an exceedingly courteous and diligent official who had discharged his duties faithfully at a time when there was little encouragement for extensive road improvements and severe restrictions on funds.

LYNAM, Patrick John (1852–1912), was a member of one of the most prominent engineering families in Ireland in the nineteenth century. His maternal grandfather was John Kelly (1791–1869) who had been one of Richard Griffith's principal assistants from 1814 onwards on geological survey and valuation work, and his father was James Lynam (1812–83), a native of Co. Carlow who lived at Raheen, Ballinasloe, from the 1850s onwards; James was at different times a land agent, a land surveyor and a valuer in the employment of the Valuation Office. As an engineer, he was employed on railway and river works and on surveys of the Shannon basin but was best known in the 1860s and 1870s for his severe criticism, in the course of a number of official inquiries, of the management by the Board of Works of the river Shannon and the failure of the Valuation Office to achieve a uniform valuation of lands throughout the country – an object which, he asserted, the office had 'totally and shamefully' failed to achieve. James and his wife, Maria Kelly, had ten children; five of their six sons became engineers and two of them – Patrick and his younger brother, **Francis Lynam** – became county surveyors. The Lynams were subsequently connected by marriage with the family of **John Kelly**, Roscommon county surveyor, and with **James Perry**, **John Moran** and **George Lee**, all of whom served as county surveyors in Galway

Patrick Lynam, the third son of James and Maria Lynam, graduated from Queen's College, Galway, with a first-class honours BE degree in 1872. He joined the Indian civil service and worked on road development for some five years with the Ceylon Public Works Department. On his return to Ireland, he competed unsuccessfully for the post of Cork city engineer in 1880 but gained appointment as county surveyor for the northern division of Mayo in July 1883, with headquarters in Ballina. This was a difficult assignment, carrying responsibility for the maintenance of some 1,300 miles of road at an annual cost of £12,000, and with only one assistant. With such an extensive area and no allowance for travel expenses, Lynam decided to limit his travel costs to £100 a year 'and let the roads take their chance'; he drove over the network once or twice each year but freely admitted that he could not properly estimate the amount or the quality of the work done by the contractors.

Although his biannual reports to the grand jury regularly contained criticism of the neglect of road contractors, Lynam held strong views on the question of replacing the traditional contract system by a system of direct labour. He told a meeting of the Institution of Civil Engineers of Ireland in December 1885 that, from his considerable experience, he believed that even the best ordinary hired labourer would not do half the work that would be done by a contractor working for himself. No overseer could be trusted with the uncontrolled supervision of men: even the most honest and well-meaning of them would become demoralized or be intimidated, idling would begin, false names of men who never worked would be entered on the pay sheets, and to combat all of this the surveyor's supervisory work would be quadrupled. 'It is quite true', Lynam went on, 'that the main object of every hired labourer is to put in his day' and thus, without adequate supervision, passers-by would have cause to make unpleasant remarks and county work would be brought into disrepute. As late as 1906, when many of his colleagues were operating substantial direct labour schemes, Lynam was still an opponent of the system, explaining that he did not trust labourers to work well unless competent gangers and overseers could be engaged to supervise them – but 'competent men cannot be found to work at incompetent wages and the county council cannot be induced to pay gangers and overseers at a sufficiently high rate to secure the services of reliable men'.

Lynam was appointed to succeed the long-serving **John Neville** in Louth in April 1886 but seems to have continued to perform his duties in Mayo until the summer assizes in July when members of the grand jury remarked on the great improvement he had brought about in the roads through the use of broken stone rather than gravel, although at considerably increased cost to the cesspayers. In Louth, he was fortunate that the roads were in a relatively good condition when he took over, in contrast to those of adjoining counties, and he maintained this situation by introducing better working methods which also reduced costs.

He was an early and ardent advocate of steamrolling as a means of improving road standards and a strong advocate of the need for a proper supervisory organization to pursue contractors. He told a conference in 1891 that if the average county surveyor saw his roads twice a year, he did very well, but in Louth he was fortunate to have five assistant surveyors at the time – a relatively large number for a county with only 600 miles of public roads. On this basis, he claimed that his county was the only one in Ireland in which the level of supervision of road work was satisfactory but he was critical of the low pay of the assistants who had to lay out at least £12 a year out of their salaries of about £80 for car-hire in the winter months, besides keeping a bicycle for summer travel. In addition, a system had grown up under which contractors who wished to claim advances on foot of work done clubbed together to pay the cost of an assistant's travel to inspect the various road sections; Lynam thought that this was open to serious objection and abuse and contended that assistants should instead be paid an allowance per mile for car-hire in the winter months and for cycling in the summer.

At the final meeting of the Louth grand jury in February 1899, a foretaste of what was to come in relation to roads and road traffic was presented in Lynam's half-yearly report. He described how the Greenore Granite Quarries Syndicate had introduced a steam traction engine, pulling two wagons, to transport their stone to the pier at Carlingford, cutting deep trenches in the roads, destroying culverts and making the roads virtually impassable for ordinary traffic. While Lynam had received legal advice to the effect that the company could operate on any road they pleased, he suggested that it was a reckless thing for them to run a locomotive like this in December when the roads were very soft, leaving the ratepayers to meet the cost of restoration.

Lynam filled the separate post of surveyor for the county of the town of Drogheda until it was abolished in 1899. His only significant building project in the town seems to have been a new courthouse – a plain two-storey building completed in 1889 in the yard of Francis Johnston's 1796 Corn Exchange complex. As town surveyor he had to deal with the replacement of St Dominick's bridge – a wooden structure completed by his predecessor in 1863 but limited to pedestrian traffic since the 1870s. When the Board of Works refused early in 1889 to allow a loan to be raised for a new wooden bridge, Lynam proposed the construction at a cost of £5,500 of a substantial iron bridge with two river piers and a central span of 120 feet. However, the grand jury would not agree to spend more than £1,500 on the project and he was forced to come up with an alternative design. This involved retaining the massive stone abutments of the earlier bridge and constructing a light structure, with steel decking 16 feet wide (10 feet less than the existing bridge) carried in four spans of 37 feet on clusters of screw-piles. Lynam acknowledged that screw-piles were no longer in general use in tidal waters and that the new bridge would not be suitable for use by traction engines, but the technique was still the cheapest available and, he believed, would give a good durable bridge on a difficult site. The contract to build the new structure was awarded to Thomas Gibson of Westminster in June 1889 but delays arose because of a dispute with the holders of the patent for the decking; this forced Lynam to design an alternative superstructure of girders and steel plates but one year later the contractor was still complaining of delay by the local firm of Thomas Grendon & Co. in supplying these. In the meantime, the grand jury had second thoughts about the width of the bridge and Lynam was forced to come up with a scheme for a wider bridge, including footpaths, at an additional cost of £1,000. Tenders were sought again in June 1892 and a contract with Messrs Brettel (who had built similar bridges elsewhere in Ireland) was signed in July 1893. One year later, Lynam was able to report that the bridge had been completed and that the piles and decking had been subjected to sixteen-ton loads with satisfactory results. The bridge is still in existence but is confined to pedestrian traffic only.

Lynam married Margaret Mary Barry, daughter of Captain Fitzwilliam Barry RM, Ballina, but his 26-year-old wife died in 1889 shortly after their marriage. He then went to live at Stapleton Place, Dundalk, with his sister Anna who was elected in 1899 to be a member of the local board of guardians, one of the first women to achieve this status. After a long and painful illness which continued for over a year, he died on 7 August 1912 when he was just over 60 years of age. A few days later his funeral went by train from Dundalk to Dublin where he was buried in Glasnevin cemetery with his wife. The chairman of the county council, in a tribute to Lynam, described him as 'a man of retiring disposition, an able, conscientious and painstaking official, a high-minded and honourable gentleman, and a truly upright and just man, who did his duty to the end'. According to the *Irish Builder and Engineer*, no death in Irish engineering circles for many years was more regretted than his; he was said to have discharged his duties in a most efficient manner and to have displayed qualities of a very high order, his work being marked by solidity, forethought and sound judgment.

LYONS, Andrew Oliver (1827–1902), was born on 14 July 1827. Having been apprenticed to **John Fraser**, Down county surveyor, he began his engineering career in 1845 when, at the age of 18, he took up a position as assistant engineer on the construction of the Waterford & Limerick Railway for which Charles B. Vignoles was consulting engineer. He was living in Templemore in 1849 when he was admitted to membership of the Institution of Civil Engineers of Ireland and subsequently worked for eleven years as district engineer under the Board of Works. He had responsibility for arterial drainage and land reclamation works in counties Galway, Longford and Tipperary, and was based at Killimor, Co. Galway, in the 1850s when his pay amounted only to twelve shillings a day. In 1861, he supervised the construction of a model school at Enniscorthy and worked with **John Bower** on a number of railway surveys in the north of Ireland.

Lyons qualified at the county surveyor examinations held in December 1851 but was not offered an appointment until nearly ten years later. When the lord lieutenant agreed in 1861 that the east riding of Cork should have a second surveyor the new post was offered to Charles Hargrave who had also qualified in 1851. However, Hargrave had gone to work in Australia and was unwilling to return and so the post was offered to Lyons who was next on the list. He took up duty as surveyor for the northern division of the east riding on 4 September 1861 and served there for thirty-seven years.

In his first report to the grand jury in February 1862 Lyons reported that the road surfaces in his new division were very poor and, although his efforts to improve the situation seem to have been more successful than those of **W.A. Treacy**, his immediate predecessor, they were not enough to deflect regular criticism in the ensuing twenty years. Lyons was responsible for a number of fine bridges in the east riding, the most substantial being the replacement bridge constructed in 1864–5 to carry the main Dublin–Cork road over the river Blackwater at Fermoy, with seven dressed limestone segmental arches and an unusually wide roadway. Further downstream, when the grand jury refused for cost reasons to authorize the construction of an iron bridge, he was forced to construct and maintain one of the last survivors of the era of wooden bridge-building – a large structure of five forty-foot spans over the Blackwater at Ballyhooley. Having periodically renewed some of the woodwork and applied treatments of creosote, he was able to advise the grand jury in 1898 that the bridge was still in sound condition.

Lyons enjoyed a considerable private practice and was, for a time, the principal waterworks engineer in the south of Ireland, regarded as an expert in all matters relating to water supply, drainage and hydraulic engineering. He designed waterworks in Mallow (1877), Fermoy (1886) and Mitchelstown (1892), by which time he had been joined in his practice by his son, Francis Oliver (b. 1867). He also undertook a considerable amount of arbitration work including assignments for the Board of Works in connection with the rush to promote light railways in the 1880s.

In opting to retire on pension in October 1898 at the age of 71 due to failing health, Lyons may also have been influenced by his view that the new county council system would involve 'a great deal of trouble for the county surveyors until things get into shape'. With his wife, Frances Margaret, he lived at Shortcastle, Mallow, and died there at the end of February 1902. His funeral took place 'amid every mark of respect' according to the *Cork Examiner*, which described him as a man 'of marked distinction in the profession to which he belonged' and who was 'well known and much esteemed by all classes throughout the county'.

McCONNELL, Edgar H. (1889–1948), was born on 15 January 1889, son of William McConnell and his wife, Jane, of Ballymena. He was educated at Ballymena Academy and the Royal College of Science in Dublin where he qualified in engineering in 1910. He was employed as an inspector by the Congested Districts Board between 1911 and 1921. He qualified for appointment as county surveyor at an examination held in 1920 and was appointed Armagh county surveyor, on an acting basis, in March 1921. He was elected to the permanent post in June of that year with a salary of £800 and was in office until 1947 when he was succeeded by James McRoberts who had served as one of his assistant county surveyors since 1926. Married to Anna Mays in 1912, McConnell died in Armagh on 26 February 1948 and was buried in the New Cemetery.

MACKENZIE, William (fl. 1787–1840), was employed as a road inspector in Roscommon in 1818 having had an earlier career in Scotland where he was employed under Thomas Telford on the construction of the Caledonian Canal. He was living at Catherine Street, Limerick, when he was appointed Wexford county surveyor on 28 May 1838 and seems to have carried out his duties to the general satisfaction of his grand jury. He got into serious difficulties, however, with the Board of Works arising from the construction of a new bridge over the river Slaney at the Deeps, between Enniscorthy and Wexford, for which a loan of £1,964 was sanctioned by the board in April 1839. When Mackenzie reported that the wooden piles were being kyanized by an entirely new process to ensure their preservation, the board asked for details and, when the surveyor failed to provide these, sent one of their own engineers, **Barry D. Gibbons**, who had been Mackenzie's predecessor in Wexford, to carry out an inspection. Gibbons quickly came to the conclusion that the reports on the new process were no more than 'fabrication and falsehood' and the board immediately demanded Mackenzie's resignation, taking the view that his conduct was 'a very gross attempt to deceive them or an instance of extreme and unjustifiable negligence'. Mackenzie pleaded that the episode should be overlooked in view of his generally meritorious conduct as county surveyor and the Board of Works agreed to allow his future to be decided in the light of the grand jury's assessment of this. But when they failed to obtain 'a satisfactory document in his favour' from the jury, the board wrote to Mackenzie in July 1840 giving him three weeks to resign or face removal from office by the lord lieutenant. The surveyor elected to resign but made a series of allegations – which were never substantiated – about Gibbons' performance as county surveyor and malpractice by the secretary of the grand jury and the treasurer.

Nothing is known of McKenzie's subsequent career. The wooden bridge at the Deeps was completed in 1844 by his successor, **J.B. Farrell**, and replaced in 1915 by a ferro-concrete structure which is still in service.

McMAHON, Thomas (1898–1979), was born at Milltown, Co. Kerry, on 1 April 1898 and was educated at Blackrock College, Dublin, and at University College, Galway, where he received BE and BSc degrees in 1918. He worked initially on mineralogical surveys in Ireland and later did similar work in India. He returned to Ireland early in 1921 and was assistant director of engineering at IRA headquarters until July 1922 when he took the republican side in the civil war. Following a period of imprisonment, he was employed for a short period in 1924 by Kildare County Council and by the Valuation Office in 1924–5. He then became an assistant county surveyor with Dublin County Council and was engaged mainly on road construction and maintenance work until 1935. In that year, he became chief engineer with the Turf Development Board, working with Dr C.S. Andrews, the board's managing director, who had been a close friend of his since their days together in the IRA.

In 1937 McMahon took up an appointment with the South of Ireland Asphalt Company, working initially on the development of the new housing area at Crumlin, Dublin. Having qualified at the county surveyor examinations held in 1931, he was an applicant for the post of Donegal county surveyor in 1939. Because of the rule under which an absolute preference was given to those with a knowledge of Irish for this particular post, McMahon, as the only candidate with a good knowledge of the language, defeated a number of strong candidates for the post, including **T.C. Courtney**, chief engineering adviser in the Department of Local Government and Public Health, and took up duty in October 1939.

McMahon was in office in Donegal for nine years, working mainly on programmes to improve the county's large network of main and county roads, some of which were still being maintained under the contract system until the mid-1940s. He pioneered the introduction of costing systems for road works, quarrying operations and capital projects, in the interests of increased output and economy, and read a paper on the subject to the Institution of Municipal and County Engineers in Dublin in May 1946.

In 1948 McMahon left the local service to set up a contracting business in Dublin (Thomas McMahon, Contractors) and later established the Municipal & General Supply Company which supplied heavy equipment and materials to local authorities. In the late 1950s he teamed up with David McIlvenna, a major building contractor, to found Wavin Pipes of which he was chairman and managing director until his retirement in 1976, three years before his death. The company's first production plant was located in McIlvenna's warehouse in Drumcondra, Dublin. Turnover increased rapidly after operations were moved in 1961 to a new plant at Balbriggan, Co. Dublin, where the company still thrives.

MAHONY, Florence (fl. 1834–61), became Westmeath county surveyor on 17 May 1834 and served the county until his death on 9 January 1861. Like the other surveyors of the period, he was obliged to take responsibility for the supervision of famine relief works on the roads in his county in 1846–7 but, while other engineers and commentators took the view that all such works were of an unproductive nature, Mahony argued that they were, in some instances, by far the most productive works that could be undertaken; the repair of roads leading to extensive and populous districts and to bogs had been productive of the utmost good, he claimed, and had assisted many large landowners who lived far from the existing public roads and given others the accommodation to which they were entitled. He reported also that, while he had been spending an average of £1,000 a year on the lowering of hills, it had proved to be impossible to carry out work of this kind during the winter without inconvenience to travellers. His relief works were operated on a task-work basis but, because of the short winter days and the weak condition of many of the workers, he had to apply relatively high rates which pushed up the cost of individual projects by 75 per cent as compared with contract prices.

Mahony was responsible for the design in 1850–2 of the sessions house at Delvin, a plain building of coursed rubble masonry; he took the view that certain parts of the work were imperfectly executed, and advised the grand jury at the spring assizes in 1852 to withhold £5 from the contractor on account of this.

MEADE, Thomas Patrick (1891–1951), was born on 10 January 1891 and educated at Rockwell College and University College, Dublin, from which he graduated with BE and BSc degrees in 1912. He was assistant to P.H. McCarthy, consulting engineer, in 1913 on the Dodder Valley sewerage scheme and between November 1913 and September 1920 was an assistant engineer with Dublin Corporation, engaged on main drainage, maintenance work at Roundwood reservoir and roads.

Meade was a member of the 4th battalion of the Dublin brigade of the IRA and with a number of others was arrested by the military in the Dublin mountains in September 1920. He was subsequently court-martialled and sentenced to two years imprisonment for illegal drilling, a sentence which he served in Mountjoy, Belfast and Cardiff jails. He was one of four applicants for the post of county surveyor in Tipperary south riding in 1920 but failed to gain the appointment which went to **T.S. Duggan**. However, after the latter had taken up duty on an acting basis, it emerged that he had not passed the qualifying Civil Service Commissioners examinations held in June 1920, whereas Meade had done so and with very high marks. Having discussing the dilemma at a number of meetings, the council decided in March 1921 to appoint Meade, who was still imprisoned, as county surveyor while continuing Duggans's appointment in an acting capacity.

On his release from jail in January 1922 Meade took up the appointment in south Tipperary and was soon faced with the problem of reconstructing or replacing some 200 bridges which had been destroyed or damaged during the civil war. He went on to serve the county effectively for almost thirty years. The centenary history of the county council, published in 1999, described him as 'a man noted for a keen sense of humour, forthright in the expression of his views on county council matters within his field of responsibility, and something of a legend for his capacity, when under fire, to turn defence into attack with his gift of subtle but good-humoured repartee ... It was said of him that there was not a road, bohereen or drain in the county that he did not know and, without ever having to think, would immediately announce where it was. He was among the most approachable of officials and is eagerly credited by the elected representatives of his days as one who gave every consideration possible to them when they had occasion to come to him with their multifarious suggestions and requests from their constituents'. Meade died in office on Christmas Eve, 1951 and was buried with full military honours on St Stephen's Day in St Patrick's cemetery, Clonmel.

MEGAW, David (1869–1927), was born in Co. Antrim on 26 August 1869, son of John Megaw and his wife, Ellen, née Dick. He was educated at Coleraine Academical Institution and at Queen's College, Belfast, from which he graduated with BA and BE degrees in 1895.

He was assistant to James Otway of Waterford on general engineering work until May 1896 and worked with Otway again in 1899–1900, laying out the railway from Waterford to Rosslare and preparing contract documents and plans. Between 1896 and 1899 he worked as assistant to **James Perry**, the Galway county surveyor, and from 1900 to 1914 was special assistant to **J.H. Brett**, the Antrim surveyor, working on a major steamrolling programme.

Megaw qualified for a county surveyor appointment at the examinations held in 1903 and, when Brett retired at nearly eighty years of age in February 1914, he was appointed to succeed him. He was one of the best paid of the surveyors at the time, with a salary maximum of £700 and an allowance of £150 a year for travelling expenses. He was known as an enterprising and progressive county surveyor whose painstaking efforts brought the roads of Antrim up to a high standard. The reconstruction of the main road from Belfast to Coleraine and Portrush was regarded as one of his finest achievements. He was one of the first surveyors to use reinforced concrete for road construction in rural areas without, as had been the practice until then, a covering of wood blocks, granite sets, or tarmacadam. For reinforcement on poor ground, he used the British Reinforced Concrete Company's fabric (familiarly known at the time as BRC), a mesh of drawn steel wire enclosed within up to seven inches of concrete.

Megaw was taken suddenly ill while on a summer holiday in Greencastle, Co. Donegal, in 1927 and died in hospital in Derry a few days later on 14 August.

MOORE, John Alexander (1881–1944), was born on 1 June 1881 and educated at Coleraine Academical Institution and at Queen's College, Galway, where he was awarded a BE degree in 1904. He was engaged until 1912 as an assistant engineer on road works in the Federation of Malay States and until 1916 was director of the Universal Machinery Company, London. He served as a second lieutenant in the Royal Engineers from 1916 onwards and worked in India until 1924, first as a military engineer on the northwest frontier and later as a state engineer. After the death in June 1924 of **C.L. Boddie**, the Derry county surveyor, Moore became deputy county surveyor in Derry and was appointed county surveyor some months later. He died in office in 1944.

MOORE, Joseph Henry (1844–1912), was born in London on 26 August 1844, the eldest son of Revd Patrick Moore MA, chaplain at the Training College, Westminster. He was educated at Drogheda Grammar School and at Lewisham Grammar School, London. In June 1862 he entered Trinity College, Dublin, where he was a scholar and gold medallist and graduated with BA and BAI degrees in 1867. He gained the MAI in 1874. Moore served a one-year pupilage to Bindon Blood Stoney, engineer to the Dublin Port and Docks Board, and later worked with Beyer, Peacock & Co. in London. Having worked as assistant county surveyor in Antrim under **Alexander Tate**, he took first place in a county surveyor examination in November 1870 and was appointed to Westmeath in the following month. He transferred to the Meath surveyorship at his own request in August 1874.

In an address to the Institution of Civil Engineers of Ireland in November 1907, Moore strongly criticized the arrangements for road maintenance and improvement which had been introduced by the Local Government (Ireland) Act 1898 and particularly the fact that the initiative in proposing any work rested with the district councils rather than the county council. This meant that the district councils not only controlled the level of expenditure but could dictate the specification for road projects, demand the use of inferior local stone, oppose the use of steamrollers and give priority to roads of purely local importance instead of the main roads: according to Moore 'their chief idea is to make every lane leading to a few houses a county road, or even to make new roads to accommodate half a dozen occupiers'.

Moore was an early user of a form of reinforced concrete for bridge works. Speaking in March 1911, he referred to bridges over the river Nanny which he had built in the early 1880s using disused rails, with nine inches of concrete on top, and which were as good, twenty-seven years later, as the day they were erected. In 1903–4, he was responsible for the design and construction of the four-span Watergate Bridge over the river Boyne, just upstream of the old bridge at Trim; the technique was a simple one, he explained: steel beams were put in place on the concrete abutments and piers, with a platform of planks beneath them so that concrete could be rammed between the beams by an ordinary labouring man. In 1879 he reconstructed and lowered Somerville Bridge over the Boyne navigation which had been erected by Richard Evans in 1792; the bridge was improved and widened in 1936 by **E.J. Duffy**, a subsequent Meath county surveyor. In 1906, he completed a fine single-arch masonry bridge at Herbertstown, near Naul, Co. Dublin.

Between 1890 and his retirement Moore carried out all of his inspections of road works using a bicycle. He was said to be 'a man of much mental culture and many activities'. He was an active member of the Institution of Civil Engineers of Ireland of which he was president in 1908–9 and subsequently honorary treasurer. He was a member of numerous other societies, including the Royal Society of Antiquaries of Ireland, the Georgian Society which he had helped to establish, and the Meath Archaeological & Historical Society of which he was secretary for a period.

In 1871 Moore married Elizabeth Jane, daughter of Revd R.W. King, rector of Portglenone, Co. Antrim, with whom he had four sons and four daughters. The family lived in Navan in the 1880s but later moved to Dublin, living first at Eccles Street and later at Brookfield Terrace, Donnybrook. Moore retired in June 1907 and died at the Meath Hospital, Dublin, on 18 March 1912 after a brief illness; he is buried at Mount Jerome cemetery with his wife who died in 1919. His younger brother, William Patrick (1846–1915) was employed as assistant mechanical engineer by the Dublin Port and Docks Board from 1881 to 1912.

MORAN, John (*c*.1866–1916), son of James Moran, was born at Ballysteen, Askeaton, Co. Limerick, and became a county surveyor in somewhat controversial circumstances. When a vacancy arose in the eastern division of Co. Galway in February 1907, the county council initially appointed James Hardiman, one of their assistant county surveyors, to fill the post on a temporary basis. A salary of £300 a year was fixed for the permanent position with travelling expenses of £100 and an allowance of £60 for office expenses. The post was duly advertised in May and, although the *Irish Builder and Engineer* advised that 'candidates not possessed of strong local influence need not trouble themselves to apply', thirteen applications were received. At a meeting of the council in May, Hardiman got twenty-one votes as against nine for Edward W. Lynam BE, of Ballinasloe, two of whose brothers were already serving as county surveyors and whose wife, Meta Kelly, was a granddaughter of **John Kelly**, a former Roscommon county surveyor. Hardiman, who held BA and BE degrees, subsequently failed to present himself for the statutory qualifying examination and when this was reported to the council in December 1908 an unusual situation arose. The chairman took a serious view of the matter, noting that the council could be surcharged if they continued to pay Hardiman's salary or to pay contractors on foot of certificates which he had signed. Hardiman claimed that he had been ill at the time of the examinations and had handed in a medical certificate to that effect and there was some support for him by councillors who felt that it would be unfair 'to treat him so shortly'. But the weight of opinion at the meeting was against him, with members asking 'are we to come here and fool the county Galway and allow Mr Hardiman to do what he likes'? The chairman finally called for Hardiman's resignation which he there and then wrote out and handed to him; he was allowed to revert to his former position as assistant surveyor.

When the county council re-advertised the position in January 1909 it specified that there would be an absolute prohibition on private practice by the new appointee. This was hailed by the *Irish Builder and Engineer* as 'a very proper rule' which should be adopted by every county council in Ireland and it was noted with satisfaction by the council of the Royal Institute of the Architects of Ireland. Although he had been placed last of the three candidates at the qualifying examinations in April, John Moran was appointed to the vacant post at a council meeting in late June, defeating the other leading contender (W.N. Binns BE, a Galway man who was later to become Galway town surveyor) by fourteen votes to ten. Moran had gained a BA degree in 1891 and a BE degree in 1898 at Queen's College, Galway, and before his appointment to Galway he had been an assistant county surveyor in Tyrone for some years under F.J. **Lynam** and town surveyor in Omagh where he was based. He had also carried on a significant private practice, dealing mainly with housing and other minor architectural projects. While he was therefore an experienced and academically well qualified engineer, the factors which seem to have influenced councillors when the appointment was being discussed were that Moran was 'a consistent nationalist' who was 'strongly recommended by the leading nationalists of the west'.

Moran's assignment was a difficult one. He had barely time to settle in to his new post when he found himself enmeshed in a controversy about 'our rotten, rutty roads' – the title given by the *Tuam Herald* to an article in September 1909 which was trenchantly critical of the 'deplorable state of ruin and disrepair to which the once fine roads of the county have been reduced'. The newspaper attributed this situation to a number of causes, including the fact that there had been so many changes in the ranks of the county surveyors within a relatively short period, with consequent interruptions in the continuity of supervision and management. In addition, the friable limestone used for road repairs was unsuitable, according to the newspaper – in a dry summer, the roads were a mass of dust and, in wet weather, a mass of mud. Moran made it clear that he planned to take immediate steps to tackle the legacy of neglect and ineffective maintenance and not long after his appointment he obtained the sanction of the Local Government Board to the suspension of one of the assistant surveyors and withheld payments from contractors in a substantial number of cases. This led to a warning by the *Tuam Herald* that many of the members of the rural district councils were interested, directly or indirectly, in road contracts to such an extent that the safety and welfare of road users 'does not matter one straw so long as certain road contractors can scamp their work and do as little as human ingenuity can devise for the upkeep of their sections ... He would be a more daring man, more so than most Irish officials are, who would face the enmity of so widespread a crowd of overseers, whose aim is to provide that their friends' work on the road is as cheaply and as badly done as it can be'.

When **P.J. Prendergast**, the surveyor for the western division of Galway, resigned due to ill-health after less than four years in office, the county council decided to appoint Moran to be surveyor for the whole county for a term of three years beginning on 26 November 1910. His salary was increased to £440 a year and he was given an allowance of £200 a year for travelling expenses, on the understanding that he would provide himself with a car. The Local Government Board doubted whether the appointment of one surveyor for the whole county was expedient but agreed reluctantly to sanction it. Moran, who was allowed six assistants, did not take long to make his presence felt in west Galway as he had done in the eastern division. At the first meeting of the Galway rural district council which he attended, he reported that the condition of the main roads in the district was extremely bad with a crust so weak that it could not carry the ordinary traffic; he therefore withheld payment from a number of the contractors and announced that he was deferring payment to most of the others who had failed to carry out their maintenance contracts to his satisfaction. A turbulent scene ensued for some ten minutes, according to a press report, with contractors surging up to the bench where Moran and the clerk sat, waving their fists, and declaring that they would throw up their contracts entirely. Moran was surrounded by a threatening crowd as he attempted to leave the courthouse and there were shouts of 'kill him', 'he wants to starve us', and 'he will destroy the county'. He escaped without injury and was escorted to a place of safety.

In November 1913 the county council decided to revert to the arrangement which had applied from 1838 to 1910 and appointed a separate surveyor for the eastern division. Moran continued to serve the western division from then until his death at the age of 50 on 14 April 1916; he had contracted acute pneumonia arising from the fact that he had been 'out and about at all hours and in all kinds of weather' inspecting the roads. He was buried at Rahoon cemetery in Galway. Moran lived with his wife, the former Miss K. Hill, at Salthill and had four sons and a daughter, Margaret, who married Charles G.J. Lynam (1889–1969), a nephew of **Francis Lynam** for whom Moran had worked as assistant county surveyor in Co. Tyrone. Charles was a captain in the Royal Engineers at the time of Moran's death and was later to become deputy director of public works in Iraq and the first chief engineer of the Iraq Petroleum Company; he and Margaret had one daughter, Shevawn (1914–98) linguist, journalist and author of *Humanity Dick: A Biography of Richard Martin MP, 1754–1834* (London, 1975).

MORRIS, Patrick Richard (1880–1909), served as county surveyor for only three months. His untimely demise 'when life seemed to be full of promise' was greatly to be deplored, wrote the *Irish Builder and Engineer*, when reporting his death at the age of 28 in March 1909 after a short illness. He had been elected county surveyor for Queen's County in the previous November and had taken up duty on 1 January – another case of success for an applicant who had received the lowest marks at the qualifying examinations held a few months earlier. His grandfather, James Feehan, and his brother, George, were both assistant county surveyors.

Morris was born on 7 January 1880, the son of Samuel Morris of Newrath, Waterford, who had been an anti-parnellite nationalist MP for Kilkenny South from 1894 until 1900. He was a graduate of Queen's College, Cork, where he had gained a BE degree in 1904, and had served a pupilage with J.H. Ryan, one of the leading Irish consulting engineers of the day, working on sewerage schemes at Greystones, Delgany and Enniskerry, Co. Wicklow. He was employed by the Congested Districts Board as clerk of works for two years before his appointment as county surveyor.

MOYNAN, John Ouseley Bonsall (1852–1932), was born in Dublin, one of the eight children, and the eldest son, of Richard Moynan, a manager with Ferrier Pollack & Co., and his wife, Harriet Nobel (daughter of a church of Ireland clergyman), who lived at Eldon Terrace in the South Circular Road area of the city. Why he was called after John Ouseley Bonsall, author of the *Essay on the Attribute of Knowledge to God* (Dublin, 1830) is not clear.

His younger and more conventionally named brother, Richard Thomas RHA (1856–1906), was to become a distinguished Irish painter, a member of the Royal Hibernian Academy and a political cartoonist with the unionist journal *The Union*; his most popular and best-known work was *Military Manoeuvres* (1891), a large painting of a group of children acting in the street at Leixlip as if they were a military band and which now owned by the National Gallery of Ireland.

Moynan was educated privately and in Nice and entered Trinity College, Dublin, in June 1869. He graduated with MA and BAI degrees in 1874 and subsequently served a pupilage with James Price, consulting engineer. He was resident engineer in charge of the works on one division of the Waterford, Dungarvan & Lismore Railway in 1876–8 and was later employed on various projects undertaken by **John H. Brett** and **Edward Glover**, both county surveyors. He was an unsuccessful candidate for the post of Cork city engineer in 1880 but went on to win an appointment as engineering inspector with the Board of Works which he held for three years.

In September 1883 Moynan was appointed county surveyor in Longford but seems not to have been very content with his position there. He regularly complained that road contracts were taken at 'absurdly low prices' even though he had introduced more stringent specifications, and urged his grand jury to consider arrangements by which good contractors for road maintenance operations could be obtained at realistic prices without unduly interfering with legitimate competition. He complained, too, that some contractors were fraudulently moving heaps of stone from one position to another for re-measurement, and about the difficulty he experienced in detecting abuses such as this with only one assistant surveyor to help him. On top of this, the assistant was so badly paid that Moynan had to give him the use of his own horses to enable him to carry out his share of the work of inspecting the 600 miles of roads for which the grand jury was responsible.

Moynan moved to Tipperary north riding in August 1891 and served there for thirty-nine years. On his transfer to the service of the new county council in 1899, his salary was fixed at £650 a year, an increase of £150 on what he had been receiving from the grand jury. According to the *Irish Builder*, he was a very popular and energetic official who won the respect and regard of all for the efficiency with which he discharged his duties. He was certainly one of the most prominent county surveyors of his era, a trailblazer in the introduction of large-scale direct labour schemes, and among the first to advocate and use steamrolling extensively. When he took up duty in Tipperary, road maintenance was carried on by means of 750 separate contracts, with individual contracts covering an average of less than two miles of road and costing an average of £25 each per year. Assuming a contractor's profit of 20 per cent, Moynan argued that this system involved the diversion of £4,000 a year from actual road work. In the early 1890s he introduced a limited form of direct labour operation on roads for which no acceptable tenders had been received and, in 1900, after one of his assistants had left to join the army during the Boer war, he persuaded the county council to seek sanction to new staffing arrangements under which there would be only one assistant surveyor and a number of overseers. Sanction for this was refused by the Local Government Board but, when the law was changed in 1901 to allow direct labour operations to be introduced, the north riding of Tipperary, together with Co. Limerick, were the first to adopt the system exclusively for all works, maintenance and otherwise, instead of the contract system. Beginning in 1903 with a scheme under which up to 350 miles of the more important roads were dealt with by direct labour, all work on the county's 1,300 miles of roads had been transferred to the new system by 1909. This involved the employment of over 1,000 men on

61 road sections, each the responsibility of a ganger, with each 5 sections grouped to form a district in the charge of a road steward, and each 4 districts forming the area of responsibility of an assistant county surveyor. It also involved the acquisition of plant and equipment on a scale previously unheard of; in January 1904, for example, Moynan's council accepted tenders for an Aveling and Porter steamroller (12½ tons) at a price of £420, another heavier steamroller made by Clayton and Shuttleworth costing £468, several sleeping vans costing £52 each, water carts at £28 each, horse rotary road brushes at £27, and horse road scrapers at £16.

In the overall national context, Moynan saw little future for the canals and very limited prospects for extensions of the railway network and he correctly forecast that roads (with the development of the internal combustion engine) would become the dominant mode of transport, especially for freight, as the twentieth century progressed; he paraphrased a well-known couplet to read: 'Canals were, Railways are, Roads will be, the most important of the three'. To cope with the anticipated growth of road traffic, he was advocating from about 1905 onwards what he called the 'nationalization' of some 25 per cent of the rural road network which then totalled about 56,000 miles in the thirty-two counties. These roads, linking one important town with another, were the only ones which he considered to be of general public utility and, in his view, they should be handed over to an executive Road Board, financed partly by central and local taxation and partly by borrowing, and brought up to a uniform standard over a period of years.

Moynan was using concrete in the construction of road bridges by the mid-1890s. Between 1910 and 1915 he supervised the construction of a major new bridge over the river Shannon at Portumna which had been designed to his specification by C.E. Stanier of London. A wooden bridge at this location, built by Lemuel Cox in 1795, had been replaced between 1838 and 1842 by a structure composed of cast-iron girders resting on piers formed of wooden piles. A bridge commission advised in October 1909 that this second bridge should be replaced and recommended a steel plate-girder bridge on cast-iron cylindrical piers, nine feet in diameter and filled with concrete. A tender of £21,500 for the work was accepted in March 1911 and the bridge was completed by English contractors at a cost of about £30,000.

In the 1890s the Ionic central portico of Nenagh courthouse (built in 1843 to a design by J.B. Keane) was rebuilt under Moynan's direction, in conjunction with the architect, Walter G. Doolin, who was engaged at the time on the construction of the town's new Catholic church. For most of his years in Tipperary, however, Moynan's partner in an extensive private practice in engineering and architecture was Robert Paul (Bertie) Gill, who had been an assistant county surveyor since the summer of 1882 and served until his death in July 1928. Gill was a brother of T.P. Gill, nationalist MP for Louth from 1885 to 1892, and from 1899, secretary of the Department of Agriculture and Technical Instruction; Bertie Gill, himself an unsuccessful candidate at a bye-election in 1915, was the father of Tomás Mac Giolla (b. 1924), who served as a Sinn Féin The Workers' Party TD from 1982 to 1992. Operating under the title Moynan & Gill, Architects, from an address at the courthouse, Nenagh, the county surveyor and his assistant attracted a wide variety of commissions; their projects included new schools at Church Road, Nenagh (1906); the Town Hall in Nenagh (1889); the Nenagh Convent of Mercy schools, completed in 1910 to cater for 520 pupils; a water supply extension scheme for Nenagh costing £3,000 (1918); housing schemes for the urban district council; dispensaries; and numerous residential development schemes.

Moynan took an active part in the affairs of the Institution of Civil Engineers of Ireland of which he was president in 1918–19. As such, he presented an address to the lord lieutenant,

Viscount French, in July 1918 congratulating him on his appointment and expressing 'sentiments of loyalty and devotion towards the sovereign'. He was a keen admirer and strong supporter of John Redmond and his policies but when his council, following Sinn Féin policy, decided to break with the Local Government Board, he continued to work with them, declaring that he would stick by his county and his country through thick and thin. Consistent with this was his appearance in December 1919 as a witness before a Dáil Éireann Commission of Inquiry into the Resources and Industries of Ireland; he urged that the state should become directly involved in a number of activities which private enterprise could not be expected to finance – the provision of economical and efficient power and transport, suitable housing, arterial drainage, peat development and re-afforestation. Although his own family, he told the commission, had about £20,000 'sunk in a mill' near Nenagh where water power was used, he was not a believer in the 'great possibilities of water-power in Ireland' about which other witnesses had spoken, and he argued that the Shannon Scheme, in particular, would not be financially or economically possible.

In November 1929, when Moynan was 77 years old and in office as a county surveyor for forty-six years, the Department of Local Government and Public Health called for his resignation on foot of an adverse report from **James Quigley**, their chief engineering inspector. Moynan duly submitted his resignation in the following month but continued to serve the council until September 1930, just before his successor took up office. His retirement was marked by the usual tributes, with a particularly lavish address from the road workers who 'with burning indignation noted, in the closing stage of your public career, the attempt to sully your great record because of our alleged indolence and inefficiency'. With Henrietta (Hennie) his wife, who predeceased him, Moynan had three sons and a daughter. He died at his home, Islandbawn House, Nenagh, in October 1932 and was buried at Kilruane cemetery near the town after an impressive funeral procession. An obituary described him as 'a man of great personal geniality and charm and of an untiring activity'; his passing, said the *Irish Builder and Engineer*, 'removes another of the fast-dwindling band of Irish county surveyors of the old regime, and a very distinguished and capable band they were'.

MULVANY, Christopher John (1852–1919), born on 4 September 1852, was a son of Christopher Mulvany (1818–95), engineer to the Grand Canal Company for forty-five years, and his wife Isabella. The young Mulvany served a pupilage to his father between 1871 and 1875 and, with seven others, including Percy French, graduated from the school of engineering at Trinity College, Dublin, in 1881. He briefly held positions as resident engineer on the Great Southern & Western Railway and under Sir Joseph Bazalgette of the Metropolitan Board of Works, London. After a competition in July 1882 at which there were sixteen candidates, Mulvany was appointed Roscommon county surveyor in the following September. He devoted a great deal of attention to the improvement of the roads of the county and made extensive use of direct labour when it became permissible to do so in the first decade of the new century. He was responsible for the construction of numerous bridges, including two early ferro-concrete bridges at Cloonagh and Floatford, and had to deal with at least three major bog slides in his area.

Mulvany was one of the surveyors who carried on a substantial practice as a consulting engineer and architect, an arrangement which attracted constant criticism from the Royal Institute of the Architects of Ireland and from the influential *Irish Builder and Engineer* in the early decades of the twentieth century. When the Local Government Board sanctioned an increase of £200 in his salary in 1900, the county council took the view that Mulvany should desist from private practice and devote his full time to official work. The surveyor, however, continued as before but, when he claimed at a public inquiry in July 1909 that it was impossible for him, with the staff at his command and without a qualified competent deputy, to give the business of the county the supervision it required, he left himself open to a particularly harsh attack in the columns of the *Irish Builder and Engineer* which described his statement as astounding. 'It is perfectly well known', the article went on 'that many county surveyors have so much free time on hands that they not merely engage in an extensive practice as architects and engineers, but enter upon public competitions for the most troublesome and varied architectural and engineering projects, involving the expenditure of much time ... It is incontrovertible that the counties where the surveyor devotes his whole time to the duties of his office are naturally the best maintained'. The subsequent report from the inspector who held the inquiry did not appear to criticize Mulvany who was fully entitled under the law as it stood to carry on private practice, but it did recommend that the entire staff of his department should be reorganized and that the salaries paid to his assistant surveyors should be increased to attract properly qualified men.

Mulvany was engineer for Roscommon, Strokestown, Tuam and Castlerea district councils and was responsible for water supply schemes at Castlebar (1896); Roscommon (1901); Castlerea (1904); and Tuam (1906). As county surveyor he undertook the restoration of Roscommon courthouse which had been destroyed by fire in June 1882, having won the commission in competition with some twenty architects. In 1915 he designed a new TB dispensary in Roscommon which was estimated to cost £1,200, much more than the cost of the standard plans issued by the Local Government Board. Mulvany told the county council that no permanent building could be erected for less than his estimate and he could not advise them to accept 'a building of tin and sticks' built with ratepayers' money. He designed several business premises in his area and, with the Westmeath surveyor, **A.E. Joyce**, designed a new water supply scheme for Mullingar in 1898.

A prominent sportsman in his early days, Mulvany rarely missed a meeting of the Roscommon hounds and on several occasions rode winners of steeplechases. His service as county surveyor was ended by his sudden death on 15 October 1919 at his home, Ranelagh House, Roscommon, at the age of 67. According to the local press, he was in his usual robust good health to the day of his death, and had been travelling up to 100 miles a day on his motorcycle in the course of his duties. The *Roscommon Herald* described him as an efficient, courteous, and painstaking official who would be difficult to replace after thirty-seven years service, and noted that the high esteem in which he was held was evidenced by the large cortege which followed his funeral procession to the new cemetery at Roscommon.

MURPHY, Cornelius (1890–1961/2) was born on 2 August 1890 near Inchigeela, Co. Cork. He was educated at Christian Brothers' College, Cork, and went on to study at the Royal College of Science in Dublin. He gained the ARCScI qualification in 1915, having been a first class exhibitioner and medallist in the course of his studies, and was also awarded a BSc degree. He worked for a year with Siemens Brothers at the Dynamo Works, Stafford, before taking first place at assistant county surveyor examinations in 1916. He took up duty with Cork county council later that year and was promoted to the new grade of chief assistant county surveyor in 1925, working under **R.F.M. O'Connor** in the northern division.

Murphy passed the county surveyor qualifying examination in 1927 and, following the first competition held by the newly established Local Appointments Commission for a county surveyorship, he was recommended to Westmeath county council for appointment to the vacant post in that county in March 1928. After a long debate, the council voted to appoint Murphy but only under protest. Some members argued that the commission's recommendation should be treated with contempt and referred back, while others criticized the sending of 'planters into Westmeath over the heads of natives' – a reference apparently to the claims to the post of a local man, Joseph Buckley, who had been an assistant county surveyor in Westmeath since 1910 and had acted as county surveyor since the death of **A.E. Joyce** in January 1927. Murphy served in Westmeath until his retirement on 31 July 1955. He subsequently lived at Stillorgan, Co. Dublin, and died in 1961/2.

MURPHY, Jeremiah Thomas (1882–1929), was born on 23 August 1882 near Timoleague, Co. Cork, and educated at Butlerstown national school and at Christian Brothers' College, Cork. He studied at Queen's College, Cork, and was awarded the BE degree with first class honours in 1904. His first appointment (1905–9) was as assistant to Francis Bergin, a prominent consulting engineer, with whom he was engaged on the design of water and sewerage schemes in Dublin, Kildare and Monaghan; his assignments also included periods as resident engineer on water schemes at Thurles and New Ross and a sewerage scheme at Naas.

When **Samuel A. Kirkby** gave notice in 1908 of his wish to resign from his post as county surveyor for the east riding of Cork, the county council decided to revert to the arrangement which had applied between 1862 and 1898 and to appoint two surveyors for the riding at salaries of £400 each, rising by annual increments to £500. When applications were invited in October 1908, the council gave notice that each candidate would have to be nominated by six councillors and pass an examination in Irish before being admitted to the statutory Civil Service Commissioners' examinations, and they went on to bind themselves to appointing the two highest-placed candidates in those examinations. Similar arrangements had been introduced by the council some years earlier for other appointments, including those of assistant county surveyors. The Local Government Board was impressed by the council's decision to make appointments on merit rather than on a patronage basis and told the council that, while their proposal to have a preliminary examination in Irish was *ultra vires*, they could give preference at the appointments stage to those who had qualified in Irish. In all, there were thirty-nine applicants for the two posts, sixteen of whom presented themselves for the examination in Irish which was conducted by three priests, including Very Revd Canon O'Leary PP, better known as An tAthair Peadar, the Irish author. Thirteen candidates qualified in Irish and the council told the Local Government Board that they 'may examine as many candidates as they think proper, but the county council would appoint none other than the two who have attained the two highest places out of the thirteen with Irish'.

By January 1909 the county council were concerned about the delay in supplying them with the list of qualified candidates and were complaining that in the absence of proper supervision the assistant surveyors were becoming too friendly with the road contractors! Having unsuccessfully pressed **R.W.F. Longfield**, the west riding surveyor, to take on the added responsibility, with additional remuneration, the council appointed one of the assistants to act as county surveyor until permanent appointments could be made. Eventually, at their meeting in mid-June, the council was informed that J.T. Murphy had gained first place in the qualifying examination held in the previous April, with **R.F.M. O'Connor** in second place. Of the original 39 applicants, 7 had been declared ineligible, 23 had withdrawn or failed to attend the statutory qualifying examinations, 4 had failed to pass those examinations leaving only 5 qualifiers. As they had already agreed to do, the council duly appointed Murphy and O'Connor with effect from 1 July while glorying, as one councillor put it, 'in the fact that the first four qualifiers were Cork men'. Murphy initially opted to take charge of the central division but, when the number of surveyors in the county was reduced to two late in 1925 following Longfield's retirement, a new division of responsibilities was arranged between him and O'Connor, with Murphy retaining what was then termed the southern division, mainly the area south of the river Lee.

Roads in Co. Cork, including the main roads and those serving tourist areas, were in poor condition when the two new surveyors took up duty, sharing with Longfield (who had been in office in the west riding since 1899) responsibility for bringing standards up to the levels already achieved elsewhere. They were fortunate that, notwithstanding opposition from some of the rural district councils, a major steamrolling programme had been decided on by the county council late in 1908, involving the purchase of an additional eight rollers. Loans were subsequently raised to fund special road works and to purchase rollers, stone crushers and other plant, and the system of assistant surveyors was reorganized, with travelling expenses being allowed for the first time. Whatever progress in roadmaking was made by the new surveyors in the following ten years was more than offset by damage caused by the additional motor traffic which had come onto the roads, the exceptional damage caused by British armoured cars and lorries between 1919 and 1921 (estimated by the surveyors to have cost £114,000), the destruction of more than 300 bridges and culverts in the months before the truce of 1921 and the destruction of an even greater number during the civil war of 1922–3. In December 1923 the Dáil was told that the estimated total cost of repairing and reconstructing damaged bridges in the county was almost £164,000 and **James Quigley**, chief roads engineer at the Department of Local Government and Public Health, reported in the following year that bridge destruction in Cork 'was on a vastly greater scale than in any other county I have seen ... the damage is far oftener in the nature of total destruction than is the case elsewhere'. But the run-down state of the county's roads was not due entirely to 'the troubles' for, according to Quigley, the council's roadmaking plant was 'wretchedly inadequate', no tar had yet been tried, and the increase of 200% in wages since 1914 was limiting the amount of work that could be financed from the council's resources. The only consolation, said Quigley, was that the situation in Cork city was worse: 'barbarous is the only adjective that comes to my mind to describe the road work there'.

Taking account of Quigley's report, approximately one-eighth of the three million pounds provided in special grants by the government in 1924–6 for road improvement works was allocated to Cork county council, enabling Murphy and his colleagues to achieve enormous improvements in the standard of the trunk routes and to provide tarred surfaces on many of them. Murphy was only 47 years old when he died from pneumonia at a Cork hospital in November 1929.

MURPHY, John Joseph (fl. 1882–1942), was born in Cork city on 28 August 1882. He was an articled pupil for three years with the Cork city engineer before becoming an associate member of the Institution of Civil Engineers, London. He was assistant engineer with Cork corporation from 1904 to 1908 and from then until 1914 was employed by various local authorities in Co. Cork on diverse projects. In 1914–16, he was engaged by the military authorities to supervise the construction of a sea-plane base at Aghada and an extension to the dry dock at Passage West. He took on a contract from Tipperary south riding county council to construct two reinforced concrete bridges in 1917 and he worked as assistant county surveyor in Wexford in 1918. He was in practice as an architect on his own account and later with Joseph O'Malley of Limerick until 1923 when he became an assistant engineer with the Board of Works.

Murphy became Longford county surveyor in unusual circumstances. Soon after the resignation of the previous holder of the office in January 1927, he was appointed by the county council as temporary county surveyor. The council were reluctant to fix the salary for the position at a level which the Department of Local Government and Public Health considered would attract 'efficient candidates' and it eventually fell to the Department to make a formal request to the Local Appointments Commissioners to hold a competition for the post. Following the usual advertisements, and interviews in January 1928, Murphy, who was the only applicant for the post, was declared by the commissioners to be unsuitable for appointment. The outcome was no different when the position was re-advertised in April 1928. In the meantime, the county council maintained their opposition to a higher salary (which might have attracted a wider field of candidates) and insisted that Murphy had given 'entire satisfaction' as temporary surveyor. When the Department advised that they also considered Murphy's temporary service to have been entirely satisfactory, the Commissioners sensibly decided to override the advice of their interview board and recommended him for appointment in January 1929. He served in Longford until his retirement in March 1942.

MURRAY, Bernard Barnaby (1823–89), was the son of Patrick Murray, a silk manufacturer of Francis Street, Dublin, and his wife, Frances Tennant of Belfast. He was educated at St Vincent's College, Castleknock, entered Trinity College, Dublin, in November 1839 and was awarded a BA in 1844 and the Diploma in Civil Engineering in 1845. He gained valuable experience of engineering in America before returning in the late 1840s to Dublin where he took up residence at 14 Ushers Island and established what was believed to be a lucrative private practice. In 1854, his offices were at Lower Gardiner Street.

The fact that Murray – who was considered to be a clever graduate and well qualified – failed to secure an appointment as Dublin district surveyor in 1856 was one of the factors which led to considerable controversy about the selection system for surveyors, culminating in the legislation of 1862 which assigned responsibility for conducting the qualifying examinations to the Civil Service Commissioners. He was subsequently certified to be qualified for appointment to positions of county surveyor generally and was eventually appointed to the southern division of Co. Down; he took up office in Newry in September 1862 and served there for the rest of his life. Apart from acting as engineer to the Newry Navigation Company, he seems to have devoted all of his time to his grand jury duties.

Murray enjoyed the respect and esteem of successive grand juries and while it was said that he was a strict task-master, he was never liable to be accused of being harsh or tyrannical. There was general agreement that he kept the roads and public buildings in the area in first-class condition. In 1905, his successor as county surveyor, **P.C. Cowan**, by then chief engineering inspector at the Local Government Board, paid fulsome tribute to Murray's skill as a road engineer; he noted in particular, that Murray was among the first of the surveyors to employ a large number of surface-men, one of whose main tasks was to cross-sweep the roads frequently both in summer and winter to remove dust and mud, instead of applying fresh layers of broken stones.

Murray married Barbara Mary Power, a native of Cork, in the Pro-cathedral, Dublin, in 1852 and, after her death in 1879, he married Mary Barry, who died in 1933. By early 1889, when he was 66 years of age, his health had begun to fail and he planned to retire on pension in the summer. He died rather suddenly, however, on 14 May 1889 at his residence, South Lodge, Newry. His funeral cortege, a few days later, to the Old Chapel yard in Newry was a large and impressive one, with his massive oak casket being carried first by four of his assistant surveyors and then in a four-horse hearse.

NERNEY, Michael (1892–1965), was born on 22 November 1892 at Culleen, Elphin, Co. Roscommon, and was educated at Summerhill College, Sligo. He gained a BE degree at University College, Galway, in 1916 and served in France as a lieutenant with the Royal Engineers from then until 1919.

Within a week of the sudden death in October 1919 of **C.J. Mulvany**, Roscommon county surveyor, the *Roscommon Herald* reported that a large number of candidates were already actively canvassing for the appointment. The paper named no less than eleven putative candidates, mentioning in many cases their connections with local families or their links with churchmen and professional people throughout the west of Ireland. The extent of the canvassing moved one of the local catholic curates, Father Michael O'Flanagan (1876–1942), who was already prominent in the national movement and in Gaelic League affairs, to preach a special sermon on the subject a few days later which the *Herald* reported in detail. 'The pulpit is not the place from which to throw as much as the weight of a feather in the scale in favour of one candidate as against another' said the priest, 'but it certainly is the place from which to lay down the general moral principles upon which the selection should be made'. There was scarcely any councillor, he went on, who would stoop so low as to accept a money bribe for his vote, but there was a far greater evil – a subtle insidious form of bribery which was commonly described by the expression 'scratch me and I'll scratch you'; and there was no more detestable thing in public life than the giving of public positions through the operation of that abominable thing which is called influence. Motives of a personal nature should not prevent any councillor from carrying out his trust, and anyone who voted for any such motive should be removed from office at the first opportunity.

Canvassing continued as the county council made the necessary arrangements to get the formal competition under way. They fixed the salary for the post at £500 a year, rising by annual increments to £600, but were pressed by the Local Government Board to allow an additional £150 for travelling expenses so that the new surveyor could provide himself with a car and give closer supervision to the road works. The post was advertised in December 1919 and shortly afterwards a list of candidates was sent to the Local Government Board so that their qualifications could be examined. The usual examinations were duly held in June 1920 and six candidates qualified, with Michael Nerney in fifth place. However, due to the disturbed conditions of the time, and possibly because the county council had transferred its allegiance from the Local Government Board to the Dáil Éireann Department of Local Government, the post in Roscommon was not filled until December 1923 when Nerney was appointed.

Nerney's term of office in Roscommon lasted thirty-five years ending with his retirement in November 1958. He carried on the roads, water and sewerage, housing and other functions of the council efficiently and to the general satisfaction of the council and his sense of humour and witty remarks were said to have often eased tense situations at council meetings. It was said of him that he knew every one of the council's large workforce; in turn, because he was small in stature, the roadworkers were known to be always on the alert when they saw a car approaching 'with no driver'. A keen fisherman, he continued to live at Abbeytown, Roscommon, until his death on 28 July 1965. He was buried locally at St Coman's cemetery.

NEVILLE, John (1813–89), was born in Co. Limerick, the son of John Neville who was in practice as an architect and a member of a family said to be the hereditary keepers of St

Patrick's Bell. He was living in Clonmel in 1836 when he became a member of the Civil Engineers Society of Ireland and was employed by the Shannon Commissioners on drainage and navigation works between then and 1838. Among other projects, he carried out a detailed survey of the river at Athlone and was involved in work on the locks between Athlone and Meelick. He also worked for a period with the land valuator who was associated with the drainage project. Almost thirty years later, he told a House of Lords select committee in 1865 that the original plans for the works on the Shannon were defective and that the works themselves were quite ineffective, had not achieved the planned reductions in flood levels and had cost much more than they should have done. He repeated his criticism in 1874 when fresh proposals for the drainage of the river basin were being considered, claiming that the government of the day 'threw £250,000 … into the Shannon without any corresponding advantage to the drainage' and that both the engineers' designs and the carrying out of the works had been defective.

Neville sat the county surveyor examinations in March 1838 but was found not to be fully competent. When he was examined again in August 1840 he was reported to be considerably improved because of the practical experience he had gained in the meantime as a road contractor in Co. Limerick, and was appointed Louth county surveyor in the following September. He proved to be a competent road engineer but was frustrated by, and always critical of, the contract system of road maintenance and the limited resources available to county surveyors for supervising the contracts. He campaigned unsuccessfully for modifications of grand jury law to make the system a more efficient one, attacking what he described as the labyrinth of contracts with which the surveyors had to deal and the lack of any provision for the systematic consolidation of these contracts. In his report to the grand jury at the summer assizes of 1842, he noted that there were 272 different contracts costing a total of £3,847 per year for the repair of 411 miles of road in the county, giving an average contract length of 1½ miles and an average contract price of just over £14; the lengths of contracts varied from 100 yards to 10,000 yards and the cost ranged from less than £1 to almost £123 per contract, with well over 50 per cent costing less than £8. Neville argued that a minimum contract of £50 would reduce the total number to 77, bring considerable administrative savings and lead to the engagement of a better class of professional contractor who could employ road labourers on a permanent basis and pay regular weekly wages. This could be achieved, under his proposals, if the surveyor himself were allowed to apply for presentments for all maintenance works which he considered necessary, on the basis of a scheme of road districts and sections; tenders would then be invited and accepted only on the basis of this scheme instead of leaving it to the cesspayers and the contractors to decide, effectively, how the work and the contracts should be organized.

Neville campaigned also for an increased number of assistant surveyors (even though he had only about 400 miles of road to maintain in 1840) and for the introduction of higher levels of qualifications and higher salaries, in order to attract a better standard of assistant. At the summer assizes in 1844 he told the jury that the fourth assistant allowed to him at the previous assizes had been engaged, but he had found it necessary 'to make an additional allowance of salary, from myself, to provide a person suitable to the views of the grand jury' and, to make matters worse, the unfortunate assistant had been killed in a fall from a horse just before the assizes – 'and I will now have to provide another'. Twenty-five years later, when his road mileage had increased to 590, Neville was still complaining about the difficulty of recruiting assistants of suitable standard and arguing that expenditure on the supervision of contractors was small compared with what could be saved by assistants of integrity and skill.

His regular refusal of certificates to contractors made him many enemies – his position, he wrote, was 'between Scylla and Charybidis', unable to satisfy any of the interests involved. A paper entitled *Grand Jury Laws and County Public Works, Ireland* which was published under the initials JN in the *Dublin University Magazine* in March 1846 but which is clearly attributable to Neville, set out the various defects in the legal and organizational framework under which the county surveyors worked; for some reason, the council of the Institution of Civil Engineers of Ireland had earlier refused to accept the paper and have it printed in their transactions. Another paper of Neville's dealing with the best form of road cross-sections had already appeared in the quarterly *Original Papers on Engineering*, published in London in 1846.

In February 1847, when he was directly responsible for a force of 6,395 workers on relief schemes, Neville was able to assert that road work was more economical and better done by daily labour than by contractors, especially when task work, with good rates, was in force. But he was critical of other aspects of the relief works on which women and children as young as 10 years of age were employed, as well as men who were weak and infirm: putting these forward for employment on the relief works was mischievous, in his view, and the system should be changed so as to allow the engineer to select the labour and raise the scale of wages instead of using destitution or virtual physical infirmity as the test of qualification for employment.

In his evidence in February 1856 to a government-appointed commissioner, Neville advocated the abolition of the turnpike road system although this would result in the loss of the salary of £25 a year which he received for acting as surveyor to the trustees of the twelve-mile-long turnpike from Dundalk to Dunleer. Staff costs, he pointed out, absorbed up to 50 per cent of the revenue from tolls leaving insufficient funds for maintenance of the road and, as a result, its condition was such that the grand jury had been forced to step in and carry out essential repairs funded from the county cess. This, and similar evidence about much of the remaining 325 miles of turnpike roads throughout the country, led to the abolition of the system in 1857 and the transfer of full responsibility for the roads to the grand juries.

On 5 October 1853 Neville was a passenger on the Great Southern & Western Railway's Cork–Dublin express train when the locomotive failed near Straffan. A following goods train collided with the stationary express causing a railway disaster which was the most serious in Great Britain or Ireland to that time. The five carriages were extensively damaged, there were sixteen deaths and many of the other forty passengers were injured, including Neville who suffered a leg fracture and other injuries. He was taken to the Imperial Hotel in Dublin where he received medical treatment and a few days after the accident he was well enough to be able to give the correspondent of *The Times* a detailed account of his experience, describing how he had been the only one left alive in the carriage in which he was travelling and which ended up nearly a quarter of a mile from the point of impact. This prompted *The Times* to comment that 'the escape of this gentleman from the terrible death which overtook so many of his fellow travellers would seem to be one of those fortunate chances which frequently occur under circumstances deplorably fatal to others'.

In addition to his role as county surveyor, Neville was surveyor for the county of the town of Drogheda until 1878, borough surveyor in Drogheda until February 1869 and town surveyor in Dundalk from 1861 until 1871 when he resigned due to pressure of work. He persuaded the town commissioners and the grand jury to co-operate in 1864-5 in financing the construction of a new road along the southern shore of Dundalk harbour, allowing a large area of slobland to be reclaimed for use as a fair green and town park; the road was originally known as Neville's Road but later became Fairgreen Road. In 1867, the town was said by the

Irish Builder to be 'the cleanliest and healthiest' town on the east coast 'thanks to the vigilance and activity of the town surveyor'. Ten years later, however, the same journal was reporting the dirty state of the streets of Dundalk and an acrimonious dispute between Neville and the town commissioners as to where responsibility for street cleansing lay. When the surveyor contended (with the support of the Local Government Board and a judgment of the queen's bench) that it was the duty of the commissioners to provide for the cleansing of the streets, he was accused of insolence and of acting as a potentate. He replied to one of the commissioners, apparently more in sorrow than in anger, pointing out that 'for the last thirty-seven years, I have been improving the formation of your streets, entrances, footpaths, drainage and sewerage and they present a very different look now to what they did then; but no one can keep the town clean and free from nuisances but your own officers, who are paid for it'. Neville's stormy relationship with the Dundalk commissioners (to which he was elected at the first elections in March 1855 but resigned after six months) continued to the end of his days as county surveyor; for example, he refused in 1885 to have anything to do with the replacement of the flagstones which surrounded the courthouse in the town claiming, even in defiance of his own grand jury, that the matter was one for the commissioners whose chairman he described as a jobbing road contractor.

Having spent over £20,000 between 1840 and 1855 on harbour development works designed by Sir John MacNeill and his son, Telford, Dundalk harbour commissioners found that difficulty was still being experienced in maintaining a navigable channel through the sandbanks in the outer harbour. When Neville was engaged in 1864 to study the problem and to recommend improvement works, he was severely critical of the earlier work, even though it had been approved by Sir John Rennie, and put forward new proposals for improving the navigation by extensive dredging and by the erection of long training walls to fix the location of the channel. Work commenced in 1864 but the scheme was criticized by the Admiralty surveyor and never fully implemented. In 1869, when there were fears that Dundalk would lose much of its trade to the new port at Greenore, Neville was authorized to spend a further £2,000 on improving the outer works of the harbour; he also drew up plans that year for a graving dock, estimated to cost a total of £13,000, to facilitate ship-building but this did not proceed. Although his engagement by the commissioners ceased in 1886, construction and renewal of the training walls continued until 1893 by which time over 46,000 tons of stone had been used to build walls almost three miles long. In 1895 a report by John Purser Griffith, the Dublin Port & Docks Board's chief engineer, was very critical of Neville's work at Dundalk, while recognizing that the shipping requirements of his day were very different and that much less was known at the time about 'this difficult branch of engineering'.

Neville was sufficiently confident of his abilities in the design of bridges to join the large number of engineers and architects who submitted designs when a competition for a new Carlisle Bridge in Dublin was announced in 1862. But Neville's £50,000 scheme was not among the prize-winners and some of the bridges which he constructed in Co. Louth were not particularly successful. For example, Aclint Bridge, on the road between Ardee and Carrickmacross, which he built soon after taking up duty in Louth, was badly built, according to an 1844 report by **Samuel Ussher Roberts**, who was then engaged on drainage works for the Board of Works in the district; however, the bridge survived until the 1980s when it was replaced as part of a major road improvement scheme. Another of Neville's bridges – St Dominick's Bridge, Drogheda, a wooden bridge with large stone abutments and piers which he completed in 1863 – had to be closed to all vehicular traffic in 1872 due to disrepair and was replaced in the early 1890s by an iron structure.

In 1861 steps were taken to provide another new bridge over the river Boyne in Drogheda in place of the existing St Mary's Bridge, built in 1722 and condemned for its inadequacy and poor condition as far back as the early 1800s. Neville's plans for a bridge with a single eighty-foot arch were approved in March 1862, tenders were received in October 1863 and work began in April 1864. By November, with the old bridge already removed, work had been suspended because of the default of the contractor and fresh tenders were sought, this time allowing the option of a one-span or two-span structure. A tender of £6,657 submitted by a Dublin firm for a two-arch bridge was accepted in January 1866 and work began again in the following April. By November 1867, when four years had elapsed since the foundation stone was laid, the bridge was still not completed, the contractor blaming the stone suppliers for his inability to finish the parapets but still having to suffer a penalty of £5 a week. By the following March, the bridge appeared to be complete, but Neville repeatedly refused on grounds of public safety to allow it to be opened until the contractor carried out some additional works to protect the foundations in the bed of the river. James Barton, on behalf of the contractor, disputed the need for these works and advised that the bridge was perfectly safe but Neville refused to yield. The matter was finally resolved at arbitration in June 1868 when it was held that the bridge had been completed in accordance with Neville's plans and specifications; the Drogheda grand jury then pointedly passed a vote of thanks to the contractor, ordered that the award and vote be printed and circulated to the ratepayers, and refused to allow Neville his fee of 5 per cent on their share of the cost. Neville's bridge had a reasonably long life (it was replaced in 1984 by a reinforced concrete structure) but the controversy surrounding its construction had an adverse effect on his relationship with contractors; matters did not improve in this respect in the years which followed and, by 1883, it was being reported that 'no contractor will work under him'.

In February 1867 a wooden bridge erected in 1856 (possibly to a design by **Alexander Schaw**) at Oldbridge, three miles above Drogheda, was badly damaged by ice in the river Boyne and it partially collapsed in the following May. As the bridge was an inter-county one, responsibility for its replacement was shared by Neville and the Meath surveyor, **Samuel Searancke**, and **Alexander Tate**, the Antrim surveyor was also involved; however, Neville's name alone, was carried on the large stone plaques at each end of the new bridge. It was reported at first that a substantial stone bridge was to be built at a location nearer Drogheda where the river was not so deep but the final decision was in favour of a single 120-foot span of wrought-iron lattice girders, nearly eleven feet deep, on limestone abutments at the old location, close to the obelisk which had been erected in 1736 on a rocky outcrop near the place where William of Orange was believed to have crossed the river in the course of the 1690 battle. The Drogheda firm of Thomas Grendon & Co. was awarded the contract for the fabrication of the bridge and its erection was completed by them in June 1869 after the heavy girders were floated upriver on pontoons from the company's works. The bridge is still in use, subject to a five-ton weight restriction, but the obelisk from which it took its name was blown up in August 1921.

The 1857 report of the Select Committee on County and District Surveyors recognized Neville as one of the few surveyors who were competent architects but the only major public building project which he completed was the Italianate county gaol, built at Dundalk between 1849 and 1854 at a cost of about £23,000; the building ceased to be used as a gaol in 1931 and part of it was reconstructed for use as a Garda station in 1945. He was also responsible for a small bridewell constructed at Ardee in 1863 (described as 'a plain square block with tiny windows and massive jambs and lintels'); an extension in 1846 of Dundalk's Greek Revival courthouse which was completed between 1813 and 1820 to designs by Edward Parke and John Bowden; extensions and alterations carried out in 1858 at the prison in Drogheda of which only the front walls now survive; and modifications in 1861–2 at the Tholsel in Drogheda containing a court room and the corporation's offices and, much later, used as a branch of the Bank of Ireland. He is credited also with the design of the O'Connell monument at Ennis, completed by a local contractor, William Carroll, in 1865; the monument dominates the centre of the town and consists of a tall column on a massive base, with a figure of O'Connell sculpted by James Cahill on top, and is believed to be one of the first such monuments raised by nationalists to a public figure.

Neville carried on a large private practice in architecture but was often in competition with another Dundalk-based architect, John Murray (1807–82), for commercial work and ecclesiastical commissions in the 1850s and 1860s. His country house projects included Shanlis House (1858), Dowdstown House and Cardistown House (1865), all in the Ardee area. He designed various shops and warehouses in Dundalk and Drogheda, including a substantial new wine store with an elaborate façade at Clanbrassil Street, Dundalk, which was noted and illustrated by the *Irish Builder* in 1868. He was responsible also for chapels and convents for the Sisters of Mercy (of which his two daughters had become members) at Mount St Vincent, Limerick (1851), Ardee (1858) and Dundalk, Omagh and Roscommon (1859). In the case of the latter, there was a substantial dispute with the contractor when Neville refused to certify proper completion of the work and withheld payment of various balances and extras; the whole matter was likely, according to one report, to be referred to 'the gentlemen of the long robe'. Neville and Murray were awarded the prizes in 1858 when designs were invited for a new corn exchange and assembly-room building at Dundalk but the commission was awarded to Bellamy & Hardy of Lincoln; however, when the winning design was found to exceed an acceptable cost limit, the commission for the building (now the town hall) went to Murray in 1859.

Unlike many of his county surveyor colleagues, Neville did not become involved in the planning and construction of railways, even when the railway mania was at its height. Instead, he rightly saw it as his duty to defend the public interest – and the interests of road users in particular – against the railway companies. He took strong action against the Dublin & Belfast Junction Company and its engineer, Sir John MacNeill, in the 1840s when he found that some of the bridges on the main line to Belfast would interfere with the free flow of road traffic; he sought and obtained court orders requiring a number of these bridges to be taken down and others to be altered. Twenty years later, when the Dundalk & Enniskillen Railway proposed to extend their line to the quays at Dundalk, Neville told the local town commissioners and the grand jury that great injury would be caused to the streets of the town and urged them to take energetic measures to prevent the company's bill from passing into law.

Neville wrote extensively on technical subjects. He appears to have had a special interest and competence in hydraulic engineering arising, presumably, from his early work on the river Shannon. His *Hydraulic Tables, Co-efficients, and Formulae for Finding the Discharge of Water from Orifices, Notches, Weirs, Pipes, and Rivers* was published in London in 1853 and, according to himself, had a considerable circulation, justifying a second edition in 1860 and a third revised and extended edition in 1875. In 1858, he was confident enough of his expertise in these matters to write and publish a pamphlet entitled *The Hydrodynamics of Blue Books and of the Metropolitan Drainage Reports* in which he commented critically on 'the new theory for the discharge of water through notches and submerged apertures propounded in the Metropolitan Main Drainage Report' which had been prepared earlier that year by some of the foremost experts in the field, including Thomas Hawksley, George Parker Bidder and Sir Joseph Bazalgette. He was also the author of a lengthy paper on earth pressure and retaining walls, containing elaborate and detailed mathematical formulae, which was published in the first volume of the *Transactions of the Institution of Civil Engineers of Ireland* in 1845; another paper on the same subject was published in the *Proceedings of the Royal Irish Academy* in 1847 and subsequently discussed in the columns of the *Civil Engineers and Architects Journal.*

In the early 1880s, Neville and his assistant, Eugene McSweeney, developed proposals for a Dundalk water supply scheme but was unsuccessful in his efforts to have these accepted by the local town commissioners; the scheme involved taking a supply from Camlough Lake which was already being used to supply the town of Newry. Although the proposals were considered by Richard Hassard, the consulting engineer who was advising the commissioners, to be the best of the thirteen schemes on offer following an 1881 competition, the scheme ultimately adopted in 1883 was one put forward by Hassard himself; because of difficulties with the contractors and other problems, this was not fully completed until 1889.

Neville married Constance Cox at Bruree, Co. Limerick, in 1849, and had three sons and two daughters. While the family lived at Roden Place, and later at Jocelyn Street, Dundalk, Neville was listed in 1876 as the owner of 776 acres of land in Co. Tipperary. A one-time worshipful master of the local masonic lodge, he was elected a member of the Institution of Civil Engineers of Ireland in January 1845 and became a member of the Royal Irish Academy in June 1844, and a fellow of the Royal Geological Society of Ireland. After almost forty-six years service in Louth, he retired on a pension of £500 following the spring assizes in March 1886, due to 'old age and bad health' and went to live in Dublin with his son, William, a gynaecologist. He died at 71 Lower Baggot Street, Dublin, on Whit Monday, 10 June 1889, aged 76. He was certainly an effective and conscientious county surveyor but not the most popular with his colleagues, his employers, his clients, the road contractors or his subordinates. An excellent biographical note by Christine Casey in the *County Louth Archaeological and Historical Journal* (1985) suggests, in fact, that he was a cold and unemotional man, confident and assured by nature, with no time for false modesty nor even at times for diplomacy, and single-minded to the point of ruthlessness in the pursuit of quality and efficiency.

NOONAN, John Joseph (fl. 1910–21), was a native of Ballyhaunis, Co. Mayo, and had been employed as an inspector by the Congested Districts Board before his appointment as one of the Mayo county surveyors. The county council had invited applications for the post in their northern division in July 1910 at a salary of £400 a year (to include all expenses) and the thirty-six applications were referred in the usual way to the Local Government Board with a view to the holding of a qualifying examination. The examination was duly held in December and in February 1911, when the list of qualified candidates was received, the number remaining had been whittled down to eight. County councillors met privately in a local hotel immediately before their March quarterly meeting and settled on the appointment of Noonan who was placed second on the list of qualified candidates. When the council formally assembled later on in the day, the appointment process was a mere formality with no candidate other than Noonan being proposed. Nevertheless, Noonan's proposer thought it necessary to put on record the reasons for his selection – he and his people were always nationalists, he was a native of the county and he was an exceptionally brilliant young man of high attainments. Noonan took up his new post on 1 April 1911 and held office for ten years.

By 1921, relations between Noonan and the county council had virtually broken down. He was accused at the annual meeting in June of 'impudent and unmanly conduct', lecturing the council, and disgracefully neglecting his business. This led to a public inquiry in September 1921 by inspectors of the Dáil Éireann Department of Local Government and a letter was issued later that month by W.T. Cosgrave, Minister for Local Government, recommending the surveyor's immediate dismissal. The council duly voted to dismiss Noonan at a meeting on 17 October, citing unjustifiable delays, repeated insubordination, gross irregularities and continuous neglect of correspondence.

O'BYRNE, Joseph Thomas (1899–1987), was born on 13 March 1899 in Co. Wicklow, and educated at Mungret and Clongowes Wood Colleges. While studying engineering at University College, Dublin, he became involved in the independence movement but successfully completed his studies for the BE degree in 1921. In 1921–3 he was resident engineer on a Wicklow town waterworks improvement scheme and in 1924–5 was employed by Dublin Corporation in connection with preliminary planning of the Liffey reservoir project. Subsequently he was a temporary assistant county surveyor in Co. Carlow and, from 1925 to 1938, held a similar post in Wicklow as well as the positions of Wicklow town surveyor and engineer to the board of health.

Following the retirement of the Wicklow county surveyor, **Stephen Gallagher**, in 1938, O'Byrne (who had passed the qualifying examinations for county surveyorships as far back as October 1927) was appointed county surveyor by vote of the elected council but without reference to the Local Appointments Commission. He served in the post on an acting basis for over two years before his appointment was regularized with effect from July 1940 following a competition arranged by the commission and in which there were twenty-nine candidates. He was the last man to attain the rank of county surveyor and he served until reaching the age limit on 13 March 1964. A keen yachtsman, he was active also in the local lifeboat service and in the preservation of national monuments. He died at the age of 88 on 9 September 1987. An appreciation written by one of his former colleagues in Wicklow, Peter O'Keeffe, described O'Byrne as a dynamic and progressive engineer who implemented a considerable programme of public works in the county.

O'CONNOR, Richard Francis Mary (1875–1940), son of John O'Connor, a Cork timber merchant, **was** born in Cork on 10 January 1875 and educated at Christian Brothers' College and at Clongowes Wood College. He attended Queen's College, Cork, between 1896 and 1899 and graduated with BA and BE degrees. He trained for two years under J.C. Inglis, chief engineer of the Great Western Railway Company, at its Paddington headquarters and worked on surveys and as resident engineer on some branch line extensions. Between 1901 and 1903 he was assistant engineer with Messrs Pauling & Co., contractors on the Northolt & High Wycombe Railway, and for the following two years was engineer and agent for Messrs C.H. Walker & Co., contractors, on the reconstruction of the harbour at Watchet in Somerset. In partnership with John Martin of Drogheda, O'Connor then moved to public works contracting in Ireland, working between 1905 and 1909 on pier and quay extensions at Baltimore, Courtmacsherry and Castletownbere, Co. Cork; the Howth sewerage scheme, including 1,100 yards of sea outfall; and a new pier at Cleggan, Co. Galway.

In June 1909 O'Connor was appointed county surveyor for the northern division of Co. Cork, centred on Mallow, while **J.T. Murphy** was appointed at the same time to take charge of the central division, based on the city. Unlike so many of the other appointments made by county councils in the previous ten years, Murphy and O'Connor were appointed solely on merit, having taken first and second places, respectively, at the qualifying examinations held by the Civil Service Commissioners in April 1909. In 1925, following the retirement of **R.W.F. Longfield**, surveyor for the west riding, a new division of responsibility was arranged, with O'Connor taking on the area north of the river Lee at an increased salary of £750. Four years later, after Murphy's death in November 1929, O'Connor assumed responsibility for the whole county – the first county surveyor to do so since the office was introduced in 1834. In 1931, he was allowed two chief assistant county surveyors, each of whom had county surveyor qualifications and who carried much of the responsibility for the northern and western areas. He was still in office as county surveyor when he died at his home, Clydaville, Mallow, after a short illness, on 2 March 1940. After his death, the engineering organization was revised again; the two chief assistants – **John T. O'Donnell** and **Gerard Fogarty** – were promoted to fill two posts of acting county surveyor, a situation which continued until the late 1940s when three deputy county engineers were appointed.

O'Connor was active in the debates on the development of road policy and investment programmes in the 1920s as evidenced by papers on *Irish Free State Transport Problems* and on *The Wear and Tear of Roads* which he delivered to the Institution of Civil Engineers of Ireland in 1927 and 1929. He argued strongly that vehicles should be designed to fit the road, rather than the reverse, that the use of containers capable of being easily transferred from road to rail should be promoted, and that six-wheeled vehicles would be less damaging to road surfaces. At the same time, he favoured the abolition of speed limits and a ban on solid tyres on larger heavier vehicles. He made substantial progress towards modernizing the huge network of roads in the county and was responsible for some unusual projects. These included the widening and reconstruction in concrete of the Carrigrohane 'straight road' on the outskirts of Cork city in the 1930s, making it a suitable location for motor cycling events, and the construction of the spectacular Healy Pass through the mountainous Beara peninsula, from Adrigole on Bantry Bay to Dereen on the Kenmare River, a distance of about eight miles in all. Efforts to build a road across the peninsula linking Cork and Kerry had begun in famine times but the work was abandoned before the difficult summit section was attempted; the completed sections were allowed to become derelict and only two miles of the 'Kerry Pass' remained in use by 1930. Tim Healy, the first Governor General of the Irish Free State and a native of Bantry, had been attempting for almost thirty years to have the route completed when, in the late 1920s, his representations eventually led to the allocation of a Road Fund grant of £7,680 for the project. Construction work began in January 1931 under the general supervision of O'Connor and his chief assistant, Gerard Fogarty, and was pressed ahead with remarkable speed. Explosives were used extensively near the summit to blast away some 20,000 cubic yards of rock which was subsequently crushed and used to make up the road surface. The breakthrough to the Kerry side was achieved in April 1931 and the first motor cars were able to use the new road in the following August. The grades on the road are generally 1 in 20 but are as steep as 1 in 10 near the summit, 950 feet above sea level, where a concrete arch spanning 40 feet was constructed over a gorge on the Cork side. The new route was named 'the Tim Healy Pass' in memory of Tim Healy who died in March 1931.

O'Connor was a man of great vision who was interested in promoting the development of Cork on many fronts, especially in the field of transport. One of his many innovations was the completion of a special motor racing circuit near the city and he was directly involved himself in the organization of races there. He also deserves to be remembered for his pioneering work towards the establishment of an airport in the Cork area. This began in 1933–4 when O'Connor presented detailed proposals to the Cork Airport Joint Local Authorities Committee for a major international airport and seaplane base on the eastern side of Cork harbour, involving the reclamation of some 460 acres of tidal mudflats. The idea was that English and European airlines could connect at Cork – the most westerly harbour in Europe – with steamship and air services across the Atlantic. O'Connor also initiated efforts to set up an Irish airline to serve routes into and out of Cork and, in this context, has been credited with originating the title *Aer Loingeas* (literally, air fleet). Interests in the Cork area campaigned for some years for government support for O'Connor's proposals but the scheme began to fall apart when Aer Lingus was set up by the government in 1936 as the Irish national airline, using a modified form of O'Connor's title, and when Foynes, on the Shannon estuary, was designated in the same year as the transatlantic seaplane base. There was a further setback that year when a British firm of consultants found that O'Connor's site at Belvelly was far from ideal and that construction of an airport there would be a very difficult and expensive undertaking. However, the consultants endorsed an alternative site near Midleton which had been put forward by O'Connor and reported that this would be more than adequate for all future airport requirements. By 1939, the government had agreed to provide a grant of 45 per cent of the cost of building an airport at the new site if the local authorities met the balance, but the plans had to be put in abeyance on the outbreak of war later that year. When eventually an airport was opened at Cork in October 1961, it was located on a different site, southwest of the city, chosen on foot of survey work in the 1940s by the Board of Works.

The development of a new county council headquarters building was another project initiated by O'Connor but which he did not live to see completed. In 1929, when there were proposals for a new city and county hall on the site of the former city hall which had been destroyed by fire in 1920, O'Connor expressed serious reservations about the suitability of the section of the building, fronting the laneway serving the Cork & Bandon Railway station, which was to be allocated to the county council. The council subsequently withdrew their support for the project and their main offices remained in the courthouse. Within a few years, however, there was growing concern about the inadequacy of the offices for transacting the council's business, with increased staff numbers accommodated in an overcrowded courthouse and in a variety of unsuitable buildings. O'Connor was then instructed to report on buildings which might be adapted for use as council offices and on possible sites for new county buildings. By 1939, a site had been acquired and, just before his death in 1940, O'Connor had moved on to the design stage and was seeking the appointment of an architect to assist him. Subsequently, a variety of factors combined to delay the project and a new county hall – a sixteen-storey building designed by county architect P.L. McSweeney – was not completed until 1968.

Obituaries described O'Connor as an ideal type of public official, with a genial and kindly personality and an enthusiastic interest in his profession. He appears to have had some marginal involvement in political matters at one stage: in 1922, in the early months of the civil war, when Eamon de Valera was travelling covertly through the southern counties, his diary for 11 August which describes the day as 'one of the most if not the most miserable days I ever spent', concludes by recording 'slept at O' Connor's, county surveyor'. Mary C. Bromage in *De Valera and the March of a Nation* (London, 1956) records that O'Connor had attempted to persuade De Valera to prevent the destruction of the large ten-arch masonry viaduct carrying the Dublin-Cork railway over the river Blackwater at Mallow but met only obduracy – 'it will save thousands of lives' was the reply. The viaduct was completely destroyed by explosives on 9 August, effectively cutting off rail connection between Cork city and the rest of the country.

O'DONNELL, John Thomas (fl. 1894–1957), was born on 24 December 1894 and educated at Mount Mellary College. He served an apprenticeship at Messrs Green's Engineering Works in Leeds from 1913 to 1915 and was employed as assistant engineer at the Royal Aircraft Factory at Farnborough and at the Royal Ordnance works at Coventry until 1918. He studied engineering at Sheffield University in 1919-24 where he gained degrees in mechanical and electrical engineering, and took up a position as assistant to the borough surveyor in Clonmel in 1925. After a period on the staff of Messrs Siemens Bauunion at the Shannon Scheme, he returned to Clonmel as acting borough surveyor in 1927 and became borough surveyor in Wexford in 1929. Having successfully sat the county surveyor examinations in 1930, he moved to Cork in March 1931 as chief assistant county surveyor. After the death of **R.F.M. O'Connor** in January 1940 O'Donnell initially became one of two acting county surveyors and, from April 1947, one of three deputy county engineers who served the county in the absence of a county engineer post. He retired in 1957.

ORCHARD, William Patterson (1850–1920), was born at Islington, London, on 29 January 1850 and gained a BE degree at Queen's College, Belfast, in 1869. He was a pupil of Charles Nixon in London until 1873 when he became an assistant to the county surveyor for West Sussex. He was appointed county surveyor for the northern division of Mayo in December 1886 and transferred to the service of the new county council in 1899. Long before that there were recurring complaints about the condition of the roads in his area, with Orchard in turn complaining about the pressure of work arising from major bridge projects at Crossmolina and Cloonacannana in the early 1890s.

Orchard was in serious difficulty with his council by 1909 because of the state of the roads in his area; he was accused of mismanagement and of allowing the roads to become virtually impassable, with the tyres on the mail-car being cut to ribbons on the road to Belmullet. In the following March the county council considered a report by one of the Local Government Board's inspectors in which the deterioration in road conditions was said to be almost wholly attributable to the failure of the county council's officials to devote proper supervision to the work of the contractors; as evidence of Orchard's neglect, the inspector noted that his diary showed that in the last eight months he had spent only twenty-three days on inspections in his district even though the assistant county surveyor was incapacitated. A covering letter from the Local Government Board was even more explicit in its criticism than the inspector had been: it expressed serious concern about the dangerous condition of the roads, attributed responsibility largely to the county surveyor and his assistants and suggested that the council should appoint a small committee to consider what changes might be necessary in the county surveyor's department. The committee reported in April 1910 and recommended that Orchard should be allowed to resign on health grounds and be awarded a pension. The surveyor agreed to this while continuing to perform the essential duties of the office until a successor was appointed. He eventually left the service of the council in March 1911 and died on 19 October 1920. He is buried at Mount Jerome cemetery in Dublin.

O'REILLY, Henry J. (fl. 1884–1929), was born on 15 May 1884 and was educated at Kevin Street Technical School in Dublin. He was awarded the diploma of the Royal College of Science in 1907 and subsequently gained BE and BSc degrees. He was an assistant county surveyor in Wexford in 1909–10 and an assistant engineering inspector with the Congested Districts Board until 1914.

When **R.N. Somerville** gave notice in May 1913 of his intention to retire from his post as Cavan county surveyor, the county council decided that the new appointee should serve on probation for two years at a salary of £400 and, if found to be efficient, would be granted increments of £20 bringing the total to £500. O'Reilly canvassed actively for the post and was elected on 28 February 1914, having gained top marks at the qualifying examinations held in the previous month. He took up the appointment on 1 May 1914 but had a difficult early period in Cavan. His relationship with the county secretary was strained and difficult, and dissatisfaction with the condition of the roads led the county council to seek to have him removed from office in April 1916. In the following month, an inquiry into the working of the county surveyor's department was held by **Dr P.C. Cowan**, the Local Government Board's chief engineering inspector who declared at the outset that he wanted to establish why 'so many roads in the county are a heartbreak to the ratepayers'. Having considered Cowan's report, the board concluded that there was not sufficient evidence to warrant dismissal of the surveyor but warned the council that they should take action to resolve the personal differences between him and the secretary.

Five years later, in July 1921, an inspector from Dáil Éireann's Department of Local Government reported that the Cavan roads were the worst in Ireland and this probably contributed to the decision of the council, at a time when many of his colleagues were being granted substantial salary increases to compensate for the price inflation of the war years, to deny O'Reilly any increase. In 1924, a report from **James Quigley**, the Department's chief roads engineer, restated the view that the roads in Cavan, with the possible exception of one or two counties, were the worst in Ireland. He went on to say that 'there is not a single mile, save a mile or two at Virginia, where a motor car can be run at normal speed without risking broken springs. The usual road … is a continuous succession of deep potholes filled with water along which there is a heavier traffic of motor cars and motor lorries than one would expect in such an inland county, remote from large towns'. Quigley had been told by the county surveyor and his assistants that the allocation of insufficient funds for maintenance work was the root cause of the problem but he questioned this, claiming that a totally inadequate return in road upkeep had been achieved for the expenditure involved. He concluded his report by advising that the county surveyor should be given a year to improve matters and if there was no real improvement, some arrangement should be made 'whereby the Cavan people will be free to try another surveyor'.

O'Reilly continued in office in Cavan until May 1929 when, after yet another adverse report from one of his engineering inspectors, **T.C. Courtney**, the Minister advised the county council that he could come to only one conclusion – 'that apart from his educational attainments, Mr O'Reilly could not be regarded as possessing the other qualifications essential for the office of county surveyor and that no success could attend the work while it remained under his control'. He demanded, therefore, that arrangements be made without delay for the appointment of a suitable replacement and warned that no further grants for road works would be allocated to the county until this was done. Councillors criticized what they described as the unprecedented course adopted by the Minister and O'Reilly initially refused to resign but he soon bowed to the inevitable and, although he was only 45 years old, was allowed to resign on pension on the basis that he was not in good health. In retrospect, it seems that O'Reilly may have been unfairly held to be personally accountable for the dreadful condition of the roads as none of his successors managed to make any real impression on the Cavan potholes until the 1990s when special state grants were made available to finance complete reconstruction of the county's network of roads.

O'SULLIVAN, Ignatius J. (fl. 1909–27), was a native of Cork and became surveyor for King's County in 1909, having previously held an appointment with the Caledonian Railway. Following the sudden death of the previous office-holder, the position had been advertised on 14 August at a salary of £400 a year, up to £150 of which was expected by the councillors to be absorbed by expenses including, as one of them put it, the upkeep of a horse and a man. The council decided also that the new appointee would not be entitled to engage in private practice and would be expected to live in or near Tullamore, and they agreed also that 'all things being equal', preference would be given to a candidate with a knowledge of Irish. There were six candidates and O'Sullivan (who had been the choice of the majority of the councillors at a preliminary private meeting) was elected unanimously when the appointment came to be made at the quarterly county council meeting on 25 August. One of the unsuccessful candidates, **W.F. Barry**, was already a county surveyor and two of the others, **J.P. Punch** and **J.J. Hannigan**, gained county surveyor appointments in the following three years.

According to a report from a Dáil Éireann inspector towards the end of 1921, O'Sullivan carried out hardly any road inspections and attended the office only two days each week, arriving by the 11.00 a.m. train from Dublin, and returning on the 4.00 p.m. train. At the end of June 1927 he applied for three months sick leave and an acting county surveyor was appointed to carry out his duties. Shortly after his resumption of duty he died on 12 November 1927 having suffered severely, according to the *Midland Tribune*, from 'the worries of office and the unjust criticism to which he had been subjected'. O'Sullivan left a widow and four children and was buried at Clonminch, Tullamore.

OTTLEY, Drewry Gifford (1845–96), eldest son of Drewry Ottley MD, was born at Bedford Place, London, on 20 October 1845. He was educated initially at Pau in southwest France, where his father was well known as a physician, and later at a private school in Bath and at King's College, London. In 1863 he joined the engineering works of the Midland Railway Company at Derby where he served a pupilage for three years and afterwards worked for a year as assistant engineer. He was appointed as assistant engineer in the Indian public works department in October 1868 and was involved from 1870 in surveys for the Indus Valley State Railway. He was subsequently placed in charge of the company's large workshops and of the locomotive and wagon stock. In 1876 he was transferred to the Punjab Northern State Railway as executive engineer and held that post until 1879 when he retired from the Indian government service.

Ottley practised privately in England and in the United States in the early 1880s, apparently without great success. He was appointed in October 1884 as Leitrim county surveyor and served uneventfully for some twelve years. In addition to his ordinary duties he acted as an arbitrator for the Board of Trade in connection with the Cavan, Leitrim and Roscommon Light Railway system, built between 1885 and 1887. His ability as an engineer was said to be 'considerable'; he was elected an associate of the Institution of Civil Engineers (London) in 1875 and was transferred to the class of member in 1885. It was also said of him that his unvarying kindness and consideration to all who worked under him, as well as his pleasant manner, made him universally popular. He was only in his fifty-first year when he died at his residence, Cartown House, Carrick-on-Shannon, on 27 September 1896 after a few hours' illness.

OWEN, Henry H. (fl. 1817–49), was born in 1817, one of the several sons of Jacob Owen, principal engineer and architect at the Board of Works, who followed their father into careers in engineering and architecture. He entered Trinity College, Dublin, in 1833 at the age of 16 but did not graduate; instead, he became a pupil in his father's office and, by 1839, was engaged by the Board of Works in the supervision of contract works on roads in Kerry. In March 1840 he took up an appointment as a drawing and measuring clerk at a salary of two guineas a week at the board's offices in Dublin.

Owen was appointed county surveyor for Queen's County in August 1841 following the dismissal a few weeks earlier of **Edward V. Forrest** and was transferred at his own request to Waterford on 16 August 1845 to replace **Charles Forth** who had died. Although the original plans may have been prepared by **John Walker** who served in Waterford from 1840 to 1844, Owen was largely responsible for the completion in 1847 of Victoria Bridge, a five-span masonry arch bridge over the river Blackwater at Cappoquin which replaced a seventeenth-century wooden structure. In September 1849 he sought a transfer to Co. Down in an exchange with John Fraser to which the latter had agreed but this did not come into effect.

Owen resigned in September 1849 and no details of his subsequent career have come to light. He married Maria O'Shaughnessy and they had two children, a son who was born at Dungarvan in May 1847 and a daughter, Mary, who married Edward Malins MD in 1869.

PATTERSON, Robert Harold Scott (1900–57), was born in Co. Fermanagh on 21 January 1900. After service with the 28th London Regiment (Artist's Rifles) in 1918–19, he graduated with a BSc Eng degree from Queen's University, Belfast, in 1922 and was later awarded a DSc. He joined the staff of Down county council, initially as resident engineer on a new road construction project, and became deputy county surveyor under **J.G. Wilkin** in 1939. At the beginning of the Second World War he joined the Royal Engineers but at the request of the county council was released from the army with the rank of major in 1941 to resume his former duties. He became Down county surveyor in 1945 and had responsibility for a major post-war road construction programme, including the Belfast outer ring road on the eastern side of the city and the widening and improvement of the Belfast–Bangor road. Patterson was regarded as an outstanding county surveyor, with a keen legal mind. He took a great interest in the training of engineers and was active in the various engineering institutions and other representative bodies. He died at Down Hospital, Downpatrick, on 31 January 1957, at the age of 57.

PERRY, James (1845–1906), son of Samuel Perry and Agnes Smith, was born into a Presbyterian family at Garvagh, Co. Derry, on 7 September 1845. He was educated at the local national school, studied at Queen's College, Belfast, and at the Royal College of Science in Dublin, and graduated from Queen's with a first-class honours degree in engineering in 1873. He was awarded the ME degree at Queen's in 1882. Perry obtained his practical training at the engineering workshops of the Belfast & County Down Railway and subsequently worked on water supply projects in Beirut (then part of Syria) and in Belfast. In October 1877 he took first place at the county surveyor examinations and was appointed to Roscommon on 18 December of that year. He transferred to the western division of Galway in July 1882 and acted also as surveyor for the county of the town of Galway. With his Scottish wife, Martha Park, Perry lived at Wellpark on the east of the city, near the race course. The couple had one son, who retired from the British army with the rank of major and five daughters, one of whom, Martha, married Edward William O'Flaherty Lynam (1885–1950), a nephew of two of her father's county surveyor colleagues, **P.J. and F.J. Lynam**, and who was superintendent of the map room at the British Museum until 1950.

In west Galway Perry assumed responsibility for the maintenance and repair of a network of more than 1,000 miles of public roads as well as numerous bridges, public buildings, piers and coast protection works, but he still found time to carry on an extensive and varied private practice in architecture and civil engineering. Among his undertakings were new or extended waterworks at Galway, Clifden, Roundstone and Bundoran, extensive alterations at the Clifden workhouse and a new dispensary and doctor's residence near Tuam. In 1901 he completed a new school at the Dominican Convent, Taylor's Hill, Galway, at a cost of about £10,000; this was a large square building, four storeys high with a balcony surrounded by a wrought-iron railing on the top floor, separate cubicles for boarders, and classrooms, science laboratory, laundry and large dining room.

Perry acted as consulting engineer for the Board of Works and assisted the Piers and Roads Commissioners in 1886–8 in the selection and design of a variety of relief work projects involving the construction of piers and harbours, roads, bridges and causeways linking islands to the mainland. Completed projects included a dry-stone granite causeway to Mweenish Island, 1,450 feet long; a causeway at Lettermullen/Goruma which included an iron revolving bridge, on a central pivot, giving two twelve-foot openings for small boats; bridges at Kylesalia, Acoonera, and Screeb; and a five-mile length of road across the neck of the Carna peninsula. For the Congested Districts Board, he designed and supervised similar works in the early 1890s, including the construction of what the board described as 'the heavy and costly work of constructing a line of causeways between the island of Gorumna and the intervening islands, and the mainland of Galway'. The entire project cost about £8,000 and gave direct access to the mainland to 3,687 people. On the Aran islands, where great difficulties arose in dry seasons when wells dried up, he built concrete water storage tanks, with piped gravity supplies to a number of centres.

Perry was a man of exceptional energy, ability and business capacity. In 1886–7 he acted as substitute lecturer in geology at Queen's College, Galway, during the illness of his brother-in-law, Richard John Anderson, professor of natural history, geology and mineralogy. In 1888, in conjunction with his younger brother, John Perry ME DSc FRS (professor of mechanics and mathematics at the Royal College of Science, London, a well-known writer on engineering and mathematical subjects and the inventor of the navigational gyroscope), he built and equipped an electric power station in Galway, initially using hydropower, to supply private consumers. In 1897 the Galway Electric Light Company was formed, with Perry as managing director and chief engineer, and the operation continued to expand, undertaking the public lighting of the city of Galway in 1905 when Perry declared that 'gas for lighting purposes is on its deathbed'. Power was supplied by the company at a cost of three pence per unit for 'motor work', and five pence for lighting – relatively inexpensive prices at the time but made possible by the low price of the coal then used for generating purposes. Some engineers in Ireland and in the rest of the United Kingdom were sceptical about this initiative by a local authority engineer but Perry argued that electrical engineering 'is forced upon us, and we have got to do our duty in regard to it as we have done our duty by the other human refinements which have preceded it'. The undertaking developed by the Perrys was absorbed by the Electricity Supply Board in 1929 after completion of the Shannon Scheme.

Perry died in office on 28 November 1906 after a brief illness and was buried at the city cemetery in Galway. Two weeks later, the *Connaught Champion* reported that a special meeting of the county council, as 'a graceful tribute' to his memory, had unanimously elected 'his brilliant young daughter', 21-year-old Alice Jacqueline, to replace him temporarily. In noting the appointment, the *Irish Builder and Engineer* suggested that 'having regard to the lack of initiative displayed by some of the county surveyors' it might be a case of 'angels tread where fools fear to venture'. After her early education at the High School, Galway, Miss Perry had entered Queen's College, Galway, in 1902 and registered for an arts degree but transferred to engineering in the following year. In her first year of study she won first rank awards for engineering and drawing and went on to graduate with first-class honours in October 1906, the first woman engineering graduate in Great Britain or Ireland. Some months after her appointment, the *Connaught Champion* reported that 'the many and arduous duties of county surveyor have never been better or more faithfully discharged than since they were taken over by Miss Perry'. Based on her own inspection visits and those conducted with her late father, she submitted an impressively detailed report to the county council in February 1907 setting out the condition of the roads, bridges, piers and courthouses in the western division and making proposals for improvement works, and she seems to have had no difficulty in defending any of her proposals which were questioned by the more experienced councillors.

When the permanent post was advertised by the county council in February 1907, the salary offered was only £300 a year, plus expenses, much less than the amount which had been paid previously. Alice Perry, together with seventeen men, applied for the post but she was not successful; four of the candidates were formally proposed and Alice gained only three votes as against seventeen for **P.J. Prendergast** who won the appointment. Leaving aside the extraneous considerations which influenced the majority to vote for Prendergast, her failure may also be attributed to the fact that she was less than the prescribed minimum age of 26 and had limited experience; she may also have had difficulty in meeting the Irish language qualification which the county council, under the influence of the Gaelic League, had laid down for all new appointees. Alice Perry was among the unsuccessful candidates again in May 1907 when a county surveyor vacancy in east Galway was filled. Nevertheless, she holds the distinction, one hundred years later, of being the only woman to serve, even temporarily, as a county surveyor or county engineer. She and her sisters moved to London in 1908 and she subsequently took up an appointment as an inspector under the Factories Acts. In 1916 she married an Englishman, Robert Shaw, who was killed in action in France in October 1917. She gave up her engineering career in 1921 and moved to Boston where she devoted herself to the Christian Science movement and achieved distinction as poetry editor of the *Christian Science Monitor*. Her own poems were published in various church journals and magazines and seven books of collected poems were published in Boston between 1930 and 1961, among them, *The Children of Nazareth and Other Poems* (1930), *The Morning Meal, Collected Poems, Mary in the Garden and More Poems, One Thing I Know and Other Poems* (1953) and *Women of Canaan and Other Poems* (1961). Most of her poems had spiritual themes, some based directly on scriptural quotations, but there were a few which reflected her former life in west Galway. In one of these, Still Life (included in *One Thing I Know and Other Poems*), she imagines herself back in the Joyce Country 'where I am at home' and demonstrates that her engineering training and experience had not been forgotten:

> ... These are good roads;
> These are good granite roads;
> Not like the roads of East Galway
> Made of limestone; white, soft and useless.

Alice Jacqueline Perry-Shaw died in Boston, aged 84, on 21 August 1969.

PRATT, James Butler (*c.*1822–86), was the eldest son of James Pratt, a retired army captain, and grandson of James Butler Pratt of Drumsna, Co. Leitrim. He entered Trinity College, Dublin, in October 1840 and gained a Diploma in Civil Engineering in 1844 as one of the second class of graduates of the new school of engineering. He was awarded a BA degree in 1846 and an MA in 1850 and was elected to membership of the Royal Irish Academy in 1854.

Pratt's first employment was with the Board of Works as an inspector of drainage during the famine years in Limerick and Clare but he resigned in early 1847. At that stage, the organization of the works was falling apart because of the pressure of numbers seeking employment, the constant threats of violence to those responsible for organizing and managing the works and, according to Pratt's reports, the reluctance of men to work on drainage as against road works. He was resident engineer on the Dublin & Bray Railway in 1854 and was later employed as contractor's engineer on the Limerick & Ennis Railway, but before completion of the line in 1859 he was appointed Leitrim county surveyor, having qualified at the 1851 examinations. He took up duty in January 1857, becoming the first county surveyor to hold an academic qualification in engineering. He seems to have continued his railway work at least until the end of 1858 when he was able to report that the line was in running order from Limerick to Clarecastle, just outside Ennis.

While Pratt was to serve as county surveyor for twenty-seven years, he is best remembered for his close association with the thirty-eight-mile-long Ballinamore and Ballyconnell Canal, described in 1880 as 'one of the most shameful pieces of mismanagement in any country'. The canal was constructed between 1846 and 1859 by the Board of Works, at a cost of some £229,000, to provide a navigation from the Erne to the Shannon, to improve drainage, to alleviate flooding, and to provide employment during the famine. Boats began to pass through the canal in the late 1850s, one of the first being a screw-steamer which Pratt himself was authorized to operate between Carrick-on-Shannon and Belturbet; Pratt used this steamer to draw barges carrying coal from Arigna, the only occasion, apparently, on which the canal was used by commercial traffic.

When the completed navigation was about to be handed over to trustees in July 1860, Pratt was appointed engineer and secretary. In the following November he sent a detailed report to the Board of Works listing the numerous defects which were already apparent in the canal and detailing the cost of putting the navigation into proper repair. The board, however, took the view that Pratt's claims were exaggerated, if not largely unfounded, and refused to accept responsibility for carrying out and financing the necessary works, and the trustees had no resources to do so. Pratt subsequently claimed to have done what he could to keep the navigation open until 1865 but maintenance and other work was run down some time after that due to the limited traffic – only eight toll-paying boats were reported to have passed through the canal between 1860 and 1869. Repeated efforts by the trustees, who had no power to operate boats themselves, failed to interest anyone in taking a lease of the waterway and operating it. One of the trustees, who had himself brought a steamer from the river Shannon to Lough Erne in 1868, told an official inquiry some years later that 'only through the kindness of the people and Mr Pratt, the engineer, was I able to get through the canal. They took the greatest trouble to get the water from one reach to another to float me down. But my coming through it is no proof, for it took me three weeks to get through'.

The canal was allowed to become derelict in the 1870s and it was 'absolutely useless' and, to all intents and purposes, closed according to the report in June 1878 of a parliamentary committee chaired by Viscount Crichton MP which had been established to inquire into the activities of the Board of Works. Although the board of trustees lapsed some years later, Pratt continued to be listed as their engineer and secretary until his death in 1886. Over one hundred years after it had been constructed, the canal was restored and reopened to traffic as the magnificent Shannon-Erne waterway.

Pratt's son, Mervyn James Butler Pratt (b. 1857), followed him into engineering. He entered Trinity College, Dublin, at the age of 15 and graduated in arts and engineering in 1879 having won a gold medal moderatorship and a prize in experimental science. After two years of laboratory work, Pratt became assistant to his father on relief works in Leitrim before going to work with J.R. Manning on improvement works on the river Shannon at Jamestown, Banagher and Killaloe. He was involved in the construction of the Killeshandra branch of the Midland Great Western Railway up to the time of his death, following surgery, on 16 March 1884 at the early age of 26.

J.B. Pratt retired from his county surveyor post in July 1884 owing to failing health, having carried out his duties 'faithfully, and to the satisfaction of the grand jury and the county at large', as the *Leitrim Advertiser* put it. He was granted a pension of £180 a year but did not live long to enjoy it. He died at 25 Lower Fitzwilliam Street, Dublin, on 1 February 1886, aged 64.

PRENDERGAST, Patrick Joseph (1853–1911), was born at Clifden, Co. Galway, in March 1853. He graduated in engineering at Queen's College, Galway, and worked initially on railway construction projects for the Great Northern and Midland Great Western Railway Companies. When he left the service of the GNR in February 1886 after nine years, he was presented by his friends on the staff with a handsome engraved gold watch and testimonials to his courtesy and genial bearing. In 1889 he was resident engineer on the construction of a bridge in masonry and steel in north Mayo and by 1892 had become town surveyor in Athlone. In addition to the usual work on roads, water and sewerage and public housing, he designed a market house to be erected at the Market Square.

When **James Perry**, county surveyor for the western division of Galway, died unexpectedly in November 1906, there was much speculation in the local press as to who his successor was likely to be in an office which was said to carry 'a splendid salary'. (The salary actually fixed by the council for the new appointee was a modest £300 a year, with £100 for travelling expenses, and £60 for clerical and office expenses). The names of several possible contenders were openly mentioned and their merits and accomplishments publicly discussed. The county council moved quickly to fill the vacancy and the names of no less than eighteen candidates (including Perry's daughter, Alice, who had been acting county surveyor since the death of her father) were before them at their meeting in February 1907. The candidates came from a number of different counties and there were four from England, but when the voting was over, it was Prendergast, the local man, who gained the appointment in April 1907. His supporters on the council argued that, at 54 years of age, he was far more experienced than any of the other candidates but added, for good measure, that he had been a nationalist all his life and had the support of the archbishop of Tuam and of other bishops in the region!

Prendergast's term of office was a short one. He was forced to resign his position in October 1910, due to continuing ill-health, and he died on 1 February 1911, aged 57. He was the father of Lt.-Col. William Hillary Prendergast DSO (1895–1957) who, after a distinguished career in the royal engineers in India and Burma and with the Bengal-Assam railway, was professor of civil engineering at University College, Galway, from 1947 to 1957. The engineering career of another of his sons, John K. Prendergast ME, was cut short by death in 1936 while he was in charge of the reconstruction of a barrage on the Nile; he had worked in the late 1920s as the government's resident engineer responsible for certification of payments to the main contractors engaged on the construction of the Shannon Scheme.

PUNCH, John Philip (1880–1967), was born on 4 December 1880 near Fermoy, Co. Cork, second son of John Philip Punch and his second wife, Ellen O'Sullivan. He was educated locally at St Colman's College and at Clongowes Wood College. He was awarded a BE degree at Queen's College, Cork, in 1903 and subsequently worked for a year with one of the most prominent consulting engineers of the day, J.H. Ryan, on a variety of water and sewerage schemes. In 1905–6 he designed and carried out water schemes for Fermoy and Conna village in east Cork, and between 1906 and 1908 was a temporary assistant engineer with Dublin Corporation, working on main drainage. He then spent some time in private practice on his own account.

In February 1911, Carlow county council advertised the vacant office of county surveyor, offering a salary of only £300 a year inclusive of all expenses. Although the closing date for applications was 11 February, the county council could not make an appointment until 28 August when the results of the qualifying examinations held in July became available; at that stage, Punch was elected to the post by twelve votes as against eight for **J.J. Hannigan** who was to become Monaghan county surveyor in the following year. Punch took up duty on 1 October 1911 and served for more than thirty years.

The years from 1919 to 1923 were difficult ones for most county surveyors, due to the disturbed political situation and the virtual collapse in the finances of local authorities. In August 1922 Punch complained that the limited funds available for road works in his county were, in effect, being used to support a kind of outdoor relief: council employees, he believed, sometimes destroyed roads and bridges by night in order to create employment for themselves. Five years later, with peaceful conditions restored, Punch was still lamenting the inefficiency with which road works were conducted and, in particular, the fact that direct labour had still not been adopted universally in his county. He continued to have difficulties with road contractors in the 1930s, complaining that even those engaged for special works sometimes made no serious effort to start or complete the work on time. As he colourfully put it, the halcyon days, when tinkering at roads was regarded as a fruitful source of income for small farmers and publicans, had not fully passed away!

Punch retired on 31 December 1942 and returned to live in his native Fermoy. He died there on 14 February 1967.

QUIGLEY, James (1869–1941), was born on 30 January 1869 at Newbliss, Co. Monaghan. After his secondary education at St Macartan's seminary in Monaghan, he worked as a clerk in an oil company in Liverpool but soon tired of office work and joined the French Foreign Legion in 1892. After some years he deserted and worked his way back to Liverpool by ship from Port Said. He became a teacher in Liverpool but tired of that too, and returned to study at Ushaw College, Durham, and Queen's College, Galway; he completed the BE course between 1902 and 1906 but received a diploma rather than a degree because he had not matriculated. He was subsequently an assistant county surveyor in Monaghan, based at Clones.

When Meath county council invited applications for the post of county surveyor in March 1907, the *Irish Builder and Engineer* remarked that 'none but candidates possessing considerable local influence, political or personal, have the slightest chance of being appointed'. In the event, there were nine candidates for the post and Quigley was voted into office by the county council at a meeting in August 1907. His remuneration was to be £500 (to include the salary of his clerk and travelling and all other expenses) with increments of £20 a year up to a maximum of £600, subject to satisfactory performance. Whether Quigley's appointment was attributable to his qualifications as an engineer, or whether his canvassing of votes was simply more effective than that of his opponents, can only be a matter of speculation at this stage; the official record shows, however, that he received fewer marks than any of the other candidates who sat the qualifying examinations in July 1907.

Quigley claimed considerable success as a county surveyor – and there is ample evidence to support this. Writing in 1923, by which time his salary had been increased to £1,000 a year, he stated that when he took up duty sixteen years earlier he found the Meath roads to be the worst in Ireland but they were then the best. At the second Irish Roads Congress in Dublin in April 1911 he described how, following the success of 'a sort of tentative scheme that certain roads were to be made by direct labour ... there was not a single voice raised against the proposal to work all the roads in the county – some 1,600 miles – with direct labour. ... After three years, there was nobody that ventured to say we did not do a lot better than the contractors had done at the same money'. One of those who worked as a road labourer under Quigley during these years was the poet, Francis Ledwidge (1887–1917) who, ironically, died on the western front while engaged in road building.

By the spring of 1920 Quigley was complaining that his operations were being adversely affected by the enormous increase in costs and in wages during the world war: it was difficult to get men to undertake laborious quarry work while high wages could be earned for much lighter work elsewhere; it was hard to get coal for the rock drilling plant; and worst of all, it was practically impossible to get explosives for the smaller quarries due to restrictions introduced by the authorities in the light of the security situation. As a result of this, Quigley feared that the council could not keep pace with the extraordinary increase in motor car, motor lorry and traction engine traffic and he reported that the roads were becoming more pot-holed and worn than they had been only a year before.

With **R.H. Dorman**, Armagh county surveyor, Quigley was appointed by the British government to act on a committee entrusted with the task of investigating claims for war damage to roads in Ireland and, following the establishment of the ministry of transport in 1919, he was appointed in January 1920 to represent the county surveyors on an Irish sub-committee of the ministry's roads advisory committee. These appointments could be taken as evidence that Quigley was fully on the side of the authorities and the establishment during the period 1916–22 but this does not seem to have been the case. He was involved in the establishment of the national volunteers in Navan in 1914 and later that year became secretary of the Co. Meath board of the organization and helped to organize a parade of 2,500 volunteers at Slane in August. According to a note which Quigley wrote in 1927, he had 'risked my position and my livelihood all the way through' and he was always 'on the side of Ireland' whereas, he claimed, none of the other county surveyors had risked anything in the national struggle. He was, in fact, fortunate to escape with his life and liberty when he unwittingly became involved in the events of Easter Week, 1916. On 28 April, the Friday of that week, when the fifth (Fingal) battalion of the volunteers under Thomas Ashe and Richard Mulcahy was in action at the battle of Ashbourne (an engagement which led to the deaths of eight members of the Royal Irish Constabulary and the wounding of another fifteen), a section of the ambushers moved to a position in a deep ditch on the side of the main road north of the village so as to cut off any attempt at escape by the police. Colonel J.V.

Lawless, one of the participants, in an account of the battle published in *Dublin's Fighting Story* (Tralee, 1948), described Quigley's involvement:

> Having noticed a motor cycle standing on the opposite side of the road, I then observed a man in civilian dress crouched near it in the hedge, and was about to fire on him when he saw me, and stood up with his hands raised. He said his name was Quigley, and that he had been trying to get in touch with us, to warn us of the coming of the police convoy. Actually I did not believe him, and, somehow, perhaps because he was a very tall man, I thought he was connected with the police, but there was a doubt in the matter. I told him to leave his machine where it was, and to clear off across the fields. I afterwards learned that Mr Quigley was the County Surveyor of Meath, and a good nationalist and supporter of the volunteer movement.

But that was not the end of the matter. Quigley was 6' 4" tall and his presence at the scene of the ambush had been noted by the police; he was arrested shortly afterwards and charged with aiding and abetting the commission of an act prohibited by the Defence of the Realm Regulations by conveying information to persons who were taking part in an armed rebellion, and with having possession of a rifle, shotgun, explosives and seditious literature which had been found at his home. The county council declared that they 'refused to believe that such a manly, generous-hearted official would have been party to the Ashbourne outrage' but the authorities believed otherwise and arranged for Quigley to be court-martialled in Dublin in June 1916.

Major Kimber, the prosecutor, outlined the Crown case. During the week of the rebellion, he said, a man named Ashe, who had since been sentenced to death, set out with armed rebels, took possession of various police barracks, made policemen prisoners, and ultimately came to a pitched battle with the constabulary at Ashbourne where a county inspector was mortally wounded, a district inspector was killed, and a number of constables were wounded. He went on to say that Quigley was seen on the road on his motorcycle and later in communication with the rebels, and a search of his premises turned up a rifle, a shotgun, some ammunition and seditious literature. In his defence, Quigley handed in a written statement declaring that he had no hand, act or part in the rebellion and explaining that the explosives found at his home were for quarrying purposes. He said that he had publicly expressed loyalty to the king and the constitution and was never a member of Sinn Féin. When the split in the volunteers arose in September 1914, the Navan corps, of which he was chairman, voted to remain loyal to John Redmond and to the Irish parliamentary party. Witnesses (including the nationalist MP for North Meath, Patrick White) confirmed in evidence that Quigley was associated with the national volunteers and Redmond's party rather than with the more militant groups, and that he had been known to express views critical of Sinn Féin. Quigley's defence was conducted by J.C.R. Lardner KC, nationalist MP for Monaghan, and by Henry Hanna KC who made what was described as a powerful address on his behalf. After a two-day hearing, Lord Cheylesmore, who presided at the court martial, declared Quigley not guilty and he was released.

Quigley was arrested again under the Restoration of Order Act in December 1920 and, according to a note which he wrote in 1924, 'was imprisoned for the whole of that year [1921] in various workhouses, jails, and internment camps, without trial or charge assigned. My house was broken up, my family scattered, and my household furniture sold by auction. I was put to hundreds of pounds of expense by the action of the British government, apart from all personal suffering'. He returned to his county surveyor duties in Meath following his release in 1922 but left again in 1923 when he was offered a post in the ministry of local government which had been established in the previous April. The ministry had taken over responsibility for roads which had been vested in the ministry of economic affairs until October 1922 but found itself in difficulty when **Dr Peter Cowan**, the local government board's experienced chief engineering inspector, who had transferred to the ministry, was removed from office in March 1923 and J.P.J. Butler, the only remaining roads inspector, opted for early retirement. The ministry was then in urgent need of what its secretary described as 'the services of an engineer of proven administrative ability and good professional standing' in view of the major programme of road work which was to be carried out and looked to Quigley to fill this need. He took up office in September 1923 as chief roads engineer but at a salary significantly below the rate at which his successor in Meath was being paid. He was regraded as chief engineering inspector in 1928 and served in that capacity until 1930 when the administrative duties of the post were taken over by Nicholas O'Dwyer. Quigley continued in the department as 'general consultant engineer', but resumed his former position in July 1932 when O'Dwyer left to establish a consulting engineering practice.

Beginning in 1924, a major programme of works was undertaken by the new state to overcome the neglect and destruction of roads during the previous few years and to bring the trunk roads of the country up to the standard required to cope with the rapidly increasing volumes of motor traffic. In February 1924, grants of £1 million were allocated to local authorities for road improvement works, a sum which exceeded the total of all the grants made for such works since the passing of the Development and Road Improvement Funds Act 1909. In the following year, a national road scheme estimated to cost a further £2 million was launched with a view to improving the worst sections of 1,667 miles of trunk roads of national importance. New techniques of road-making were introduced, including the laying of reinforced concrete, asphalt, tar and bitumen macadam. With only a very limited staff, Quigley played a leading part in planning and supervising the implementation of this programme. He insisted on high standards of performance by the county surveyors and his highly critical reports on road conditions in a number of counties led to the enforced resignation of several of his former colleagues. His detailed reports on the progress of the works throughout the country and on trials conducted with different road surfaces on the main roads out of Dublin were published as appendices to the department's reports for 1922–5 and 1925–7. By 1930, he claimed that in the previous six years 'all the main roads of the Saorstát – all roads on which there is any considerable motor traffic – have been put in good condition … our roads and city streets are now well up to the British standard for such traffic as obtains thereon, and we are ahead of what has yet been accomplished on the continent and elsewhere'. Through his chairmanship of an interdepartmental committee on the subject, Quigley also had a significant impact on the modernization and development of road traffic law to take account of the growth in motor traffic; his committee's 1928 report was a fully comprehensive one which led to the introduction of compulsory motor insurance and to the enactment of the Road Traffic Act 1933 which was the basis of road traffic law until the 1960s.

Quigley retired at the end of September 1934. He was nominated by the Institution of Civil Engineers for election to Seanad Éireann on the Industrial and Commercial panel in July-August 1938 but was one of seventeen candidates on the nominating bodies sub-panel who did not receive even a single vote. He died on 23 October 1941 at his residence in Leinster Road, Rathmines, aged 72, and his funeral went by rail a few days later to Killevan,

Co. Monaghan, where one of his brothers was parish priest. With his wife, Linda Hynes, whom he had met at Lisdoonvarna in 1911, Quigley had four sons: Richard, a priest of the Dublin diocese; Edward, a doctor; and James and Paul who were both engineers; the latter had a distinguished career in the public service and was for twenty-five years general manager of the Shannon Free Airport Development Company.

QUILTON, Edward Tankerville (*c*.1842–1920), son of Edward Quilton, a Dublin gentleman, graduated from Trinity College, Dublin, with a BA degree in 1862, an LCE in 1864 and an MA in 1872. He gained first place at the county surveyor examinations held in September 1876 and was appointed to Co. Roscommon on 27 October to replace **John Kelly** who had retired after forty-two years service.

At the spring assizes in March 1877 the foreman reported that a committee of the grand jury had met the new surveyor to discuss the question of whether he intended to carry on a private professional practice in addition to his official duties. Quilton, however, refused to bind himself to give all of his time to grand jury work, even though he had been warned that his service would not be reckonable for pension purposes unless and until he announced his intention of doing so. At the same assizes a more serious issue arose for Quilton when his predecessor objected to the payment to him of the full half-year's salary of £200; Kelly argued that while he had retired on pension at the summer assizes in July 1876, he had been obliged to continue on an acting basis until his successor took up duty in October and was therefore entitled to the salary payable for that period. Quilton contended that Kelly had done very little work after his retirement and had, in fact, told the under-secretary that his health would not allow him to discharge the duties of the office. After much discussion, it was agreed to pay Kelly's salary for two months, leaving Quilton with the balance of the £200 which he had expected to receive.

Quilton found the roads in Roscommon to be 'in fair order' when he took up duty but he reported that the very severe winter of 1876–7 caused substantial damage to some road sections and left others, between Boyle and Carrick-on-Shannon, dangerously flooded. He was strict in his supervision of the road contractors, refusing certificates in many cases because the materials supplied were far short of the quantities required or were of poor quality.

At his own request, Quilton was transferred to Carlow in August 1877 and continued in the service of the new county council after 1899. He was in difficulty with the council at an early stage about the number of his assistants whose work and abilities were questioned by councillors. The fact that he himself lived at 24 Leinster Road, Rathmines, Dublin, was also a regular cause for complaint. In 1910, the council demanded that he should keep a diary showing the number of inspections he carried out in the county, but he took offence at this, pointing out that no neglect of duty could be proved against him and that it was an indignity after thirty-four years as a county surveyor to make such a demand on him. The matter was referred to the Local Government Board which expressed the opinion that the council's request was not unreasonable. This seems to have been the last straw for Quilton who decided to resign at the end of 1910, citing ill-health, as well as the attitude of the council towards him, as his reasons for doing so.

The *Irish Builder and Engineer* supported Quilton in his dispute with the council, arguing that while the requirement to keep a diary would not be inappropriate in the case of a junior or new official, 'it is rather an indignity to make such an order on an old official who has served for so many years. The council should be in a position to trust their senior officials more. If there is not more trust, there can be very little mutual co-operation'. But these fine sentiments did nothing to resolve the conflict between Quilton and his council; he was replaced in 1911 by **John P. Punch** and lived in retirement at his home in Dublin until his death at the age of 78 on 26 April 1920. He is buried at Mount Jerome cemetery with his first wife, Laura who died in 1912, and his second wife, Kathleen Marianne, who died in 1914.

RICHARDS, Richard (fl. 1810–42), worked as a surveyor for the bogs commissioners between 1810 and 1813 and subsequently came to notice as an architect-contractor when he was living in Dublin. In association with Thomas Colbourne, also of Dublin, he submitted designs for a new Roscommon county gaol in 1814 to replace the building – constructed in the 1740s – which was then in use. A design by Richard Morrison was preferred but Richards and Colbourne were chosen as contractors and completed the £20,000 project in 1818. The gaol was demolished around 1945, leaving its predecessor still standing in the Market Square in the centre of the town.

Richards and Colbourne again tendered unsuccessfully in 1821 for the construction of Armagh lunatic asylum, the first of a series designed by Francis Johnston who considered their tender price of nearly £14,000 to be too low. In 1820 Richards submitted proposals for remodelling the east front of Tullynally castle in Co. Westmeath for Lord Longford but, while his plans were not adopted, they influenced some subsequent alterations to the castle. His work for the Church of Ireland on the construction of a number of churches and glebe houses gave great satisfaction, according to the archbishop of Tuam, and his design and construction of a bridge at Longford was praised by Lord Longford.

Richards was appointed as the first county surveyor for Clare on 17 May 1834. He immediately set out on an inspection of the roads and bridges in the county and, in his first report to the grand jury at the following summer assizes, he noted the extreme narrowness of many of the roads, both between fences and in their metalled parts. This situation, he reported, was exacerbated by nuisances and encroachments on the roads 'to an extent almost incredible'; watercourses were obstructed and timber, stone, peat, dung, soil, compost, sand, gravel, sea-wrack, lime and even limekilns were to be found on some of the roads. In addition, almost all of the bridges needed repairs, with the stones in the arch-rings exposed in many cases.

According to the *Clare Journal*, the presentments brought before the assizes in July 1834 were discussed with great patience and fairness and Richards' opinion was 'deferred to in every case'. The *Journal* noted that, considering the limited time since his appointment, the surveyor's performance was quite remarkable and reported that all parties were satisfied with the fair and consistent manner in which he acted. But notwithstanding this impressive beginning, Richards' term of office in Clare was a short and an unhappy one: the grand jury sought his removal in October 1834, only five months after his appointment, and he was promptly asked by the Dublin Castle authorities to resign or face dismissal by the lord lieutenant. Richards elected to resign and did so on 19 November 1834 – the first county surveyor to suffer effective dismissal – but he and his wife sent a series of memorials to the lord lieutenant in the following years seeking an appointment as county surveyor in another county. He argued that his difficulties in Clare had arisen because he had not been prepared to authorize payments where contracts had been executed in the most disreputable manner by influential magistrates and their associates and he refuted the grand jury's allegations that he had 'associated on terms of intimacy with some of the most degraded characters in society' and was occasionally in a state of intoxication.

When consulted by the under-secretary, Thomas Drummond, about Richards' case in 1836–7, the Board of Works noted that he had given up a business connection of some extent and respectability to become a county surveyor, that he had performed his duties very competently and that no dishonesty had been alleged against him. He had suffered enough, and while he was a violent man and had written some intemperate letters, he would probably be more cautious in the future. In these circumstances, and 'under the difficulties of procuring persons for the situation', the board advised that 'it may be worth reconsidering how far his case may admit of a renewed appointment' while adding that it would be 'impolitic to continue a person in the service of the grand jury who was displeasing to that body'.

While it appears that Richards was considered for appointment as county surveyor in Tipperary north riding in May 1839, he had to wait until August 1840 to be restored to the ranks as county surveyor for Tyrone. On that occasion, the Board of Works advised that Richards 'is more competent in professional knowledge than many of the present county surveyors or than any candidate that could possibly be obtained now'. They went on to say that although he had been compelled to resign because he had not given satisfaction 'by his personal habits to a number of influential gentlemen' in Clare, he had not been allowed the formal investigation which he solicited. He had resided in Roscommon since leaving Clare, and had numerous and strong testimonials of good conduct there.

Richard's new appointment in Tyrone lasted for little more than a year: the grand jury decided at the summer assizes in 1841 that his salary should be withheld until certain charges brought against him had been investigated, and he was dismissed in the following December.

ROBERTS, Samuel Ussher (1821–1900), was one of a family of Welsh descent which had a long association with architecture and building in the south of Ireland and was distinguished in arts and in arms. 'Honest John' Roberts (1714–96) was the architect of many important Waterford buildings, including the Catholic and Church of Ireland cathedrals, and one of his sons, Thomas Sautelle Roberts (1760–1826), who initially trained as an architect, became a painter of some distinction. Samuel Ussher Roberts was born into this family on 21 May 1821 at Weston, near Waterford, a large house built by his father, Edward Roberts JP, a solicitor (and a grandson of 'Honest John') and his wife, Martha Ussher. Another great-grandson of 'Honest John' was Field Marshal Earl Roberts VC (1832–1914) who distinguished himself in India and Afghanistan in his younger days, commanded the British forces in the Boer War and was to be the last commander-in-chief of the British army.

Roberts was educated at Burney's Royal Academy in Gosport and, despite the family tradition of architecture, his early training was in engineering: he served a pupilage under **Thomas Kearney**, one of the Limerick county surveyors, and worked, initially as a temporary drawing clerk, under Jacob Owen at the offices of the Board of Works from 1841 to 1844. He was assigned in 1844 to report on Ardglass Pier which had been badly damaged by storms some years earlier, and in May of that year became district engineer for the Ardee, Glyde and Fane drainage districts in counties Monaghan, Louth and Meath where he was to have charge of extensive arterial drainage works. One of his assistants, initially as accountant and draftsman and later as assistant engineer, was Robert Manning (1816–97), who became chief engineer at the Board of Works in 1874 and was the author in 1889 of the formula (which is still in use by engineers throughout the world) for calculating the flow of water in open channels and pipes. Roberts' extensive reports in 1844–5 provide detailed descriptions of the proposed works, including the widening and deepening of river courses, the construction of embankments, and the replacement or reconstruction of numerous stone bridges. The reports also describe the methods employed (including the successful use of turbines worked from the head of rivers) to pump water out of river channels while bridge works and channel deepening was in progress; a paper by Roberts on this subject was published in the *Transactions of the Institution of Civil Engineers* in 1849.

Reports from Roberts in 1846–7 are even more interesting for the light they throw on conditions in the Monaghan and Louth areas at the time of the famine. In February 1847, for example, Roberts wrote that he could not describe the poverty and destitution among the people in the barony of Farney in Co. Monaghan – men came to work starving; they could not earn even eight pence a day, sufficient to enable them to survive outside the workhouse; and yet, he noted, they were peaceful and quiet, with no riots or disturbances. Roberts administered some fifty different relief work schemes in that one barony, providing employment for men, women and children, and he raised the daily rates of wages to allow the weak and infirm to earn enough to buy turnips. By the end of January 1847 the average numbers employed each day had risen to 9,608 men, 7,042 women and 3,606 boys. When, in March 1847, the relief works had to be closed down by order of the Treasury, Roberts wrote of people being thrown out of work with no recourse at all: he found it horribly painful to look at 'their deserted and wretched cabins, their forlorn and distressing look, with pain and sickness so dreadfully depicted in every countenance' and complained that 'no proper efforts are being made to relieve the burden'.

Roberts was transferred to Galway in 1848 where, as Board of Works district engineer, he took charge of the Loughs Corrib, Mask and Carra drainage district, one of the largest in the country. The works which were to be undertaken at a cost of some £200,000 comprised the improvement of the drainage of Lough Corrib and its tributaries, the opening of a navigable canal from Galway city to Lough Corrib, improvement of the water-power of the Corrib in and near the city, and the connection of Loughs Corrib and Mask by a locked canal. Most of the work was successfully carried out: Roberts claimed that the level of Lough Corrib had been lowered by three feet and that 13,685 acres of land had been relieved from injurious flooding except under extraordinary circumstances; the canal from Galway Bay to Lough Corrib was opened by – and named after – the lord lieutenant, the earl of Eglinton, in August 1852; the famous salmon weir was reconstructed; and a new west bridge was completed in the same year. It was, however, the failure to complete one part of the project – the ill-fated Cong canal which was to connect Loughs Corrib and Mask for both navigation and drainage purposes – that attracted most attention in the annals of engineering.

By 1850, some 500 men were employed on the canal works at different points along the four-mile route. In 1852, difficulty was being experienced in deep excavations, with numerous caverns being intersected in the limestone. In the following year, lining walls were being built and puddling of the bottom and sides was in progress. By April 1854 the works were said to be in an advanced state, with several of the locks and some fine stone bridges completed. However, considerable additional expenditure was still required to complete the canal because of the porous nature of the limestone rock through which it was being cut. The official attitude towards drainage works had altered considerably by that stage and the pressure for relief works had eased. In this situation, Roberts received sudden and unexpected instructions from his superiors in Dublin to suspend all navigation works and to complete only such works as were necessary for flood relief purposes. The incomplete canal has often been described as a disastrous and costly engineering failure – so dry, in 1872, that Sir William Wilde saw 'little boys … playing marbles on the bottom' – but it seems to be at least arguable that Roberts was coping well with the undoubted difficulties and that the abandonment of the project was due primarily to shortage of funds, the changed circumstances since the work was undertaken and a realization that there were no real prospects of traffic through the canal.

In 1853 Roberts had a salary of £530 from the Board of Works, with an allowance of £150 for travel and other expenses, making him one of the highest-paid engineers on the staff. He took on the post of surveyor for the county of the city of Galway in August 1855 (having qualified at the examinations held in 1841 and declined several offers of appointment in the intervening years) and left his full-time position with the board in July 1858 when he was appointed also as county surveyor for the western division of Galway. He was, however, retained by the board as consulting engineer and inspector on land improvement and drainage projects in Galway and elsewhere, and he served as engineer to the Lough Corrib Drainage and Navigation Trustees, the Galway relief committee (which funded public works schemes in the 1860s) and the Galway harbour commissioners. In this latter capacity, he became heavily involved in efforts in the late 1850s and 1860s to develop the harbour as a packet station to cater for passenger and freight services to north America. To facilitate this, he prepared plans for a long causeway linking Mutton Island to the mainland and a pier extending from the island into the deep water of Galway Bay, but the campaign in Galway and in London to raise funds for the project (estimated at different times to cost from £70,000 to £150,000) was unsuccessful. Roberts, however, rebuilt the enclosed Galway docks and their gates in the early 1870s at a cost of some £5,000, and prepared plans for a new graving dock.

Roberts was actively involved for more than ten years in efforts to provide a railway between Clifden and Galway city, a distance of about fifty miles. He presented plans and cost estimates in 1861 for a line generally following Alexander Nimmo's Galway–Oughterard–Clifden road, built more than thirty years earlier. The scheme was revived in 1866 and in 1870–1, with Roberts again acting as promoter and engineer but a more southerly route was also canvassed locally. The Galway, Oughterard & Clifden Railway was eventually approved by parliament in August 1872 but, in the absence of government grants or loans, or active support from the Midland Great Western Railway, construction work could not get underway; it was not until 1895 that the line originally proposed by Roberts was opened to traffic.

As county surveyor, Roberts built a new Galway–Headford road in the early 1870s and new roads on reclaimed land in the Claddagh and Salthill areas. In Connemara, he completed an attractive new road by Derryclare Lough and Lough Inagh towards Kylemore in 1867, incorporating several nicely finished stone arch bridges. He was responsible also for completing the coast road from Letterfrack to Tully and on towards Leenane (begun in the famine years) with a thirty-two foot span bridge over the short river Culfin which drains Lough Fee to the sea. He also replaced Nimmo's masonry-arch bridge at Maam (destroyed by a flood in the 1860s) by a lattice-girder structure on the original abutments; this bridge, in turn, was replaced in the 1980s by the present concrete structure.

In 1873 Roberts prepared plans for a new town sewerage scheme in Galway where sanitary matters were said by the *Irish Builder* to be in a lamentable condition, but he left Galway before these could be put into effect. Well over 100 years later, with the same unsatisfactory sewerage disposal arrangements still prevailing in Galway, his proposal to construct a causeway to Mutton Island was revived, this time to permit the construction and operation of a sewerage treatment plant on the island, linked by a trunk sewer and causeway 890 metres long to the mainland. He put forward plans for a water supply scheme for the city of Galway in 1862 taking water from the Corrib and using a large storage reservoir on Prospect Hill; two large water-wheels powered the pumps when the works opened in 1867. Roberts had already designed and constructed a large wrought and cast-iron water wheel, eighteen feet in diameter and twelve feet wide, at the Newcastle Distillery.

Roberts developed a substantial private architectural and engineering practice, mainly in the Galway area, during his years as county surveyor. He was responsible for the construction of Kylemore Castle (now Kylemore Abbey, the convent of the Benedictine Nuns) for Mitchell Henry MP between 1867 and 1870, at a cost of about £29,000, although J.F. Fuller, who carried out some alterations at the castle in the 1880s also claimed credit for the work. Every modern convenience was incorporated in the large Gothic Revival style building which was finished to a very high standard on a difficult and remote site. Roberts also appears to have worked with George Ashlin on the rebuilding of Ashford Castle for Sir Arthur Guinness (later Lord Ardilaun) after Ashlin had replaced Fuller as the lead architect. He had earlier designed Gurteen Le Poer, a spacious baronial style mansion at Kilsheelan, near Clonmel, which was completed in 1866 at a cost of about £10,000. In 1862, he designed a new hotel at Ballinahinch and a new town hall, incorporating a ballroom and a supper-room, at Loughrea. In the following year, his commissions included a dispensary and lying-in hospital at Galway and a new gate-house and entrance gates for Lord Dunsandle. In that year, too, his Gothic church of Saint Ignatius at Salthill – possibly an adaptation of earlier plans by the prolific church architect, J.J. McCarthy – was completed and consecrated for the Jesuits. In 1874, a large new house for Captain Knox was being completed under Robert's direction at Creagh, near Ballinasloe, and he carried out improvements at Curraghmore, seat of the marquis of Waterford, in 1876. Two years earlier, at Faithlegg, six miles east of Waterford on the Suir estuary, he had extended and renovated a house dating from 1783 to form a large victorian mansion for the Power family; the house was restored and enlarged in 1998 to form a hotel.

When Sir Richard Griffith retired as chairman of the Board of Works in 1864 at the age of 80, the board was effectively reduced to two members – his successor Colonel J.G. McKerlie and W.R. Le Fanu, a former consulting engineer. Complaints of maladministration and inefficiency and controversy about the role and value of the organization, led the Treasury in 1872 to appoint a committee chaired by Lord Lansdowne to examine the board's procedures and to make recommendations. The report was never printed and published and the only real result of the exercise was a decision, announced in September 1873, to appoint Roberts as assistant commissioner. The *Galway Vindicator* carried reports over a number of weeks of the great loss that Roberts' departure from the county would represent. He was a universal favourite, according to an editorial in one issue, efficient, popular and straight-forward, liberal in character and a man in whom there was unlimited public confidence. There was general satisfaction that Roberts arranged to defer taking up his new appointment for a few months so as to allow him to complete important harbour and other projects on which he was engaged. He eventually left his post in Galway in mid-November 1873 and was the recipient of a generous testimonial some weeks later.

Roberts' salary as assistant commissioner was initially £800 a year. He was the board's first architect-commissioner and had direct charge of all of the organization's building and architectural work. He became a full commissioner in 1878, with a salary of £1,200, following the death of Sir Richard Griffith (who had continued to be a nominal member of the board since 1864) and the publication in the same year of the report of the Crichton committee which, among other things, recommended that the strength of the board should be restored to three. He retired on pension in 1886 and immediately resumed private practice. He became consulting engineer and architect to the Asylums Board of Control in 1891 and in this capacity was involved in the planning and supervision of extensions and improvement works at a variety of locations around the country. He worked also for the Congested Districts Board for which he designed a new bridge over the Gweebarra estuary in Co. Donegal to complete the road through the Rosses and the coastal link between the north and south of the county. Roberts' original design, prepared early in 1893, involved a lattice-girder bridge 288 feet long, consisting of five 52-foot spans and one of 28 feet, supported on piers formed of clusters of steel screw-piles, each 10 inches in diameter and driven up to 16 feet into the sandy bed of the estuary. As built in 1896, the bridge had 9 spans of 52 feet each, the additional spans being substituted for a long causeway. With nearly 2,000 yards of approach roads, the whole project cost about £9,000, but the bridge was not a success. At its last meeting in March 1899, only three years after the bridge had been completed, the grand jury learned of a failure in the foundations due to defective construction but their hopes that the Congested Districts Board would put matters right were in vain and it was left to the county council to attempt to repair the bridge in the following years. By 1915 the bridge was seriously defective, according to the Local Government Board's inspector, but little was done to improve matters and the iron girders and decking continued to deteriorate due to the action of sea spray. Weight restrictions were introduced in 1931, with tar barrels on the deck to reduce speeds, and the bridge survived in this condition until its was replaced, after long delays by the contractor, by the present reinforced concrete structure which was completed at a cost of more than £133,000 at the end of 1953.

In 1887 Roberts was created a Companion of the Order of the Bath in the jubilee honours list. He was for long an active member of the Royal Dublin Society, a member of the society's committee of agriculture and its vice-president at the time of his death. Said to be one of the best judges of a horse in Ireland, he was one of the chief organizers and judges at the annual horse shows. In the 1880s he designed the new jumping course at the society's premises in Ballsbridge, with double bank, ditch and bank and water-jump, which remained unaltered until 1935. A perpetual challenge cup for the best light-weight hunter which was presented by some of his friends is still competed for at the horse show. His interest in horses also led to an involvement in the All-Ireland Polo Club for which he arranged to have their ground at the Phoenix Park railed off and drained to the lake in the Zoological Gardens.

Bright and active to within a few days of the end, Roberts died at his home at Upper Pembroke Street, Dublin, after a short illness, on 11 January 1900, the day after his more famous second cousin arrived at Capetown to take over as commander-in-chief of the British forces which had suffered a series of disastrous defeats in the first months of the war against the Boers; one wonders if the 70-year-old field marshal had called to Upper Pembroke Street to say his goodbyes before leaving his headquarters in the Royal Hospital, Kilmainham, in mid-December to begin the long journey with his staff to the Cape. Roberts' body was taken by train from Kingsbridge station to Waterford a few days after his death for burial in the cemetery at John's Hill; a memorial tablet was later erected to Roberts and his wife, Emily Isabella Forster (d. 1907) in the ruins of the medieval French Church in the city where 'Honest John' had been buried over 100 years earlier. It was said of him that his sterling and genial character caused him to be mourned, not only by his family but also by many friends and that his marked ability, firmness and judgment made him greatly missed as a public man. As against that, a contemporary, Robert Dobbyn, solicitor, wrote of him as a 'vain extravagant man, always in debt to keep up a show at the Castle and with the Ward and Kildare Hounds'.

RORKE, John (1886–1948), was born at Clondalkin, Co. Dublin, on 2 January 1886 and was educated at Belvedere College. He was awarded a BE degree at Queen's College, Galway, in

1908 and subsequently served for two years as assistant county surveyor under **James Quigley** in Co. Meath. From 1911 to 1914 he was an assistant engineering inspector with the Congested Districts Board before his appointment as county surveyor for Queen's County in April 1914 at a salary of £350 a year. He transferred to Co. Kildare in the following December and served there for nearly thirty years.

In the 1920–3 period Rorke, like the other surveyors, found himself in a difficult position in attempting to maintain the county's network of roads. In the first half of 1921 local units of the IRA set out to disrupt military and police activity by trenching roads, felling roadside trees and damaging bridges – and even greater damage was caused during the civil war of 1922–3 bringing the total estimated cost of road restoration in the county to more than £4,000. Because the financial resources of the council were limited arising from the withholding of government grants and difficulties in rate collection in the disturbed conditions of the time, there was little that Rorke could do to improve matters until grants from the new government became available in 1923.

In 1924, when it was announced that further grants totalling one million pounds were to be made available to local authorities for the improvement of the trunk roads using new methods of surface treatment, Rorke pleaded for an immediate programme of research to determine which treatments would be most suitable and cost-effective in Irish conditions. 'It is not that any great difficulty lies in the task of writing the theoretically best specification', he wrote, 'or even in carrying this out if one has sufficient money to expend'; the need was to find the solution which was best for particular traffic conditions having regard to the kind of materials that could readily be procured, at least cost, and with the workmen available. Arguing that there could be as many as fifty different combinations of methods and materials, Rorke called for the laying down immediately of as many trial lengths of road as possible, with a range of surface treatments, in order to compare the behaviour of the materials, and the costings, before decisions came to be made on practical and standard methods to be applied.

Rorke was an acknowledged expert in various aspects of roadwork and a pioneer in the introduction of super-elevation and other techniques. An obituary recorded that he had earned the reputation of being 'the foremost road-maker of his time'; he had developed a methodical system of road reconstruction, with up-to-date plant and materials and the roads in his county, particularly the main arteries serving the south and west, were said to have been improved beyond recognition. The concrete road which Rorke built in the late 1920s across the bog between Kildare and Monasterevan was the subject of much attention at the time and was regarded as a tribute to his ability and skill; until the completion of a motorway section eighty years later, it was carrying traffic volumes that its designer could never have imagined. A paper on road-making practice in Ireland, which he presented to the Institution of Civil Engineers of Ireland in 1936 provides a good description of the state of the art at the time. However, while modern road-making methods were being used on the trunk roads in Kildare at the time, over 500 miles of the county roads were still being maintained under the traditional contract system which was favoured by councillors and ratepayers because of lower costs

Forced to resign his position because of ill-health in April 1944, Rorke died in Dublin on 11 May 1948. An obituary described him as a man who was enthusiastic about his work and a kindly, unselfish, honourable gentleman.

ROWAN, Frederick J. (fl. 1842–47), appears to have been a native of Belfast and was probably a nephew of Field-Marshal Sir William Rowan (1789–1879) who served in the

Peninsular War and at Waterloo and was later commander-in-chief in Canada, and of Sir Charles Rowan (1780–1852), who became first Chief Commissioner of London Police in 1829. Having worked for the Midland Great Western Railway Company, he was appointed county surveyor in Tyrone on 3 January 1842 following the dismissal of **Richard Richards**. According to a statement which he submitted to the Devon Commission in 1845, he found that there had been great negligence and mismanagement of the grand jury's affairs before he took up duty and that grievous malpractices had been allowed to exist. He had attempted, with the 'most ready support' of his grand jury, to improve matters by establishing strict discipline among his assistants, by requiring the assistants to submit monthly reports on their activities, by prosecuting defaulting contractors and by attempting to discourage excessive competition for contracts which could only result in inadequate prices and work of a poor standard. But, he told the commission, there would always be occasional cases of fraud which even the utmost diligence of a public official would not be able to remedy.

In 1842, when the Poor Law Commissioners' architect, George Wilkinson, certified that the Omagh poor law guardians were liable to pay the contractor who had completed their 800-place workhouse a sum of £2,174 over and above the contract price of £6,657 to cover extras and fittings, Rowan was engaged by the guardians to review the matter. He reported that the sum claimed by the contractor should be reduced by £837 but Wilkinson robustly defended his original figures adding that while Rowan might be 'well qualified for the duties of his position [he] is not professionally familiar with those of an architect which in many respects differ from the qualifications required of an engineer'. The dispute was reviewed by a royal commission in 1843–4 but, notwithstanding the fact that Rowan was 'a person of considerable reputation', the view was taken that the guardians were legally bound to pay the amount certified by the architect.

Like his contemporaries, Rowan was heavily involved in the supervision of famine relief works in 1846–7 under the general direction of the Board of Works. At the summer assizes in the following July, the judge cautioned him and the grand jury about accepting responsibility for the completion and repair of roads which had been cut up or left in an unfinished state when the works programme was summarily terminated a few months earlier by order of the authorities in Dublin. Members of the jury were reluctant to adopt this approach, pointing out that even the main road from Omagh to Derry was almost impassable and that the Board of Works had told them that they had no funds to complete the works. However the judge was adamant – it should be left to the board to finance the work and the grand jury should not bring the county to a state of bankruptcy by taking on the responsibility.

Rowan resigned in September 1847 and, according to the Board of Works, then left Ireland. While details of his subsequent career have not been established, it seems likely that he was the same Frederick Rowan who worked as an engineer with Messrs Peto, Brassey and Betts on the Grand Trunk Railway of Canada in the mid 1850s. He was the father of Captain Frederic Charles Rowan (1844–92) who, after service in the army and with the police in New Zealand, practised in Melbourne as a civil engineer and became managing director of the Australasian Electric Light, Power and Storage Company in 1882.

RYAN, Joseph Albert (1888–1971), was born in Dublin on 18 August 1888, son of a barrister, and educated at Castleknock College. He was an apprentice with the Kerry Electricity Supply Company at Killarney for two years before entering Trinity College, Dublin, in May 1907 where he obtained a BA degree in 1910 and a BAI in 1911. He worked for a short time as assistant surveyor with Pembroke urban district council before setting out for Canada where he was employed on construction works by the Grand Trunk Pacific and the Pacific Great Eastern Railways in British Columbia between 1913 and 1914.

Ryan was appointed county surveyor for Queen's County in April 1915, having passed the qualifying examination earlier that year. Three years later, he was reporting that the road maintenance programme had been disrupted due to an inadequate supply of materials which, in turn, was caused by the confiscation by the RIC of the explosives which were stored for use for at the council's quarries. Like many of his contemporaries, he also had to deal with serious damage caused to roads and bridges during the war of independence and the civil war: he reported in April 1922 that sixty bridges in the county had been damaged and was authorized to begin a repair and reconstruction programme estimated to cost £7,000. In the following September, he told the council that an additional nineteen bridges had been destroyed or rendered impassable and that workmen and contractors were being prevented by threats or by armed men from carrying out repairs.

When the post of Dublin county surveyor, which carried a salary of £800, came to be filled by the members of the county council in July 1924, there were twelve applicants, four of whom already held county surveyor posts. After several divisions, Ryan was selected to fill the post having gained fifteen votes as against five for E.M. Murphy of Bray. His term of office of almost thirty years was notable for the major programmes of road development, housing, and sanitary services which were carried through under his supervision. In 1924, he contributed notes on road construction to the *Irish Builder and Engineer* in which he dealt in some detail with the factors to be taken into account in designing road schemes and the materials and methods that might be employed in surfacing. He was responsible for conducting a traffic census in 1924 on the main arteries in the county and, in 1925, for the conduct of experiments with different forms of road construction and different materials on nine short sections of the Naas road between the city boundary and Newlands Cross under the general supervision of **James Quigley**, the Department's chief roads engineer; the results were summarized in a paper which he presented in January 1926 to the Institution of Civil Engineers of Ireland.

Ryan was president of the Institution of Civil Engineers of Ireland in 1938–40. He retired from his county surveyor post in July 1953 and then worked for some years as sales manager with Moracrete Limited of Dublin. He died on 16 September 1971 and is buried at Deansgrange cemetery with his wife, Maimie, who died on 30 December 1995, in her ninety-ninth year.

RYAN, Thomas Francis (1885–1948/49), was born on 15 July 1885 near Limerick Junction. He was educated at Clongowes Wood College and studied privately for the diploma in engineering of the Royal University. He served a pupilage with **E.A. Hackett**, Tipperary south riding surveyor, and subsequently worked for London county council on main drainage schemes. He then became an assistant engineer with the Port of London Authority and worked at the Victoria and Albert Docks and on the construction of new ferro-concrete quays at the West India Docks.

Ryan became county surveyor for the western division of Co. Limerick in June 1914 when, following the retirement of **John Horan**, the county council decided to employ two county surveyors at reduced salaries of £300 each; he had been placed first at the qualifying examinations held in September 1913. He had a difficult time in the 1920s arising from the demands of unemployed labourers and their unions who sought to maximize employment by the county council and objected to the use of machinery. At the same time, he and his council faced criticism because of the relatively high wage rates paid to the road workers and the fact that average expenditure per mile of road was higher in the county than elsewhere – an unnecessary expense, in the view of a local government inspector who held a sworn inquiry into the affairs of the council in 1927.

With an increased salary of £1,150 a year, Ryan took over responsibility for the entire county in 1929, when his colleague in the eastern division, **Robert Davison**, retired. In an unusual move, he was appointed to act also as Kerry county surveyor for some months in 1943 when the post in that county fell vacant. He was among the first of the county surveyors to embark on turf production programmes during the emergency years and, until his retirement in 1945, he was noted for his energy in dealing with the various other problems he was called on to handle. Many tributes were paid to him on his retirement when his county manager described him as 'a splendid official and a good colleague'. He died in 1948–9.

SANDERS, Richard Barnsley (1845–1900), was born in Lisburn on 6 October 1845 and educated at Queen's College, Belfast, where he gained an honours diploma in civil engineering in 1866 and a BE degree in 1878. He served a pupilage under James Thompson, afterwards professor of civil engineering at Glasgow University, and subsequently worked on a variety of projects. He was contractor's engineer on the Dundrum & Newcastle Railway in 1868–9 and on the Belfast Central Railway in 1870–2. He then went into practice on his own account in Belfast and was involved in the design of waterworks and docks in the city.

Sanders took second place at the county surveyor examinations in June 1874 and was appointed in August 1874 as surveyor for King's County. By 1883 he had become seriously concerned about the damage being caused to all classes of roads in his county by steam engines and their trains of wagons. In a paper entitled *Steam Traction on Common Roads* which he presented to the Institution of Civil Engineers of Ireland, he complained strongly about the other disadvantages of the new form of traction: there was nuisance and danger to the public through noise, smoke and vibration; the engines were unsightly and hideous in appearance; they occupied too much of the roadway making it difficult for others to pass; their drivers were guilty of incivility and carelessness; and there was real danger arising from frightened horses who could not be trained to face 'such unsightly, dangerous and horrible-looking machines'. For Sanders, the solution lay in limitations on the size, power, weight and width of engines and wagons, and restrictions on the roads which could be used by particular classes of engines; the alternative would be to undertake the entire reconstruction of three-quarters of the road network at enormous cost. He argued also for changes in the design and appearance of traction engines to reduce the risk of frightening horses; in particular, the naked fly-wheels, pistons, and other working parts should be covered by wooden or iron casings to improve their appearance and deaden the noise, and the engines should be made to consume their own smoke.

In 1898 Sanders contributed a paper on the management and maintenance of public roads in Ireland to the Institution of Civil Engineers (London) of which he was elected a member in April 1892. He described in some detail his own methods of road construction, the system of road repair generally adopted in Ireland which he described as 'darning' or patching, and the staffing and other arrangements which he had in place to deal with the maintenance of the considerable mileage of roads for which he had direct responsibility in the absence of local contractors willing to undertake the work. But, he complained, he was not free from difficulties often caused by interested and unscrupulous opposition, and noted that he and his colleagues were 'at times liable to serve as buffers between the ratepayers and interested or unprincipled parties who have considerable power over them, and with whom they may not infrequently come into collision'.

In 1893, when reform of local government at county level was already under consideration, he dealt extensively with the grand jury system in a paper delivered to the Incorporated Association of Municipal and County Engineers. As he saw it, the Irish system was a unique one, almost an exact miniature of the British constitution, with proposals being considered by representative bodies at two levels, and final approval being in the hands of a judge appointed by the crown; thus 'with efficient and conscientious county officers, there could scarcely be a better system for the conduct of the business entrusted to it'. The system had had worked well and while it had its defects, these could easily be remedied; thus, replacing it 'by county council boards, or dividing the county business amongst a large number of local bodies, would be most disastrous to the best interests of the country'.

Sanders' view did not, of course, prevail and the county councils that he seemed to fear came into office in April 1899. Like his colleagues in other counties, he transferred to the new county council at that stage and was actively involved in his duties until his rather sudden death less than a year later on 13 February 1900. His body was taken by train from Birr to his native Lisburn where he owned a substantial house and farm, and he was buried at the City cemetery, Belfast, on 17 February. In reporting the death, the *King's County Chronicle* said that the late surveyor 'was never very robust but very abstemious and careful in regard to his eatables at all times and systematic in open air exercise' (he was, in fact, one of the surveyors who travelled by bicycle on his inspection visits around the county). It added that 'there never breathed a more conscientious official', a man who rigidly enforced the road contractors' obligations, guided by his own sense of what was right and proper and who was 'careless as to whom he might give umbrage'.

Sanders was unmarried and lived in lodgings at Cumberland Square, Birr, where he also carried on a private practice involving engineering work of various kinds. He left a substantial fortune of some £30,000 and, in his will, left specific legacies to a long list of charities, with the residue going to Dr Barnardo's Homes in London. The will was challenged by some of his relatives in the probate court in July 1900 but was held to be valid.

SCHAW, Alexander (1825–54), (sometimes Shaw), was born in Dublin on 11 December 1825, the son of Lt. John Sauchie Schaw of the Royal Artillery and his wife, Frances Anne, daughter of Archdeacon Henry Irwin of Armagh. Schaw's mother died soon after his birth whereupon his father married Catherine Louisa Sirr, daughter of Major Henry Charles Sirr of Dublin Castle, best known for his part in the capture of Lord Edward Fitzgerald and of Robert Emmet; the couple had three children one of whom became Major-General Henry Schaw (1829-1902), commander of the Royal Engineers.

Alexander Schaw began his career as a pupil of Sir John MacNeill and was appointed in 1849 as resident engineer at Newry on the Dublin & Belfast Junction Railway of which MacNeill was engineer-in-chief. In 1850 he worked with MacNeill and James Barton, the assistant engineer, in developing the plans for the viaduct which was to carry the railway 90 feet above high-water level in the Boyne estuary at Drogheda. This was the first large lattice-girder bridge in the world; it had three spans of wrought-iron girders, the longest of them 267 feet between bearings, and 15 masonry arches, each 60 feet wide. At the detailed design stage in early 1851 Schaw spent up to twenty hours daily at his desk engaged on what Barton described as the 'tedious and complicated calculations' required for proper analysis of the stresses and strains of the proposed structure and its components; he was awarded a bonus of £75 by the board of the railway company for his efforts. In addition, according to MacNeill, who engaged in furious controversy with Barton between 1858 and 1861 as to which of them was entitled to be credited with the detailed design and construction of the bridge, Schaw acted as his resident engineer on the actual construction of the bridge for some eighteen months in 1851-3 'until it was three-fourths done'. He is credited also with the design of the Newfoundwell viaduct crossing a ravine just north of the main viaduct at Drogheda, with five massive stone arches, castellated parapets and semi-circular towers, apparently in imitation of the nearby thirteenth-century St Laurence's Gate. It seems possible that he was also the designer of the 'Egyptian' Craigmore viaduct at Mullaghglass, Co. Armagh.

Schaw was well regarded by both Barton and MacNeill but, as construction work progressed, he was regularly in trouble with the board of the railway company because of his frequent absences from duty – partly due to his strict Sunday observance – as well as his failure to answer correspondence from the board and the tone of his replies. On one occasion, he was told by the board to receive orders with respect and not to make them the subject of frivolous and absurd comment. When he left the service of the company in 1853, soon after trains had begun to cross the viaduct on a temporary wooden superstructure, Barton told the board that he much regretted the loss of 'one who, whatever his eccentricities or failings, yet has decided cleverness'. He was succeeded in his post at Drogheda by Bindon Blood Stoney (1828–1909) who remained until the bridge was opened for traffic in April 1855; Stoney then became assistant engineer, and later chief engineer, with the Dublin Port and Docks Board.

Schaw was a keen early amateur photographer and, according to a letter of his to Sir John MacNeill in August 1853, he had 'by photography procured very accurate views of the works at all periods from their commencement to the present day'. He went on to say that he had retained all of his 'calculation papers' relating to the design of the lattice-girder work and had produced a large series of accurate drawings and views of the works at various stages and of the plant and machinery employed, and was proposing to use these, together with memoranda, diaries and notes which he had preserved to produce a book to be called *The History of the Boyne Viaduct by Sir John MacNeill*. In seeking MacNeill's approval for the proposed publication, Schaw concluded that 'it seems a pity to let all the memoranda and plans and pictures I have accumulated lie dormant' but this, unfortunately, is what seems to have happened; the book never reached publication stage and there is no record today of any of the Drogheda photographs or other materials. Some of Schaw's other photographs, however, mainly portraits, have survived in an album (now in private ownership) compiled by Francis Edmund Currey, agent of the duke of Devonshire at Lismore Castle, and himself a pioneer of photography in Ireland.

On 15 July 1853 Schaw took up the positions of Waterford city and county surveyor, having qualified at the examinations held in 1851. In the following November he had to cope with a major emergency when all but the central arch and one adjoining arch of Thomas Ivory's bridge and eight-span approach viaduct over the river Blackwater at Lismore, dating from the 1770s, was carried away by a huge flood. A temporary wooden bridge was quickly put in position and, within a month, plans for the renewal of the bridge, but with larger arches, had been prepared by Schaw. A contract valued at £2,000 for the necessary work was awarded to Albert Williams of Dublin at the summer assizes in July 1854 and work began shortly afterwards. Schaw was due to visit the site on 1 September to settle details of the piers with the contractor, and he set out from Clogheen, some ten miles away, to travel by car through the Knockmealdown mountains to Lismore. He left the car and the road after a few miles and took a direct route uphill until he reached Bay Lough, near the famous 'Vee' section of the road which had been built in the early 1830s. When the car driver found Schaw's clothes, watch and other possessions on the lakeshore some time later, he was presumed drowned and a search organized by the duke of Devonshire's agent, using grappling irons and boats, was unsuccessful. However, the body was found floating in the lake by passers-by on 19 September 1854 and a subsequent inquest returned a verdict of accidental death. According to the *Waterford Mail*, Schaw had earned the good opinion of all classes during his short term of office in the county and his melancholy end occasioned much sympathy. He was buried with his mother at Sandford church, Ranelagh, Dublin.

SCOTT, Harold Kyle (fl. 1912–77), was a native of Derry. He gained a BSc (Eng) at Queen's University, Belfast, in 1932 at the age of 20 and served as assistant county surveyor in Tyrone from then until 1940 when he joined the British army. He was initially garrison engineer in the Banbridge and Portadown areas where he was in charge of the construction of army camps for occupation by American troops. Early in 1941 he was granted a commission in the Royal Engineers and was posted to the Middle East in 1942. After a short stay in Beirut, he served for two years in command of the airfield construction unit in Cyprus and was later on the staff of the general of engineers for the entire Middle East, with the rank of lieutenant-colonel.

Scott was appointed Derry county surveyor in April 1946 and took up duty in September. In the following twenty-five years he was responsible for planning and carrying out a large programme of improvement works on the trunk roads in his county, evading efforts by the authorities in Belfast to curtail his expenditure. He and his team introduced a number of innovations and techniques to Northern Ireland, such as aerial surveys to expedite the design of major schemes, prestressed concrete bridges, mechanical surfacing of carriageways, plastic white-lining, snow blowers, roadside picnic areas and tree planting. His standing as a road engineer was marked by his election as president of the Institution of Highway Engineers in 1970, the first Irish engineer to gain this honour.

With the reorganization of local government in Northern Ireland in October 1973, county councils ceased to exist, the office of county surveyor was abolished and all road staff became civil servants in the roads service of the Ministry of Development. Scott subsequently wrote that, for him, this was 'a most unwelcome and upsetting occurrence … county surveyors … were relegated and given the new title of divisional road managers – not even engineers or surveyors!' He retired in 1977.

SEARANCKE, Samuel Stephen (1811–85), was born in England on 15 October 1811, son of Francis Searancke, a brewer, of St Albans Abbey, Hertfordshire, and his wife, Harriet Smith. Having served his time as an apprentice civil engineer with the prominent consulting engineer, Charles Blacker Vignoles, he was appointed first county surveyor for Co. Meath on 17 May 1834 and went to live initially in Trim, later moving to Violet Hill, Navan.

With some 1,700 miles of roads to maintain, Searancke was allowed only one assistant by his grand jury in the 1840s, and as he considered the assistant to be 'a better class than usual', he paid him £10 out of his own pocket, in addition to the official salary of £50, in order to retain him. The assistant, in fact, was his younger brother, William Nicholas (1817–1904) who emigrated in 1842 to New Zealand where he worked in succession as a government land surveyor, district commissioner and resident magistrate and, according to the *Dictionary of*

New Zealand Biography 'exploited every opportunity to further his career in the administration of land and justice'.

During the famine years, Searancke was called on to assist in the planning and supervision of relief works in parts of his county and encountered the same difficulties as in other areas in enforcing task work; he reported in early 1847 that the labourers in one area had 'turned out' against the system and had 'perambulated the country adjoining, calling on gentry and farmers, on whom by the intimidation of numbers, they levied blackmail'.

Searancke had constructed a total of eighty-five miles of new roads in his county by 1868. He complained that there was frequently a private understanding among the bidders for a road contract and that some bidders were simply stalking-horses for their friends. He reported that he was frequently offered presents by road contractors and that country people often delivered gifts of poultry to his servants; on one occasion, a sheep was put into his paddock, but he felt that this was going too far and so the animal was turned out onto the road instead of ending up on the surveyor's table.

Searancke seems to have had only a very limited private practice. He served in Meath for forty years, but was seriously ill for some years before being forced to retire for health reasons in July 1874, just a year before the law was amended to allow pensions to be paid to surveyors; he died in Wales in 1885. He married Caroline Isabella Jackson Provis in 1835 and had at least one son, Samuel (b. 1846) and a daughter, Ellen Isabella.

SIMINGTON, Thomas Aloysius (1905–94), was a native of Dublin and graduated with BA and BAI degrees at Trinity College, Dublin, in 1927. He joined Great Southern Railways in the same year and gained valuable experience in the design and maintenance of steel structures. He subsequently worked on the design of concrete structures with Messrs Tilesman & Co. of London and in the early 1930s was contractor's site agent and engineer for a succession of British and Irish firms on a variety of bridge, marine and other works. These included a new deep-water jetty at Foynes; extensions of the Limerick docks; and Butt Bridge, Dublin – a three-span reinforced concrete cantilevered structure which was completed in 1932 at a cost of £65,000. He was contractor's engineer also on the replacement of the old suspension bridge at Kenmare (completed in 1841 and probably the first of its kind in Ireland) by a reinforced concrete bow-string girder bridge which was opened by Seán T. O'Kelly, then Minister for Local Government, on 25 March 1933; the new bridge was designed by L.G. Mouchel & Partners and was completed in one year at a cost of approximately £10,000. Simington was again contractor's engineer on the construction of the Cusack Stand at Croke Park, Dublin, completed in 1937 and demolished in 1993 to make way for a new stand.

Simington, by his own account, 'took refuge' during the Second World War, when contract work was scarce, in the local authority service. He worked for Dublin Corporation in 1940–4 on a variety of projects and was, for a time, engineer in charge of relief schemes. He became acting county engineer in Kildare in June 1944 and took up duty as county engineer in Clare on 17 September 1945. When construction activity picked up after the war he opted to return to contracting; he resigned his post in Clare at the end of October 1949 and began a new career as joint managing director of the newly-formed John Paul Construction Company whose first major project involved the civil engineering work for the Lee hydro-electric scheme in the early 1950s, including three major bridges in reinforced concrete. Simington's subsequent work related mainly to marine and bridge structures, including the reconstruction of Fenit Harbour where the long timber approach viaduct was replaced by reinforced concrete. He retired in 1972.

Tom Simington was president of Cumann na nInnealtóiri in 1959–60 and of the Institution of Civil Engineers of Ireland in 1961–2; his presidential address to the institution was devoted to tracing the history of bridges and bridge-building in which he had a lifelong interest, describing himself as a 'pontist'. This culminated in co-authorship with Peter J. O'Keeffe of a major work on *Irish Stone Bridges* (Dublin, 1991). He died on 17 June 1994, in his ninetieth year, and was buried at Deansgrange cemetery.

SMITH, John (1865–1907), was born on 19 April 1865. He trained as a contractor's engineer for six years after which he was awarded a Whitworth scholarship – an award made under a scheme founded by Joseph Whitworth in 1868 to enable outstanding young engineers to pursue degree-level studies in engineering. Smith's scholarship was tenable at the Royal College of Science in South Kensington, London, where he studied until 1889. He was then employed in England as a surveyor by the Royal Engineers before taking first place in the Autumn 1891 county surveyor examinations. He was appointed to the eastern division of Galway in the following December and went to live in Ballinasloe where he also carried on a private practice.

Smith transferred to the service of the new county council in 1899 and was in office until his death from pneumonia on 15 February 1907 at his home, Ardcairne, Ballinasloe. He was buried locally at Creagh cemetery. According to the *Connaught Champion*, he had 'enjoyed a marked popularity and earned to a striking degree the entire confidence of the county council'. The *Connaught Leader* added that he was second to none as an architect but this latter may be open to question because, apart from some works at the lunatic asylum at Ballinasloe in 1904 and a number of private houses and shops, none of his projects seem to have attracted notice.

SMYTH, Henry (*c*.1820–94), a native of Dublin, began his career as a pupil of Jacob Owen in the offices of the Board of Works at the Custom House, Dublin. He was employed by the board as clerk-of-works on the construction of Mountjoy gaol in Dublin when he was placed first on the list of qualifiers at the county surveyor examinations held in 1847. He became Donegal county surveyor on 9 September 1850 but, when a second surveyor was appointed a year later, Smyth's responsibilities were limited to the new northern division of the county. After a further year, he transferred to Co. Down on 24 September 1852 (exchanging positions with **John Fraser**) and went to live in Downpatrick where, for many years, he was a prominent member of the local freemason's lodge. Smyth had responsibility for the entire county until September 1862, when **Bernard B. Murray** was appointed to take over the southern division.

Although he had nine assistant surveyors in Co. Down, Smyth, like his colleagues elsewhere, had difficulties in ensuring that road contractors fully met their obligations in maintaining the extensive mileage of roads in the county and his experience led him to argue in 1856 that skilled labour and expert supervision should replace the contract system. Contracts in the county generally specified that that particular quantities of materials were to be provided and made available for measurement before being spread on the roads but there were allegations that stones, when measured at one location, were moved to another to be measured again. Another strategy was exposed by Smyth in 1869 when he brought proceedings against a contractor on the Belfast–Newtownards road for having fraudulently presented heaps of stones for measurement although the interior of the different heaps was made up of mud, with only an external coating of stones.

Smyth became a fellow of the Royal Institute of the Architects of Ireland in 1849 and carried on a substantial private practice in architecture, employing a number of assistants. One of his articled pupils in 1867, and subsequently one of his assistants, was Robert Cochrane (1844–1916) who went on to become principal surveyor and architect to the Board of Works, inspector of national monuments, and president of the Institution of Civil Engineers of Ireland, 1904–6. Smyth's projects included a number of national schools and dwelling houses. In the mid-1860s he was responsible for the provision of a large complex of buildings, including a chimney 172 feet high, for an extensive lime and brick works at Castle Espie near Comber, Co. Down, where some of the first Hoffman kilns used in Ireland were introduced for the burning of lime and the firing of brick. In the public sector, Smyth undertook a number of significant projects. He rebuilt and enlarged the county courthouse at Downpatrick after a fire during the spring assizes in 1855, earning the thanks and congratulations of the grand jury for his skill and taste in providing such a convenient and handsome building. He carried out improvements at Newry courthouse in 1862 and designed a small courthouse at Banbridge which was completed in 1872 at a cost of about £2,500. He was the architect of the Co. Down lunatic asylum (later known as the Downshire Hospital), a large complex of red-brick buildings completed at a cost of some £60,000 between 1865 and 1869. With a main frontage of nearly 1,000 feet, a clock-tower nearly 100 feet high, and separate male and female accommodation on each side of the central administrative bay, the building catered for over 300 patients, with all ancillary services located on its sixty-acre site; an extension designed by Smyth some years later brought the capacity up to 420. A regular worshipper at Downpatrick cathedral and a member of the cathedral board from 1873 onwards, Smyth was asked in 1856 by the chapter to prepare plans for substantial alterations to the building but, while these were not carried out due to lack of funds, he remained actively involved in subsequent programmes of maintenance and improvement at the cathedral where a memorial window was later erected in his honour.

With **Charles Lanyon**, Smyth designed Ormeau Bridge, Belfast, built between 1861 and 1863 at a cost of some £15,000, and his Lisburn Bridge was completed in 1885. With Luke L. Macassey CE, Belfast, he proposed a major water supply scheme for Lurgan, Portadown, Banbridge and Downpatrick in 1885 but the project fell through because of the inability of the authorities concerned to agree on joint action.

Having retired on pension in March 1890, he became a JP and in 1893 was a member of the Ulster Defence Union which as set up to oppose home rule. He died at his home in Newcastle, Co. Down, on 21 April 1894.

SOMERVILLE, Richard Newman (fl. 1871–1914), was a graduate of Queen's College, Belfast, where he gained a BA in 1871 and a BE in 1873. He served a pupilage with **Samuel Ussher Roberts**, Galway county and town surveyor, and became town surveyor and harbour engineer in Galway when Roberts took up a position in the Board of Works in September 1873. During his term of office, the new east dock was constructed at a cost of about £40,000, enclosing an area 385 feet by 180 feet with a water depth of 28 feet at high spring tides.

Somerville became Cavan county surveyor on 15 October 1889 and transferred to the new county council in 1899. He seems to have taken a hard line with defaulting road contractors, two of whom sued him at Cavan quarter sessions in November 1907 for payments which he had disallowed. He had to withstand aggressive cross-examination from the plaintiff's solicitor who reminded the surveyor at one stage that he could not 'walk over us now, like long ago' and declared that it 'would take a pick-axe to get at your intelligence'! Somerville, however, successfully defended his position and the contractors lost their case. In May 1913, when he gave notice of his intention to retire, he was granted a retiring allowance of £275 a year by the county council 'as a testimony of their feelings on the severance'. He continued in office until 1 January 1914 to allow time for a successor to be appointed.

Somerville was among the earliest Irish motorists. He imported a 3.5 h.p. two-seater Benz in 1897 – Ireland's second petrol-engined car – and christened it *Judy*. In 1898, when he drove from Cavan to Bailieborough, a distance of less than twenty miles, it was reported that the first thing he did on reaching the town was to send a telegram to his wife in Cavan to announce his safe arrival. By 1906, he had graduated to a more powerful machine – a 10 h.p. four-seater manufactured by the Alldays & Onions Pneumatic Engineering Company of Birmingham and with a 1.6 litre engine.

St LEGER, Noblett Rogers (1808–72), whose surname was, apparently, pronounced as 'Sellenger', was a son of Heyward St Leger of Heyward's Hill, Rathcooney, just outside Cork city, and his wife, Matilda, daughter of Noble Rogers of Lota. The family was descended from Sir Warham St Leger, chief governor of Munster, who caused James, brother of the earl of Desmond to be hanged at Cork in 1580 during the Munster rebellion. Over 200 years later, the family were indirectly associated with another rebellion: the legendary Sarah Cullen was staying with them in Cork when her fiancé, Robert Emmet, was captured and sentenced to death in 1803.

St Leger worked with Alexander Nimmo in 1826–9 as surveyor and superintendent on the Killary to Westport road and on new roads in Achill and Erris. He worked also with the Shannon Commissioners and seems to have been in partnership at one stage with **Samson Carter** who had also been employed by Nimmo and was himself to become a county surveyor: describing themselves as 'engineers and hydrographers', they prepared a detailed survey of Waterford harbour in 1834–5 for the Commissioners for Improving the Port. He was 26 years old when he was appointed to be the first Leitrim county surveyor on 17 May 1834 but was transferred to Co. Sligo at his own request on 14 October 1836. He also took on responsibility for the borough of Sligo and was credited with bringing about substantial improvements both there and in the county, particularly in relation to the road network. He had built a total of 150 miles of new roads by 1868, including the Albert Line into Sligo from the south, now known as Pearse Road, and the Victoria Line from the north, now Markievitz Road. With **Sir John Benson**, he designed the town's five-arch bridge over the river Garavogue, originally named Victoria Bridge but since renamed Hyde Bridge. Plaques on the river-facing parapets of the bridge attribute it solely to St Leger, apparently because of a disagreement with Benson (who became county surveyor in Cork in March 1846) and date it to 1846, but the bridge was not, in fact, completed until 1852. St Leger built some further substantial bridges throughout Co. Sligo including those at Easky, Aclare and Curry and he was responsible for the town's first main drainage scheme.

St Leger and his grand jury were active in the 1850s in the development of proposals for a railway from Athlone to Sligo; these were taken up by the Midland Great Western Railway Company which constructed a line from its Longford branch to Sligo between 1859 and 1862. His limited private practice included the carrying out of surveys and valuation assignments under the Incumbered Estates Acts in the 1850s.

Shortly before his transfer to Sligo in 1836 St Leger married Elizabeth Catherine (Bessie) Cullen, the handsome, amusing and vivacious second daughter of Lieutenant-Colonel John James Cullen, a country gentleman and land agent who lived near Manorhamilton, Co.

Leitrim. The newly-married couple lived initially at Quay Street, Sligo, before moving to Wine Street and finally to a large house at The Mall. A son born to them in 1837 died of a cold within a few months and they had no other children. The memoirs of Bessie's youngest sister, Kate, published as *The Sligo-Leitrim World of Kate Cullen* (Dublin, 1997), provide a vivid picture of the personalities and lifestyle of the St Legers and of the society in which they lived. St Leger comes across as a congenial and widely respected man whose house in Sligo was a centre of sociability, with regular parties, card games and musical evenings attended by the landed gentry and professional classes of the area. He died suddenly at his home on 8 April 1872, aged 64, and was buried in an impressive mausoleum close to the main entrance to Sligo cemetery at Cleveragh Road. His wife was 72 years old, and had been an invalid for some years, when she died in 1887.

STEADMAN, James Hardie (fl. 1862–1933), was born in Edinburgh, the son of David Steadman, a draper, and entered Trinity College, Dublin, as a sizar – the holder of an entrance scholarship for a student of limited means – in June 1882. He became a scholar in 1885 and was awarded BA and BAI degrees in 1887 and an MAI in 1893.

Steadman became county surveyor for the southern division of Donegal at the end of 1898 and transferred to the service of the new county council in the following April. In September 1912, after the resignation of his colleague, **J.R.A. Ferguson** in the northern division, the council agreed to Steadman's request for a transfer to that post and this made it possible for him to take up residence in Derry. He continued in office until his retirement in 1933, after which the council agreed, under strong pressure from the Minister for Local Government and Public Health, that one county surveyor, with two senior assistant surveyors, should take on responsibility for all of the engineering work of the county.

In 1906, Steadman prepared plans for a gas generating station and distribution mains at Donegal for the rural district council. In the same year, he was appointed to prepare plans for a water supply scheme for the lunatic asylum at Letterkenny, the committee of management taking the view that no better engineer could be found in Donegal and that nobody could criticize his work.

STEEDMAN, John (fl. 1790–1868) (sometimes Steadman), Donegal's first county surveyor, is the only one of the first group of county surveyors to merit an entry in the *Biographical Dictionary of Civil Engineers in Great Britain and Ireland, 1500–1830* (London, 2002). It appears that he was originally an architect, builder and land drainage contractor in Edinburgh and became the first secretary of the Society of Scottish Surveyors in 1826. The dictionary records Steedman's work as assistant engineer with Robert Stevenson, the eminent Scottish engineer best known for erecting the Bell Rock lighthouse off the coast of Scotland in 1811; every rock, pool and ledge was given a name by Stevenson, including 'the engineers' ledge, named as a compliment to Steedman and the other assistant engineers. Steedman was associated with a great variety of other important projects in Scotland, including surveys of the harbour of Dundee and of the Firth of Tay, a survey for a proposed canal between Edinburgh and Glasgow, and several early Scottish railway surveys. He became a corresponding member of the Institution of Civil Engineers in 1829 and was contractor for Stevenson's Hutcheson Bridge in Glasgow, built between 1831 and 1834 at a cost of £24,000.

Steedman took up duty in Donegal in May 1834 and set up an office in Letterkenny. As the third largest county in Ireland, with substantial areas of mountainous terrain, the maintenance and improvement of the county's extensive network of roads presented a major challenge. By 1847, the grand jury had come to the view that Steedman, because of his age, was unequal to the task and decided to apply to the lord lieutenant to form two divisions in the county and to appoint two younger engineers to replace him. At the summer assizes in July that year they decided to dismiss Steedman, apparently on the understanding that he would be appointed as county surveyor in a smaller county, but this did not happen. Instead, Steedman found himself unemployed and as he put it himself 'advanced in years, his country and his friends given up, his early commissions in his professional life gone' and with no relations in Ireland and 'those left in Scotland almost all dead'. As there was no provision for pensions for county surveyors until 1875, a 'gratuitous fund' was raised by some of his former colleagues and some other local gentlemen from which Steedman was paid £3 per month from September 1847 to May 1860 and £2 a month thereafter. In April 1861 when serving county surveyors were pressing for the introduction of pension arrangements, Steedman lodged a petition with the House of Commons complaining of his removal from office without compensation and praying for redress. Claiming that he had barely one shilling and four pence a day to live on, he addressed a series of petitions and memorials to the grand jury, the cesspayers, the county surveyors and the lord lieutenant in the years 1865–7 seeking financial support, but to no avail. In 1868, he was still, according to one of his former colleagues, 'thrown upon the world' and being supported in his old age by some of these colleagues and friends.

STOKES, Henry (1808–87), was born in Dublin into a family whose members had distinguished themselves in many fields in the century before his birth and who continued to achieve prominence throughout the nineteenth century. His great-grandfather was Gabriel Stokes (1682–1768), a Dublin mathematical instrument maker, map-maker and surveyor who became Deputy Surveyor General of Ireland in 1748. His grandfather, Revd Gabriel Stokes (1726–1806), was professor of mathematics at Trinity College, Dublin, and his father, Whitley Stokes MD (1763–1845), was regius professor of medicine at the College from 1830 to 1843. Henry's eldest brother, Sir William Stokes (1804–78), succeeded his father as professor of medicine in Trinity and was the leading Irish physician of his time and one of the greatest contemporary physicians in Europe; he was also the friend of George Petrie, whose life he wrote, and patron of Benjamin Woodward (1816–61), one of the most important Irish architects of the early Victorian era. Henry's first cousin, Sir George Gabriel Stokes (1819–1903), became professor of mathematics at Cambridge, president of the Royal Society and a baronet in 1889. Anther first cousin, William Stokes (1793–1864) was an engineer and land surveyor, practising initially in and around Dublin but operating in the 1840s as engineer and contractor on the Shannon navigation works and on other government and grand jury commissions; William's father, John Stokes, had been the Grand Canal Company's engineer from 1804 until his death in 1843. Margaret Stokes (1832–1900), Ireland's greatest contemporary archaeologist, was a niece of the future county surveyor and his nephew, Whitley Stokes (1830–1909), was an important celtic scholar who edited and published numerous gaelic manuscripts and texts.

Henry Stokes told a royal commission in 1841 that he had learned his engineering as an apprentice to a Mr McKenna but no significant engineer of this name seems to have been in practice in Ireland at the relevant time. It is possible that the reference should have been to **William Mackenzie** who was to become Wexford county surveyor in 1838 or, more likely perhaps, William Mackenzie (1794–1851), an important English civil engineering contractor and railway builder. Stokes was only 26 years old when he became Kerry county surveyor on 17 May 1834 and he went on to serve the county for forty-two years. He was probably the most active of the early surveyors in building up a network of new main roads in his county and under his direction the total mileage of roads for which the grand jury had responsibility increased from 520 in 1834 to over 1,200 in 1847, with a further 175 miles under the charge of the Board of Works. In that thirteen-year period, some £120,000 had been made available by the government in loans and grants for construction and improvement work on roads and bridges in Kerry. The collection of OPW Engineering and Architectural Drawings held in the National Archives includes up to fifty drawings of new roads and bridges produced by Stokes between 1835 and 1843, many of them relating to roads which are now major tourist routes – for example, the roads from Tralee to Dingle and to Listowel; the road from Killarney to Castlemaine and on to Dingle via Inch Strand and Anascaul; and the road from Kenmare to Sneem and Cahirciveen. In all, Stokes claimed to have constructed 383 miles of new roads between 1834 and 1868, when the total mileage under his charge had grown to 1,654.

Stokes also had a substantial involvement in famine relief work in Kerry where an expenditure of £322,000 was sanctioned by the Treasury in 1846 under the Labour Rate Act, much of it intended to finance road works. It was claimed by a committee of the grand jury in July 1847, after the works had been closed down, that 370 of the 450 miles of road which had been partly formed at a cost of £346,000 remained 'useless and totally unavailable to the public'. Based on returns and reports from Stokes the committee estimated that it should have been possible to complete the 450 miles for little more than half of the expenditure which had been incurred. Nevertheless, they urged that additional funds of about £80,000 should be made available immediately by the government so that the more important works could be completed and so that a 'wholesome system of employment' could be provided instead of the 'present ruinous and demoralizing system of outdoor relief'.

Stokes was the only one of the first group of county surveyors who did not become a member of the Civil Engineers Society of Ireland at its foundation meeting in August 1835 but he subsequently attended meetings of the society and seems to have been very active in promoting the interests of the surveyors themselves. In September 1840, for example, he acted as secretary at a meeting of the surveyors at which **Charles Lanyon** presided and which adopted a long list of recommendations for the amendment of grand jury law, including provision for increased salaries for the surveyors. In transmitting these to the lord lieutenant, with a request for the establishment of a committee of inquiry, Stokes argued that 'if any mischief has been done by bad surveyors, the cause of such being appointed is that better educated men could not be had for the salary of £300 per annum'. The government, some months later, established a commission to revise the grand jury laws but its 1842 report was shelved and Stokes and his colleagues did not receive increased salaries until nearly twenty years later after a campaign of lobbying in which he was again particularly active.

In 1857 Stokes estimated that up to half of his official salary was absorbed by travel and other expenses but, on the other hand, he had to admit that his earnings from private practice were roughly equal to his salary. He was one of a small number of the early county surveyors who were prepared not only to undertake assignments in architecture and engineering but also to act as contractor on building works. He tendered for the building of the large workhouse at Tralee in 1840 but his price of £11,600 was 25 per cent greater than that of William Hill of Cork who was awarded the contract. Some twenty years later, his tender of £9,000 for building a school at Kenmare (as against the lowest tender of £3,853 submitted by William Martin Murphy of Bantry) attracted critical comment in the *Dublin Builder*: it was said to be an illustration of one of the problems which completely puzzled the public and a 'pons asinorum, not to be solved by the most accomplished of our modern mathematicians'.

Stokes carried out some work on public buildings in Kerry, including the construction of a courthouse at Listowel for which a loan of £1,500 was raised in 1842, and others at Dingle and Cahirciveen. In his early years he also carried on a private surveying practice, charging 6 pence per acre for land surveys and up to £2 12s. per mile for new road surveys, in addition to a fee for supervision of the works. He was engaged by the Board of Works in the late 1830s and 1840s on the maintenance of the 'grant roads', earning about £100 a year from this and, in the 1860s and 1870s he acted as inspector for the board under the Landed Property Improvement Act; in this capacity, he earned a fee of one guinea a day for dealing with applications from counties Cork and Kerry for loans for the improvement of farm land and buildings and for the construction of dwellings for small farmers and agricultural labourers. His reports (often printed in the annual reports of the Board of Works) provide interesting commentaries on the state of agriculture, current farming practices and housing conditions. In 1877 he was advocating that the cost of labourers' dwellings should be kept below £50 – and this could be achieved anywhere, in his view; building such houses at a cost of up to £160 would only deter prudent people from engaging in what was already considered by many to be a waste or unprofitable use of public money. Interestingly, Stokes reported that the use of concrete roofing had been tried with success in his area – and it was cheap, strong, weatherproof and safe from decay.

After his retirement in 1876 with a pension of £400 a year, Stokes produced a number of papers containing unorthodox suggestions on a variety of engineering subjects which were published in the *Transactions of the Institution of Civil Engineers of Ireland* and in the *Irish Builder*. There were papers on the ventilation of coal mines, incombustible materials for building and a paper on *Railway Sleepers and Rails*; the latter advocated the use of longitudinal sleepers tied together by cross-bars, in place of the transverse arrangement of sleepers which was then the well-established norm, and argued for very large flanges on the wheels of locomotives and rolling stock to guard against derailments. In another paper dealing with water supplies for provincial towns and private houses, Stokes suggested that two separate piped supplies should be made available, the smaller one to be filtered for drinking and cooking and the larger one for general use and for fire-fighting. He expressed strong criticism of the water supply for Cork city – the worst he had ever seen – and of Dublin Corporation's Vartry scheme; he suggested that in both cities each house should be supplied with a charcoal filter so as to purify for cooking and drinking purposes the water delivered from the pipes and which he regarded as polluted.

Having lived in Tralee until 1865, Stokes moved with his wife Letitia (neé Bland) to Askive Cottage on a small estate near Sneem owned by his wife's family. He died there on 27 March 1887, aged 79. Letitia subsequently left Askive Cottage (which was burned out in 1922 and later demolished) and went to live with one of her sons in Dublin where she died in 1893. Two of the couple's other sons, Henry Edward and Gabriel, were knighted for services to the government of Madras.

TARRANT, Charles (1815–77), was born in Dublin into a family which was already prominent in Irish engineering. His grandfather, Major-General Charles Tarrant (1730–1818), worked on the Grand Canal and other Irish canals from the 1770s onwards and for the Dublin Wide Streets Commissioners in the 1780s, and was distinguished also for his work as an architect, artist, cartographer and surveyor. His father, one of five illegitimate children and the only son of the Major-General, was Charles Tarrant junior, engineer to the Royal Canal Company and to the Midland Great Western Railway and, for some years, engineer/surveyor to Dublin Corporation; he died in April 1854 when struck by a railway locomotive at the Broadstone station in Dublin.

After a private education, the third Charles Tarrant was apprenticed to his father and worked with him on the maintenance of the Royal Canal and on private commissions. He went on to become assistant engineer to the canal company under his father, and later went to Scotland where he worked on the construction of the Edinburgh & Glasgow Railway. He returned to Ireland on completion of that project and worked for some time as an engineer on the staff of Dublin Corporation. He became assistant engineer to the Waterford & Kilkenny Railway Company under Captain W.S. Moorsom, chief engineer. In that capacity, he was directly responsible for supervising the construction between 1846 and 1850 of a large railway viaduct near Thomastown which was formed of latticed woodwork fabricated and erected by Robert Mallet; it had a span of 200 feet between heavy masonry abutments and a height of nearly 80 feet above the river Nore. Described by the *Illustrated London News* as 'the largest work of its kind in the three kingdoms', the bridge superstructure had to be replaced in 1876–7 by the present bowstring girder bridge in wrought-iron, using the same abutments.

In 1850 Tarrant accepted an engagement on the Susquehanna & Reading Railway in the United States and successfully pressed on with the construction works using large teams of both German and Irish navvies. The climate, however, did not agree with his family and he returned to Ireland in 1851 to take up an appointment as district engineer under the Board of Works in Armagh. He became Monaghan county surveyor on 12 September 1854, having originally applied for such a post in a letter from Glasgow in September 1840 and having qualified at examinations held late in 1851. He was transferred to the Waterford post on 30 March 1855 and also held the position of surveyor to the city grand jury for which a fee of £20 was payable; the latter was effectively a sinecure as the corporation of Waterford repaired all of the streets and roads in the city.

Under Tarrant's management, the roads in Co. Waterford were said to be among the best in Ireland but the growth in the mileage under maintenance – from 833 in 1854 to 1068 in 1868 – and the annual cost of £10 per mile caused problems; he suggested to the grand jury in 1868 that while the additional mileage was obviously beneficial to particular localities, the large additional cost to the cesspayers was difficult to defend. Tarrant was responsible for the construction of a significant number of bridges including, in 1860–2, a large three-arch stone structure with piers standing up to fifty-two feet above the river Dalligan at Ballyvoyle, east of Dungarvan; the central arch of this bridge was destroyed in the civil war in 1922 and rebuilt the following year by **J.K. Bowen**. He also undertook some architectural and drainage projects and constructed sea walls at a number of locations along the coast. He reconstructed the Waterford city and county gaol and designed the landmark Gothic clock tower which was completed with the aid of public subscriptions in 1861 and still stands on the quay at Waterford.

When Tarrant took up duty in Waterford, work on the rebuilding of Thomas Ivory's bridge over the river Blackwater at Lismore after the disastrous floods of November 1853 was well under way and the bridge was due to be opened to traffic at the end of 1855. However, the whole of the new work – five arches and a causeway – collapsed on 26 December 1855 leaving only the main arch and one adjoining arch standing amid an immense mass of rubble. The bridge contractor was held not to be responsible for the collapse and Tarrant set about preparing fresh plans which would eliminate the design faults which had led to the disaster. These were approved at presentment sessions in May 1856, subject to the incorporation of such ornamental features as might be agreed with the local agent of the duke of Devonshire whose castle overlooked the bridge, and who had sought the advice of Sir Joseph Paxton on the matter. Paxton, who was originally the duke's head gardener at his Derbyshire home at Chatsworth, had come to fame as architect of the unique Crystal Palace built for the Great Exhibition of 1851, but the extent to which the modifications he suggested (at an estimated cost of £1,400) may have affected Tarrant's plans for the new bridge is far from clear. However, the fact that the surveyor's March 1856 estimate of £3,000 had increased to £4,000 by the following July suggests that at least some of Paxton's ideas, possibly relating to the parapets and coping stones, were incorporated in the final plans. Work began again on the rebuilding of the bridge in the summer of 1856 and was completed in 1858. The old foundations were used, but six higher arches with spans of up to fifty feet were built to replace all but the central arch of Ivory's original bridge which had been completed in 1779.

Before his Waterford appointment Tarrant had acted as engineer to the seven-mile long Waterford & Tramore Railway, completed in 1853. Twenty years later, in conjunction with Wellington Purdon, a prominent London-based railway engineer, he was engineer for the Waterford, Dungarvan & Lismore Railway which was constructed between 1874 and 1878; the line included substantial viaducts at Ballyvoyle and a number of other locations, a tunnel near Kilmacthomas, and a pratt-truss bridge, 1,200 feet long with an opening span, crossing the river Suir near Waterford. Tarrant's involvement attracted some local criticism because of the damage caused by the works to county roads which it was his duty to protect. He was one of a number of engineers who put forward proposals in 1867 for a water supply scheme for Waterford city but his low-cost scheme, using groundwater from Lisduggan and surface-water from other areas, was not accepted. He seems to have been involved as joint engineer with Thomas Hawksley in the 1870s in developing plans for an alternative scheme but the assignment was terminated before construction stage was reached.

In 1877 the commissioners who owned and operated the wooden toll bridge which had been completed by Lemuel Cox at Waterford in 1794 engaged Tarrant to investigate the condition of the 834-foot-long structure and, particularly, its 40 sets of oak piers. He reported in March 1877 that the superstructure of oak and memel beams was, for all practical purposes, thoroughly sound. He believed that all of the oak piles below the level of the water (which at low tide was up to thirty-three feet deep) were sound and serviceable even after eighty-three years; that the bridge as a whole was safe and stable; and that – unlike the bridge built by Cox at New Ross, also in the 1790s – it had not been affected by the large amounts of floating ice which the river had brought down during the severe winter of 1867. However, to ensure the safety of the bridge, and particularly the central portcullis (installed in 1854 and the weakest section of the entire structure), Tarrant recommended the imposition of a four-ton weight limit which was generous for the time and, in case replacement was decided on, he put forward plans for a new iron girder bridge with seven spans of 85 feet carried on cylindrical piers, and an 80-foot-wide opening span. In the event, the wooden bridge at Waterford remained in service until 1911, notwithstanding a maintenance cost of about £1,000 a year; it was replaced in 1913 by a ferro-concrete structure on the same site.

With William Forsyth of the Board of Works and the county surveyor for the eastern division of Cork east riding, **Samuel A. Kirkby**, Tarrant was appointed by the lord lieutenant in 1876 to report on the condition of the wooden bridge over the river Blackwater at Youghal which had been designed by Alexander Nimmo and built between 1829 and 1832. Plans prepared by Tarrant and the others for the replacement of the decayed wooden structure by a stone and iron bridge on the same site were unanimously approved by the Cork and Waterford grand juries at the spring assizes in March 1877. However, it fell to Kirkby to complete the project after Tarrant's death at his home at Belvedere Terrace, Tramore, on 29 July 1877; he had heart disease of long standing but the immediate cause of his rather sudden death was pneumonia arising from a severe wetting on a yachting trip a few days before. He was buried in the churchyard of Christ Church, Tramore, with his wife, Jane, who predeceased him in January 1871.

Elected to membership of the Royal Irish Academy in February 1848 and a founding member in 1849 of the Kilkenny Archaeological Society, Tarrant was described in an obituary as 'a thoroughly upright, single-minded man'. His death at the age of 61 brought an end to more than a century of involvement by three generations of the family in building up Ireland's canal, railway and road infrastructure.

TATE, Alexander (1823–1904), born in Dublin in 1823, was one of the many early surveyors who received their training in engineering and architecture under Jacob Owen in the offices of the Board of Works at the Custom House, Dublin. He took up a position as assistant county surveyor under **Stewart Gordon** in Derry in March 1844 when the latter sought a competent assistant to superintend the building of Coleraine Bridge across the river Bann; the three-arch bridge was built by John Lynn and Gordon Maxwell at a cost of some £14,650 using Scottish granite. Tate was still occupied on this assignment when the district surveyor system was established in Co. Dublin under legislation passed in August 1844. Qualifying examinations were arranged some months later without any prior public advertisement and Tate was fortunate that a friend in Dublin advised him of the situation. Despite arriving one day late for the three-day examinations at the Custom House, Tate took first place from twenty-four candidates and was appointed as one of the initial three Dublin surveyors. He took up duty at the beginning of 1845 and was initially assigned to the northern district where he lived at Santry Lodge. At a later stage he took on responsibility for the central district comprising the baronies of Coolock, Castleknock and Newcastle.

In 1843–4 Tate designed the courthouse at Balbriggan and the attractive courthouse at North Street, Swords, both of which are still in use today by the district court. The two-storey main block at Swords and the entrance wings on each side are built of coursed rubble masonry with limestone dressings and the main roof is hidden behind a substantial cornice. In both cases, the grand jury refused to grant presentments to meet the construction costs and the projects had to be taken over by the Board of Works for whom the buildings were completed by contractors in 1845; Tate therefore had no involvement with the works but still claimed that the buildings were completed 'mainly according to my designs'. He designed no other public buildings in Co. Dublin and he seems to have had only a very limited involvement in private practice, either as an architect or as an engineer.

Tate gained experience of both the contract system of road maintenance and the direct labour system because, in addition to his normal duties on county roads, he served as surveyor to the trustees of the Malahide turnpike road and those of the Dublin-Dunleer road, from which he earned additional amounts of £30 and £50 a year, respectively. The Malahide road – eighteen miles long, including its branches – seems to have been particularly well managed and profitable: in 1853, receipts from tolls amounted to £1,700, or nearly £350 more than the outlay on maintenance. In his evidence in 1854 to the commissioners who were considering the future of the Dublin turnpike roads, Tate argued that the system by which the trustees maintained their roads was far more efficient and less costly than the system followed on the county roads, where there were no regular staff and maintenance contracts were given to the lowest bidder. By contrast, on the Malahide road, there was 'constant scientific supervision' by a professional person, trained and skilful overseers and labourers, and certain and prompt payment at the end of each fortnight. On the Dunleer road, where there had been numerous complaints about the standard of maintenance, Tate claimed that matters had improved somewhat after 1846–7 when direct labour was introduced; he was forced to admit, however, at the 1854 inquiry that he had not been on the road himself for as much as six months! The fact that all of the major roads in his district, except the Naas road, were managed by trustees meant that Tate, in his capacity as county surveyor, had no first-class roads in his charge for ten years after his appointment. However, this situation ended in January 1856 when full responsibility for the ninety-two miles of turnpike roads in Co. Dublin was transferred to the grand jury and the district surveyors who then became liable for their maintenance in the ordinary way.

In order to ensure that he would be eligible for transfer to any county surveyor position which might arise in the future, Tate was a candidate at the county surveyor examinations held in 1847 but, in accordance with practice at the time, was not advised of the outcome. It was only when a return of those who had qualified was presented to parliament in July 1856 that he discovered that his name was not included. He took the matter up with Thomas Larcom, the under-secretary at Dublin Castle, and was told that his name was indeed on the list of qualifiers where, according to a letter from Larcom 'it stands next in succession to Mr Garrett' but had been omitted in error from the list presented to parliament. It seems to have been assumed that because Tate held a district surveyor position in Dublin he was not a serious contender for a county surveyor post and so six candidates whose names followed his in the 1847 list were appointed to county surveyor posts between 1847 and 1856 when the error came to notice.

Following the resignation of **Charles Lanyon** from the post he had held for twenty-five years, Tate was appointed to succeed him as Antrim county surveyor in December 1861, bringing him a salary of £600 a year as against the £350 he had been receiving in Dublin; he also took on the separate but not very demanding post of surveyor to the county of the town of Carrickfergus. With a corps of up to twelve assistant surveyors, each paid £80 a year, he continued the substantial road-building programme which Lanyon had initiated. One of his assistants in 1870–1 was a young Welshman, John Purser Griffith (1848–1938), a graduate of the Trinity College school of engineering, who later became chief engineer of the Dublin Port and Docks Board, gained a knighthood and was one of the most prominent personalities in Irish engineering in the early decades of the twentieth century.

Tate was a member of the Royal Institute of the Architects of Ireland but few of his building projects are recorded; the most significant during his years in Antrim seems to have been a new town hall completed at Larne in 1876. He was engineer for a water supply scheme at Portrush in 1880 but found it difficult to recover his fee of £480 for the work from the local board of guardians. He was active from an early stage in the campaign for increased salaries and better conditions for county and district surveyors and, as honorary secretary of a committee of the surveyors which was formed in 1850, he gave evidence before a select committee of parliament in July 1857. He was elected a member of the Institution of Civil Engineers of Ireland in April 1845 and served as its honorary secretary and treasurer from 1855 until 1861, and as editor of some of the early volumes of the institution's transactions. But for his transfer to Belfast, he would probably have graduated to the presidency of the institution at a later stage. At the time of his death, he had been a member for nearly sixty years and was regarded as 'father' of the institution.

Tate held office in Antrim until November 1885. His last report to the grand jury at the summer assizes in July that year was a very detailed one; it contained a summary of his activity in the county in the previous twenty-four years during which he had undertaken works with a gross cost of more than £1.5 million, or an average of over £64,000 a year. Awarded a pension of £280, Tate lived in retirement at his home at Whitehouse, near Belfast, for some twenty years maintaining his interest in science, antiquities and astronomy and in the affairs of many learned societies. He died on 29 July 1904, aged 81.

TOWNSEND, Horace Uniacke (1816–97), sometimes known as Horatio Townsend, was born on 28 March 1816, the third son of Thomas Townsend of Clyda, near Mallow, Co. Cork, and his wife, the former Martha Uniacke. He was admitted to Trinity College, Dublin, in 1833 but left to serve a five-year apprenticeship to **Thomas Kearney** who became county surveyor for the eastern division of Limerick in 1837. In December 1838 he sat the county surveyor examinations which lasted for ten days, with written tests from 9.00 a.m. to 5.00 p.m. each day; years afterwards, he admitted that he had 'become knocked up by the close confinement and the excitement [and] had to stay away some time under a doctor's care and go back again for further examination'. Notwithstanding his temporary indisposition, Townsend's performance must have impressed the board of examiners for he was the first of the 1838 candidates to be offered a county surveyor post. He took up duty in Nenagh as surveyor for the newly-constituted north riding of Tipperary on 7 May 1839 and was transferred at his own request to Queen's County on 16 August 1845. The move was much regretted by the Tipperary grand jury which declared in a resolution passed at the summer assizes that they had every reason to be satisfied with Townsend's activity and industry in attending to the duties of his office and with his integrity, ability and attention to the public interest.

Townsend served in Queen's County until September 1883, living at Rathmoyle, Abbeyleix, and enjoying, as he put it, the private practice incident to a local civil engineer; this included the carrying out of surveys of proposed new roads for which his charge in the 1840s was £4 a mile. In 1872, when he had been in office for twenty-seven years, he came under pressure from some members of the grand jury to transfer his office from Abbeyleix to the courthouse in Maryborough (Portlaoise) so that the allowance of £50 paid to him for his clerk and office could be reduced. Townsend claimed that there was no room at the courthouse, pleaded that some indulgence should be shown him after his long service and hoped that he would not be put to the trouble and expense of moving at a time when his service was probably drawing to a close. The grand jury's demand was castigated by the *Irish Builder* which described it as 'sharp practice against a public servant of good repute and long standing' and they decided not to press the matter.

There were more problems for Townsend at the spring assizes in 1872 when he and his grand jury came in for severe criticism from the lord chief justice because of the condition of the county courthouse which had been completed some seventy years before to a design by Sir Richard Morrison. Townsend explained that a presentment for some of the necessary works had been thrown out by the presentment sessions but this only provoked a further outburst of criticism from the bench: 'I am put here in a pit I will not continue to sit in … This building is an antediluvian one … The room I have in here is a filthy one … This courthouse is like an old barn, and a very bad barn … It is a shabby, shabby, courthouse, as shabby as any courthouse you could find in the three Kingdoms'. Two years later, the grand jury acted. A committee was appointed to arrange for the necessary improvement works and since Townsend was obviously not considered to be sufficiently competent, advertisements were placed seeking the services of an architect to prepare designs. J. Rawson Carroll of Dublin was appointed and in April 1875 a tender of £1,880 was accepted for the works.

By 1879 a significant proportion of the road maintenance work in Queen's County – over 50 per cent in some baronies – was being carried out under Townsend's direct supervision and effectively on a direct labour basis, because of failure to obtain tenders or because of the default of contractors. Even then there were serious difficulties with many of the remaining contractors. Townsend complained of their failure to provide adequate quantities of materials in many cases and of the fraudulent removal of measured materials from one section of road to another. He repeatedly threatened to prosecute contractors and to withhold the certificates without which they could not claim payment for their work but there was no real improvement in their performance.

Townsend had been in office as county surveyor for forty-four years when, in May 1883, he told the county at large presentment sessions that he regretted the time had come when his health and age disqualified him for the discharge of his duties. If he felt himself able to go on any longer he would not throw up his post but, he said, 'I am not able and I don't think it would be honest for me to undertake any longer duties I am not able to discharge'. The magistrates and cesspayers accepted Townsend's resignation with regret and recommended that he should receive the maximum pension of £347, two-thirds of his salary. They noted his long and faithful performance of his laborious duties and decided to set up a fund to finance a suitable testimonial, with individual subscriptions limited to £1. Townsend was reported to have been 'much affected' by the tributes paid to him, which he said were calculated to touch a person's feelings.

After his retirement the Board of Works commissioned Townsend to conduct an enquiry under the Tramways Acts into the merits from an engineering point of view of proposals from a local group for a tramway, nearly fifteen miles long, from Skibbereen to Schull. His report of 7 March 1884 advised that the scheme was feasible subject to modifications, and his former colleague, **Nat Jackson**, surveyor for Cork west riding, agreed. The line was built in 1885–6, with yet another county surveyor, **Samuel A. Kirkby**, as joint engineer, but was never a success.

Townsend married Louisa Jane Clarke of Wilfield, Co. Dublin, in 1841 and the couple had five sons and seven daughters. After his retirement, he lived at Waltham Terrace, Blackrock, Co. Dublin and died on 15 February 1897, aged 81. He was buried at Mount Jerome cemetery in one of a cluster of graves containing the remains of several generations of the Townsend family, including his father and mother who had come to live in Dublin in the 1830s and two of his daughters. His grave adjoins that of another county surveyor, **Frederick Gahan**, to whom he was related by marriage.

TREACY, William Augustus (1818–86), (sometimes Tracey) was the only son of John Treacy of Brigadie House, Ballymena, Co. Antrim. He was living in Highgate, London, in June 1835 when he was elected to associate membership of the Institution of Civil Engineers. The fact that he was only 17 years of age at that stage and that his proposer was Sir John MacNeill suggests that Treacy was then a pupil or junior assistant of MacNeill. By the mid-1840s, he had established a private practice in surveying and engineering in Cork city, with an office at the Grand Parade, and was working as an engineer for Cork Corporation. He also had some limited involvement in the development of local railway proposals during the railway mania of 1845–6.

Treacy became county surveyor for the west riding of Cork on 10 April 1846 and some years later went to live at Kilbrogan Street, Bandon. Immediately following his appointment he was obliged to become heavily involved in the organization and supervision of famine relief works and, like most of the other surveyors, he came in for criticism from various quarters; there were allegations that works were badly designed, that there was poor supervision and abuse in the management of the works and that the payments system was grossly inefficient. On one occasion in October 1846 major controversy arose from a report in a morning newspaper that a young person had died from starvation while engaged on works in the Skibbereen area for which wages were long overdue but, when called on for a report, Treacy was able to point out that an inquest had found that the young man had 'died a natural death'.

When the situation returned to normality after the famine years, Treacy seems to have served satisfactorily in the west riding before transferring to the east riding on 31 March 1855 following the resignation of **Sir John Benson**. He had replaced a large number of bridges in the west riding after the serious floods of November 1853 and he went on to build some substantial bridges in the east riding also. These included Colthurst's Bridge over the river Blackwater, completed in 1859, and Heard's Bridge, a long wooden bridge across the estuary of the river Bandon above Kinsale; this was completed at a cost of over £6,000 in 1861 and named in honour of the local MP, John Isaac Heard. Emergency repairs to the bridge cost £1,000 as early as 1869 and notwithstanding further repair work in subsequent years, the bridge collapsed in 1881, a year before a replacement bridge was completed, the red pine eaten away by the sea and by barnacles, as one contemporary account put it.

When Treacy's transfer from the west riding was announced, the magistrates and cesspayers at several of the presentment sessions expressed great regret at the loss of his valuable services. In the east riding, however, things were very different after a few years. There were complaints about the condition of the roads generally and specific charges about the state of some of the main roads such as the road from Rathcormack to Fermoy which was said to be virtually impassable by 1861. At the spring assizes in March that year it was further alleged that Treacy was guilty of having 'disreputably' prepared a specification for a sewer in Mallow and that he had failed to provide adequate supervision of roads and public works in different parts of his area. A motion to dismiss him was defeated by eleven votes to four, with some abstentions, but the grand jury censured him for the 'considerable want of care and attention on his part and that of his assistants'. In his defence, Treacy pointed to the size of his area which had a greater mileage of roads than any other surveyor's district and to the problems caused by the heavy traffic in and out of Cork city. He would try to carry on 'as far as my health and abilities admit' but he repeated the call he had made in earlier years for the establishment of two divisions in the riding, with a separate surveyor for each. The grand jury set up a committee which endorsed this arrangement. By the summer assizes Treacy was reported to be very ill, with his colleague in the west riding having to act for him, and while the lord lieutenant had agreed to the appointment of a third surveyor, the person to whom the post was offered was in Australia and was unlikely to take up the offer. A new surveyor, **A.O. Lyons**, eventually took up duty in September 1861 and Treacy, whose health had obviously recovered, exchanged places in the following November with **F.G. Deverell** who had been serving in the southern division of Mayo since 1856.

Treacy was seeking a transfer from Mayo to 'a first class county with a good school' from early 1867 and managed to achieve this in July 1868. When he gave notice at the summer assizes that year of his transfer to the southern division of Tyrone, in an exchange which he had agreed with **W.H. Deane**, the grand jury expressed their regret at the loss of his services and, while acknowledging the financial gain for the surveyor, protested that the timing and suddenness of the transfer would cause great disruption to their business. Treacy had explained to the chief secretary's office that he wished to leave Westport because of 'the impossibility of obtaining tuition for his children in this place'. The *Mayo Constitution* noted that he had earned the good opinion of all classes and the highest testimonials from the leading gentry of the county but matters were obviously very different in Tyrone where Treacy was dismissed in July 1869 after little over a year in office. He made a number of applications to the chief secretary's office in the years that followed seeking re-appointment as county surveyor in another county but these were unsuccessful.

Treacy married Agnes, daughter of John J. Thompson, manager of the Bandon branch of the Provincial Bank of Ireland, at Rathclarin church on 3 August 1852. He was a widower when he died in Ballymena on 18 February 1886, aged 68.

TURNER, Thomas (*c.*1820–91), was born in Dublin. He was a son of the ironmaster Richard Turner (1798–1881) who was responsible, with Decimus Burton, for the design and construction of the famous Palm House at London's Kew Gardens (1844–8), the Winter Gardens at Regent's Park, London (1845–6), the great curvilinear range of glasshouses at the National Botanic Gardens in Glasnevin (1843–8), and the magnificent Palm House at the Belfast Botanic Gardens on which he collaborated with **Charles Lanyon** in 1839–40. Like many of the early county surveyors, the young Turner began his career as a pupil of Jacob Owen in the offices of the Board of Works at the Custom House. He moved to Belfast in the 1840s as Lanyon's chief assistant and the association continued until 1851. Lanyon's architectural and engineering practice was a particularly busy one during that period and, while it seems likely that Turner made a significant contribution to many of the designs produced by the office, it is not possible to associate him definitively with particular projects. Turner did, however, sign some of the drawings for sections of the Larne-Glenarm road in 1840, including a tunnel, and his signature also appears on drawings for the fifteen churches that the office designed in 1843 for the Down and Connor Church Accommodation Society.

With his father, Turner produced one of the 245 entries in the 1850 competition for the Great Exhibition Building for which Joseph Paxton's famous Crystal Palace design was eventually chosen. Around the same time he designed Craigavad House on Belfast Lough, now the clubhouse of the Royal Belfast Golf Club, and following an open competition he was commissioned in 1851 to provide an elaborate new screen of Glasgow sandstone, ninety-five feet wide, across the forecourt of St Patrick's pro-cathedral, Dundalk. He had set up in private practice on his own account in Belfast, with commissions for a number of houses in Co. Down and elsewhere when, on 6 February 1852, he accepted appointment as Cavan county surveyor, having qualified for the post at the examinations held in 1847. It appears that Turner never took up duty in Cavan but appointed **Frederick Gahan**, who had also qualified at the 1847 examinations, as his deputy. He submitted his resignation at the assizes in March 1853, indicating that he could not come to reside in Cavan because of his architectural business, leaving the way open for Gahan to succeed him. In the course of the following twenty years, Turner sought reappointment as county surveyor in 1860, 1865, 1867 and again in 1869, even on a *locum tenens* basis, and seems to have acted as deputy for **Richard Williamson**, the Londonderry surveyor in 1873.

As an architect Turner was in partnership at various times with **Stewart Gordon** and Richard Williamson who successively held the post of Derry county surveyor between 1834 and 1874 and, in addition, he carried on an architectural practice from a separate office in

Glasgow at intervals throughout the 1860s. He was in partnership briefly with the 23-year-old Thomas Drew (later Sir Thomas Drew) in 1861 when they were one of six architectural practices invited by the Royal College of Physicians in Ireland to enter a competition for the design of new premises at Kildare Street, Dublin. They submitted two proposals but the scheme of William George Murray was considered to be the best and most suitable by the building committee. In the 1870s and early 1880s Turner was in partnership with Hume Babington in Derry.

Turner's buildings include the council offices in Derry; an imposing Italianate railway station at Foyle Road, Derry for the Great Northern Railway (now demolished); the Northern Bank building in Shipquay Place, Derry (1866), probably his finest single work in the city and 'a noble monument to his memory' according to the *Londonderry Sentinel*; the now demolished Academical Institution in Derry (in partnership with Williamson); Coleraine Town Hall (1859); and the Irish Society's Schools in Coleraine (1867), also with Williamson. He designed the courthouse in Lurgan (1874) in partnership with **Henry Davison**, the Armagh surveyor and in the early 1870s, the courthouse in Magherafelt, in this case, perhaps, in partnership with Williamson. He was a prolific designer of country houses, villas and gate lodges throughout the northern counties (including Kintullagh, Co. Antrim, 1863), as well as schools and a number of churches. He was responsible for additions to the county courthouse in Bishop Street, Derry and, in conjunction with Henry Davison, for the remodelling of Francis Johnston's courthouse at Armagh between 1859 and 1863. In 1858 he was commissioned by the Cleeland family to transform their Georgian-type country house outside Belfast, dating from the 1830s, into an elaborate Scots Baronial castle, with turrets, stepped battlements and bartizans with conical caps. The result was Stormont castle which, with its 235-acre demesne, was acquired in 1921 as the site for the Northern Ireland parliament buildings; the castle itself served as the official residence of the prime minister and later as the headquarters of the secretary of state for Northern Ireland.

After his early success in the Dundalk pro-cathedral architectural competition, Turner's subsequent competition entries brought mixed results. In 1857 he was an unsuccessful candidate in the competition for the design of a corn exchange building in Dundalk, now the Town Hall. Three years later he was placed fifth out of fourteen competitors in the competition for St Andrew's Church, Suffolk Street, Dublin which was won by the firm of Lanyon, Lynn and Lanyon – but he was entitled to take some comfort from the fact that his entry was judged to be almost as meritorious as that submitted by the famous firm of Deane and Woodward which was placed fourth. In 1863 he won the premium of £200 in a competition for the design of a new head-office building in Manchester for the Lancashire Insurance Company – a compliment to his well-known but hitherto somewhat tardily recognized ability, according to the *Dublin Builder*. In 1866 he designed the podium in front of the principal entrance to Dublin's City Hall, replacing the stone plinth of the building – originally the Royal Exchange – which was completed to a design by Thomas Cooley in 1779. In 1866 work was also in progress on the erection of a tower and spire, 185 feet high, which he had designed at the Mariner's Church, Kingstown (Dun Laoghaire), together with other alterations and improvement works affecting the entire building.

Late in 1864, after the foundation stone had already been laid, he was one of the many leading architects and sculptors who submitted sixty designs in a competition for the national monument which was to be erected to Daniel O'Connell close to the then Carlisle Bridge in Dublin. Turner's entry – one of only seven submitted in model form – was remarkable for its clever and artistic grouping, in the opinion of the *Dublin Builder*, and was pleasing in outline and general effect; it consisted of a substantial square pedestal, with bas-reliefs of events from O'Connell's life and radiating pedestals on which were groups intended to represent each province under the protection of its patron saint. At the centre was a circular pedestal, with four allegorical figures representing O'Connell's highest attributes and the whole was surmounted by a figure of O'Connell; the latter was reported at the time of the exhibition of the entries at the City Hall to have been described by 'an excited spectator of the unwashed class' as a figure of 'a Virginian Negro in a night-shirt'. The selection committee found themselves unable to recommend any of the original competition entries and the commission for the monument was eventually given to John Henry Foley in 1867 after a second competition had again failed to produce an acceptable design. Foley's model for the monument was on view in Dublin and approved by the committee in December 1867 but he died in 1874 before completing the assignment. The monument was eventually completed by his principal assistant, Sir Thomas Brock and unveiled in August 1882, thirty-five years after the death of O'Connell and nearly twenty years after the commemoration committee was established.

There was another disappointment for Turner in 1875 when he entered an architectural competition for extensions and alterations at the Royal College of Surgeons at St Stephen's Green in Dublin. Turner and other prominent architects had been led to believe that the College architect, W.J. Symes, whose earlier plans had been rejected by the college, would not be an entrant but Symes was allowed to compete and one of the three designs he submitted was awarded the premium of £100. In a letter to the *Irish Builder* in August 1875 Turner was highly critical of the outcome, complaining that he and his colleagues would not have spent time and thought and money on the competition had they known of Symes' involvement; he found it difficult to understand how 'a number of gentlemen, individually beyond reproach, could, even under the cloak of the whole council, feel justified in adopting a course which is so plainly in opposition to fairness and propriety'.

In November 1861 a citizens' committee which had been formed in Dublin advertised for plans for the replacement of the narrow and hump-backed Carlisle Bridge which had been completed to a design by James Gandon in 1794 but which, with increased traffic, was considered dangerous and inconvenient. Over sixty entries were received by the summer of 1862 and placed on public exhibition but the referees did not announce their verdict until March 1864. The winning entry was a single-span cast and wrought-iron structure submitted by Thomas Turner's father, Richard, working with George Gordon Page of London and a proposal for a two-arch stone bridge, submitted by the firm of Lanyon, Lynn & Lanyon was placed second. Thomas Turner submitted a design for a three-span iron structure on stone piers but this did not find favour with the judges. No definite action was taken to rebuild Carlisle Bridge for some years but it was reported by the *Irish Builder* in 1870 that the Lanyon Lynn & Lanyon design was then favoured by Dublin Corporation. When the issue was revived in 1875 a committee of Dublin Corporation agreed after considerable discussion to recommend another scheme prepared by Turner; this involved the erection of new iron structures on each side of the existing bridge so as to give a roadway the full width of Sackville Street, with the old bridge being taken down and replaced in iron once the new side spans were in place; the existing piers were to be preserved and extended, giving three arches as before; the ironwork was to be of an ornamental character with elaborate turrets at each of the corners; and the whole project was to cost only about £35,000. But opposition to Turner's proposals soon emerged. Parke Neville, the city engineer, strongly favoured a stone bridge the cost of which he estimated at £80,000; so too did Bindon Blood Stoney, the Port and Docks Board's engineer and some members of the corporation. The view that a stone structure would be more in harmony with its surroundings and the public buildings of the city eventually prevailed, and Stoney himself was commissioned to provide plans for the replacement masonry bridge which was opened in May 1880 and renamed O'Connell Bridge.

Turner's return to live at Raheny House in his native Dublin in the early 1880s is difficult to explain but may have been connected with the death of his father in October 1881. His decision in May 1883 to take up, at the age of 63, the badly paid position of county surveyor for the northern division of Co. Dublin is even more surprising but he held the post until his death on 10 October 1891. He never married and was buried in the family tomb in Mount Jerome cemetery, close to the granite sarcophagus which stands over the family vault of the Morrisons, William Vitruvius and his father, Richard, who led the architectural profession in Ireland until his death in 1849. An obituary in the *Londonderry Sentinel* described Turner as 'an architect of great ability and a man of varied scientific attainments' whose 'kindly face and genial good nature' would long be remembered by many friends in Derry where so much of his work was carried out.

VERNON-HARCOURT, Leveson Francis (1839–1907), was one of thirty-five candidates who competed for two vacancies when the first Civil Service Commissioners examinations for county surveyorships were held in November 1869. He gained first place and **Henry Temple Humphreys** took second. In reporting the results in January 1870, *The Engineer* hoped that both gentlemen had insured their lives and went on: 'we congratulate them on no small display of personal courage; upon the whole, we think that a surveyorship in Sierra Leone would be the safest (*sic*) of the two posts'!

Vernon-Harcourt was born in London on 25 January 1839, the second son of Admiral Frederick Edward Vernon-Harcourt, a grandson of Edward Harcourt, archbishop of York, and a cousin of Sir William Harcourt (1827–1904), a prominent liberal MP from 1868 until the end of the century, home secretary in the 1880s, and chancellor of the exchequer in Gladstone's government in 1886 and again in 1892–4. The future county surveyor was educated at Harrow school and at Balliol College, Oxford where he graduated with a BA in mathematics and natural sciences in 1862 and an MA in 1865. He served a pupilage to Sir John Hawkshaw, a prominent consulting engineer, between 1862 and 1865 and remained as an assistant in the office when his training was completed. He was resident engineer on a major development project at the West India Docks in London until 1870.

Following his success in the November 1869 examinations, Vernon-Harcourt was appointed Westmeath county surveyor and took up duty in January 1870. He resigned the appointment on 11 August 1870, a week after his marriage to Alice, daughter of Lt.-Col. R.H. Brandreth of the Royal Engineers, to take up a position as resident engineer on harbour works at Alderney. He was back in Ireland in 1872–4 as resident engineer on the Rosslare railway and harbour works for which **J.B. Farrell**, Wexford county surveyor, was one of the contractors. This assignment was followed by practice in London as consulting engineer, specializing in river, harbour and dock works, as well as in water supply.

In 1882 Vernon-Harcourt was appointed to the chair of civil engineering at University College, London, and retained the professorship until 1905 when he was elected emeritus professor. During his years as an academic he continued to act as consulting engineer for a variety of public authorities and was called on to report by the government on major projects at home and abroad. Well known as a writer on technical subjects, he developed a reputation as an authority on the design of structures for inland waterways and on the development of harbours on the coast and in estuaries. He contributed no less than eighteen papers, mainly on marine engineering matters, to the *Proceedings of the Institution of Civil Engineers* and was a frequent participant in the discussions on papers presented by others. In addition, a number of his papers were published in the *Proceedings of the Royal Society* and he was the author of important textbooks, including *Harbours and Docks* (1885); *Achievements in Engineering* (1891); *Rivers and Canals* (1882, 1886); *Tunnels, Bridges, Canals, Lighthouses, etc described by a Civil Engineer* (1892); *Civil Engineering as applied in Construction* (1902); and *Sanitary Engineering with respect to Water Supply and Sewerage Disposal* (1907). He contributed (through the secretary) to the discussion at the Institution of Civil Engineers of Ireland in January 1890 on the internationally famous paper by Robert Manning on *The Flow of Water in Open Channels and Pipes*.

After a few weeks' illness, Vernon-Harcourt died at Swanage on 14 September 1907. An obituary described him as a man who had laboured incessantly but unobtrusively for the advancement of his profession and a strong and consistent advocate for the thorough scientific training of engineers. He was said to have been imbued with a high sense of duty, a man who did his work without expectation of honours or rewards and whose integrity, simplicity and gentleness of character was recognized no less by his professional colleagues than by his personal friends. On foot of a bequest which he made to the Institution of Civil Engineers, the series of prestigious Vernon Harcourt lectures began in 1920; a lecture is held every other year at the institution's headquarters in London on a subject relating to river, canal or marine engineering to which he had particularly devoted himself as a consulting engineer.

WALKER, John (fl. 1834–54), was in practice as an engineer and surveyor, mainly in the Carlow area, before his appointment as the first county surveyor for Limerick on 17 May 1834. He claimed that he had 'by much perseverance and labour in his profession, obtained the confidence of a great majority of the gentry in the line of country extending from Baltinglass to Waterford'. His designs and surveys were well received by grand juries and landed proprietors and he believed that he would have continued to be well remunerated but for the passing of the Grand Jury Act 1833 under which he foresaw that his 'most creditable and profitable line of business would be superseded by the new county surveyors'. He had hoped for an appointment to a county near Carlow so that he could continue to attend to his existing business but on his appointment to Limerick he was forced to transfer his engagements in Leinster to others.

From the outset, Walker had difficulty in meeting the demands imposed on him by his Co. Limerick post and the additional, though not very exacting, post of surveyor for the county of the city of Limerick. He complained particularly about the time and expense involved in travelling, including the support of his 'establishment of horses' which, he claimed, had amounted to more that his salary in the first half of 1835. With thirteen baronies in the county, attendance twice each year at the presentment sessions and adjourned sessions was itself an onerous undertaking. Added to this was the need to inspect beforehand each of the proposed works and the completed works, scattered over a network of about 1,000 miles of roads and some 1,500 square miles. In describing his activities in connection with the sessions held in the autumn of 1834, Walker told his grand jury at the following spring assizes that 'in the performance of these engagements and the attendance at the sessions, it became necessary to travel nearly 1,500 miles in six weeks, at set times and hours, in all the cross and indirect roads and consequently without the aid in a single instance of the public cars and coaches'.

At the summer assizes in 1835 Walker was again protesting about his working conditions. He reported that he had been 'induced to persevere and to struggle with his difficulties in the hope that the legislature would have removed the impossibilities and have obviated some of the difficulties but that hope having proved delusive, and having made exertions and suffered privations beyond his power of endurance, an engagement on his part to continue would be a want of justice towards himself and a want of candour towards the public'. He suggested to the grand jury that the county should be divided into two parts, with a surveyor for each, and that travelling and other expenses should be paid to the surveyor in addition to his salary but, although the grand jury were impressed by his zeal and skill and were willing to implement these suggestions, they were unable to do so without the approval of the authorities in Dublin and new legislation. When the case for higher salaries and travelling expenses was not accepted by a select committee which reported in August 1836 the prospect of any improvement in his conditions must have seemed to Walker to be very dim indeed. The last straw was a letter from under-secretary Drummond stating his opposition to the creation of two county surveyor posts in Limerick; on receipt of this letter, Walker submitted a letter of resignation on 11 November 1836 on the grounds that his pecuniary resources would not allow him to continue in office.

After his resignation Walker continued to reside in Limerick where he was engaged by the Board of Works to supervise the construction of an extension to the lunatic asylum. In 1837, following the dismissal of his successor, **William Horne**, a decision was made to divide Limerick into two divisions with a separate surveyor for each – the first county to be treated in this way. Walker, who was held in high regard by the Board of Works, was appointed on 1 July 1837 to one of the new posts and served in the western division for over three years, with his residence and office at Rathkeale.

After the death in July 1840 of the Waterford county surveyor, it was reported in the local press that Edgar Clements, a prominent road contractor who was 'in every way qualified', was to succeed him. However, it was Walker who was appointed to the vacant post on 12 August 1840 – 'it having originally been my desire'. The appointment was welcomed by the *Waterford Mail* which described the new surveyor as a man of urbane manner, gentlemanly demeanour, experience and first-rate ability. Walker took on the separate post of surveyor for the city of Waterford from which he earned an extra £100 but pressure of work prevented him from carrying on a private practice as he had attempted to do in Limerick.

Walker moved again on 31 July 1844, this time to Carlow in an exchange with **Charles Forth**. At the spring assizes in 1849 several members of the grand jury demanded that he should in future give his evidence in relation to particular presentments on oath, as other witnesses before the grand jury were required to do. Walker strenuously objected to this, asserting that in his fifteen years as a county surveyor 'my word was considered as valid as my oath'; besides, what was now being sought was illegal, unjust and personally insulting. After some further acrimonious exchanges, he agreed, reluctantly, to take the oath but told the jury that he would immediately apply for a transfer to another county.

One year later, Walker was still in office in Carlow, reporting on the damage to roads and bridges which had been caused by the floods of the winter of 1849–50 and presenting proposals for alleviating serious flooding in Carlow town which had arisen from the restricted channel of the river Burren which joined the Barrow in the town centre. Walker's solution was that a combined drainage and navigation scheme should be undertaken on the seven-mile stretch of the river between Carlow and Tullow at an estimated cost of £90,000; he suggested that the navigation would be of major benefit to Tullow, which had neither rail nor canal communication with the rest of the county, and that navigation fees would bring a return of 2 per cent on the outlay. Walker's scheme did not find favour with the grand jury. A few months later at the 1850 summer assizes a major dispute erupted between him and the jury arising from their rejection of yet another set of proposals, this time involving the renewal and improvement of the network of sewers and drains in Carlow town as a step towards mitigating flooding by the rivers Barrow and Burren. In a letter to the *Carlow Sentinel* a few days later, Walker defended his proposals trenchantly, explaining that they were based on his opinion as a professional man, experienced in the care of cities and towns, and clear of every local influence; he went on to criticize the 'irresponsible and ill-considered interference' which had defeated proposals whose value should be obvious 'to any intelligent and unprejudiced mind'. On 7 August 1850, ten days after this attack on his own grand jury, Walker was, not surprisingly, transferred from Carlow to Monaghan.

In his reports to his new grand jury in 1851 and 1852 Walker noted that, because of a 'general anxiety to economize the public funds', the county works programme was a very limited one, confined essentially to repair work and other work of absolute necessity, and little progress was being made even with this limited programme because so many of the local labourers were engaged in agricultural activity. He and his grand jury were seriously concerned about the condition of many of the bridges constructed by Board of Works engineers in the previous five years in conjunction with the major land drainage programmes carried out in the area; settlement was reported in some cases, there were complaints about unsafe parapets and some bridges, like the one at Aclint, were said to be in danger of being undermined because of inadequate pavement of the river bed.

At the spring assizes in February 1853 Walker told the grand jury that the main roads were then in an excellent state of repair but that his efforts to prosecute defaulting contractors had been frustrated because of irregularities in the service of summons. After the formal business had been completed the jury decided that Walker was guilty of neglect of his duties but, rather than dismiss him and suspend payment of his salary, they allowed him to submit a letter of resignation. However, when the jury had been discharged, Walker withdrew his resignation and much to their annoyance insisted on his right to attend the following assizes in July and the assizes in March 1854. On that occasion he brought the matter before the assizes judge, Baron Greene, insisting that he wanted an opportunity to defend himself in public and to retire to private life after twenty-eight years public service without the stigma of a dismissal. However, the grand jury adhered to their position and the judge ruled that it was not for him to interfere and that Walker had been properly dismissed. A subsequent appeal by Walker to the lord lieutenant to be retained in office, at least until the summer assizes, was unsuccessful and **William Carroll** was appointed to succeed him a few weeks later. A return submitted to parliament in July 1856 noted that Walker had died before then.

WALSH, Thomas (1875–1955), was born on 7 March 1875 near Gortnahoe, Co. Tipperary, and was educated at Rockwell College. After a pupilage with T.R. O'Donoghue ARCScI, he attended Queen's College, Galway, from which he graduated with a BE degree in 1905. Having served as an assistant county surveyor in Tyrone, Monaghan and Roscommon and with the Congested Districts Board, he became county surveyor for Queen's County in July 1909 on his second attempt; he had been an applicant for the post when it was previously filled in November 1908 but was defeated in a county council vote on that occasion by **Patrick Morris** who died after only a few months in office. Walsh served in Queen's County for nearly five years and earned the respect and admiration of his council for his endeavours to improve the county's roads. He introduced steamrolling on some sections of the main roads and, while he opposed the introduction of formal direct labour schemes, he continued to have direct responsibility for the maintenance of a substantial mileage of roads which for many years had not come within the contract system.

When **Patrick J. Lynam**, Louth county surveyor, died in office in August 1912, the county council decided to fix the salary of his successor at only £275 instead of the £480 paid to Lynam. Some members of the council were opposed to the role of the Local Government Board in the process of appointing a successor and felt that they should be allowed make an appointment themselves as the surveyor's salary was met entirely from local funds. They put forward the name of one of their assistant surveyors, J.M.C. Lyons, for appointment to the vacant post but, although Lyons had a degree in engineering and twenty-eight years service as assistant county surveyor, he was ruled ineligible by the Local Government Board, because of his age, to sit the qualifying examination. The county council thought this unfair and resolved that as Lyons had a complete practical knowledge of the requirements of the county, he should be recognized as a competent candidate. Matters dragged on until September 1913 when the council decided to go ahead unilaterally and voted by a large majority to appoint Lyons. The appointment was a very popular one but the Local Government Board still refused to sanction it on the grounds that Lyons was above the maximum age limit of 45 laid down in regulations made in February 1887. The county council finally accepted the inevitable: they advertised the position again in November 1913 at a slightly improved salary of £300, with £100 for travelling expenses, and in the following month elected Thomas Walsh – one of nineteen candidates – to the office. He took up the position in Louth in April 1914.

With the aid of substantial government grants, Walsh built long stretches of concrete roads between Drogheda and Dundalk in the 1920s as well as stretches of new roads in the Ardee, Carlingford and other areas. In 1930–1 he planned and supervised the construction of additional offices for the county council on part of the site of the old county gaol at Crowe Street, Dundalk; as this was not considered to be part of the normal duties of the county surveyor at the time, Walsh was paid the appropriate architect's fee for his work. He was given delegated responsibility for the turf production programme in the hilly north Louth area during the emergency years; the work was carried out on a direct labour basis and involved both the construction of access roads and bog drainage works. Because of the shortage of material and equipment and the diversion of road labourers to turf production, the roads of the county suffered, necessitating the preparation of a post-war road restoration and improvement programme and the allocation of substantial additional funds for which Walsh received the approval of the council in 1946. He retired on 31 March that year when he was 71 years of age and died, unmarried, on 25 March 1955 at his residence, Seatown Place, Dundalk; he was buried locally at St Patrick's cemetery.

WARNER, Claud Alphonsus (1911–2003), was born in November 1911 in Galway city where his father Joseph J. Warner was employed by the Midland Great Western Railway Company as station master. He was educated at St Joseph's College and gained a first class honours BE at University College, Galway, in 1932 and an ME in 1942 for which he submitted a thesis on 'Fundamentals of Highway Engineering'. He was an assistant county surveyor in the western division of Galway until June 1933 and worked subsequently for short periods in a private practice and as assistant to the Galway borough surveyor. He was employed by the Department of Local Government and Public Health between 1934 and 1936 on housing inspection work in Galway, Mayo and Roscommon.

In July 1936 Warner joined the newly-established Turf Development Board, working initially in the west of Ireland, and became assistant chief engineer in 1941 when the work of the board was expanded considerably in an effort to compensate for the wartime scarcity of other fuels. In his autobiography *Man of No Property* (Dublin, 1982), Dr C.S. Andrews who was managing director of the board, describes how Warner 'a particularly bright young engineer' contributed to the drawing up of a post-war development programme which was submitted to the government in December 1944. Andrews noted that Warner had 'unusual technical and mathematical skills' and dealt with the 'engineering calculations and financial costings' of a programme which was to produce one million tons of machine turf each year; apart from this 'he was something of an original … his taste in dress, which ran to green trousers, yellow pullovers and linen jackets, foreshadowed in the forties the product of Carnaby Street in the sixties'.

Following the first competition arranged by the Local Appointments Commissioners for county engineer positions, Warner was appointed Leitrim county engineer with effect from 29 May 1944, becoming the first person to take up appointment in this new grade. He transferred to Kerry in December 1944 and moved again in January 1959 to his native Galway where he served until retirement in November 1976. He died in Galway on 31 December 2003 and was buried at the new cemetery in Bohermore.

WEBSTER, Henry (fl. 1862–1930), was born at Monkstown, Co. Dublin, in October 1862, the son of John Webster, a merchant. He entered Trinity College, Dublin, in November 1879 as a pensioner, paying a fixed annual fee for his education, and obtained a BA degree in 1883 and a BAI in 1885, having also been a gold medallist. He was employed from 1885 to 1889 in the Board of Works under Robert Manning, mainly on harbour work in the west of Ireland, before his appointment as county surveyor for south Kerry in October 1889. In response to a resolution passed by the grand jury at the summer assizes in 1891, the lord lieutenant agreed that Kerry should revert to having only one surveyor, as was the case from 1834 to 1887; Webster was therefore transferred to Wexford in August of that year, leaving the way open for **Singleton Goodwin**, his colleague in North Kerry, to take on responsibility for the entire county.

After his transfer to the service of the new Wexford county council in April 1899 Webster had a series of long and bitter conflicts with his new employers on a variety of issues. His salary on taking up appointment in Wexford in 1891 had been set at £500 a year and he had been advanced to the legal maximum of £600 in 1894. He sought an increase of 50 per cent in 1899 to bring his salary to a new total of £900 a year, claiming that the increase was justified by the additional work arising under the Local Government (Ireland) Act 1898 and by the fact that his new duties were so great that he could not attend to his private practice. The council were not willing to concede any increase and Webster then appealed to the Local Government Board, as did many of his colleagues who were not happy with their new situations. In February 1900 the board fixed new salary levels for counties of different classes, differentiating mainly by reference to road mileages, and directed that Webster's salary on this new basis should be increased to £800. But the council still held that no increase was justified and, when the board refused to reconsider the matter, initiated *certiorari* proceedings in the courts seeking to have the decision quashed. In December 1900 the queen's bench division ruled that the board had acted lawfully but the county council refused to accept this. The issue was taken to the court of appeal as a test case and, on 25 February 1901, Lord Chief Baron Palles ruled that the board had indeed acted *ultra vires* by determining the surveyor's total remuneration rather than the increase, if any, he was due on foot of the 1898 Act; as Lord Justice FitzGibbon put it, the board instead of simply protecting the interests of the transferred staff had turned the proceedings into an occasion for raising salaries all-round, and this it had no jurisdiction to do. But this ruling was not the end of the matter; Webster maintained his claim for increased remuneration and the issue was the subject of an inquiry held by an inspector of the Local Government Board in Wexford in April 1901. The county council was represented by T.M. Healy KC who had also acted for them in the court of appeal and who mounted a savage attack on Webster's case and subjected him to a severe cross-examination. In the event, the board decided that Webster was entitled to an increase which would bring his salary close to the level determined by them two years earlier, but the county council succeeded in establishing a new salary figure just less than £700.

The dispute over salary levels clearly contributed to Webster's ultimate fate but it was the condition of the bridges over the river Slaney at the Deeps (Killurin), Ferrycarrig and Wexford that led the council in May and June 1901 to consider a motion calling for his dismissal. The wooden bridge at Ferrycarrig, built by Lemuel Cox in 1794, seemed to be a particular cause of concern at that stage; the bridge had originally been operated by commissioners as a toll bridge but was taken over by the grand jury in 1851. As far back as the 1840s, Mr and Mrs S.C. Hall, on one of their Irish tours, found it to be 'long, narrow and rickety' and it was in a seriously deficient condition by the turn of the century. Webster had declared in August 1900 that £40 would make Ferrycarrig perfectly safe but J.H. Ryan, a prominent Dublin consulting engineer who had been called in by the council, advised that an outlay of £500 was needed. Similarly, Webster had suggested that the iron bridge at Wexford dating from 1866 was in perfect order, whereas Ryan advised that it would take an outlay of £105 to restore it to a satisfactory condition. In these circumstances, the council took the view that Webster had shown incompetence and neglect of duty of such a character as deserved dismissal but he countered that the action of the council was shabby and that it arose solely from their defeat on the question of his salary. The council called on the Local Government Board to dismiss Webster and to take immediate action to enable them to appoint a successor so that work on Ferrycarrig and the other bridges could get under way, but the board refused, holding that the facts were not such as to warrant dismissal. They pointed out that Webster had been certified by the Civil Service Commissioners ten years earlier to be fit to discharge the duties of a county surveyor, that he was a member of the Institution of Civil Engineers (London) and that they felt obliged to accept these facts as evidence of his professional qualifications and of his competence to undertake the renovation of Ferrycarrig bridge. But none of this was accepted by the county council which adhered to the view that there was ample evidence to warrant Webster's dismissal.

Although he had completed a bridge, 300 feet long with an opening-span, over the river Slaney at Edermine, allegations by the council about Webster's negligent supervision of bridges and road works generally led to a continuing deterioration in relations between him and his council. In August 1908 matters came to a head; after a horse driven by Colonel Walker, a member of the council, had put his hind legs through the planks on the bridge at the Deeps, a committee of inspection reported that the bridge, as well as the bridge at Wexford, was in an absolutely rotten and dangerous state and that a great many of the planks in the bridge at Ferrycarrig were so worn that a stranger in a motor car who might cross the bridge at full speed could lose his life. Webster admitted that the condition of Ferrycarrig had been a great worry to him for the previous two or three years but pleaded that he had intended to apply for funds to repair it, and expressed the hope that the council would not do anything harsh. The discussion ended when the council shrank from calling for Webster's dismissal, as some members demanded, and resolved instead that, having lost confidence in their surveyor, the Local Government Board should be asked to send down **P.C. Cowan**, their chief engineering inspector, to advise them as to the best course to adopt.

A public inquiry was duly held by Cowan in September 1908 and his findings were published some months later. He concluded that all three of the bridges (as well as the bridge at Edermine which had been built only in 1897–8) were defective and 'discreditable' to the county and while the board ruled that the county council were justified in attributing the blame for this to their surveyor, they left it to the council to decide what action to take in relation to him and his assistants. At a council meeting on 17 February 1909 it was resolved that Webster and his principal assistant surveyor, T.J. Ryan, should be asked to resign. They were summoned to come before the council and both of them tendered their resignations, although Webster agreed to serve until the end of June to allow time for a successor to be appointed. In submitting his resignation Webster declared that he had not the slightest doubt that he could get an appointment and make a better living elsewhere. He was, however, forced to seek employment abroad, taking up an appointment in Vancouver as surveyor for the province of British Columbia in November 1910. Some members of the council felt that the disgraced surveyor should be replaced by two surveyors, each responsible for part of the county, but the majority view was that a single surveyor should be sought at a salary of £500 a year. **Stafford Gaffney**, a native of Wexford, was appointed to the post in June 1909.

Webster married Lucie Margaret, younger daughter of Luke White of Chandos, Co. Longford, in 1891 and lived at Cliff House, Enniscorthy, during his years in Wexford. He was living in Victoria, British Columbia, at least until 1930.

WHITE, Henry Vincent (fl. 1848–1934), was born in Wexford on 13 February 1848, entered Trinity College, Dublin, in November 1866 and graduated LCE in 1869. He worked for some time as resident engineer for the corporation of Kingston-upon-Thames before becoming Longford county surveyor in August 1874. He moved to Queen's County in September 1883 and soon began to take strong action against defaulting road contractors and their sureties. A case which he brought at the Abbeyleix quarter sessions in April 1885 attracted considerable interest and was regarded as a test case since the object was to establish some definite principle regarding the liability of a defaulting contractor for damages. After evidence by White about his inspections of the road within the previous twelve months, the contractor's neglect of it and the amount that had to be spent to bring it up to a satisfactory standard, the judge ruled that the contractor and his sureties were liable not only to forfeit the contract sum but also to pay damages equivalent to the additional amount needed over and above the contract price to restore the road to good repair. White's tactics soon had an effect – but not the one he intended. Many contractors were unwilling to bid for contracts at all and others increased their prices considerably. As a result, members of the grand jury expressed great concern at the spring assizes in 1885 about a 10 per cent increase in the sums presented for road maintenance; some of them felt that the county surveyor's estimates were too high and others felt that he was not conversant with local labour rates and that he was too ready to take roads from contractors. White denied these allegations and argued that the high proportion of roads in his hands – approximately one-third of the total – was due solely to the fact that no tenders had been received at the specified prices.

At the summer assizes in July 1885 the question of retaining the services of three assistant surveyors at salaries of £70 a year each came up for consideration. Some grand jurors took the view that, since a high proportion of the county roads were then in the surveyor's hands, the third assistant should be dispensed with. White contended that he could not do without three assistants and pointed out that he relied on reports from the assistants with regard to the roads in his own hands just the same as if they were contract roads. Besides, the assistant surveyors were needed as a check on the five overseers. In 1889, when nearly 50 per cent of the roads were in the hands of the county surveyor, the number of his assistants came up for consideration again. Some jurors felt that he could easily dispense with one of the assistants and there was renewed criticism of the extent to which White's actions resulted in the taking of roads from contractors who, some members felt, generally did a better job. White agreed that contractors often put out more material, because they had their own horses and family assistants, but he contended that, on the whole, he and his staff did the best they could with the roads in their charge and strongly opposed reversion to the old system of small contracts which were designed to promote greater competition from local men.

In his report to the summer assizes in 1890, White noted that the proportion of the roads in direct charge had begun to fall and said that he would be extremely glad 'to get rid of all our roads and revert to the old method of having the work done. These roads in our charge involve an immense amount of extra labour. To this I have never objected, or don't object, but I find it difficult sufficiently to express my strong and increasing personal dislike to have anything to do with them as part of a regular organized system, and I will only be carrying out the law as it was intended by making special efforts gradually to reduce or get rid of them altogether. I should deeply regret many of our deserving trained men thus losing their employment with us for so many years, but difficulties that cannot possibly be avoided may occur at any time to which contract roads are not liable'. White's paper on *County Work in Ireland*, presented to the Institution of Civil Engineers of Ireland in December 1885, is one of the few which gives an accurate, comprehensive and penetrating view of the overall system of road maintenance in Ireland towards the end of the nineteenth century.

White seems to have had only a limited private practice. He provided plans for a new water supply scheme for Carlow town, involving a fifteen-million-gallon reservoir at Killeshin, on which work began in 1894 and in the same year he prepared a preliminary report on a possible water supply scheme for Portlaoise. Some years earlier, in a paper read to the Institution of Civil Engineers of Ireland, he presented a detailed assessment of the Leinster coalfield and put forward proposals for a light railway network in the Carlow–Kilkenny area to facilitate development of the coal and culm deposits near Castlecomer, Co. Kilkenny and Crettyard, Co. Carlow, and the development of lime-burning, brick-making and the fire-clay industries. With these objects in view, he suggested a light railway from Castlecomer to Carlow and onwards to Tullow, with possible extensions southward to Kilkenny and eastward to Shillelagh, Co. Wicklow. He estimated that the twenty-one mile section from Castlecomer to Tullow could be completed at an all-in cost of £120,000 and that a shorter tramway, running along the public roads from Castlecomer to Carlow would cost £45,000. White's proposals were not acted on although a seven-mile branch railway to Castlecomer was constructed in 1919 under the Defence of the Realm Acts, the last branch railway to be built in Ireland.

Like his contemporaries, White transferred to the service of the new county council in 1899 and early the following year, the Local Government Board recommended that his salary should be increased to £680 to take account of his increased responsibilities; however, the county councillors felt that £600 would be an equitable sum. He continued, however, to serve his new masters loyally and well – never, as he said himself, taking leave or receiving temporary relief since the day of his first appointment. In 1903, he was heavily involved in preparations for the Gordon Bennett Cup motor race; he told one of the district councils in

April that in the Mountmellick and Athy areas, it would be necessary for 'road bumps to be levelled … and corners rounded off' and three miles of road would need to be steamrolled. All of this work was to be financed by the race organizers and so the councillors 'heartily approved' of the plans.

In January 1908 when he had served as surveyor for almost thirty-four years and was nearing 60 years of age, White told the county council that he had been advised to take a long rest from professional duty and gave notice of his intention to retire in the following August. He had lived in Portarlington during much of his service as county surveyor but moved to Dublin after his retirement. He was listed as a resident of St Helen's, Ailesbury Park, until 1934.

WILKIN, James Gilbert (1887–1958), was born on 10 June 1887 and graduated in engineering at Queen's University, Belfast, in 1910. He joined the Down county surveyor's department in 1907 and spent his entire career there. By 1917 he had become chief assistant to the county surveyor, **James Heron**, and acted as county surveyor on the latter's retirement in 1920. The post was advertised later that year and a special meeting of the county council in May 1921 unanimously elected Wilkin to fill it. His salary on appointment was £1,200 a year, with the use of a motor car, one of the best – if not the best – remuneration packages available to a county surveyor at the time. His term of office was noteworthy for the efforts he made to respond to the demands for road improvements to cope with the rapid growth in road traffic; the widening and realignment of the Belfast–Bangor road was one of his major projects. He was in office until 1945 and died on 29 January 1958.

WILLIAMSON, Richard (fl. 1847–74), began his career as a pupil of Jacob Owen in the offices of the Board of Works at the Custom House, Dublin. He became county surveyor for the eastern division of Tyrone on 18 December 1847 when the county was first divisionalized and transferred to Derry in July 1860. At a function in Dungannon some months later he was presented with a testimonial 'most numerously signed' and a handsome service of plate, which cost £150, to mark what was regarded as a promotion.

Williamson was responsible for improvement works costing about £2,000 at the courthouse in Derry in 1866–7. As an architect, he was in partnership in the 1860s with **Thomas Turner** with whom he designed the Irish Society's schools in Coleraine and the Academical Institution in Derry; described at the time as a building 'in the Italian style, remarkable for the simplicity of treatment and for the almost total absence of any more ornamental features', the latter building has since been demolished. Williamson's role as surveyor to the Irish Society and his private architectural practice were criticized at the spring assizes in 1869 when there were complaints about the condition of the roads and lack of attention to them by the surveyor and his assistants. Williamson defended himself vigorously, claiming that he employed an assistant who did most of the society's work, that he had already given up agencies in Donegal to allow him to devote three or four days each week to his county surveyor duties and that while he was a partner in the architectural practice he gave none of his time to that business.

Williamson travelled to London towards the end of April 1874 to give evidence to a committee of the House of Lords on the Londonderry Port and Harbour Bill. He received a severe wetting on landing from the steamer at Fleetwood and this led to acute bronchitis from which he died on 27 April at the Victoria Hotel, Euston Square, London.

WILLSON, Frederick Richard Thomas (fl. 1848–98), was born in Co. Longford, the son of Revd Frederick Willson. He was educated by a Mr Stackpoole before entering Trinity College, Dublin, in May 1867 when he was aged 19. He became a scholar in 1870 and gained BA and BAI degrees in summer 1873. He was employed by the Board of Works for a year before winning first place in the county surveyor examinations of June 1874. He was appointed to Co. Westmeath in August of that year but was transferred to Fermanagh at his own request in October 1876 'in consequence of being threatened for the discharge of public duties'.

A paper which Willson presented to the Institution of Civil Engineers of Ireland in March 1886 provides a valuable record of the state of the art of road construction and maintenance at the time and includes details of the design and construction of Drumain Bridge (1883) between Belturbet and Enniskillen at a cost of £1,500; this was a masonry arch structure of two thirty-foot spans on piled foundations. He was involved in the construction of Inishmore viaduct in the 1890s but had to tell the grand jury at the spring assizes in 1898 that despite his best efforts no work on the structure had been carried out by the contractor, J.G.V. Porter, for several years.

Willson retired on health grounds in July 1898 and was awarded a pension of £217 a year. The grand jury paid generous tribute to him, referring particularly to the fact that he had kept the network of nearly 900 miles of county roads in an admirable condition during his twenty-two years service in the county.

WOODHOUSE, Thomas Jackson (1793–1855), was reported to have been placed first on the list of qualified candidates which was presented to the lord lieutenant in 1834 following the first competitive examinations for appointments as county surveyor. Woodhouse was allowed his choice of the appointments on offer and, as the county surveyor system did not then apply in Dublin, he selected the Antrim post as the most important one because it included responsibility for the rapidly developing city of Belfast. Woodhouse took up duty on 17 May 1834 but soon found that it was 'impossible for any one person to manage efficiently what is required and ... the salary, after deducting the expenses incurred, is not an adequate remuneration for the labour and anxiety attendant upon the office'. He gave evidence to this effect to the Select Committee on County Cess in May 1836, pointing out that the cost to him of making the necessary six circuits of his extensive county each year amounted to £200, including the keep of his two horses. He pleaded also for assistants – practical honest men in each barony and others who would be competent to supervise bridge construction – but assistants were not allowed for by law at the time. Woodhouse resigned in October 1836 to return to a railway engineering career in England and further afield, and was replaced by **Charles Lanyon**.

Woodhouse was born at Bedworth in the English county of Warwick on 9 December 1793, the eldest son of Dorothy Jackson and John Woodhouse, a civil engineer whose own father had been a mining engineer. He was educated at Towcester in Northamptonshire. By 1809, when the time came for him to select a profession, England had been at war with France for six years and against the wishes of his parents the young Woodhouse decided to go to sea as a midshipman on the *Broxonbury*, one of the East India Company's armed ships. On his return from his first voyage to Calcutta he was persuaded by his parents to abandon his seagoing career and to embrace the profession of civil engineering, the rudiments of which he had already learned under his father. He was first engaged on a number of projects for which his father was either engineer or contractor; these included the construction of the Grand

Junction Canal, including the Blisworth tunnel which was one of the longest canal tunnels of the day; the Worcester and Birmingham Canal; and between 1822 and 1824 Telford's Gloucester and Berkeley Ship Canal in south Wales. Turning to railway engineering in 1825, in conjunction with Josias Jessop, he surveyed a line for a railway from Bristol to Birmingham but this was not proceeded with because of the downturn in the investment cycle. In 1826 he became resident engineer on the Cromford and High Peak Railway, one of the earliest railways in England, passing through wild, mountainous country and with gradients so steep that the line had to be worked by means of stationary engines and with horses.

Woodhouse first came to Ireland in 1832 as resident engineer under Charles Vignoles on the design and construction of Ireland's first railway, the Dublin & Kingstown; the detailed plans of the line were drawn under his direction and bear his signature, and he was said to have gained great credit and to have displayed great talent, as well as zeal and assiduity, in bringing the project to fruition. He was also involved in 1833 in planning an extension of the line to Dalkey but the Bill to authorize this failed in parliament when first presented. He was assisted in the work by his brother, George, who was later to become principal agent in Ireland for William MacKenzie, the contractor who was engaged in an extensive programme of works to improve the Shannon navigation.

On moving to Belfast in 1834 Woodhouse had to undertake a number of substantial projects as county surveyor in addition to the routine duties of the office. With William Bald, a government engineer, he was responsible for the Antrim coast road on which construction had begun in 1832. In conjunction with **John Fraser**, the Down surveyor, he undertook the design of Queen's Bridge in Belfast to replace the old Long Bridge of twenty-one arches which had been built in 1685 to link Belfast to Co. Down; the new bridge was completed in 1842 by his successor, Charles Lanyon. Woodhouse also accepted an appointment as resident engineer to the Belfast Ballast Board in which capacity he prepared reports in September 1834 on the improvement of Belfast harbour. The board had, since 1820, had been considering various schemes to improve the inadequate quays and the notoriously difficult passage through the winding river Lagan; the question had been referred to a number of prominent engineers, including John Killaly, Thomas Rhodes, John Rennie and Thomas Telford, all of whose schemes had to be rejected on cost grounds. A scheme put forward by James Walker, involving the formation of two cuts across the bends in the river, and the building of new quays, was finally adopted by the board and sanctioned by legislation in 1831. But objections to the scheme continued to be raised by various interests and, in an effort finally to resolve the matter, the Board of Works instructed yet another eminent engineer, William Cubitt, to examine and report on the merits of the different schemes. In consultation with Woodhouse, Cubitt reported that Walker's scheme was the most suitable and should be proceeded with as regards the cuts and the quays. Woodhouse had, however, long ceased to be associated with the problem when eventually, in April 1839, William Dargan began the excavation of the first cut or channel to eliminate the bend nearest the city; the work was finished in 1841 and the second cut to reach the deep water at Garmoyle was completed between 1847 and 1849, with Dargan again acting as contractor.

Woodhouse continued his railway engineering activities during his years in Belfast. Initially, he worked with Bald in 1834–5 on plans for the thirty-six-mile-long Ulster Railway from Belfast to Armagh which was authorized by parliament in May 1836 and went to construction stage under William Dargan early in 1837. He designed the original stations at Great Victoria Street and at Lisburn and some of his fine red-sandstone bridges on the section of the line between Belfast and Lisburn are still in use as part of the Belfast–Dublin main line. In 1836 Woodhouse and Bald again worked together on surveys and preliminary plans for a proposed North Eastern Railway from Belfast to Ballymena, thirty-three miles in length, but this was not proceeded with.

After his retirement from the Antrim surveyorship in the autumn of 1836, Woodhouse took up an appointment under Charles Vignoles as chief resident engineer on the Midland Counties Railway, connecting Derby, Nottingham, Leicester and Rugby. Construction of this sixty-mile line through hilly country was a difficult operation but Woodhouse saw it through successfully at a cost of one million pounds. In recognition of his services, over one thousand guineas was subscribed by those connected with the line and used to purchase a handsome silver dinner service which was presented to Woodhouse at a public dinner in Derby in May 1841. The railway company itself presented its warmest thanks to Woodhouse for his constant exertions, and in acknowledgement of his eminent services, resolved that he should be requested to accept a perpetual free ticket for passage over the railway. One of his pupils and assistants on the railway was James Beatty (1820–56), a native of Fermanagh, who went on to become chief engineer with the contracting firm of Peto, Brassey and Betts and head of the civil engineering corps which built the Balaklava railway in record time in 1855.

William Mackenzie (1794–1851), one of the giants of early railway contracting, had been one of the two main contractors on the Midland Railway. He had known Woodhouse since their work together in the 1820s on the Gloucester and Berkeley Canal and had developed an excellent working relationship with him on the Midland contract. Thus, after some further railway work in England, Woodhouse went to work in France in 1842 with Mackenzie who by then had teamed up with Thomas Brassey to create the most formidable civil engineering contracting firm of the day. Woodhouse worked initially on the construction of the canal linking the Rhine and Marne, and from 1843 onwards was the contractors' chief engineer on the £2.5 million contract to build a railway between Orleans, Tours and Bordeaux. With the line completed between Orleans, Tours and Poitiers, he resigned from this assignment in 1848 on the outbreak of revolution in France, and returned to England for a time.

Other European engagements were soon to follow. In January 1849 Woodhouse went to Spain to review on behalf of Mackenzie the plans for railways in the Asturias and to prepare estimates and tenders. However the firm did not go ahead with the scheme due, possibly, to Mackenzie's failing health. Later that year, Woodhouse was in Belgium working on estimates for the construction of a railway from Charleroi to the French border but, again due to Mackenzie's illness, the work was abortive. Early in 1850 when Brassey became alarmed at his partner's continuing illness, he sent Woodhouse to discuss with Mackenzie the dissolution of the ten-year old partnership; this was sorted out by arbitration some months later and the unfortunate Mackenzie died in October 1851.

Woodhouse continued to work with Brassey by whom he was engaged to go to Italy to build a fifteen-mile section of railway from Pistoia to Prato, near Florence. He then undertook for Brassey the construction of the line from Turin to Novara and had completed about sixty miles of this when he was struck down by illness in mid-September 1855. He died in Turin on 26 September and was buried there. It was reported that, so great was the respect and esteem he had gained among the citizens, his remains were followed to the grave by a greater number of people than had ever paid this last mark of respect to a protestant living in the city.

A memoir in the *Proceedings of the Institution of Civil Engineers* noted that whether acting for the promoters of large construction projects or as a contractor's engineer Woodhouse seemed to have had the ability to achieve his objectives without conflict or litigation. His patent honesty and impartiality, as well as his unassuming and conciliatory demeanour, were said to be of great value in many difficult negotiations and arbitrations. In addition, those who knew him best considered him to be an excellent judge of work and of its fair value to a contractor and they placed great reliance on the accuracy of his estimates. In both his character as a man and his ability as an engineer, he obviously had few equals.

WORRAL, John Leslie (1817–90), was born in June 1817 in Co. Limerick where his father, William, was a solicitor. He entered Trinity College, Dublin, in November 1837 but left after two years, without graduating, to take up an apprenticeship with an experienced practitioner in land surveying. He went on to become one of Sir John MacNeill's assistants and worked under him for nearly five years on railway surveys and construction. His initial assignment was on the Dublin & Drogheda line, where he was involved in supervising the construction of a long wooden viaduct across the estuary at Malahide. In 1843 he worked with W.R. Le Fanu, later to become one of MacNeill's senior assistants, on some of the early surveys for the proposed Great Southern & Western line from Dublin to Cork for which work he was paid eight shillings a day. From 1846 onwards Worral was in private practice on his own account, engaged mainly on land drainage work and on the survey and valuation of land; he was heavily involved in the 1850s in work arising under the Incumbered Estates Act in the Limerick area.

When Worral gave evidence to the Select Committee on County and District Surveyors in August 1857 he was very critical of the manner in which the county surveyor examinations and the appointments process had been conducted. He complained that he had been seeking an appointment for fifteen years and while he had been examined on three occasions there was great difficulty in finding out the results. He had never been able to get any information on the outcome of the 1847 examinations at which he was one of thirty-four candidates and it was not until 1854 that he learned from Thomas Larcom, the under-secretary, that he was one of the sixteen candidates who had qualified when examinations were again held in 1851.

The *Dublin Builder* reported in August 1860 that Worral was to be appointed to the county surveyorship of Derry which was vacant following the death of **Stewart Gordon** but when this turned out to be inaccurate, the journal chose in its next edition to attack its sources rather than apologise to Worral: 'the information relative to Mr Worral being appointed was conveyed to us as *authentic* … and we must observe that gentlemen ought to be more cautious in communicating as a fact to the public press that which, though probable, is not actually accomplished'. Worral was appointed county surveyor for Tipperary south riding later that month. Soon afterwards, he undertook extensive restoration works on the trackway along the Suir navigation between Clonmel and Carrick-on-Suir which was then in poor condition. Some years later, Worral reported that 'certain persons' who had been cleaning the bed of the river had removed sections of the foundations of the river walls; as a result, more than 300 feet of wall was carried away by a flood and further heavy expenditure had to be incurred to make good the damage.

Worral was active in road and bridge building. He designed and built a new bridge in 1866–7 at Glengool on the Newcastle–Clogheen road where a forty-foot arch had been carried away by a flood. His report for the 1869 spring assizes listed a number of other bridge projects: a new bridge at Ballybane on the boundary with the north riding was complete as were bridges at Ballyvistea and at Cromwell on the Limerick boundary; as well, a small bridge at Croone was open for traffic, and a new bridge at Bishopswood, near Dundrum, was finished. His major bridge, designed with Robert Manning of the Board of Works, was Dillon Bridge – a lattice-girder structure, with three ninety-two-foot spans and masonry piers, land arches and abutments, which was completed at Carrick-on-Suir in 1882 at a cost of about £13,500.

Worral was responsible for a number of architectural projects in his area. He carried out improvement works at the asylum at Clonmel, including the provision in the 1870s of a tunnel under the intervening public roadway to link the different buildings; this was in use until the 1930s. Some years before, he designed a new barracks in Clonmel for the South Tipperary Artillery; the foundation stone was laid on the site of the old county gaol in November 1868 with great pomp and ceremony and the barracks was completed by 1870. The plans provided for three-storey buildings, 246 feet long and 27 feet in width, together with the usual storehouses, parade grounds and other facilities.

In 1878, when he had been in office there for eighteen years, Worral was reported to have said that Tipperary south riding was 'the most difficult county in Ireland to manage, the two county surveyors before him having been dismissed and he having often been threatened'. Nevertheless, he carried on for another eleven years before retiring due to failing health in July 1889. He died on 26 January 1890, aged 73, and although he had lived at Stream Vale, Clonmel, his body was brought to Dublin for burial at Mount Jerome cemetery. His wife, Anna, née Bradshaw, died in December 1891 and was buried with him at Mount Jerome.

YEATS, John (1808–65), was born in January 1808, the second son of Revd John Yeats, rector of Drumcliffe, Co. Sligo, and thus was a grand-uncle of the poet, William Butler Yeats (1865–1939) and of the painter, Jack B. Yeats (1871–1957). He served a five-year apprenticeship to Richard Griffith during which he was employed as an assistant draftsman at the Boundary Survey Office. In 1834 he was engaged by the Board of Works to work under William Bald on surveys of the Drogheda–Navan section of the Boyne navigation which had been constructed between 1759 and 1800 and was then in the control of the board. As resident engineer, he carried out some limited improvement works on the navigation in the following two years.

Yeats became Louth's first county surveyor on 17 May 1834 but, in his own words, 'did not pull very well with some of the gentry'. He protested that members of the grand jury found fault with everything he did and complained that nothing was done right even though he devoted his full time to his official duties. To make matters worse, the jury refused to grant him the full salary allowed by law – initially, he was paid at the rate of £200 a year, later increased to £250.

Following the transfer to Antrim of **Charles Lanyon**, Yeats was appointed on 14 October 1836 to succeed him in Co. Kildare where his younger brother, Matthew, was secretary of the grand jury. During his service of nearly twenty-nine years in the county, he was said to have commanded the highest respect for a character of singular amiability and kindness of heart, and for unswerving integrity in the discharge of his duty.

Towards the end of his career, however, his treatment by the grand jury in a dispute about completion of a new bridge over the river Liffey at Clane was poor recompense for his years of faithful service. The existing bridge, said to be of great antiquity, had six arches and a very narrow roadway and it was reported in February 1864 that it was to be rebuilt at a cost of £2,100. At the spring assizes in March 1865 Yeats reported that the new bridge had been completed in the previous November but he went on to say that he was not completely happy with one of the arches which was imperfect, at least in appearance, due to settlement when the centring was removed. Although the contractor had refused to put matters right, claiming that the imperfection did not affect the durability of the bridge, Yeats admitted under questioning by the grand jury that he had certified the payments which had been claimed and that only £45 of the contract sum remained unpaid. The grand jury reacted strongly to this and some of them wanted the surveyor to be dismissed immediately even though he committed himself either to having the defects made good before the next assizes, or to resign. Eventually the jury voted to require Yeats to bind himself to have the defects made good to the satisfaction of an architect to be approved of by them. The unfortunate surveyor became seriously ill some weeks later but he arranged with **James Bell**, one of the Dublin surveyors, that he would insect the bridge with **Henry Brett**, the Wicklow surveyor, and the Dublin architect, Thomas Drew. Bell reported at the summer assizes that, while the bridge was not a very handsome one and there had indeed been some settlement in the centre of one of the arches, only a comparatively trifling amount of work had been necessary to put matters right. The contractor had already reconstructed the arch and the bridge was open again, although the centring would remain in place for some weeks as a precautionary measure. The bridge at Clane still stands as a monument to Yeats: it is a fine masonry structure, with three main arches and a small flood arch, and carries a plaque attesting to its design by him. It was named Alexandra Bridge, having been formally opened by Princess Alexandra, later Queen Alexandra of England.

Yeats died at his home in Monasterevan on 29 July 1865 after a long and painful illness brought on, according to the death notice in the *Leinster Express*, 'by anxiety in the discharge of his arduous duties' which he had 'faithfully, zealously and indefatigably' carried out for twenty-nine years with integrity and uprightness. He had married his first cousin, Ellen Sophia Terry, daughter of Captain Henry Terry of the 99th Regiment, and his wife, Mary Yeats, in 1842. The couple had three sons, all of whom emigrated to the United States, and three daughters.

APPENDIX I

Assistant County Surveyors

The Grand Jury (Ireland) Act 1836 made provision for assistants to the county surveyor with salaries of up to £50 a year.[1] The number of assistants in any county was to be decided by the grand jury, having regard to the size of the county and the extent of the duties to be performed, and the appointments themselves were to be made by the county surveyor from among persons certified to be fit and competent by the same board of engineers which examined candidates for county surveyorships. The Board of Works, which in practice, was entrusted with the task of examining the qualifications of applicants for the new posts, felt that 'it would be very desirable if these situations could be filled by men of a certain degree of education and well conversant with the practical superintendence and mechanical trades adapted to public works ... but such a class of men, thoroughly qualified, are rare in Ireland, and it would be impolitic and unjust to introduce them in any degree from England and Scotland, even if they could be had for the remuneration allotted, which however is not to be expected'.[2] For these reasons, they announced that in certifying candidates they were prepared to rely largely on the recommendation of the county surveyor but they thought it necessary to lay down some basic requirements: in general, candidates were to be between 24 and 45 years of age, tolerably robust and of good character, especially as to sobriety; they should at least be able to read and write, have some knowledge of arithmetic and accounts, and some practical or theoretical knowledge or experience that would be relevant to the duties to be performed. Subsequently, in the light of experience, the board drew attention to the need to appoint practical men instead of inexperienced young men with no knowledge whatever of the business,[3] and the need to avoid as far as possible appointing persons who lived in the barony in which they were to serve – they felt that such persons might be more concerned with increasing their income than with zealous service to the grand jury and that they might have partialities in favour of friends and neighbours.[4] A development of this was a decision in 1843 that certificates would not be issued for persons connected with families or individuals of influence in the area.[5]

---

1 An Act to consolidate and amend the Laws relating to the Presentment of Public Money by Grand Juries in Ireland (Grand Jury (Ireland) Act 1836), 6 & 7 Wm IV, c.116, section 43; in Co. Dublin, provision for the appointment of assistant surveyors was not made until 1897. 2 NAI, OPW 1/1/9, circular to county surveyors, 26 Oct. 1836. 3 NAI, OPW 1/1/9, circular to county surveyors, 3 Nov. 1845. 4 NAI, OPW 1/1/9, circular to county surveyors, 20 Jan. 1837. 5 NAI, OPW 1/1/9, circular to county surveyors, 16 Dec. 1843.

From November 1836 onwards, county surveyors presented large numbers of candidates for the qualifying examinations held by the Board of Works at 12 o'clock on Friday mornings at their offices in the Custom House, Dublin.[6] While certificates were quickly granted in most cases, the early assistants were generally of poor calibre and only three of them – R.J. Hampton, Alexander Tate and Peter Burtchaell – ever graduated to the county surveyor or district surveyor rank. Nominees were sometimes rejected. John Steedman of Donegal, who submitted fourteen names, was told to resubmit them, a few at a time, having first checked the qualifications himself. Farmers, when nominated, were declared to be unsuitable because, as was explained in one case, 'there will naturally be a strong persuasion that he cannot give up the time he ought to the duty of an assistant to the surveyor'. Cess-collectors were rejected for the same reason. Sanction was generally refused for the appointment of boys less than 21 years of age, without any practical knowledge; and other candidates attracted adverse comment because of their bad writing, lack of knowledge of measurement, and failure to understand the use of mathematical instruments. In general, however, the Board of Works seems to have taken a rather lenient view as it believed it was 'seldom that men can be procured fully qualified at the time of examination for the office' and they usually gave the benefit of the doubt to the surveyors and their nominees. Candidates who were found to be intelligent and promising, but without much practical knowledge of road works, were often qualified on the basis that it was primarily a matter for the surveyor himself to see to it that they became fully competent. In one such case, when a surveyor responded that he had 'no desire to become the assistant's instructor', he was told by the board in a strongly worded letter that if he did not take some pains with his assistants in teaching them their duty, the service would suffer and the works would not be executed in a respectable manner. Several of the surveyors submitted some of their apprentices for certification – a practice which was acceptable in principle according to the board, but not where the apprentice was only 15 or 16 years old. Others succeeded in obtaining certificates for members of their own families, including sons and brothers.

In the early years, the discretion left to the grand juries about the appointment of assistants was 'exercised very variously', according to an 1842 report.[7] At that stage, the Antrim grand jury, which had 1,900 miles of road under its charge, allowed twelve assistants but Meath, with 1,700 miles of road, allowed only one. Armagh employed six assistants for 917 miles of road, while there was no assistant in Galway with 957 miles. Kilkenny had five assistants for 816 miles; Kildare had three for 806 miles; and Louth had three for 400 miles. Five surveyors were allowed no assistant at all – those for east Galway, west Galway, Longford, Roscommon and Westmeath. To remedy this situation, county surveyors argued that the number of assistants should be fixed by law in proportion to the number of miles of road in each county and a standard of one assistant for every 100 miles was sometimes suggested. In 1857, when 121 assistants were employed, a select committee drew attention to the 'disparity found to exist in reference

---

6 At a later stage, the task of examining candidates seems to have fallen mainly to Jacob Owen, the board's engineer and architect; by the 1840s, candidates were also being examined locally by inspectors of the board when they were visiting particular areas on tours of inspection.   7 *Report of the Commissioners appointed to Revise the Several Laws under or by virtue of which Moneys are now raised by Grand Jury Presentment in Ireland*, HC 1842 xxiv (386).

to many counties' and recommended that counties 'should be divided into districts according to some common principle' and that there should be an assistant for each district.[8] But in this matter, as in so many others, it continued to be a matter for each individual grand jury to decide what arrangements should be made. In a number of counties, refusal to allow assistants to be appointed, or to increase the number of assistants, was linked to the grand jury's disapproval of the extensive private practice carried on by their surveyor.

The 1857 select committee agreed that 'the present salary of £50 is insufficient to secure the services of competent and trustworthy persons' as assistant surveyors and recommenced that it should be 'somewhat increased'. The assistants campaigned strongly at the time for an increase to £100 and for the payment of retiring allowances, instead of allowing them to be 'cast away' after being 'worn down to uselessness' by a lifetime of arduous public service.[9] These demands were not met, however, and while provision for an increased salary of up to £80, at the discretion of the grand jury, was made by law in 1861, many of the grand juries were slow to grant increases. In 1868, only 85 of the 146 assistants serving at that stage were receiving the maximum salary and assistants in some counties (Galway, Roscommon and Queen's County) were still being paid at the 1836 rate of only £50 a year.[10]

When responsibility for examining candidates for county surveyorships was transferred to the Civil Service Commissioners in 1862, the system of examination and appointment of assistant surveyors was not altered notwithstanding the fact that the select committee of 1857 had recommended that the examinations should be 'rendered more effective', possibly being held once a year, with appropriate advance publicity, and leading to the publication of panels of qualified candidates ranked in order of merit. The bill which was to become the 1862 Act had initially proposed the transfer of the power to appoint assistants to the lord lieutenant, but the clause was struck out at committee stage in the commons following opposition by some of the Irish members. The issue was reviewed in 1877 by a committee chaired by Viscount Crichton MP which had been appointed to inquire into the constitution and duties of the Board of Works.[11] In his evidence to the committee, William R. Le Fanu, a member of the board since July 1863, explained that the chairman, Colonel McKerlie, the board's architect, J.H. Owen, and himself constituted the board of examiners for assistant surveyors. But Le Fanu went on to reveal that in practice 'one of our surveyors writes a series of questions as to the practical carrying out of works, the repairing of roads, etc … the candidates give their answers and they are brought up to me and I go over them'. Le Fanu considered that the arrangements were anomalous but argued that 'a great weight would have to be given to practical work' in any alternative system. The committee reported in June 1878 that there was no reason why assistants should not be examined by the Civil Service Commissioners, provided the examination had a practical character. But no change was made and candidates for assistant surveyor posts continued to be examined by the Board

8 *Report from the Select Committee on County and District Surveyors etc. (Ireland)*, HC 1857 (Sess. 2) ix (270). 9 *DB*, 1 May 1861, 501. 10 *Returns from County Surveyors in Ireland of Number of Assistants and their Salaries, etc.*, HC 1867–68 lv (240). 11 *Report of the Committee appointed by Treasury to Inquire into the Constitution and Duties of the Board of Works in Ireland*, [c. 2060] HC 1878, xxiii (1).

of Works into the 1890s when about ten candidates, on average, were being certified each year.[12] The system was to survive in that form until 1899.

At the beginning of the 1870s, when there were about 170 assistant surveyors, many of them had yet to be allowed the maximum salary of £80 set by the 1861 Act and some of them were still receiving no more than the salary of £50 fixed originally in 1836.[13] An Assistant County Surveyors' Association had been formed in 1869 to campaign for higher salaries, travelling expenses and pensions,[14] but letters to the *Irish Builder* from dissatisfied individuals in 1870 and 1871 suggest that the association was not very active in those years. Deputations met Lord Hartington, the chief secretary, in early 1872 and his successor, Sir Michael Hicks Beach, in November 1874 to press their claims and many grand juries passed resolutions of support.[15] A private member's bill which provided for a salary scale rising annually from £120 to a maximum of £200 at the discretion of the grand jury, as well as for the payment of travelling expenses and pensions for those who retired because of old age or infirmity was introduced at Westminster in March 1887 but not enacted. Broadly similar bills were introduced in 1891, 1892, 1893, 1896 but none of them reached the statute book. As a result, the maximum salary remained at its 1861 level until 1899 when the new county councils were given power to determine salary levels, subject to the agreement of the Local Government Board; there was no provision for pension payments for assistant surveyors until 1925.

The qualifications required for appointment as assistant county surveyor were gradually stepped up although many of the appointments continued to be held on a part-time basis, leaving the assistant free to engage in other work. Young graduates begin to appear as assistants in 1880s although, under orders made by the Local Government Board, it continued to be possible after 1900 to appoint persons who did not hold recognised degrees or diplomas but who produced satisfactory evidence of training and passed the qualifying examination. Most county surveyors in the latter half of the century had pupils or apprentices some of whom were sent on to be certified as assistants; a small minority of these were accepted as members of the Institution of Civil Engineers of Ireland and a few of them went on to become county surveyors themselves. Two assistants appointed in Co. Tyrone in 1899 had university degrees and by the early years of the twentieth century the position appeared to have achieved a level of respectability judging by the annual reports of university presidents which recorded successes by their graduates in gaining appointments.

Under the 1898 Act the function of prescribing qualifications for candidates for assistant surveyor positions, and conducting qualifying examinations, was taken on by the Local Government Board. Under an order made in February 1900, candidates were to be between 21 and 45 years of age and were deemed to be qualified without further examination if they held a degree or diploma in engineering, or associate membership of the Institution of Civil Engineers (London), or of the Institution of Civil Engineers of Ireland, or had already been certified to be qualified for appointment as a county

---

12 The annual reports of the Board of Works give the numbers (but not the names) of assistant county surveyors whose qualifications were certified at the request of particular county surveyors each year. 13 *Galway Vindicator and Connaught Advertiser*, 8 Mar. 1873.   14 *IB*, 1 Sept. 1869, 200.   15 *IB*, 15 Feb. 1872, 58; 1 Dec. 1874, 326.

surveyor; others had to have at least two years satisfactory training or experience in engineering and pass a qualifying examination in English composition, arithmetic, mensuration, building construction, road construction and maintenance, and chain surveying and levelling.[16] Initially, at any rate, the examinations were of an elementary character according to P.C. Cowan who set the papers, as the position required 'little special skill, but chiefly probity, intelligence and industry'; there was no shortage of candidates, but the marks obtained were generally low, and passing the examination was 'in no way evidence of qualifications for general engineering or architectural work'.[17] Qualifying examinations were conducted annually in Dublin from 1900 onwards, with up to thirty candidates presenting, and special examinations were held occasionally at the request of particular county councils at provincial centres. In Co. Cork, where the council had decided to fill vacancies by competitive examination, a special examination of candidates nominated by them was held in March 1906 and the first-placed candidates were appointed.[18] Similar arrangements were made in subsequent years.

Following representations from the engineering schools of the universities, the qualifications for appointment were modified in November 1906.[19] They were again revised in 1913 by providing for the appointment of assistant surveyors of two classes for one of which 'special qualifications' were required to cope with 'the important developments which have taken place in the maintenance and improvement of roads'.[20] When the first examinations under the new regulations were held in March 1914, twelve candidates attended the examination for qualification as 'ordinary' assistant surveyors, and seven qualified; the standard was higher in the case of an appointment where the prescribed special qualifications were required, and the only candidate who sat that examination failed to qualify.[21] When examinations were held again in 1915 nine of the fifteen candidates qualified and in October 1918, eleven of the fourteen candidates qualified.[22]

Some county councils increased the salaries of their assistants in 1899 and the Local Government Board, acting under the 1898 act, set new national scales in February 1900, graduated according to the road mileage in the assistant's district; a new minimum of £120 was to apply where the district had less than 250 miles of road, and a maximum of £150 was to be paid where the district had 400 miles.[23] Strong objection was taken to the new figures by many county councils, and legal proceedings were instituted by the councils of Fermanagh, Monaghan and Wexford, among others, seeking to have the board's decisions quashed. The queen's bench division of the high court held in December 1900 that the board had not exceeded their authority,[24] but Wexford county council succeeded in February 1901 in overturning this ruling in the court of appeal.[25]

16 Order of 9 Feb. 1900 made by the Local Government Board for Ireland, published in *Annual Report of the Local Government Board for Ireland*, 1899–1900; for specimens of examination papers, see *TICEI* xxxi (1906), 67.   17 Remarks by P.C. Cowan at a meeting of ICEI, 4 November 1903, *TICEI* xxxi (1906), 33; samples of the examination papers are at pp 143–148 of that volume.   18 *Annual Report of the Local Government Board for Ireland, 1905–6*.   19 *Annual Report of the Local Government Board for Ireland, 1906–7*.   20 *Annual Report of the Local Government Board for Ireland, 1912–13*.   21 *Annual Report of the Local Government Board for Ireland, 1913–14*.   22 *Annual Reports of the Local Government Board for Ireland, 1915–16* and *1918–19*.   23 *Annual Report of the Local Government Board for Ireland, 1900–01*.   24 *The Irish Law Times Reports*, xxxiv (1900), 196.   25 *R (Wexford County Council) v. The Local Government Board for Ireland (Leary's Case)*, [1902] 2 Irish Reports, 348.

As a result, the remuneration of assistant surveyors in many counties continued to be relatively low; the average for the 188 assistants serving in 1913 was only £123 a year and in 1918, when wages and salaries generally had increased to take account of the increased wartime cost-of-living, 120 of the 185 assistants were still being paid less than £150 a year, inclusive of travelling expenses.[26] Nevertheless, the positions were much sought after by junior engineers, as time spent in the position counted towards gaining the practical experience which was required for admission to the county surveyor examinations. Of itself, however, such experience was not sufficient to secure admission to these examinations and, in practice, most of the assistants who served before 1922 had no prospect of promotion to a county surveyor position.

In November 1922 an order was made by the Minister for Local Government and Public Health introducing a degree in civil engineering (or the diploma in civil engineering of the Royal College of Science for Ireland) as an essential requirement for appointment as assistant county surveyor, except where a person had qualified by examination for a county surveyorship or for corporate membership of one of the English engineering institutions.[27] In addition, the Department began to insist that new appointments should be made on a whole-time basis. These steps transformed the grade of assistant surveyor and, by the 1930s, led to a situation in which service as an assistant became the normal route to appointment as county surveyor.

26 *IBE*, 19 July 1913, 464; 2 Nov. 1918, 466.   27 Department of Local Government and Public Health, *First Report*, 1922–5.

Careers of Surveyors by County

COUNTY	NAME OF SURVEYOR	DATE APPOINTED	TENURE
Antrim	Thomas J. Woodhouse	May 1834	Resigned, October 1836
	Charles Lanyon	October 1836	From Kildare; retired, December 1861
	Alexander Tate	December 1861	From Dublin; retired, November 1885
	John H. Brett	November 1885	From Kildare; retired February 1914
	David Megaw	February 1914	Died, August 1927
	Joseph N. Beatty	March 1928	Retired, 1934
	William Grigor	1935	Retired, 1962
Armagh	Henry L. Lindsay	May 1834	Resigned, February 1846
	Henry Davison	February 1846	Retired, October 1886
	Richard H. Dorman	January 1887	From Mayo ND; retired, 1921
	Edgar McConnell	June 1921	Retired, 1947
Carlow	Charles G. Forth	May 1834	To Waterford, July 1844
	John Walker	July 1844	From Waterford; to Monaghan, August 1850
	Alexander Harrison	August 1850	From Monaghan; died, March 1851
	Peter Burtchaell	March 1851	To Kilkenny, December 1860
	John Bower	December 1860	Resigned, July 1877
	Edward T. Quilton	August 1877	From Roscommon; retired, December 1910
	John P. Punch	August 1911	Retired, December 1942
	Gerard P. Fogarty	December 1944	To Louth, October 1946
Cavan	Alexander Armstrong	May 1834	Died, February 1852
	Thomas Turner	February 1852	Resigned, March 1853
	Frederick Gahan	March 1853	To Donegal (SD), July 1864
	John Fraser	July 1864	From Donegal (ND); resigned March 1867
	Alfred Gahan	May 1867	From Tyrone (WD); died, October 1871
	Frederick J. Deverell	November 1871	From Cork (ER-SD); retired, 1889
	Richard N. Somerville	October 1889	Retired, December 1913
	Henry J. O'Reilly	May 1914	Resigned, May 1929
	Denis Coffey	August 1930	To Cork, April 1947
Clare	Richard Richards	May 1834	Resigned, 1834
	Richard B. Grantham	December 1834	From King's County; resigned, September 1836
	James Boyd	September 1836	To Wicklow, December 1845
	John Hill	December 1845	To King's County, June 1855
	Arthur C. Adair	August 1855	To Tyrone (WD), May 1867
	John Hill	May 1867	From King's County; retired, March 1893
	P.L.K. Dobbin	April 1893	From Mayo (SD); retired, August 1922
	Francis Dowling	November 1923	Died, February 1944
	Thomas A. Simington	September 1945	Resigned, October 1949

ER = East Riding	ND = Northern Division	ED = Eastern Division
WR = West Riding	SD = Southern Division	WD = Western Disivision

COUNTY	NAME OF SURVEYOR	DATE APPOINTED	TENURE
Cork	Patrick Leahy (ER)	May 1834	Dismissed, March 1846
	Edmund Leahy (WR)	May 1834	Resigned, January 1846
	John Benson (ER)	April 1846	Resigned, March 1855
	W.A. Treacy (WR)	April 1846	
	(ER)	March 1855	To Mayo, November, 1861
	Nathaniel Jackson (WR)	March 1855	Retired, July 1898
	Frederick G. Deverell (ER–SD)	November 1861	From Mayo; to Cavan, November 1871
	A. Oliver Lyons (ER–ND)	September 1861	Retired, October 1898
	Charles B. Jones (ER–SD)	February 1872	To Sligo, May 1872
	Charles F. Green (ER–SD)	August 1872	Resigned, June 1874
	Samuel A. Kirkby (ER–SD)	July 1874	
	Entire East Riding	October 1898	Retired, January 1909
	Richard W.F. Longfield (WR)	December 1898	From Donegal (SD); retired, September 1925
	Jeremiah T. Murphy (SD)	June 1909	Died, November 1929
	Richard F.M. O'Connor (ND)	June 1909	
	Entire county	November 1929	Died, March 1940
	Gerard P. Fogarty (deputy)	1940	To Carlow, December 1944
	John T. O 'Donnell (deputy)	1940	Retired, 1957
Donegal	John Steedman	May 1834	Dismissed, July 1847
	Robert Garrett	November 1847	Resigned, August 1850
	Henry Smyth	September 1850	
	(ND)	September 1851	To Down (ND), September 1852
	William Harte (SD)	September 1851	
	(ND)	July 1864	Died, March 1895
	John Fraser (ND)	September 1852	From Down; to Cavan, July 1864
	Frederick Gahan (SD)	July 1864	From Cavan; retired, December 1891
	James R.A. Ferguson (SD)	January 1892	
	(ND)	August 1895	Retired, July 1912
	Richard W.F. Longfield (SD)	August 1895	To Cork (WR), December 1898
	James H. Steadman (SD)	December 1898	
	(ND)	September 1912	Retired, 1933
	John Caffrey (SD)	March 1913	
	Entire County	October 1934	Retired, March 1938
	Thomas McMahon	October 1939	Resigned, 1948
Down	John Fraser	May 1834	To Donegal (ND), September 1852
	Henry Smyth	September 1852	From Donegal
	ND	September 1862	Retired, March 1890
	Bernard B. Murray (SD)	September 1862	Died, May 1889
	Peter C. Cowan (SD)	June 1889	From Mayo (SD)
	Entire county	March 1890	Resigned, January 1899
	James Heron	March 1900	From Monaghan, retired 1920
	James G. Wilkin	May 1921	Retired, 1945
	Robert H.S. Patterson	1945	Died, 1957
Dublin	Alexander Tate (ND)	January 1845	To Antrim, December 1861
	Robert J. Hampton	January 1845	Died, August 1855
	Richard H. Frith	January 1845	Retired, 1865
	Richard A. Gray (SD)	1856	Retired, 1896
	James Bell (ND)	1862	Resigned, 1883
	Thomas Turner (ND)	May 1883	Died, October 1891
	William Collen (ND)	December 1891	
	Entire county	1897	Retired, June 1924
	Joseph A. Ryan	July 1924	From Queen's County; retired, July 1953

COUNTY	NAME OF SURVEYOR	DATE APPOINTED	TENURE
Fermanagh	Roderick Gray	May 1834	Retired, 1876
	William H. Deane	August 1876	From Mayo (SD); to Kerry, October 1876
	Frederick F.R. Willson	October 1876	From Westmeath; retired July 1898
	James P. Burkitt	December 1898	Retired, April 1940
	James W. Charlton	1940	Retired, 1956
Galway	Henry Clements	May 1834	
	WD	May 1838	Dismissed, July 1858
	James F. Kempster (ED)	May 1838	Retired, 1891
	Samuel U. Roberts (WD)	July 1858	Resigned, November 1873
	Henry T. Humphreys (WD)	December 1873	From Limerick (WD); resigned, January 1878
	Carter Draper (WD)	May 1878	To Wicklow July 1882
	James Perry (WD)	July 1882	From Roscommon; died, November 1906
	John Smith (ED)	December 1891	Died, February 1907
	Patrick J. Prendergast (WD)	April 1907	Resigned, October 1910
	John Moran (ED)	June 1909	
	Entire county	November 1910	
	WD	November 1913	Died, April 1916
	Michael J. Kennedy (ED)	November 1913	
	WD	June 1916	Died, December 1938
	George Lee (ED)	August 1919	
	Entire county	February 1939	Retired, June 1956
Kerry	Henry Stokes	May 1834	Retired, 1876
	William H. Deane	October 1876	From Fermanagh; died, April 1887
	Edward A. Hackett (ND)	September 1887	To Tipperary SR, September 1889
	Singleton Goodwin (SD)	September 1887	
	(ND)	October 1889	
	Entire county	July 1891	Retired, May 1920
	Henry Webster (SD)	October 1889	To Wexford, August 1891
	Valentine D. Doyle (ND)	September 1920	
	Entire county	May 1938	From Longford; resigned, October 1943
	Thomas J. Delahunty (SD)	1921	Died, May 1938
	Claud A. Warner	December 1944	From Leitrim; to Galway, January 1959
Kildare	Charles Lanyon	May 1834	To Antrim, October 1836
	John Yeats	October 1836	From Louth; died July, 1865
	Charles P. Cotton	August 1865	Resigned, October 1865
	Thomas Brazill	October 1865	Resigned, July 1869
	John H. Brett	December 1869	From Limerick (WD); to Antrim, November 1885
	Edward Glover	January 1886	From Mayo (SD); retired, June 1914
	John Rorke	December 1914	From Queen's County; retired, April 1944
Kilkenny	Samson Carter	May 1834	Died, December 1860
	Peter Burtchaell	December 1860	From Carlow; died, June 1894
	Alexander M. Burden	November 1894	Retired, February 1923
	Richard F. Bowen	October 1923	Died, May 1938
	Jeremiah G. Coffey	April 1940	Retired, 1974
King's County (Offaly)	Richard B. Grantham	May 1834	To Clare, December 1834
	Henry Brett	December 1834	To Mayo, April 1836
	Thomas Barclay	April 1836	From Mayo; died, May 1855
	John Hill	June 1855	From Clare; to Clare, May 1867
	Theophilus B. French	May 1867	To Tyrone, June 1874
	Richard B. Sanders	August 1874	Died, February 1900

COUNTY	NAME OF SURVEYOR	DATE APPOINTED	TENURE
King's County (Offaly) *(contd)*			
	James Delaney	October 1900	Died, June 1909
	Ignatius J. O'Sullivan	August 1909	Died, November 1927
	Thomas S. Duggan	August 1928	Retired, March 1954
Leitrim	Noblett R. St Leger	May 1834	To Sligo, October 1836
	Thomas D. Hall	October 1836	Died, January 1857
	James B. Pratt	January 1857	Retired, July 1884
	Drewry G. Ottley	October 1884	Died, September 1896
	Eugene O'Neill Clarke	1897	Retired, February 1936
	Patrick J. Haugh	April 1936	To Sligo, January 1943
	Claud A. Warner	May 1944	To Kerry, December 1944
Limerick	John Walker	May 1834	Resigned, November 1836
	William Horne	November 1836	Dismissed, June 1837
	Thomas Kearney (ED)	June 1837	Died, June 1862
	John Walker (WD)	July 1837	To Waterford, August 1840
	Richard Lanauze (WD)	August 1840	Resigned, July 1863
	Thomas Fosbery (ED)	June 1862	Retired, March 1893
	John H. Brett (WD)	July 1863	To Kildare, December 1869
	Henry Temple Humphreys (WD)	January 1870	To Galway (WD), December 1873
	W.E.L. Duffin (WD)	February 1874	To Waterford, October 1877
	John Cox (WD)	December 1877	Retired, March 1884
	John Horan (WD)	October 1884	
	Entire county	September 1893	Retired, March 1913
	Robert Davison (ED)	April 1914	Retired, 1929
	Thomas F. Ryan (WD)	June 1914	
	Entire county	1929	Retired, 1945
Londonderry (Derry)	Stewart Gordon	May 1834	Died, June 1860
	Richard Williamson	July 1860	From Tyrone; died, April 1874
	Arthur C. Adair	May 1874	From Tyrone (ND); retired, March 1900
	Charles L. Boddie	March 1900	Died, June 1924
	John A. Moore	1924	Died, 1944
	Harold K. Scott	April 1946	Post abolished, October 1973
Longford	James Bell	May 1834	Died, May 1872
	Samuel A. Kirkby	August 1872	To Cork, East Riding, July 1874
	Henry V. White	August 1874	To Queen's County, September 1883
	John O.B. Moynan	September 1883	To Tipperary (NR) August 1891
	John W. Gunnis	December 1891	Retired, January `1914
	Valentine D. Doyle	June 1915	To Kerry (ND), September 1920
	Bernard J. Kilbride	August 1924	Resigned, January 1927
	John J. Murphy	March 1927	Retired, March 1942
	Michael G. Ahern	June 1944	To Limerick, July 1948
Louth	John Yeats	May 1834	To Kildare, October 1836
	Francis Dubourdieu	October 1836	From Sligo; resigned, August 1840
	John Neville	September 1840	Retired, March 1886
	Patrick J. Lynam	April 1886	From Mayo (ND); died, August 1912
	Thomas Walsh	April 1914	From Queen's County; retired, March 1946

COUNTY	NAME OF SURVEYOR	DATE APPOINTED	TENURE
Mayo	Thomas Barclay	May 1834	To King's County, April 1836
	Henry Brett	April 1836	From King's County; to Waterford, October 1849
	Henry Brewster (2nd surveyor)	August 1848	
	Entire county	October 1849	
	(ND)	March 1856	Died, June 1883
	Frederick G. Deverell (SD)	March 1856	To Cork (ER–SD), November 1861
	William A. Treacy (SD)	November 1861	From Cork (ER–SD); to Tyrone (ED), July 1868
	William H. Deane (SD)	July 1868	From Tyrone (ED); to Fermanagh, August 1876
	Edward Glover (SD)	November 1876	To Kildare, January 1886
	Patrick J. Lynam (ND)	July 1883	To Louth, April 1886
	Richard H. Dorman (ND)	July 1886	To Armagh, January 1887
	William P. Orchard (ND)	December 1886	Resigned, March 1911
	Peter C. Cowan (SD)	February 1886	To Down (SD), June 1889
	Peter Le Fanu K. Dobbin (SD)	October 1889	To Clare, April 1893
	Edward K. Dixon(SD)	September 1893	Retired, October 1924
	J.J. Noonan (ND)	March 1911	Dismissed, October 1921
	Thomas P. Flanagan	October 1924	Retired, March 1960
Meath	Samuel S Searancke	May 1834	Resigned, July 1874
	Joseph H. Moore	August 1874	From Westmeath; retired, June 1907
	James Quigley	August 1907	Resigned, September 1923
	Edward J Duffy	October 1923	Died, February 1947
Monaghan	Edward V. Forrest	May 1834	To Queen's County, May 1839
	Alexander Harrison	May 1839	From Queen's County; to Carlow, August 1850
	John Walker	August 1850	From Carlow; dismissed, March 1854
	William Carroll	March 1854	To Waterford, September 1854
	Charles Tarrant	September 1854	To Waterford, March 1855
	William Carroll	March 1855	From Waterford; retired, July 1891
	James Heron	January 1892	To Down, March 1900
	William F. Barry	June 1900	To Wexford, January 1912
	James J. Hannigan	February 1912	Retired, April 1947
Queen's County (Laois)	Alexander Harrison	May 1834	To Monaghan, May 1839
	Edward V. Forrest	May 1839	Dismissed, July 1841
	Henry H. Owen	August 1841	To Waterford, August 1845
	Horace U. Townsend	August 1845	From Tipperary (NR); retired, September 1883
	Henry V. White	September 1883	From Longford; retired, August 1908
	Patrick R. Morris	January 1909	Died, March, 1909
	Thomas Walsh	July 1909	To Louth, April 1914
	John Rorke	April 1914	To Kildare, December 1914
	Joseph A. Ryan	April 1915	To Dublin, July 1924
	Malachi A. Feehan	1924	Retired, 1947
Roscommon	John Kelly	May 1834	Retired, July 1876
	Edward T. Quilton	October 1876	To Carlow, August 1877
	James Perry	December 1877	To Galway (WD), July 1882
	Christopher J. Mulvany	September 1882	Died, October 1919
	Michael Nerney	December 1923	Retired, November 1958
Sligo	Francis Dubourdieu	May 1834	To Louth, October 1836
	Noblett R. St Leger	October 1836	From Leitrim; Died, April 1872
	Charles B. Jones	May 1872	From Cork (ER–SD); retired, May 1906
	Richard J. Kirwan	February 1907	Retired, November 1942
	Patrick J. Haugh	January 1943	From Leitrim; retired, July 1957

COUNTY	NAME OF SURVEYOR	DATE APPOINTED	TENURE
Tipperary NR	Samuel Jones	May 1834	Tipperary (SR) from May 1839
	Horace U. Townsend	May 1839	To Queen's County, August 1845
	Andrew H. Crawford	September 1845	Died, August 1891
	John O.B. Moynan	August 1891	From Longford; retired, September 1930
	Thaddeus C. Courtney	September 1930	Resigned, October 1934
	William J. Chadwick	June 1935	Died, September 1951
Tipperary SR	Samuel Jones	May 1834	Dismissed, March 1852
	William H. Deane	March 1852	Dismissed, July 1860
	John L. Worral	August 1860	Retired, July 1889
	Edward A. Hackett	September 1889	From Kerry (ND); retired, May 1920
	Thomas S. Duggan	March 1920	Appointment terminated, March 1922
	Thomas P. Meade	January 1922	Died, December 1951
Tyrone	James Clarke	May 1834	Died, September 1838
	James B. Farrell	March 1839	To Wexford, August 1840
	Richard Richards	August 1840	Dismissed, December 1841
	Frederick J. Rowan	January 1842	Resigned, September 1847
	Alfred Gahan (WD)	October 1847	To Cavan, May 1867
	Richard Williamson (ED)	December 1847	To Londonderry, July 1860
	William H. Deane (ED)	August 1860	From Tipperary; to Mayo (SD), July 1868
	Arthur C. Adair (WD)	May 1867	From Clare; to Londonderry, May 1874
	William A. Treacy (SD)	July 1868	From Mayo (SD); dismissed, July 1869
	James A. Dickinson (SD)	January 1870	From Westmeath; retied, July 1895
	Theophilus B. French (ND)	June 1874	From King's County; retired, July 1895
	Francis J. Lynam (ND)	November 1895	Retired, September 1924
	John W. Leebody (SD)	November 1895	
	Entire county	October 1924	Retired, December 1934
	G.B.L. Glasgow	January 1935	Retired, 1958
Waterford	William Johnston	May 1834	Died, July 1840
	John Walker	August 1840	From Limerick; to Carlow, July 1844
	Charles G. Forth	July 1844	From Carlow; died, July 1845
	Henry H. Owen	August 1845	From Queen's County; resigned, October 1849
	Henry Brett	October 1849	From Mayo; to Wicklow, July 1853
	Alexander Schaw	July 1853	Died, September 1854
	William Carroll	September 1854	From Monaghan; to Monaghan, March 1855
	Charles Tarrant	March 1855	From Monaghan; died, July 1877
	William E.L. Duffin	October 1877	From Limerick (WD); retired, October 1914
	John K. Bowen	April 1915	Died, November 1944
Westmeath	Florence Mahony	May 1834	Died, January 1861
	James A. Dickinson	January 1861	To Tyrone (SD), January 1870
	Leveson F. Vernon-Harcourt	January 1870	Resigned, August 1870
	Joseph H. Moore	December 1870	To Meath, August 1874
	Frederick R.T. Willson	August 1874	To Fermanagh, October 1876
	Arthur E. Joyce	December 1876	Died, January 1927
	Cornelius Murphy	March 1928	Retired, July 1955
Wexford	Barry D. Gibbons	May 1834	Resigned, April 1838
	William Mackenzie	May 1838	Resigned, July 1840
	James B. Farrell	August 1840	From Tyrone; retired, July 1891
	Henry Webster	August 1891	From Kerry (SD); resigned, June 1909
	Stafford Gaffney	July 1909	Resigned, January 1912
	William F. Barry	January 1912	From Monaghan; retired, December 1942
	Thomas Kelly	June 1944	To Kildare, November 1945

COUNTY	NAME OF SURVEYOR	DATE APPOINTED	TENURE
Wicklow	William Hampton	May 1834	Died, November 1845
	James Boyd	December 1845	From Clare; died, July 1853
	Henry Brett	July 1853	From Waterford; died, May 1882
	Carter Draper	July 1882	From Galway (WD); died, December 1902
	Stephen G Gallagher	August 1903	Retired, May 1938
	Joseph T. O'Byrne	July 1940	Retired, March 1964

Bibliography

In the absence of official records of the activities of the grand juries and of the proceedings of the early county councils, the preliminary essays in Parts I and II and the biographical notes in Part III are based on a wide range of other sources listed below. Footnotes indicate the main sources relied on in preparing the essays but because of the multiplicity of sources used in compiling the individual biographical notes in Part III, footnoting is impracticable in these cases.

For basic information about dates of appointment, dismissal and retirement of county surveyors, the Chief Secretary's Office Registered Papers dating from 1818 and held at the National Archives of Ireland have been heavily drawn on; unfortunately, some of the relevant papers are not extant but, in a number of these cases, the relevant entries the annual index volume can provide valuable information. Additional information on appointments and details of some of the activities of the surveyors can be found in the Board of Works letter books and other papers (including a substantial collection of architectural and engineering drawings) also held at the National Archives. Documentation held at The National Archives (formerly the PRO) at Kew, London, including files created by the Civil Service Commissioners, provides information about appointments from 1870 onwards. The British Parliamentary Papers contain a substantial volume of additional source material; they include reports presented to parliament at different stages listing surveyors then in office; reports of commissions and select committees set up to investigate different aspects of the operation of the grand jury system; and reports of other inquiries at which surveyors gave evidence. The annual reports of a number of public authorities are also relevant, particularly the reports of the Board of Works from 1834 onwards, reports of the Poor Law Commissioners in the 1840s, reports of the Civil Service Commissioners from 1870 onwards, the annual reports of the Local Government Board for Ireland from 1899 to 1920 and the annual reports of the Department of Local Government and Public Health from 1922 to 1940.

The county directories for each year, included in the volumes of *Thom's Directory*, contain useful listings of the county officers, including county surveyors and assistant surveyors, and in many cases, their year of appointment. The provincial newspapers circulating in Ireland in the nineteenth century generally carried detailed reports on the proceedings of the grand juries at the spring and summer assizes although some of them devoted far greater attention to the more news-worthy criminal trials conducted before the jury than to the more mundane fiscal business. Many of the newspapers printed in full the biannual reports which each county surveyor was required to present on the ongoing programme of county works and the general condition of the county's stock of roads, bridges and public buildings; these reports often provide a good indication of the personality of the individual surveyor as well as a fund of information on work in progress and plans for new projects. In the later decades of the nineteenth century and subsequently, substantial obituaries of surveyors can also be found in the local newspapers.

Apart from their biannual reports, most of the early surveyors have left no written record of their operations or of their approach to the engineering and other issues of the day. A small number published pamphlets or books and some – particularly from the 1880s onwards –

contributed papers to the journals of professional organisations or delivered presidential addresses which touched on a wide range of issues relevant to the engineering profession.

The *Biographical Dictionary of Civil Engineers in Great Britain and Ireland, Vol. 1: 1500 to 1830* (London, 2002) contains an entry for only one of the surveyors, John Steedman, who served in Co. Donegal from 1834 to 1847; a second volume, now in preparation, covering the years from 1831 to the end of the century is likely to include a small number of others. The *Directory of British Architects, 1834–1914* (London, 2001) contains entries for Sir John Benson, Edward Glover, Sir Charles Lanyon and Thomas Turner. The *Dictionary of Land Surveyors and Local Map-makers of Great Britain and Ireland, 1530–1850* (London, 1997) which lists over 13,700 individuals, contains entries for thirty-two of the surveyors but these are generally very brief and incomplete. *British Engineers and Allied Professions in the Twentieth Century* and other publications edited and published by W.T. Pike between 1908 and 1911 contain short entries for a number of the surveyors. The *Oxford Dictionary of National Biography* (Oxford 2004) includes entries for Sir John Benson, Sir Charles Lanyon, and Levenson Francis Vernon Harcourt and the forthcoming *Dictionary of Irish Biography* (Cambridge, 2009) is expected to include a small number of others. The Irish Architectural Archive's Biographical Index of Irish Architects contains information on Irish architects and other architects who worked in Ireland between 1720 and 1940; it includes biographical details on some of the surveyors and lists projects on which they are known to have worked.

It was the practice of the Institution of Civil Engineers (London) to include obituaries or memoirs of deceased members in the *Proceedings* which were published annually; notices relevant to county surveyors are listed below. Unfortunately, the Institution of Civil Engineers of Ireland, of which most of the surveyors were members, rarely published obituaries or memoirs in its annual *Transactions*.

1. PUBLICATIONS BY COUNTY SURVEYORS

(a) Papers in Transactions of the Institution of Civil Engineers of Ireland
County surveyors constituted a significant proportion of the membership of the Institution of Civil Engineers of Ireland but the transactions of the Institution which began publication in 1845 contain a relatively small number of papers by surveyors in the early years. After 1869, however, when a growing number of the surveyors were university graduates, they began to figure more prominently in the activities and publications of the Institution even though they then formed a smaller proportion of the total membership than in previous years.

Benson, John, An account of the Carrigaline (*sic*) Bridge, over the River Lee, on the new Cork and Macroom road, iii (1849), 57–8.
Benson, John, An account of the Skew Bridge built on the New Western Entrance to Cork, iv (part 1) (1851), 1–2.
Bowen, J., The Salvage of the Main Road Bridge at Ballyvoyle, Co Waterford, liv (1929), 191–203.
Bower, John, Description of a method of setting out Curves with the Theodolite and a chain of chords, ii (1848), 49–53.
Bower, John, Single and Double Lines of Railway, v (1860), 11–36.
Bower, John, The Monster Contract System; its influence on the Cost of Works, and the position of the Engineer, v (1860), 65–104.
Chadwick, W.J., Turf Production in County Tipperary (North Riding), lxviii (1942), 145–53.
Coffey, J.G., Presidential Address, lxxxvii (1961), 1–19.
Collen, William, Loughshinney Pier, County Dublin, xxxvi (1910), 165.

Collen, William, Presidential Address, xl (1913), 1–25.

Cotton, C. P., On the Discharge of Tidal Sluices, x (1875), 88–94.

Cotton, C.P., Loans for Sanitary Purposes, xvi (1886), 236–45.

Cotton, C.P., Notes on Concrete and on Some Experiments with Concrete Arches, xi (1877), 96–108.

Cotton, C.P., On a Novel Means of Transit for Minerals in the County of Sligo, x (1875), 1–5.

Cotton, C.P., On the Castleisland and Gortatlea Railway, xi (1877), 41–54.

Cotton, C.P., On the Failure of Beacon on the Aldern or Alderman Rocks in Crookhaven Bay in the year 1865, x (1875), 95–100.

Cotton, C.P., On the Modification of the Prismoidal Formula, xii (1880), 26–8.

Cotton, C.P., On the Reconstruction of Innoshannon Bridge, Cork and Bandon Railway, vii (part 1) (1866), 1–11.

Cotton, C.P., Presidential Address, x (1875), 101–23.

Cotton, C.P., The Tansa Waterworks, Bombay, xxii (1893), 67–72.

Courtney, T.C., Presidential Address, lxx (1944), 1–14.

Courtney, T.C., Road reconstruction, lxiv (1938), 99–123

Cowan, P.C., Notes on Modern Road-Making, xxxv (1909), 130–56.

Cowan, P.C., Presidential Address, xxxviii (1912), 1–26.

Cowan, P.C., The Improvement of the Main Roads of Ireland, xxxix (1913), 143–75.

Duffin, W.E. L'Estrange, The Encroachment of the Sea; Beaches and Groynes, xv (1885), 62–82.

Duffin, W.E. L'Estrange, Two Small Waterworks, xlii (1916), 23–40.

Duffy, E.J., Cement Bound Macadam Roads, lviii (1932), 305–57.

Duffy, E.J., Notes on Turf Production, lxviii (1942), 93–110.

Duffy, E.J., The Treatment of Shear in Reinforced Concrete Beams, xlix (1924), 160–94.

Farrell, James B., On the Design and Details of Construction for a proposed New Bridge over the Estuary of the Slaney, at Wexford, iv (part 2)(1851), 1–7.

Flanagan, T.P., Emergency Fuel : Some Aspects of the Problems of production and Transport in Mayo, lxviii (1942), 111–44.

Frith, R.H., On the Economy of Scavenging and Repairing the Streets of Dublin, compared with that of Scavenging and Repairing the Streets of Lambeth, in London, v (1860), 105–15.

Gallagher, Stephen Gerald, Presidential Address, lviii (1932), 3–8.

Glover, Edward, Presidential Address, xxviii (1902), 94–123.

Glover, Edward, The Bicycle and Common Roads, xxvi (1897), 80–93.

Green, C.F., On Light Railways, or Remunerative Railways for Thinly-Populated Districts, xiii (1882), 158–94.

Green, C.F., On Railway Bridge Platforms, xii (1880), 86–118.

Hill, John, On the Maintenance of Macadamised Roads, exemplified by the repairs of the Central District of Roads and Bridges in charge of the Commissioners of Public Works in Ireland, iv (part 1) (1851), 12–23.

Hill, John, On the Maintenance of Macadamised Roads: Sequel to paper read before the Institution of Civil Engineers of Ireland, v (1860), 46–51.

Kelly, T., Presidential Address, xci (1965), 1–20.

Moore, J.H., Presidential Address, xxxiv (1908), 1–26.

Moynan, J. Ouseley, Road Transport in Ireland, xlvi (1922), 187–94.

Moynan, J. Ouseley, On the Maintenance and Repairs of County Roads in Ireland, xvii (1887), 180–204.

Moynan, J. Ouseley, Presidential Address, xlv (1921), 1–21.

Moynan, J.O., A Short Description of the Existing Bridges at Waterford and Portumna, and of the Proposed New Structures to replace them, xxxvi (1910), 225–40.

Neville, John, An Investigation of some formulae for finding the maximum amount of resistance required to sustain banks of earth, or other materials, and the position of the fracture requiring that resistance, i (1845), 108–37.

O'Connor, R.F., Irish Free State Transport Problems, liii (1928), 210–31.

Quigley, James, Road-Making in the USA and in Ireland, lvii (1932), 192–228.

Roberts, Samuel U., An Account of an Iron Breast Wheel, 36 Horse Power, recently erected at the Newcastle Distillery, in Galway, iv (part 2) (1851), 53–7.

Roberts, Samuel U., Description of means adopted for unwatering works of river drainage, iii (1849), 1–8.

Rorke, J., Road making Practice in Ireland, lxii (1936), 159–80.

Ryan, Joseph A., Presidential Address, lxv (1939), 1–12.

Ryan, Joseph A., Roads and Road Construction, lii (1927), 62–82.

Sanders, R.B., Steam Traction on Common Roads, xiv (1884), 181–211.

Simington, T.A., Presidential Address, lxxxviii (1962), 1–11.

St Leger, Noblett R., On a Method adopted to raise a Roof in Ballina, ii (1845), 66–7.

Stokes, Henry, Incombustible Materials for Building, xiv (1884), 55–9.

Stokes, Henry, On the Ventilation of Coal Mines, xii (1880), 120–5.

Warner, C.A., Concrete Design, lxvii (1941), 169–96.

Warner, C.A., Peat Production: Some Mechanical Methods, lxix (1943), 39–58.

White, Henry V., County Work in Ireland considered in relation to Grand Jury Laws, xvii (1887), 19–81.

White, Henry V., On the Development of the Leinster Coalfield by Railway Communication, xviii (1888), 29–46.

White, Henry V., On the Measurement of Distances in Levelling and Survey Operations, xiii (1882), 82–89.

Willson, Frederick, County Works in Ireland, xvii (1887), 144–79.

(b) Papers in the Proceedings of the Institution of Civil Engineers *(London)*

The small number of county surveyors who became members of the Institution of Civil Engineers contributed only a few papers to the institution's *Proceedings* which were published from 1831 onwards. The following are the main papers of Irish interest:

Forth, Charles, An Account of the Alterations to Tullow Bridge, 2 (1842), 165.

Grantham, Richard Boxall, On Arterial Drainage and Outfalls, 19 (1859), 53..

Hill, John, Account of Two Drainages in Ireland, 59 (1879–80 (pt. 1), 265.

Sanders, Richard Barnsley, The Management, Maintenance and Cost of Public County Roads in Ireland under the Irish Grand Jury System, 135 (1898–9), 258–271.

(c) Papers in the Proceedings of the Incorporated Association of Municipal and County Engineers

After the Association of Municipal and Sanitary Engineers and Surveyors (founded in 1873) was transformed into the Incorporated Association of Municipal and County Engineers, many of the county surveyors were admitted to membership. From the 1890s onwards, a number of them contributed substantial papers to the association's proceedings.

Dorman, R.H., Road Maintenance, xvii (1891–2), 45.

Dorman, R.H., Light Railways and Tramways, xxii (1895–6), 34.

Dorman, R.H., Main Roads under County Councils in Ireland, xxv (1898–9), 28.

Dorman, R.H., Portadown and Banbridge Water Supply, xxxiii (1906–7), 133.

Dorman, R.H., Some Notes on Road Maintenance in County Armagh, xxxix (1912–13), 106.

Leebody, J.W., County Road Maintenance in Ulster, xxiii (1906–07), 120.

Leebody, J.W., The Road Board Circular from the Irish Point of View, xxxvii (1910–11), 198.

Perry, James, Municipal Electricity, xxii (1895–6), 17.

Sanders, R.B., Irish Grand Jury Surveyorships and the Grand Jury System of Ireland, xx (1893–4), 251.

(d) Papers in the Irish Builder/Irish Builder and Engineer

The *Irish Builder* (established in 1859 as the *Dublin Builder* and from 1903 onwards titled the *Irish Builder and Engineer*) occasionally published the text of papers or addresses delivered by county surveyors. The following items are of interest:

Brett, J.H., County Courthouses and County Gaols in Ireland, xvii, 15 Jan. 1875, 25–6; 1 Feb. 1875, 41.

Cowan, P.C., Address at the Congress of the Royal Sanitary Institute, Belfast, liii, 19 Aug. 1911, 565.

Cowan, P.C., Presidential Address to the Engineering Students Association TCD, xlvii, 30 Dec. 1905, 951.

Cowan, P.C., Roads: their making and their maintenance, xlvii, 30 Dec. 1905, 930–42.

Cowan, P.C., The Difficulties of the Housing Problem and Some Attempts to solve it, lviii, 26 Feb. 1916, 87–93.

Cowan, P.C., The Science of Health, lvii, 30 Jan. 1915, 46–8.

Dixon, E.K., Rural District Road Bridges of Small Span: some practical notes, lii, 14 May 1910, 302.

Moore, J.H., Irish Roads – past and present, lii, 16 Apr. 1910, 236.

Ryan, J.A., Some Notes on Road Construction, lxvii, 24 July 1926, 581–5.

(e) Other publications by county surveyors

Bower, J.C., *Reports & Observations on the Subject of a Supply of Water to the Town and Borough of Newry* (Dublin, 1870).

Brazill, Thomas, *Report to the Corporation of Dublin on the proposed supply of the city and suburbs with pure water at high pressure* (Dublin, 1854).

Brett, Henry, *The Reclamation of the Waste Lands of Ireland* (Dublin, 1881).

Clarke, James, *Practical Directions for Laying Out and Making Roads* (Dublin, 1818).

Cotton, C.P., *Loans for Sanitary Purposes* (1886).

Cotton, C.P., *A Manual of Procedure by Provisional Order (1887)*.

Cotton, C.P., *Manual of Railway Engineering in Ireland* (Dublin, 1861).

Cotton, C.P., *The Housing of the Working Classes Act, 1890* (1890).

Cotton, C.P., *The Irish Public Health Acts, 1878–90* (1891).

Cotton, C.P., *The Irish Sanitary Acts* (1892).

Cowan, P.C., *Reconstruction: A Trilogy* (Dublin, 1919).

Cowan, P.C., *Report on Dublin Housing* (Dublin, 1918).

Dubourdieu, Francis R.H., *Instructions for the Choice, Fortifying, Occupying, Defence, and Attack of Military Fortifications* (1830).

Dubourdieu, Francis R.H., *Wild Flowers from Germany* (Belfast, 1850).

Frith, R.H., *Macadamized Streets compared with Paved Streets* (Dublin, 1857).

Frith, R.H., *The Drainage and Water Supply of Dublin* (Dublin, 1867).

Grantham, R.B., A Treatise on Public Slaughter-Houses (London, 1848).

Grantham, R.B., *On Arterial Drainage and Outfalls* (London, 1862).

Hackett, E.A., *Economical Steam Rolling of Irish Country Roads* (Dublin, 1904).

Leahy, Edmund, *A Practical Treatise on Making and Repairing Roads, Illustrated by engravings and tables* (London, 1844, 1847).

Leahy, Patrick, *New and General Tables of Weights & Measures with ample calculations and suitable examples under each head* (Dublin, 1826).

Lindsay, Henry L., *An Essay on the Agriculture of the County Armagh* (Armagh, 1836).

Lindsay, Henry L., *The Present State of the Irish Grand Jury Law Considered as it respects the Promotion and Execution of Public Works; and an Improved Plan of Jurisprudence Recommended* (Armagh, 1837).

Neville, John, 'Grand Jury Laws and County Public Works, Ireland', *Dublin University Magazine*, xxvii: clix (March 1846).

Neville, John, *Hydraulic Tables, Co-efficients, and Formulae for Finding the Discharge of Water from Orifices, Notches, Weirs, Pipes, and Rivers* (London, 1853).

Neville, John, 'On the maximum amount of resistance required to sustain banks of earth', *Proceedings of the Royal Irish Academy*, 1847, pp 80–8.

Neville, John, 'Road Cross Sections', *Original Papers on Engineering*, iii (pt. vi) (1846).

Vernon-Harcourt, Leveson Francis, 'Civil Engineering' in W.T. Pike (ed.), *British Engineers and Allied Professions in the Twentieth Century* (Brighton 1908).

Vernon-Harcourt, Leveson Francis, *Harbours and Docks* (1885).

Vernon-Harcourt, Leveson Francis, *Rivers and Canals* (2nd ed.), (1895).

2. MEMOIRS/OBITUARIES

(a) Minutes of the Proceedings of the Institution of Civil Engineers (London)

Bell, James, 79 (1884–5), p. 360.

Benson, Sir John, 40 (1874–5), p. 251.

Bower, John, 92 (1887–8), p. 379.

Cotton, Charles Philip, 157 (1905–6), p. 373.

Cowan, Peter Chalmers, 231 (1930–1), p. 372.

Draper, Carter, 151 (1902–3), p. 410.

Farrell, James Barry, 113 (1892–3), p. 344.

Garrett, Robert, 18 (1859), p. 188.

Grantham, Richard Boxall, 108 (1891–2), p. 398.

Green, Charles Frederic, 88 (1886–7), p. 437.

Hill, John, 116 (1893–4), p. 360.

Humphreys, Henry Temple, 108 (1891–2), p. 403.

Lanyon, Sir Charles, 98 (1888–9), p. 390.

Ottley, Drewry Gifford, 127 (1896–7), p. 382.

Patterson, R.H.S, 7 (May-Aug 1957) p. 678.

Perry, James, 169 (1906–7), p. 392.

Prendergast, Patrick Joseph, 184 (1910–11), p. 360.

Roberts, Samuel Ussher, 140 (1899–1900), p. 279.

Sanders, Richard Barnsley, 142 (1899–1900), p. 380.

Tarrant, Charles, 52 (1877–8), p. 280.

Vernon-Harcourt, Leveson Francis, 171 (1907–8), p. 420.

Wilkin, J.G., 10 (May-Aug 1958) p. 604.

Woodhouse, Thomas Jackson, 16 (1856–7) p. 150.

(b) Transactions of the Institution of Civil Engineers of Ireland
Bowen, John Kingston, lxxi (1945), p. 249.
Brett, Henry, xiv (1884), p. 2.
Cotton, Charles Philip, xxxi (1906), p. 289; xxxii (1907), p. 26.
Gibbons, Barry Duncan, vii (pt. 2) (1864), p. 168.
Moore, J.H., xxxviii (1912), p. 206.
Rorke, John, lxxiv (1948), p. 370.
Tate, Alexander, xxxii (1907), p. 28.

3. NEWSPAPERS

Anglo-Celt
Ballyshannon Herald
Belfast Telegraph
Boyle Gazette
Carlow Sentinel
Clare Champion
Clare Journal
Connacht Telegraph
Connacht Tribune
Connaught Champion
Cork Constitution
Cork County Eagle & Munster Advertiser
Cork Examiner
Derry (Londonderry) Journal
Derry (Londonderry) Standard
Donegal Indepndent
Drogheda Argus
Drogheda Independent
Dublin Evening Mail
Dublin Evening Post
Enniskillen Chronicle and Erne Packet
Freeman's Journal
Galway Vindicator and Connaught Advertiser
Impartial Reporter
Irish Times
Kerry Evening Post
Kerry Sentinel
Kilkenny Journal
Kilkenny Moderator and Leinster Advertiser
Kilkenny People

King's County Chronicle
Leinster Express
Leitrim Observer
Limerick Chronicle
Limerick Reporter & Tipperary Vindicator
Limerick Standard
Limerick Star & Evening Post
Longford Independent
Longford Journal
Longford Leader
Mayo Constitution
Meath Herald and Cavan Advertiser
Midland Reporter
Midland Tribune
Mourne Observer
Nenagh Guardian
Northern Standard
Roscommon and Leitrim Gazette
Roscommon Herald
Roscommon Journal and Western Reporter
Sligo Champion
The Kerryman
The Times
Tipperary Free Press
Tuam Herald
Tyrone Constitution
Ulster Gazette
Waterford Mail
Wexford Independent

4. JOURNALS

British Birds
The Builder (London)
Building News (London)
The Countryman
Dublin Builder (1859–66)
Engineer
Engineers Journal
Irish Builder (1867–1903)
Irish Builder and Engineer (1903–)
Irish Cyclist
Irish Law Times Reports

Irish Naturalist
Irish Railway Gazette
Irish Wheelman
Motor News
Proceedings of the Incorporated Association of Municipal and County Engineers
Proceedings of the Institution of Civil Engineers (London)
Thom's Directory
Transactions of the Institution of Civil Engineers of Ireland

5. MANUSCRIPT MATERIALS

National Library of Ireland

De Vesci papers
Larcom papers
Mayo papers
Smith O'Brien papers
Vesey Fitzgerald papers
Manuscript map collection

National Archives

Chief Secretary's Office Registered Papers
Chief Secretary's Office Unregistered Papers
Official Papers
OPW Architectural and Engineering Drawings
Public Works Letter Books

Other repositories

Entrance books and admissions registers (Trinity College, Dublin).
Membership application forms, 1852–1916 (Institution of Engineers of Ireland).
Minute Books of Civil Engineers Society of Ireland, 1835–44 (Institution of Engineers of Ireland).
Patterson Kempster & Shorthall Collection (Irish Architectural Archive).

6. ANNUAL REPORTS

Calendars of the Queen's Colleges and of Queen's University
Civil Service Commissioners, 1870–1920
Commissioners of Public Works in Ireland (Board of Works), 1834–

Department of Local Government and Public Health, 1922–1940
Local Government Board for Ireland, 1899–1920
Reports of the Presidents of the Queen's Colleges (Belfast, Cork, Galway).

7. BRITISH PARLIAMENTARY PAPERS

Abstract of the Accounts of Grand Jury Presentments in 1834, HC 1835 xxxvii (220).

Civil Departments: Returns of the Names and Offices of all Persons whose Salary and Emoluments exceed £250 per annum, HC 1830 xvii (480).

Correspondence relating to measures for relief of distress in Ireland (Board of Works Series), July 1846 to January 1847, HC 1847 l (764).

Evidence taken before the Commissioners appointed to Inquire into the Occupation of Land in Ireland, HC 1845 xxi (657).

Final Report from the Board of Public Works, Ireland, relating to measures adopted for the relief of Distress in July and August 1847, HC 1849 xxxiii (1047).

First and Second Reports of the Commissioners of Inquiry into the state of the Irish Fisheries, HC 1837 xxii (77), (82).

First Report from the Select Committee on Irish Grand Jury Presentments, HC 1816 ix (374).

First Report from the Select Committee on Public Works, Ireland, HC 1835 xx (329).

First Report of the Inspectors General on the General State of the Prisons of Ireland, HC 1823 x (342).

Lough Erne, copy of the Report of the Board of Works in Ireland upon the plan submitted to the Board by the County Surveyor of Fermanagh relative to the flood drainage and navigation of Lough Erne, and the works connected therewith, HC 1852–3 xciv (694).

Piers and Roads: Counties of Galway and Mayo, Report by Colonel Fraser RE [c. 5729] HC 1889 lx (949).

Report from the Committee on Grand Jury Presentments in Ireland, HC 1819 viii (387).

Report from the Select Committee of the House of Lords appointed to consider the State of the Lunatic Poor in Ireland, HL 1843 (193).

Report from the Select Committee of the House of Lords on the River Shannon (Navigation and Drainage), HL 1865 xi (400).

Report from the Select Committee on County & District Surveyors etc. (Ireland), HC 1857 (Sess. 2) ix (270).

Report from the Select Committee on Grand Jury Cess (Ireland), HC 1836 xii (527).

Report from the Select Committee on Grand Jury Presentments (Ireland), HC 1867–8 x (392).

Report from the Select Committee on Grand Jury Presentments, Ireland, HC 1826–7 iii (555).

Report from the Select Committee on Scientific Instruction; together with the Proceedings of the Committee, Minutes of Evidence and Appendix, HC 1867–8 xv (432).

Report of the Board of Civil Engineers which sat at No. 21 Mary Street, Dublin, from 23rd December 1817 to the 19th January, 1818, HC 1818 xvi (2).

Report of the Commissioner appointed to Inquire into the Turnpike Trusts, Ireland, HC 1856 xix (2110).

Report of the Commissioner for inquiring into the Execution of the Contracts for Certain Union Workhouses in Ireland, HC 1844 xxx (562).

Report of the Commissioners appointed to revise the Several Laws under or by virtue of which Monies are now raised by Grand Jury Presentment in Ireland, HC 1842 xxiv (386).

Report of the Committee appointed by the Treasury to Inquire into the Constitution and Duties of the Board of Works in Ireland, [C. 2060] HC 1878 xxiii (1).

Report of the Royal Commission on Irish Public Works, [c. 5264–1] HC 1888 xlviii.

Report of the Select Committee on the State of the Poor in Ireland (Summary Report), HC 1830 vii (667).

Reports from the Select Committee on Grand Jury Presentments, Ireland, HC 1822 vii (353) (413) (451).

Return ... specifying the Name, Salary and Situation of each Person receiving upwards of £800 per annum, HC 1835 xxxvii (609).

Return of all Piers and Harbours built under the Board of Works in Ireland since the passing of the Act 9 Vict., c. 3, HC 1884–85 lxx (167).

Return of County Surveyors and their Assistants in each County in Ireland for the Year ending with the Summer Assizes 1877, HC 1878 lxvi (218).

Return of Officers in the Service of the Grand Jury in each County in Ireland, HC 1898 lxxiv (237).

Return of the Names of Each Engineer Employed under the Commissioners of Public Works in Ireland ... in the years 1832, 1833 and 1834, HC 1835 xxxvii (536).

Return of the Number of County Surveyors, and of their Deputies and Clerks in each county in Ireland etc., from 1834 to 1839, HC 1840 xlviii (291).

Return of the Number of Days appointed by the Sheriff for Transacting the Fiscal Business in each County in Ireland, HC 1844 xliii (130).

Return Showing all Sums of Money ... voted or applied ... in aid of Public Works in Ireland since the Union ..., HC 1839 xliv (540).

Return showing the Names ... of all Officials transferred by virtue of the provisions of the Local Government (Ireland) Act, 1898 to County Councils, District Councils or Boards of Guardians etc., HC 1901 lxiv (331).

Return showing the Names of the County Surveyors in each County in Ireland, and their Assistants, with their respective Salaries, etc., HC 1863 l (277).

Returns from County Surveyors in Ireland of Number of Assistants and their Salaries, etc., HC 1867–8 lv (240).

Returns from County Surveyors in Ireland ... of the Number of Miles of Roads under charge etc., HC 1867–8 lv (446).

Returns from the Authorities of the Harbours etc. of the United Kingdom, HC 1903 lxiii (325).

Returns of Local Taxation in Ireland for the Year 1898, HC 1899 lxxxiii (pt. 1) (360).

Returns of the Dates of Commissions issued by the Irish Government for the Examination of Candidates for the Office of County Surveyor in Ireland, etc., HC 1856 liii (335).

Second Report from the Select Committee on Irish Grand Jury Presentments, HC 1816 ix (435).

Select Committee on Advances made by the Commissioners of Public Works in Ireland, HC 1835 xx (573).

Statements by County Surveyors on the Condition of Piers and Harbours etc., HC 1884–85 lxx (266).

Tenth Report of the Inspectors General on the General State of the Prisons of Ireland, HC 1831–2. xxiii (152).

8. OTHER PUBLICATIONS

A Catalogue of Graduates who have proceeded to degrees in the University of Dublin, 1595–1866 (Dublin, 1869).

Andrews, J.H., 'Maps and Mapmakers' in William Nolan (ed.), *The Shaping of Ireland: the Geographical Perspective* (Cork, 1986).

Andrews, J.H., *Plantation Acres* (Belfast, 1985)

Andrews, J.H., 'Road Planning in Ireland before the Railway Age', *Irish Geography*, 5:1 (1964), 17–41.

Aston, Gordon, *One Hundred Years of Quantity Surveying: the Annals of Patterson & Kempster, 1860–1960* (Dublin, 2007).

Bandon Navigation, Report of the Proceedings of the Committee appointed on 17 December 1841, Cork Archives Institute, U 140/D.

Barry, Michael, *Across Deep Waters: Bridges of Ireland* (Dublin, 1985).

Bendall, Sarah (ed.), *Dictionary of Land Surveyors and Local Map-makers of Great Britain and Ireland, 1530–1850* (2 vols) (London, 1997).

Boyle, Tom, *Kilkenny County Council: A Century of Local Government* (Kilkenny, 1999).

Brett, C.E.B., *Court Houses and Market Houses of the province of Ulster* (Belfast, 1973).

Brett, C.E.B., *Buildings of County Armagh* (Belfast, 1999).

Broderick, David, *The First Toll Roads* (Cork, 2002).

Brodie, Antonia et al. (eds), The *Directory of British Architects, 1834–1914* (London, 2001).

Burgoyne, Sir John Fox, *Remarks on the Maintenance of Macadamised Roads* (Dublin, 1843).

Burtchael, G.D. and T.U. Sadleir, *Alumni Dublinenses* (Dublin, 1935).

Cadogan, Tim, *Cork County Council 1899–1999, A Centenary Souvenir* (Cork, 1999).

Carbery, Lord, *Observations on the Grand Jury System of Ireland with Suggestions for its Improvement* (London, 1831).

Carroll, J.S., 'Assistant County Surveyors in Ireland', *Engineers Journal*, 8:8 (1955), 317.

Casey, Christine, 'John Nevile: Louth County Surveyor', *County Louth Archaeological and Historical Journal*, xxi:1 (1985), 23–30.

Charlton, T.M., *Civil Engineering in Queen's, 1849–1963* (Belfast, 1964).

Cochrane, Robert, 'Remarks on Some Examinations, Competitive and Qualifying, as Avenues of Employment for Engineers', *TICEI xxxi* (1906), 5–148.

Colvin, Howard, *A Biographical Dictionary of British Architects, 1600–1840* (London, 1995).

Commission on Reconstruction and Development, Interim Report on Reconditioning and Improvement of Roads (Stationery Office, 1923).

Country Gentleman (A), *Hints on the system of Road Making* (Dublin, 1829).

Cox, R.C., *A Record of the School of Engineering, 1841–1966* (4th ed) (Dublin 1977).

Craig, Maurice, *The Architecture of Ireland from the Earliest Times to 1880* (London, 1982).

Curtis, L.P. Jr, *Coercion and Conciliation in Ireland, 1880–1892* (Princeton, 1963).

Daly, Mary E. (ed.), *Country and Town: One Hundred Years of Local Government in Ireland* (Dublin, 2001).

Daly, Mary E., *The Buffer State: The Historical Roots of the Department of the Environment* (Dublin, 1997).

Daly, Mary E., *The Famine in Ireland* (Dundalk, 1986).

De Beaumont, Gustave, *Ireland: Social, Political and Religious*, edited and translated by W.C. Taylor (London, 1839); repr. with an introduction by Tom Garvin and Andreas Hess (Harvard, 2006).

De Courcy, Catherine, *The Foundation of the National Gallery of Ireland* (Dublin, 1985).

Dixon, Robert V., *A Farewell Address delivered to the Students of the School of Engineering* (Dublin, 1854).

Donnelly, Brian, *For the Betterment of the People: A History of Wicklow County Council* (Wicklow, 1999).

Edgeworth, Maria, *Tour in Connemara and the Martins of Ballinahinch* (ed. H.E. Butler), (London, 1950).

Fennelly, Teddy (ed.), *Laois County Council: the first 100 years* (Portlaoise, 1999).

Ferriter, Diarmaid, *Cuimhnigh ar Luimneach, A History of Limerick County Council, 1898–1998* (Limerick, 1998).

First Irish Roads Congress: Record of Proceedings (Dublin, 1910).

Fitzpatrick, David, *Politics and Irish Life, 1913–1921* (Dublin, 1977).

Fraser, Murray, *John Bull's Other Homes: State Housing and British Policy in Ireland, 1883–1922* (Liverpool, 1996).

Greig, William, *Strictures on Road Police* (*sic*) (Dublin, 1818).

Griffin, Brian, *Cycling in Victorian Ireland* (Dublin, 2006).

Griffith, Arthur, *The Resurrection of Hungary* (Dublin, 1905).

Griffiths, A.R.G., 'The Irish Board of Works in the Famine Years', *Historical Journal*, xiii (1970).

Heaney, Henry (ed.), *A Scottish Whig in Ireland, The Irish Journals of Robert Graham of Redgorton* (Dublin, 1999).

Hughes, Noel J., *Irish Engineering, 1760–1960* (Dublin, 1982).

Important Articles on Road-Making and Maintenance, published for the Irish Roads Improvement Association (Dublin, 1906).

Justitia, *Observations on the General Grand Jury Act ... in a Letter to the Right Hon. E. Horsman, MP, Chief Secretary for Ireland* (Dublin, 1857).

Keating, Rev M.J., *Suggestions for a Revision of the Irish Grand Jury Laws* (n.d.).

Laffan, Michael, *The Resurrection of Ireland: The Sinn Féin Party, 1916–1923* (Cambridge, 1999).

Larmour, Paul, *Belfast: An Illustrated Architectural Guide* (Belfast, 1987).

Leahy, P., 'History of Engineering Training in Ireland', *Engineers Journal* 15 (April 1962), 147–151.

Lee, J.J., 'Centralisation and community', in *Ireland: Towards a Sense of Place* (Cork, 1985), p.84.

Lincoln, Colm, *Steps and Steeples: Cork at the Turn of the Century* (Dublin, 1980).

Long, Brendan, *Tipperary S.R. County Council 1899–1999: A Century of Local Democracy* (Clonmel, 1999).

Lyons, F.S.L., *Ireland since the Famine* (London, 1971).

Lyons, Mary Cecelia, *Illustrated Incumbered Estates, Ireland, 1850–1905* (Ballinakella, 1993).

Maclean, Sally, 'Searancke, William Nicholas 1817?–1904' in *Dictionary of New Zealand Biography* II (1993).

MacNeill, Sir John, *The Boyne Bridge: Its History* (1860).

Mansergh, Nicholas, *The Irish Free State: Its Government and Politics* (London, 1934).

Maume, Patrick, *The Long Gestation: Irish Nationalist Life 1891–1918* (Dublin, 1999).

Maynard, Henry Nathan, 'The New Ross Bridge', *PRICE* 32, 146.

McCullough, Niall, 'Courthouses – The Mirror of Society', in Mildred Dunne and Brian Phillips (eds), *The Courthouses of Ireland* (Kilkenny, 1999).

McCutcheon, W.A., *The Industrial Archaeology of Northern Ireland* (Belfast, 1980).

McDermott, Joe, *Public-Spirited people: Mayo County Council 1899–1900* (Castlebar, 1999).

McDowell, R.B. and D.A. Webb, *Trinity College Dublin, 1592–1952 An Academic History* (Cambridge, 1982).

McDowell, R.B., *The Irish Administration, 1801–1914* (London, 1964).

McNally, Brian and Maurice McHugh, *Comhairle Chontae Phortláirge, 1899–1999* (Dungarvan, 1999).

McNamara, T.F., *Portrait of Cork* (Cork, 1981).

Meghen, P.J., 'The County Surveyors of Limerick', *Engineers Journal* 7:5 & 6 (1954), 184, 229.

Micks, W.L., *History of the Congested Districts Board* (Dublin, 1925).

Mitchell, Arthur, *Revolutionary Government in Ireland* (Dublin, 1995).

Murphy, Donal A., *Blazing Tar Barrels & Standing Orders: Tipperary North's First County & District Councils, 1899–1902* (Nenagh, 1999).

Murphy, Donal A., *The Two Tipperarys* (Nenagh, 1994).

Murphy, Michael et al., *Grand Jury Rooms to Áras an Chontae: Local government in Offaly*, (Tullamore, 2003).

Murphy, Walter, *Remarks on the Irish Grand Jury System* (Cork, 1849).

Newenham, Thomas, *A View of the Natural, Political and Commercial Circumstances of Ireland* (1809).

Nolan, David M., 'The County Cork Grand Jury, 1836–1899' (MA thesis, UCC, 1974).

O Donoghue, Brendan, *In Search of Fame and Fortune: The Leahy Family of Engineers, 1780–1888* (Dublin, 2006)

Ó hOgartaigh, Margaret, 'Women Engineers in Early 20th Century Ireland', *Engineers Journal*, (Dec. 2002), 48–9.

O'Brien, Emmet, 'The Architecture of Bank Buildings in Ireland, 1726–1910' (PhD thesis, UCD, 1991).

O'Donnell, Kevin C., Dublin's Water Supply, *TICEI* 112 (1987–88), 105.

O'Driscoll, John, *Views on Ireland*, 2 vols (London, 1823).

O'Dwyer, Frederick, 'The Architecture of the Board of Public Works 1831–1923', in Ciaran O'Connor, and John O'Regan (eds), *Public Works. The Architecture of the Office of Public Works 1831–1987* (Dublin, 1987).

O'Halpin, Eunan, *The Decline of the Union* (Dublin, 1987).

O'Keeffe, Peter, *Ireland's Principal Roads, AD 1608–1898* (Dublin, 2003)

O'Keeffe, Peter and Tom Simmington, *Irish Stone Bridges: History and Heritage*, (Dublin, 1991).

O'Mahony, Canice, 'Iron Rails and Harbour Walls: James Barton of Farndreg', *County Louth Archaeological and Historical Journal*, xxii:2 (1990), 134.

O'Sullivan, Harold, *A History of Local Government in the County of Louth* (Dublin, 2000).

O'Sullivan, Harold, 'Dundalk Harbour Improvements in the Nineteenth Century', *County Louth Archaeological and Historical Journal* xxiv:4 (2000), 504–530; xxv:2 (2002), 129–50.

Parker, W., *An Essay on the Employment which Bridges, Roads and other Public Works may afford the Labouring Classes, etc.* (Cork, 1819).

Parnell, Sir Henry, A *Treatise on Roads* (London, 1833).

Pike, W.T. (ed.), *Belfast and the Province of Ulster in the Twentieth Century: Contemporary Biographies* (Brighton, 1909).

Pike, W.T. (ed.), *British Engineers and Allied Professions in the Twentieth Century: Contemporary Biographies* (Brighton, 1908).

Pike, W.T. (ed.), *Cork and County Cork in the Twentieth Century: Contemporary Biographies* (Brighton, 1911).

Pike, W.T. (ed.), *Ulster: Contemporary Biographies* (Brighton, 1909).

Prunty, Jacinta, *Dublin Slums, 1800–1925* (Dublin, 1999).

Register of the Alumni of Trinity College, Dublin (9 vols), Dublin, 1928–70.

Reports on Inspection, Assessment and Rehabilitation of Masonry Arch Bridges and of Concrete bridges (Dept. of the Environment, Dublin, 1988 and 1990).

Robinson, Sir Henry, *Memories: Wise and Otherwise* (London, 1923).

Rolt, L.T.C., *Thomas Telford* (London, 1985).

Rothery, Sean, *A Field Guide to the Buildings of Ireland* (Dublin, 1997).

Rowan, Alastair, *The Buildings of Ireland – North West Ulster* (London, 1979).

Rynne, Colin, *Industrial Ireland 1750–1930: An Archaeology* (Cork, 2006).

Scally, Robert James, *The End of Hidden Ireland: Rebellion, Famine and Emigration* (Oxford, 1995).

School of Engineering, Trinity College, Dublin: A Record of Past and Present Students (Dublin, 1909).

Second Irish Roads Congress: Record of Proceedings (Dublin, 1911).

Skempton, A.W. et al. (eds), *A Biographical Dictionary of Civil Engineers in Great Britain and Ireland, Volume 1:1500–1830* (London, 2002).

Townsend, Horatio, *A General and Statistical Survey of the County of Cork* (Cork, 1815).

Trevelyan, G.M., *British History in the Nineteenth* Century (London, 1922).

Trollope, Anthony, *An Autobiography*, 2 vols (London, 1883).

Vaughan, W.E. (ed.), *A New History of Ireland V, Ireland under the Union I, 1801–1870* (Oxford, 1989).

Viliers-Tuthill, Kathleen, *History of Kylemore Castle & Abbey* (Kylemore, 2002).

Villiers-Tuthill, Kathleen, *Alexander Nimmo & the Western District* (Clifden, 2006).

Villiers-Tuthill, Kathleen, 'Rags to Riches in Clifden "Society"', *Journal of the Clifden & Connemara Heritage Group*, 1:1 (1993), 81–4.

Webb, Sidney & Beatrice, *English Local Government: Statutory Authorities for Special Purposes* (London, 1922).

Williams, J.D., *Donegal County Council: 75 Years* (Lifford, 1974).

Index of Personal Names

COUNTY SURVEYORS (*continued*)

COUNTY SURVEYORS (*continued*)

ARCHITECTS

ENGINEERS

ENGINEERS (*continued*)

ENGINEERS (*continued*)

ADMINISTRATORS, POLITICAL FIGURES AND OTHERS

ADMINISTRATORS, POLITICAL FIGURES AND OTHERS (*continued*)

Index of Works

GALWAY *(continued)*

Galway golf club, 226
Galway harbour and docks, 130, 184, 289–90, 301
Galway town sewerage, 289
Galway–Clifden road, 217
Garbally Court, 225
Gort courthouse, 224
Gorumna Island, 277
Kylemore Castle, 290
Lismany House, 224
Loughrea town hall, 290
Lunatic asylum, Ballinasloe (St Brigid's Hospital), 78, 225
Maam Bridge, 289
Mutton Island, 289–90
Mweenish Island, 277
Newcastle distillery, 290

Oughterard courthouse, 131
Portumna Bridge, 224, 225, 257–8
Railway schemes, 112, 147, 213, 219, 230, 245, 289
Relief works in 1886–8, 68–9, 277
Road conditions in 1910, 254–5
Road development in 1860–70, 289; in 1880s, 277
Salthill promenade, 240
St Ignatius church, 290
Tuam dispensary, 277
Tuam post office, 230
Tuam workhouse, 230
Water supply schemes, 230, 259, 277, 289–90
Wolfe Tone Bridge, 226
Woodlawn House, 224

KERRY

Cahirciveen courthouse, 306
Dingle courthouse, 306
Famine relief works, 305
Fenit harbour, 299
Great Foze Rock lighthouse, 184
Kenmare Bridge, 165, 299
Kenmare–Killarney road, 189
Killorglin Bridge, 146

Listowel courthouse, 306
Listowel workhouse, 238
Road conditions in 1910, 188–9; in 1920s, 158
Road development in 1834–44, 305
Tralee & Dingle railway, 122, 189
Tralee workhouse, 306

KILDARE

Alexandra Bridge, Clane, 331–2
Athy–Wolfhill railway, 168
Barrow drainage, 107, 187, 207
Maynooth courthouse, 116
Naas courthouse, 116

Naas sewerage scheme, 114, 116
Naas workhouse, 116
Road development in 1900–10, 186–7; in 1920s, 292
Victoria Bridge, 232

KILKENNY

Callan courthouse, 126
Castlecomer collieries, 168
Kilkenny gaol, 126
Kilkenny water supply, 119
Kilkenny–Inistioge canal, 119
Road development in 1834–44, 126; in 1920–40, 104

St John's Bridge, 118, 121
St Joseph's convent, 114
Thomastown railway viaduct, 307
Urlingford courthouse, 126
Waterford & Kilkenny railway, 307